AF845633

VOLUME FIVE HUNDRED AND NINETY

METHODS IN ENZYMOLOGY

NanoArmoring of Enzymes: Rational Design of Polymer-Wrapped Enzymes

METHODS IN ENZYMOLOGY

Editors-in-Chief

ANNA MARIE PYLE
Departments of Molecular, Cellular and Developmental Biology and Department of Chemistry
Investigator, Howard Hughes Medical Institute
Yale University

DAVID W. CHRISTIANSON
Roy and Diana Vagelos Laboratories
Department of Chemistry
University of Pennsylvania
Philadelphia, PA

Founding Editors

SIDNEY P. COLOWICK and NATHAN O. KAPLAN

VOLUME FIVE HUNDRED AND NINETY

METHODS IN ENZYMOLOGY

NanoArmoring of Enzymes: Rational Design of Polymer-Wrapped Enzymes

Edited by

CHALLA VIJAYA KUMAR
University of Connecticut, Storrs, CT, United States

ELSEVIER

ACADEMIC PRESS
An imprint of Elsevier

Academic Press is an imprint of Elsevier
50 Hampshire Street, 5th Floor, Cambridge, MA 02139, United States
525 B Street, Suite 1800, San Diego, CA 92101–4495, United States
The Boulevard, Langford Lane, Kidlington, Oxford OX5 1GB, United Kingdom
125 London Wall, London, EC2Y 5AS, United Kingdom

First edition 2017

Copyright © 2017 Elsevier Inc. All rights reserved.

No part of this publication may be reproduced or transmitted in any form or by any means, electronic or mechanical, including photocopying, recording, or any information storage and retrieval system, without permission in writing from the publisher. Details on how to seek permission, further information about the Publisher's permissions policies and our arrangements with organizations such as the Copyright Clearance Center and the Copyright Licensing Agency, can be found at our website: www.elsevier.com/permissions.

This book and the individual contributions contained in it are protected under copyright by the Publisher (other than as may be noted herein).

Notices
Knowledge and best practice in this field are constantly changing. As new research and experience broaden our understanding, changes in research methods, professional practices, or medical treatment may become necessary.

Practitioners and researchers must always rely on their own experience and knowledge in evaluating and using any information, methods, compounds, or experiments described herein. In using such information or methods they should be mindful of their own safety and the safety of others, including parties for whom they have a professional responsibility.

To the fullest extent of the law, neither the Publisher nor the authors, contributors, or editors, assume any liability for any injury and/or damage to persons or property as a matter of products liability, negligence or otherwise, or from any use or operation of any methods, products, instructions, or ideas contained in the material herein.

ISBN: 978-0-12-810502-3
ISSN: 0076-6879

For information on all Academic Press publications
visit our website at https://www.elsevier.com/books-and-journals

ELSEVIER Book Aid International
Working together to grow libraries in developing countries
www.elsevier.com • www.bookaid.org

Publisher: Zoe Kruze
Acquisition Editor: Zoe Kruze
Editorial Project Manager: Helene Kabes
Production Project Manager: Magesh Kumar Mahalingam
Cover Designer: Alan Studholme

Typeset by SPi Global, India

CONTENTS

CONTRIBUTORS

Dhamodaran Arunbabu
Madanapalle Institute of Technology and Science, Madanapalle, India

Jason A. Berberich
Miami University, Oxford, OH, United States

Aoneng Cao
Institute of Nanochemistry and Nanobiology, Shanghai University, Shanghai, China

Sheiliza Carmali
Center for Polymer-Based Protein Engineering, ICES, Carnegie Mellon University; Carnegie Mellon University, Pittsburgh, PA, United States

Abhijit Chakrabarti
Saha Institute of Nuclear Physics, HBNI, Kolkata, India

Leslie Coats
Oklahoma State University, Stillwater, OK, United States

M. Rita Correro
School of Life Sciences, University of Applied Sciences and Arts Northwestern Switzerland, Muttenz, Switzerland

Philippe F.-X. Corvini
School of Life Sciences, University of Applied Sciences and Arts Northwestern Switzerland, Muttenz, Switzerland

Chad Cummings
Center for Polymer-Based Protein Engineering, ICES, Carnegie Mellon University; Carnegie Mellon University, Pittsburgh, PA, United States

Anindita Das
Indian Institute of Science, Bangalore, India

Puspendu K. Das
Indian Institute of Science, Bangalore, India

Melissa L. Dougherty
Miami University, Oxford, OH, United States

Rebecca Falatach
Miami University, Oxford, OH, United States

Henry Fischesser
Miami University, Oxford, OH, United States

Jun Ge
Key Lab for Industrial Biocatalysis, Ministry of Education, Tsinghua University, Beijing, China

Antonella Grigoletto
University of Padua, Padua, Italy

Karsten Haupt
Sorbonne Universités, Université de Technologie de Compiègne, CNRS Enzyme and Cell Engineering Laboratory, Rue Roger Couttolenc, Compiègne Cedex, France

Miao Hou
Key Lab for Industrial Biocatalysis, Ministry of Education, Tsinghua University, Beijing, China

Tushar Jana
School of Chemistry, University of Hyderabad, Hyderabad, India

Rajeswari M. Kasi
University of Connecticut; Polymer Program, Institute of Materials Science, University of Connecticut, Storrs, CT, United States

Priya Katyal
University of Connecticut, Storrs, CT, United States

Andreas Küchler
ETH Zürich, Zürich, Switzerland

Dominik Konkolewicz
Miami University, Oxford, OH, United States

Sadagopan Krishnan
Oklahoma State University, Stillwater, OK, United States

Challa V. Kumar
University of Connecticut; Polymer Program, Institute of Materials Science, University of Connecticut, Storrs, CT, United States

Konda R. Kunduru
School of Chemistry, University of Hyderabad, Hyderabad, India

S.N. Raju Kutcherlapati
School of Chemistry, University of Hyderabad, Hyderabad, India

Melanie Leurs
TU Dortmund, Dortmund, Germany

Yao Lin
Polymer Program, Institute of Materials Science, University of Connecticut, Storrs, CT, United States

Yi Liu
Institute of Nanochemistry and Nanobiology, Shanghai University, Shanghai, China

Katherine Makaroff
Miami University, Oxford, OH, United States

Katia Maso
University of Padua, Padua, Italy

Krzysztof Matyjaszewski
Center for Polymer-Based Protein Engineering, ICES, Carnegie Mellon University; Carnegie Mellon University, Pittsburgh, PA, United States

Matthew McCauley
Miami University, Oxford, OH, United States

Paulina X. Medina-Rangel
Sorbonne Universités, Université de Technologie de Compiègne, CNRS Enzyme and Cell Engineering Laboratory, Rue Roger Couttolenc, Compiègne Cedex, France

Anna Mero
University of Padua, Padua, Italy

Daniel Messmer
ETH Zürich, Zürich, Switzerland

Hironobu Murata
Center for Polymer-Based Protein Engineering, ICES, Carnegie Mellon University; Carnegie Mellon University, Pittsburgh, PA, United States

Ramanathan Nagarajan
Natick Soldier Research, Development and Engineering Center, Natick, MA, United States

Matthew Paeth
Miami University, Oxford, OH, United States

Richard C. Page
Miami University, Oxford, OH, United States

Jose M. Palomo
Institute of Catalysis (CSIC), Madrid, Spain

Gianfranco Pasut
University of Padua, Padua, Italy

Gayan Premaratne
Oklahoma State University, Stillwater, OK, United States

Kishore Raghupathi
University of Massachusetts, Amherst, MA, United States

Caterina M. Riccardi
University of Connecticut, Storrs, CT, United States

Alan J. Russell
Center for Polymer-Based Protein Engineering, ICES, Carnegie Mellon University; Carnegie Mellon University, Pittsburgh, PA, United States

A. Dieter Schlüter
ETH Zürich, Zürich, Switzerland

Patrick Shahgaldian
School of Life Sciences, University of Applied Sciences and Arts Northwestern Switzerland, Muttenz, Switzerland

Jerry Shepherd
Miami University, Oxford, OH, United States

Jacob Stapleton
Miami University, Oxford, OH, United States

Nisaraporn Suthiwangcharoen
Natick Soldier Research, Development and Engineering Center, Natick, MA, United States

Sabine Sykora
School of Life Sciences, University of Applied Sciences and Arts Northwestern Switzerland, Muttenz, Switzerland

Sankaran Thayumanavan
University of Massachusetts, Amherst, MA, United States

Joerg C. Tiller
TU Dortmund, Dortmund, Germany

Bernadette Tse Sum Bui
Sorbonne Universités, Université de Technologie de Compiègne, CNRS Enzyme and Cell Engineering Laboratory, Rue Roger Couttolenc, Compiègne Cedex, France

Olga Vinogradova
University of Connecticut, Storrs, CT, United States

Peter Walde
ETH Zürich, Zürich, Switzerland

Cameron Williams
Miami University, Oxford, OH, United States

Jingjing Xu
Sorbonne Universités, Université de Technologie de Compiègne, CNRS Enzyme and Cell Engineering Laboratory, Rue Roger Couttolenc, Compiègne Cedex, France

Yongkun Yang
Polymer Program, Institute of Materials Science, University of Connecticut, Storrs, CT, United States

Omkar V. Zore
University of Connecticut; Polymer Program, Institute of Materials Science, University of Connecticut, Storrs, CT, United States

PREFACE

Enzymes are nature's workhorses. These complex enzymes have a wide variety of essential roles in nature that include catalyzing reactions, transporting molecules, and DNA replication. Many studies in recent years have taken advantage of the specific properties of enzymes in order to design new materials and devices. There are many advantages of using an enzyme as the basis for a functional material: enzymes are highly specific to their substrates, they are biodegradable, and they have many functional groups on their surface, which can be chemically modified to tune the enzyme's interactions with its environment. Compared to traditional organic synthesis reactions, enzymes are capable of catalyzing reactions in a more economic and green manner, and they also have very high activities and excellent selectivities. Enzymes, however, are often limited in their industrial and practical applications because of their poor stability—they are only stable in their physiological conditions (such as pH and temperature) and their poor recyclability and operational stability also limit their practical application. There are numerous methods developed over the decades to stabilize enzymes while maintaining their activity and selectivity.

Stabilization of enzymes is a fundamental thermodynamic issue. The free energy of denaturation (ΔG) of an enzyme decreases with increasing temperature. Above the denaturation temperature of the enzyme, the denatured state is thermodynamically more favored, and the enzyme unravels. Similarly, other factors such as pH and solvent also affect the free energy change and contribute to the enzyme's instability. In order to control the free energy of denaturation, one important thermodynamic component to consider is the entropy of denaturation (ΔS). When the enthalpy of denaturation is constant, ΔG can be increased if ΔS is decreased and stability of the enzyme enhanced (Scheme 1). Many methods have been developed for the stabilization of enzymes via this approach. One such method includes the conjugation of enzyme to a synthetic polymer, where the conformational entropy of the enzyme–polymer conjugate is reduced, and as a result the enzyme is armored with a polymer shell and its stability enhanced (Mudhivarthi et al., 2012; Riccardi et al., 2014; Thilakarathne, Briand, Zhou, Kasi, & Kumar, 2011). The nanoarmor provides the enzyme with protection from the harsh conditions encountered during various situations. Additionally, the armor also protects the enzyme from other degrading components such as bacteria

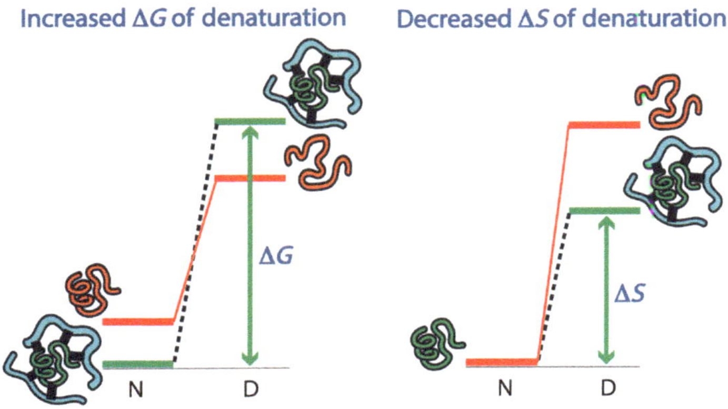

Scheme 1 Thermodynamic interconversion between the native (N) and denatured (D) states of an enzyme (*red lines*) are inhibited when the enzyme is conjugated to a nanoarmor such as a polymer (*green lines*). This is typically achieved by lowering the conformational entropy of the enzyme (*right graphic*).

and proteases as well as prevents the dissociation of the subunits of multimeric enzymes.

There are several experimental protocols, which have been reported for nanoarmoring of enzymes, such as encapsulation in nanoparticles, incorporation in metal-organic frameworks, modification with polymers, and immobilization in organosilica (Scheme 2). One example is to encapsulate enzymes inside nanoparticles, where the pore size of the nanoparticle can be tuned while also being able to tune the size of the overall particle for specific applications such as cellular imaging (Scheme 2, 1) (Cao et al., 2010). Immobilization of enzymes onto the surface of magnetic nanoparticles (Scheme 2, 10) is also used to impart stability and easy recovery of the enzyme and thus enhance its industrial value (Premaratne et al., 2016). Entrapment and immobilization are sometimes even used together, where the enzyme is first immobilized on the surface of one nanoparticle, and then other nanoparticles are then attached around it to create an enzyme armor (Correro et al., 2016). The enzyme–nanoparticle interactions are important in this design and several factors that may contribute are the binding free energy, binding stoichiometry, and structure distortion upon binding (Scheme 2, 2) (Das, Chakrabarti, & Das, 2016).

Similar to the encapsulation method in nanoparticles, metal-organic frameworks are also reported for armoring enzymes—as they are structurally diverse and can be tuned to the desired functionality (Scheme 2, 3–4) with retention of activity (Ge, Lei, & Zare, 2012).

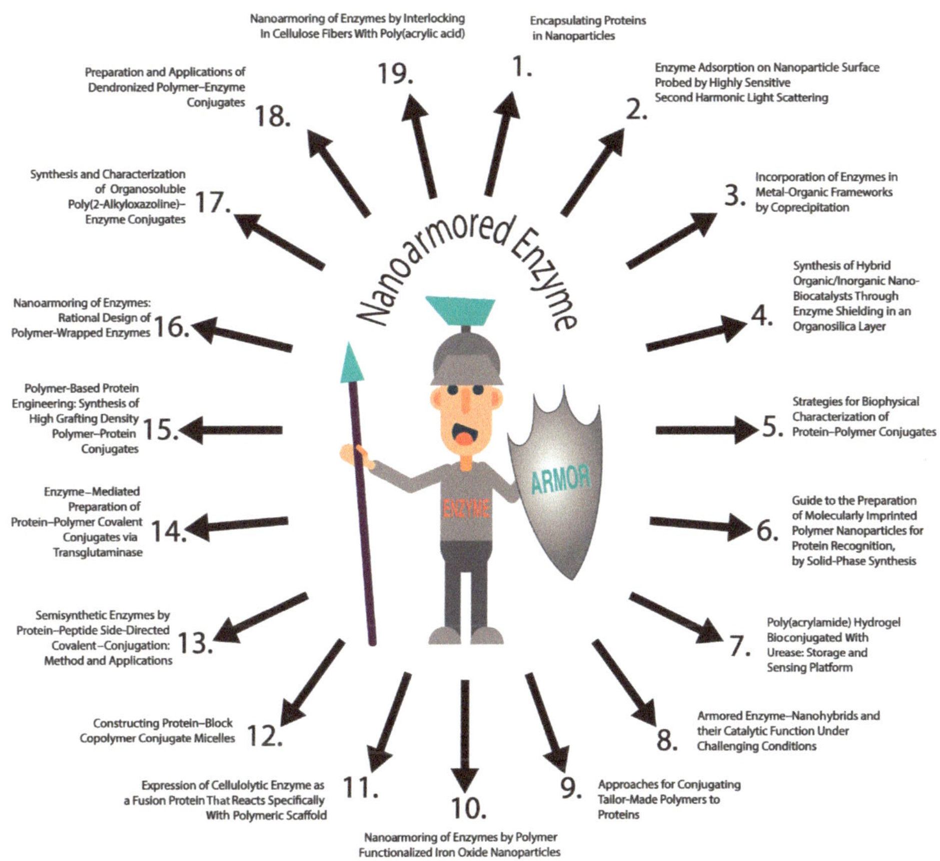

Scheme 2 Nanoarmoring of enzymes by the various methods illustrated in this book. Each number corresponds to the chapter number and title.

Polymer-based armoring (Scheme 2, 5–9) is another widely studied method. Specific functional groups on the polymers are used to covalently conjugate to the enzyme, e.g., polyethylene glycol, poly(acrylic acid), poly(acrylamide), cellulose, and polyamide. The conjugation of enzyme to the polymer can tune the preferred properties of enzymes. For example, the operational temperature of urease was controlled so that it can be used effectively at room temperature (Kutcherlapati, Yeole, & Jana, 2016). In addition to using the polymer alone, hybrid methods of using poly(acrylic acid) and graphene oxide sheets further stabilized enzymes (Scheme 2, 8) so that they were active in organic solvents, a condition in which an non-armored enzyme would otherwise denature (Zore, Lenehan, Kumar, & Kasi, 2014). Interlocking of the enzyme–polymer conjugate, as a further step, has been done in the fibers of ordinary paper, without any need for activation of the cellulose support (Riccardi et al., 2016).

Another advantage of using polymer as a support for enzyme armoring is that the polymer hierarchical structure can also be modified, such as in dendronized polymer–enzyme conjugates (Scheme 2, 18). The more branching points a dendronized polymer has, the stiffer the polymer becomes, thus potentially decreasing the free energy of the immobilized enzyme (Gustafsson, Küchler, Holmberg, & Walde, 2015). The composition and sequencing of block copolymers can also be modified so that the polymer itself can assemble into specific shapes such as micelles (Suthiwangcharoen & Nagarajan, 2014). Highly dense polymer shells around an enzyme can also serve like armor, as achieved by the "grafting-to" and "grafting-from" methods, where the nature of the polymer can be designed to impart stimuli responsiveness to the conjugate (Cummings, Murata, Koepsel, & Russell, 2014).

From nanoparticles, to metal-organic-based frameworks, to polymer-based materials, etc., these nanoarmors have all shown to enhance the stabilities of enzymes. There is still much to explore and understand this approach, but it is clear that these methods are crucial for developing and influencing the development of new biocatalysts, biomaterials, and devices on a global scale.

REFERENCES

Cao, A., Ye, Z., Cai, Z., Dong, E., Yang, X., Liu, G., … Liu, Y. (2010). A facile method to encapsulate proteins in silica nanoparticles: Encapsulated green fluorescent protein as a robust fluorescence probe. *Angewandte Chemie, International Edition*, *49*(17), 3022–3025. https://doi.org/10.1002/anie.200906883.

Correro, M. R., Moridi, N., Schützinger, H., Sykora, S., Ammann, E. M., Peters, E. H., … Shahgaldian, P. (2016). Enzyme shielding in an enzyme-thin and soft organosilica layer. *Angewandte Chemie, International Edition*, *55*(21), 6285–6289. https://doi.org/10.1002/anie.201600590.

Cummings, C., Murata, H., Koepsel, R., & Russell, A. J. (2014). Dramatically increased pH and temperature stability of chymotrypsin using dual block polymer-based protein engineering. *Biomacromolecules*, *15*(3), 763–771. https://doi.org/10.1021/bm401575k.

Das, A., Chakrabarti, A., & Das, P. K. (2016). Probing protein adsorption on a nanoparticle surface using second harmonic light scattering. *Physical Chemistry Chemical Physics*, *18*(35), 24325–24331. https://doi.org/10.1039/C6CP02196D.

Ge, J., Lei, J., & Zare, R. N. (2012). Protein-inorganic hybrid nanoflowers. *Nature Nanotechnology*, 7(7), 428–432. https://doi.org/10.1038/nnano.2012.80.

Gustafsson, H., Küchler, A., Holmberg, K., & Walde, P. (2015). Co-immobilization of enzymes with the help of a dendronized polymer and mesoporous silica nanoparticles. *Journal of Materials Chemistry B*, *3*(30), 6174–6184. https://doi.org/10.1039/C5TB00543D.

Kutcherlapati, S. N. R., Yeole, N., & Jana, T. (2016). Urease immobilized polymer hydrogel: Long-term stability and enhancement of enzymatic activity. *Journal of Colloid and Interface Science*, *463*, 164–172. https://doi.org/10.1016/j.jcis.2015.10.051.

Mudhivarthi, V. K., Cole, K. S., Novak, M. J., Kipphut, W., Deshapriya, I. K., Zhou, Y., … Kumar, C. V. (2012). Ultra-stable hemoglobin–poly(acrylic acid) conjugates. *Journal of Materials Chemistry*, *22*(38), 20423–20433. https://doi.org/10.1039/C2JM34434C.

Premaratne, G., Nerimetla, R., Matlock, R., Sunday, L., Koralege, R. S. H., Ramsey, J. D., & Krishnan, S. (2016). Stability, scalability, and reusability of a volume efficient biocatalytic system constructed on magnetic nanoparticles. *Catalysis Science & Technology*, *6*(7), 2361–2369. https://doi.org/10.1039/C5CY01458A.

Riccardi, C. M., Cole, K. S., Benson, K. R., Ward, J. R., Bassett, K. M., Zhang, Y., … Kumar, C. V. (2014). Toward "stable-on-the-table" enzymes: Improving key properties of catalase by covalent conjugation with poly(acrylic acid). *Bioconjugate Chemistry*, *25*(8), 1501–1510. https://doi.org/10.1021/bc500233u.

Riccardi, C. M., Mistri, D., Hart, O., Anuganti, M., Lin, Y., Kasi, R. M., & Kumar, C. V. (2016). Covalent interlocking of glucose oxidase and peroxidase in the voids of paper: Enzyme–polymer "spider webs". *Chemical Communications*, *52*(12), 2593–2596. https://doi.org/10.1039/C6CC00037A.

Suthiwangcharoen, N., & Nagarajan, R. (2014). Enhancing enzyme stability by construction of polymer-enzyme conjugate micelles for decontamination of organophosphate agents. *Biomacromolecules*, *15*(4), 1142–1152. https://doi.org/10.1021/bm401531d.

Thilakarathne, V., Briand, V. A., Zhou, Y., Kasi, R. M., & Kumar, C. V. (2011). Protein polymer conjugates: Improving the stability of hemoglobin with poly(acrylic acid). *Langmuir*, *27*(12), 7663–7671. https://doi.org/10.1021/la2015034.

Zore, O. V., Lenehan, P. J., Kumar, C. V., & Kasi, R. M. (2014). Efficient biocatalysis in organic media with hemoglobin and poly(acrylic acid) nanogels. *Langmuir*, *30*(18), 5176–5184. https://doi.org/10.1021/la501034b.

C.M. Riccardi, C.V. Kumar
University of Connecticut, Storrs, CT, United States

CHAPTER ONE

Encapsulating Proteins in Nanoparticles: Batch by Batch or One by One

Yi Liu, Aoneng Cao[1]
Institute of Nanochemistry and Nanobiology, Shanghai University, Shanghai, China
[1]Corresponding author: e-mail address: ancao@shu.edu.cn

Contents

Abstract

Encapsulation of proteins in nanoparticles (NPs) can greatly improve the properties of proteins such as their stability against denaturation and degradation by proteases, and branches out the applications of natural proteins from their intrinsic localizations and functions in living organisms for biomedical and industrial applications. We recently developed several methods to armor proteins in NPs with sizes from nanometers up to >100 nm, batch by batch or one by one, covalently or noncovalently, for a wide range of applications from biocatalysis to bioimaging and drug delivery. In this chapter, we provide detailed protocols on these methods. Key steps of specific protocols are explained with particular examples to help other laboratories to adopt and modify these methods for their own purposes. The advantages and disadvantages of each method are summarized, and guidelines for choosing the right method for a given application, as well as the current challenges and future directions of this field, are discussed.

Methods in Enzymology, Volume 590
ISSN 0076-6879
http://dx.doi.org/10.1016/bs.mie.2016.12.001
© 2017 Elsevier Inc.
All rights reserved.

1. INTRODUCTION

Proteins in living organisms incessantly conduct a vast array of essential functions, including catalyzing metabolic reactions (Benkovic & Hammes-Schiffer, 2003), DNA replication (Kornberg & Baker, 1992), and transporting molecules (Rothman, 1994). In the past decades many protein products, such as enzymatic catalysts, monoclonal antibodies (Liu, 2014), and fluorescent proteins (FPs) (Chalfie, 1994; Tsien, 1998), have been launched in biomedical and industrial fields with remarkable successes. However, proteins are intrinsically unstable under nonphysiological conditions, limiting their applications under normal laboratory conditions. Encapsulating or armoring proteins/enzymes in nanoparticles (NPs) is an effective way to address these issues. Enzyme binding has long been explored to improve their stability, recyclability, activity, and even selectivities (Barbosa et al., 2013; Fernandez-Lafuente, 2009; Garcia-Galan, Berenguer-Murcia, Fernandez-Lafuente, & Rodrigues, 2011; Mateo, Palomo, Fernandez-Lorente, Guisan, & Fernandez-Lafuente, 2007; Min & Yoo, 2014; Rodrigues, Ortiz, Berenguer-Murcia, Torres, & Fernández-Lafuente, 2013; Secundo, 2013). Some examples are also highlighted in several chapters of this volume.

Encapsulation of proteins and enzymes inside NPs or intercalation in nanosheets is a special method that can offer extra protection when compared to surface binding and could provide control over substrate diffusion or product release or keep the inhibitors out. Wrapping the enzymes completely by NPs (nanoshells) provides a strong shield for the delicate proteins against tough environments that are encountered in different applications. The nanoshells not only serve as shields to keep deleterious factors such as bacteria and proteases away but also prevent the dissociation of the subunits of protein assemblies and thus, enhance stabilities of multimeric proteins against heating, denaturation, and protease digestion (Cai et al., 2011; Cao et al., 2010; Yang et al., 2012; Yang, Liu, Wang, & Cao, 2013). When the nanoshells are porous and pore size can be tuned, then the shell can be made selectively permeable to the substrate but not for inhibitors. In addition, encapsulation methods usually result in smaller particle size than surface binding, and smaller size is usually preferred for many applications including enzymatic reactions, bioimaging, and drug delivery due to the possibility of achieving higher loadings of the enzyme per unit mass of the carrier matrix.

In this chapter, we describe some of the recent methods developed in our group to armor proteins (including enzymes and peptides) in NPs for different applications. These methods can be divided into two categories based on the number of proteins armored in each NP. Methods in the first category bind many protein molecules in each NP, and result in relatively large particle, and most literature methods belong to this category. The major advantage of this "batch-by-batch" approach is the high protein loading in NPs, which is preferred for applications such as delivering protein drugs.

Methods in the second category bind protein molecule individually, i.e., one protein in one particle, thus resulting in small NPs with sizes comparable to that of the proteins. The major advantages of this "one-by-one" approach are their rapid mass transfer and large specific surface area which can maximize the function of every single protein molecule. In Section 1, we demonstrate two of the category one methods: a noncovalent method (Cao et al., 2010) and a covalent one (Cai et al., 2011; Yang et al., 2012), to armor multiple protein molecules in silica NPs based on the common reverse microemulsion method.

In the common reverse microemulsion methods (Eslamian & Shekarriz, 2009; Qi, 2010), high pH, organic solvents, and surfactants are used, and all of these factors may have impact on the structure and function of bound proteins (Cao, Wang, Tang, & Lai, 2002; Lai, Cao, & Lai, 2000; Lu, Cao, Lai, & Xiao, 2006; Lu, Cao, et al., 2005; Lu, Xiao, et al., 2005). To address these specific issues, in Section 2, we demonstrate two of the category two methods to armor protein molecules individually in silica (Yang, Xiang, et al., 2013) and chitosan (CS) NPs (Liu et al., 2015), while both are carried out under mild conditions that have less influence on the structure and function of the bound proteins. Finally, we summarize the advantages and disadvantages of each method and discuss some current challenges and future directions in this field.

2. "BATCH-BY-BATCH" ENCAPSULATION OF PROTEINS IN SILICA NPs

Silica NPs are among the most extensively studied and massively produced NPs for many applications due to their biocompatibility and chemical inertness. They have also been widely applied to armor proteins/enzymes covalently or noncovalently for many bioapplications, such as biosensors, bioreactors, bioimaging (De, Ghosh, & Rotello, 2008;

Kim, Grate, & Wang, 2008; Korzeniowska, Nooney, Wencel, & McDonagh, 2013), and as described in this volume by other authors. Most current methods bind proteins "batch by batch," i.e., each silica NP bind many protein molecules. Covalent "batch-by-batch" methods can easily achieve high protein loading. For some applications, including protein and peptide drug delivery, it is preferred to bind the proteins/peptides non-covalently so that they could be released in vivo.

For noncovalent "batch-by-batch" methods, previous studies have shown that the p*I* value of a protein plays a critical role in the encapsulation process (He, Chen, Wang, Qin, & Tan, 2007). Generally, it is less facile to bind negatively charged proteins (p$I < 7$) in silica NPs due to the electrostatic repulsion between the protein and the negatively charged silica surface. To address this issue, we developed a facile His-tag-based method to non-covalently bind proteins/peptides in silica NPs that worked well for both negatively charged and positively charged proteins/peptides (Cao et al., 2010). The idea is to introduce Ca^{2+} into the silica shell to bind His-tagged proteins/peptides inside silica NPs via the strong interaction between the Ca^{2+} ions of the silica matrix and the His-tags of the proteins/peptides.

2.1 A Facile General Noncovalent Method to Armor Proteins and Peptides in Silica NPs

2.1.1 Protocol

2.1.1.1 Chemicals and Materials

Tetraethoxysilane (TEOS) (Guaranteed Reagent, GR) and Triton X-100 (GR) were purchased from Sigma-Aldrich, USA. Cyclohexane (Analytical Reagent, AR), *n*-hexanol (AR), calcium chloride, acetone (AR), and ammonia solution (GR) were obtained from Sinopharm Chemical Reagent Co., Ltd.

2.1.1.1.1 Equipment U3010 UV/Vis spectrophotometer (Hitachi, Japan), F-7000 fluorescence spectrophotometer (Hitachi, Japan), BS224S electronic scale (Sartorius, Germany), Forma Orbital Shaker (Thermo, USA), HITACH CR21G II high speed freezing centrifuge (Hitachi, Japan), JEM 200CX microscope (JEOL, Japan), JEM-2010F microscope (JEOL, Japan), nanosizer (dynamic light scattering [DLS], Malvern Nano ZS90, Malvern, UK), FV1000 confocal fluorescence microscope (Olympus, Japan), Zeeman atomic absorption spectrometer (Perkin–Elmer 5100PC, USA), and ÄKTA purifier 10 (GE, USA).

2.1.1.2 Procedures

1. Water-in-oil microemulsion is prepared by mixing 7.50 mL of cyclohexane, 1.77 mL of Triton X-100, 1.80 mL of *n*-hexanol, 300 μL of His-tagged protein or peptide (up to 100 mg/mL) in 0.2–2.0 of mg/mL calcium chloride aqueous solution, and 100 μL of TEOS.
2. Then, 60 μL of 25% ammonia is added to the above mixture to initiate the silica polymerization reaction. Keep stirring the mixture at room temperature for 24 h to complete the reaction.
3. After the above reaction, 20 mL of acetone is added to the system to break the microemulsion and precipitate the protein-encapsulated silica NPs.
4. Finally, the above solution is centrifuged at 12,000 rpm for 15 min to obtain the NPs, which are washed several times with ethanol and then with deionized water.

2.1.1.2.1 Tips

1. This method can be used to bind both negatively charged and positively charged proteins/peptides in the silica matrix.
2. Ca^{2+} ions are essential for the encapsulation of negatively charged proteins, and the concentration of Ca^{2+} ions has a major influence on the porous structure of the silica matrix.
3. The concentration of the protein/peptide determines the protein/peptide load in NPs.
4. The size of the protein-bound silica NPs can be tuned by altering the water-to-surfactant molar ratio in the reverse microemulsion system.
5. Usually, the higher the stirring rate, the smaller the particle size. However, in the presence of high concentrations of proteins/peptides, bubbles might be generated by high speed stirring, which should be avoided.
6. Even for the relatively stable proteins like enhanced green fluorescent protein (EGFP) (see the example later), it is still recommended to keep EGFP@silica in wet form to prevent possible aggregation of silica NPs. And after long-time storage as dry powder, EGFP inside EGFP@silica will be unfolded and its fluorescence will disappear. Though the unfolded EGFP can be simply refolded by resuspending EGFP@silica in aqueous solution to recover the fluorescence, other unstable proteins may not be refolded easily.

2.1.2 Application Examples: Armord EGFP as a Robust Fluorescent Probe

Green fluorescent protein (GFP) and its engineered variants (Chalfie, 1994; Tsien, 1998) have been widely used in biological sciences, especially for in vivo probing due to their efficient bioluminescence (Chudakov, Lukyanov, & Lukyanov, 2005; Wang, Tan, Zhang, Fan, & Wang, 2006). FPs might be the safest fluorescence probes that have been developed so far (as compared to organic dyes and quantum dots (QDs)), as demonstrated by their wide in vivo applications. However, free FPs are intrinsically not stable for long-term application; thus, they are generally used as gene-expression markers rather than direct probes for bioimaging. To extend the application of FPs, we encapsulated and armored FPs by optically transparent silica nanoshells, and developed a novel biosafe fluorescent nanoprobe. FPs have p*I* values around 6 and are negatively charged at neutral pH. Therefore, FPs also serve as an ideal model to develop a general method to noncovalently encapsulate proteins and peptides.

Inspired by the His-tag technique that has been widely used for protein purification (Yokoyama, 2003), we developed a facile method to encapsulate the His-tagged EGFP within silica nanoshells, as depicted in Scheme 1 (Cao et al., 2010). The resulting EGFP-armored silica NPs were denoted as EGFP@silica (in this chapter, we use @ to indicate that the substance ahead of it locates inside the substance behind it). Our method is only modified

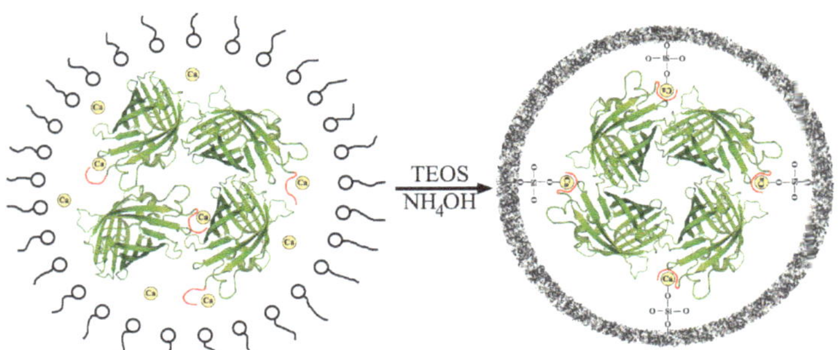

Scheme 1 Procedure for encapsulating His-tagged proteins noncovalently. Enhanced green fluorescent proteins (EGFP) (*green*) with His-tags (*red*) are linked to the silica shell through coordinate bonds between Ca^{2+} ions (*yellow*) and the histidine residues of the His-tags. *Reproduced with permission: Cao, A., Ye, Z., Cai, Z., Dong, E., Yang, X., Wang, H., … & Liu, Y. (2010). A facile method to armor proteins in silica nanoparticles: Armord green fluorescent protein as a robust fluorescence probe.* Angewandte Chemie International Edition, *49, 3022–3025, Copyright 2010, Wiley-VCH Verlag GmbH & Co. KGaA.*

slightly from that of the conventional reverse microemulsion method (Eslamian & Shekarriz, 2009; Qi, 2010).

The distinguished feature of our method is the use of His-tagged proteins, which can be easily obtained by protein engineering, and a small amount of Ca^{2+} ions are added in the reaction system. The presence of Ca^{2+} ions is critical, because Ca^{2+} ions function as anchors to link the His-tagged EGFP to the silica shell through coordinated bonds. In the presence of Ca^{2+} ions, the loading of His-tagged EGFP in silica NPs could reach up to 20 wt% as determined by elemental analysis, with almost 100% loading efficiency of the protein.

The TEM images of the EGFP@silica show an apparently hollow center (Fig. 1), demonstrating a large amount of proteins armord inside the silica shell (proteins mainly consist of light elements of H, C, N, and O, so proteins are more "transparent" for electrons than silica). As a comparison, in the absence of Ca^{2+} ions, there would be almost no encapsulation of EGFP in silica NPs. On the other hand, if negatively charged proteins/peptides do not have His-tags, Ca^{2+} ions are not helpful in encapsulating them in silica NPs. Adding EDTA in the reaction system could suppress the encapsulation of His-tagged proteins, due to the strong coordination between EDTA and Ca^{2+} ions. This also demonstrates the key role of His-tag/Ca^{2+} interaction for the encapsulation of negatively charged proteins.

Interestingly, after encapsulation in NPs, EGFP shows much higher fluorescence yield than the free, at the same concentration. The confocal fluorescence microscopy image of EGFP@silica also shows a strong

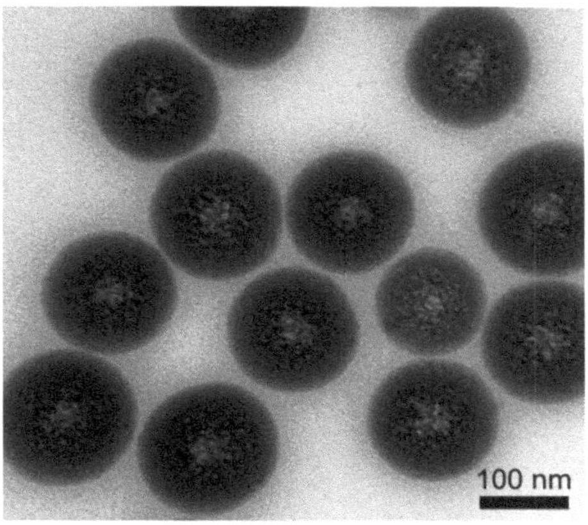

Fig. 1 TEM image of EGFP-armored silica NPs prepared in the presence of Ca^{2+}, the hollow centers demonstrate high protein load.

fluorescence response, suggesting that the EGFP@silica NPs have a promising application in bioimaging. Moreover, the encapsulation of EGFP in silica NPs can significantly increase the stability of EGFP against chemical denaturants, proteases, and heat denaturation. For instance, the pure EGFP quickly became denatured in 6 *M* guanidine hydrochloride (Gd·HCl) within 10 min, whereas EGFP@silica maintained more than 60% of its fluorescence in the same solution, after 20 min. In addition, the EGFP@silica can be easily centrifuged and dried at 40°C under vacuum and then stored at room temperature for several months. During this time, the green color of EGFP@silica decreased gradually due to the unfolding of the protein. However, the unfolded EGFP can be simply refolded by adding aqueous buffers and the fluorescence is recovered (Fig. 2).

To test the general applicability of this method, many other His-tagged proteins or peptides, including FITC-labeled His-tagged BSA, and FITC-labeled His-tagged polypeptides were also encapsulated in silica NPs following the same protocol. In all the cases, proteins or peptides can be efficiently encapsulated with good stability. This noncovalent method is generally applicable for protein/peptide drug delivery, bioimaging, and other applications, because the noncovalently bound protein could be

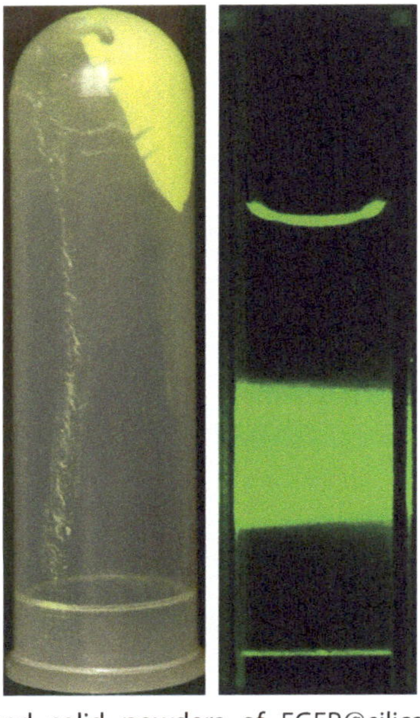

Fig. 2 (A) The centrifuged solid powders of EGFP@silica. (B) The image of the resuspension of EGFP@silica NPs under UV irradiation, showing strong fluorescence.

released from the NPs when the shell is disrupted. In fact, our current ongoing study shows a new and very interesting controllable release behavior of this His-tag-based encapsulated peptide drugs. Given that silica is generally recognized as safe (GRAS) material, this encapsulation method may have potential application for oral protein/peptide drug delivery.

2.2 An Efficient Covalent Method to Armor Proteins in Silica NPs

Covalent encapsulation of proteins in silica NPs is required to achieve long-term stability for many applications. For this purpose, we developed an efficient covalent encapsulation method to wrap protein/enzyme molecules in silica NPs (Cai et al., 2011; Yang et al., 2012). The main idea is to introduce covalent bonds between proteins and the silica nanoshells by functionalizing the proteins first with (3-aminopropyl) trimethoxysilane (APTS) groups, which will be hydrolyzed later and incorporated into the silica matrix, thereby attaching the protein to the silica shell via one or more covalent linkages.

2.2.1 Protocol

2.2.1.1 Chemicals and Materials

1-Ethyl-3-(3-dimethylaminopropyl) carbodiimide hydrochloride (EDC·HCl) and *N*-hydroxysuccinimide (NHS) were obtained from GL Biochem Ltd. (Shanghai, China); cyclohexane (AR), *n*-hexanol (AR), acetone (AR), and ammonia solution (GR) were obtained from Sinopharm Chemical Reagent Co., Ltd. (China). TEOS (GR), APTS (GR), and Triton X-100 (GR) were purchased from Sigma-Aldrich, USA.

2.2.1.1.1 Equipment U3010 UV/Vis spectrophotometer (Hitachi, Japan), F-7000 fluorescence spectrophotometer (Hitachi, Japan), BS224S electronic scale (Sartorius, Germany), Forma Orbital Shaker (Thermo, USA), HITACH CR21G II high speed freezing centrifuge (Hitachi, Japan), JEM 200CX microscope (JEOL, Japan), JEM-2010F microscope (JEOL, Japan), nanosizer (DLS, Malvern Nano ZS90, Malvern, UK), Vario EL-III elemental analyzer (Elementar, Germany), automated surface area, and pore size analyzer (Quadrasorb™ SI, USA), FV1000 confocal fluorescence microscope (Olympus, Japan), Zeeman atomic absorption spectrometer (Perkin–Elmer 5100PC, USA), and ÄKTA purifier 10 (GE, USA).

2.2.1.2 Procedures

1. The protein (enzyme) to be encapsulated is first functionalized with APTS by the conventional EDC/NHS method. Briefly, 1 mL of

10 mg/mL protein is mixed with 2 mL of 50 m*M* phosphate buffer (pH 7.5) at room temperature. Then, 40 mg EDC·HCl and 50 mg NHS are added to the mixture and stirred for 30 min. Next, 500 μL of APTS is added, and the mixture is stirred vigorously for 24 h to complete the APTS functionalization.

2. The APTS-functionalized protein (APTS-Protein) is purified by eluting on a G-25 column with slightly acidic buffer (to prevent the hydrolysis of APTS groups) and then concentrated by ultrafiltration on Millipore Amicon Ultra-50 filters.
3. The freshly concentrated APTS-Protein is then encapsulated in silica NP following the common reverse microemulsion method, during which the APTS groups on the surface of proteins are hydrolyzed and integrated into the silica NPs. In a typical procedure, the water-in-oil microemulsion was prepared by mixing cyclohexane (7.50 mL), Triton X-100 (1.77 mL), *n*-hexanol (1.80 mL), concentrated APTS-Protein solution (300 μL), and TEOS (100 μL) at room temperature. Then, 25% ammonia (65 μL) was added to initiate silica polymerization.
4. After stirring the mixture for 24 h for the completion of the polymerization, acetone (20 mL) was added to break up the microemulsion and precipitate the Protein@silica NPs.
5. Pure Protein@silica NPs were obtained by centrifugation at 12,000 rpm for 15 min and washed several times with ethanol and then with deionized water.

2.2.1.2.1 Tips

1. For multimeric proteins/enzymes, it is critical to keep the proteins/enzymes in native assembly during the APTS-functionalization step. Otherwise, the interfaces between the protein subunits might be also functionalized with APTS groups, and result in the disassembly of the multimers in the NPs.
2. The purification of APTS-Protein is to reduce potential harmful interference in the following encapsulation step. This step might be not necessary in some cases.
3. APTS-Protein should be kept under acidic conditions before the encapsulation step to reduce its potential hydrolysis/polymerization. Do not store it for long time!
4. The size of the protein-encapsulated silica NPs can be tuned by altering the water-to-surfactant molar ratio in the reverse microemulsion

system. And usually, the higher the stirring rate, the smaller the particle size.

5. The silica shell is porous, and allows the diffusion of small substrates and products.

2.2.2 Application Examples

2.2.2.1 Case I: In Situ Synthesis of Porous Silica NPs for Immobilization of Catalase

Currently, porous silica has been developed as a widely used inorganic material for enzyme encapsulation. A common approach is to immobilize enzymes into premade porous silica, usually by linking enzymes directly onto the inner surface of the silica shell (He, Li, Evans, Duan, & Li, 2000; Wang, Dai, Waezsada, Tsao, & Davison, 2001). This conventional approach generally has very low enzyme loading efficiency, because only the inner surface of the big pores can be occupied by the large enzyme molecules (Kim et al., 2008). A "ship-in-a-bottle" method (Lee et al., 2005) has also been reported that it could introduce enzymes to occupy most of the porous silica volume, and then cross-link the enzymes within the pores by treatment with glutaraldehyde. The enzyme loading is significantly improved by this method; however, the cross-linking and aggregation of crowded enzymes also prevent fast diffusion of the substrates and products in the silica matrix. To solve the problem, we applied the newly developed covalent method to immobilize enzymes in silica NPs (Yang et al., 2012).

We chose catalase, an enzyme that has been widely used in food industry to catalyze the decomposition of hydrogen peroxide to water and oxygen (Aebi, 1984), as a model system. Scheme 2 shows the in situ immobilization method (Yang et al., 2012). It is common that the enzymes lose some of their activities after encapsulation in NPs due to the unfolding effects of synthesis condition and/or the interaction with NPs. Luckily, in this study, the catalase@silica was found to maintain about 90% activity of the free catalase determined by the Goth method (Goth, 1991), indicating that there is little unfolding of the catalase during the encapsulation process and after encapsulation (Fig. 3A).

Besides, the stability of the armored catalase against protease K is greatly increased. For the catalase@silica, only less than 1/3 of the catalase was digested after 2 h of treatment with protease K, while the free catalase has been digested very quickly under the same conditions (Fig. 3B). In addition, we investigated the diffusion effect on the catalase@silica at a wide range of concentrations of the substrate H_2O_2 from 65 to 325 mM (Fig. 3A).

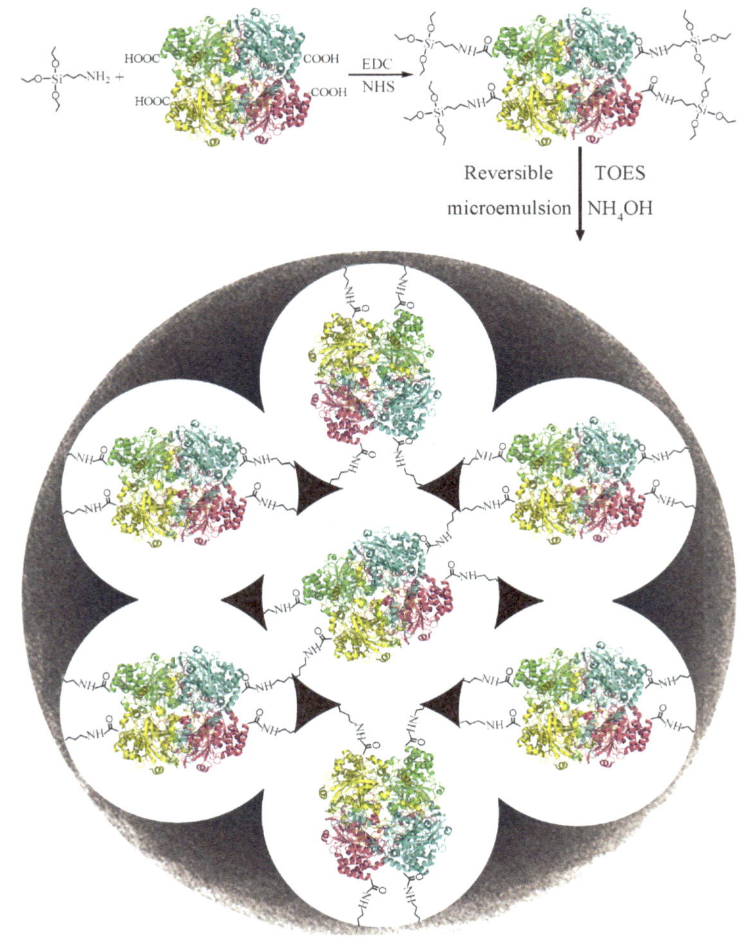

Scheme 2 The procedure to encapsulate the multimeric catalase in silica NPs. *Reproduced with permission: Yang, X., Cai, Z., Ye, Z., Chen, S., Yang, Y., Wang, H., … & Cao, A. (2012). In situ synthesis of porous silica nanoparticles for covalent immobilization of enzymes.* Nanoscale, 4, *414–416, Copyright 2012, The Royal Society of Chemistry.*

No enzymatic activity difference was observed, strongly suggesting that the substrate (H_2O_2) can diffuse fast into the active site of the encapsulated enzyme inside the silica NPs. After encapsulation, the catalase@silica was easily separated from the reaction mixture by centrifugation. Moreover, the catalase@silica can be recycled and repeatedly used for many times without substantial loss of activity. In short, through in situ synthesis method to covalently encapsulate catalase, the catalase@silica NPs were obtained with many improved properties for industrial applications.

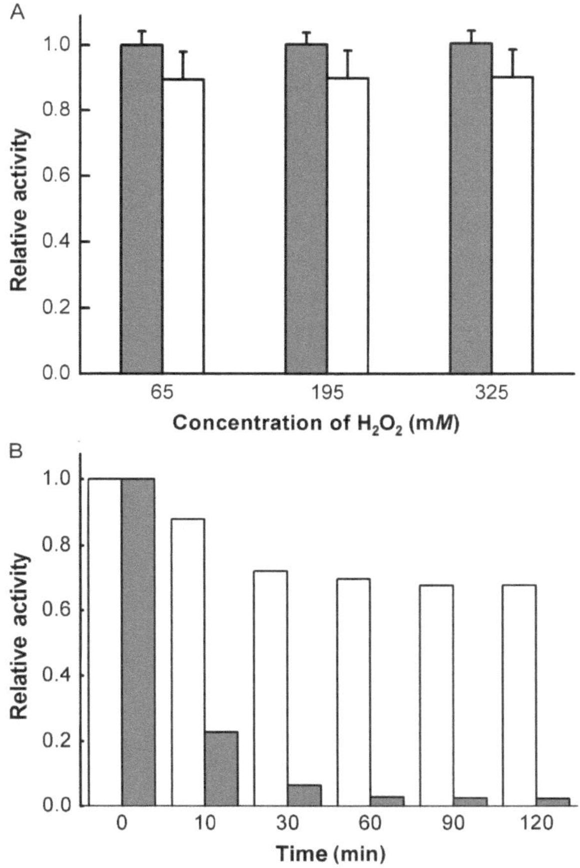

Fig. 3 (A) The enzymatic activity of catalase@silica NPs (*gray*) maintains about 90% of that of the free catalase (*white*), and there is no activity difference among different substrate concentrations (65, 195, and 325 m*M* H_2O_2). (B) The relative enzymatic activity changes of free catalase (*white*) and catalase@silica NPs (*gray*) after protease K digestion as a function of time. *Reproduced with permission: Yang, X., Cai, Z., Ye, Z., Chen, S., Yang, Y., Wang, H., … & Cao, A. (2012). In situ synthesis of porous silica nanoparticles for covalent immobilization of enzymes.* Nanoscale, *4, 414–416, Copyright 2012, The Royal Society of Chemistry.*

2.2.2.2 Case II: Encapsulated EGFP for Cellular Imaging

In this work, the EGFP was covalently encapsulated for the long-term application as a robust and safe nanoprobe (Cai et al., 2011). One of the great advantages of the reverse microemulsion method is that the size of the EGFP@silica NPs can be simply tuned by altering the water-to-surfactant molar ratio. Herein, the particle size of EGFP@silica can be easily controlled by changing the volume of 25% ammonia used, and the particle-size distribution has been found to be narrow. The covalently armored EGFP has many improved properties, including increased stabilities against protease digestion, chemical denaturants, and heat denaturation, as well as excellent storage performance. The improvements are greater than those seen with the EGFP encapsulated by the previous noncovalent method (Cao et al., 2010). This improvement could possibly due to just one coordination site in the first method, while multiple covalent bonds can be formed between each EGFP and silica shell in the second method.

After encapsulation, the fluorescence intensity of EGFP was nearly doubled when compared to that of free EGFP (Fig. 4A). This result is significant, considering that the free EGFP has a high quantum yield near 0.60 (Tsien, 1998). The fluorescence quantum yield increase by NPs entrapping has been reported for many fluorescent materials (Kester et al., 2008; Kumar et al., 2008; Muddana, Morgan, Adair, & Butler, 2009), mainly due to the nanoconfinement of the fluorophore and shielding from the solvent that may cause nonradiative decay of the fluorescent state. The RAW264.7 cell was treated with the EGFP@silica for 5 h, and confocal microscopy (Fig. 4B) indicated bright green spots in RAW267.4 cells, which indicated that the EGFP@silica can be applicable in cellular imaging, at least in this one particular cell line.

3. "ONE-BY-ONE" ENCAPSULATION OF PROTEINS IN NPs

In Section 1, we described three methods to efficiently armor multiple protein molecules in silica NPs. These methods are more suitable to synthesize NPs with large size (e.g., size >40 nm). On the other hand, there are pressing needs to synthesize small size NPs (e.g., size <20 nm) to entrap protein/enzyme, of particular interest is to realize unimolecular encapsulation. Besides, all methods in Section 1 are based on the reverse microemulsion that contains organic solvents, surfactants, and pH values far away from physiological conditions. It is well known that pH (Cao et al., 2002), organic solvents (Lai et al., 2000), and surfactants (Lu et al., 2006; Lu, Cao, et al., 2005; Lu, Xiao, et al., 2005) may induce structural change and even unfolding of proteins (especially for multimeric proteins and unstable enzymes). Therefore, we developed two novel approaches to coat each protein molecule individually in a NP at mild

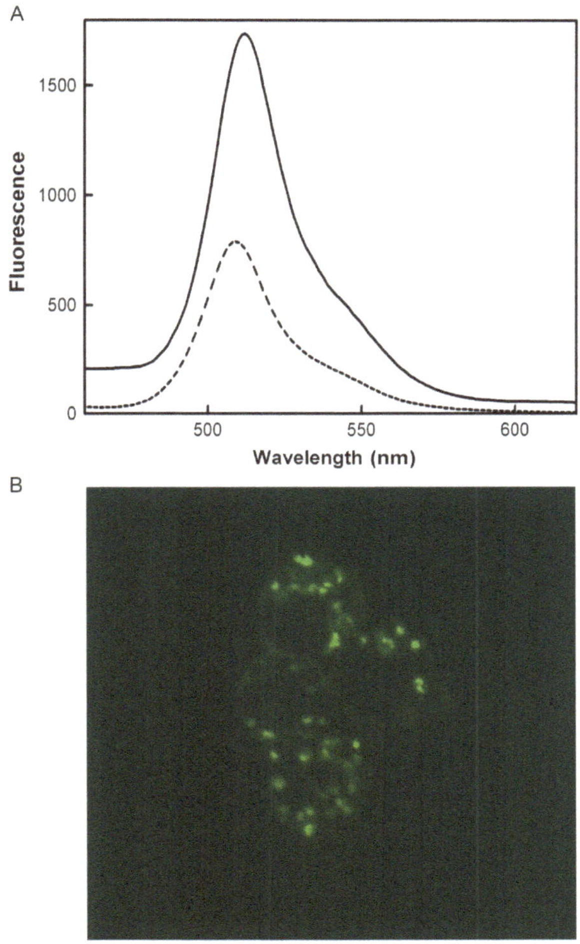

Fig. 4 (A) The fluorescence spectra of 0.12 mg/mL EGFP@silica NP (*solid line*) and free EGFP (*dashed line*). (B) Confocal microscopy images of cells treated by EGFP @silica NP (excited at 488 nm). *Reproduced with permission: Cai, Z., Ye, Z., Yang, X., Chang, Y., Wang, H., Liu, Y., & Cao, A. (2011). Armord enhanced green fluorescence protein in silica nanoparticle for cellular imaging.* Nanoscale, 3, *1974–1976, Copyright 2011, The Royal Society of Chemistry.*

conditions, which is greatly helpful to maintain the structure and function of the armord protein. The material we used are silica (Yang, Xiang, et al., 2013) and CS (Liu et al., 2015), both are GRAS material. Dramatically different from the "batch-by-batch" encapsulation methods that are inevitable to have some proteins immobilized on the surface of NPs, the "one-by-one" methods are basically growing NPs from each protein molecules. So theoretically, there would be no protein molecule left outside the NPs.

3.1 Individually Coating Protein Molecules in Silica NPs

3.1.1 Protocol

3.1.1.1 Chemicals and Materials

1-Ethyl-3-(3-dimethylaminopropyl) carbodiimide hydrochloride (EDC·HCl) and NHS were obtained from GL Biochem Ltd. (Shanghai, China). Cyclohexane (AR) was obtained from Sinopharm Chemical Reagent Co., Ltd., TEOS (GR), APTS (GR), and L-lysine (AR) were purchased from Sigma-Aldrich, USA.

3.1.1.1.1 Equipment U3010 UV/Vis spectrophotometer (Hitachi, Japan), F-7000 fluorescence spectrophotometer (Hitachi, Japan), BS224S electronic scale (Sartorius, Germany), Forma Orbital Shaker (Thermo, USA), HITACH CR21G II high speed freezing centrifuge (Hitachi, Japan), JEM 200CX microscope (JEOL, Japan), JEM-2010F microscope (JEOL, Japan), nanosizer (DLS, Malvern Nano ZS90, Malvern, UK), Xenogen IVIS@-SPECTRUM (Caliper Life Sciences, Alameda, CA, USA), FV1000 confocal fluorescence microscope (Olympus, Japan), and ÄKTA purifier 10 (GE, USA).

3.1.1.2 Procedures

1. Dissolve the protein to be encapsulated (~10 mg) in phosphate buffer (15 mL, 200 m*M*, pH 7.5), and then mix it with EDC·HCl (200 mg) and NHS (250 mg).
2. Stir the above mixture for 30 min at room temperature, and then add APTS (200 μL).
3. Stir the above mixture vigorously at 4°C for 12 h to complete the reaction and obtain APTS linked protein (APTS-Protein).
4. Desalt the above reaction mixture on a G-25 column with slightly acidic buffer (to prevent the hydrolysis of APTS groups), and then concentrate the pure APTS-Protein by ultrafiltration on Millipore Amicon Ultra-50 filters.
5. Add L-lysine (75 mg) to the freshly obtained APTS-Protein solution (17.1 mL, 0.3 g/L) in a 50 mL vial.

6. Add TEOS solution (30 μL) (in cyclohexane, volume ratio: $V_{TEOS}/V_{cyclohexane} = 5/4$) slowly to the above solution under magnetic stirring. Keep stirring the solution for 12 h to coat the protein with silica shell.

7a. To synthesize Protein@silica NPs with thin silica shell (i.e., small size NPs), the above mixture is simply stirred continuously for another 12 h.

7b. To synthesize larger size Protein@silica NPs (thicker silica shell), repeat step 6 for certain rounds.

8. The synthesized Protein@silica NPs can be easily purified by ultra-centrifugation at 4000 × *g* by a Millipore MWCO (molecular weight cutoff) 100 kDa ultrafilter.

3.1.1.2.1 Tips

1. For multimeric proteins/enzymes, it is critical to keep the proteins/enzymes in native assembly during the APTS-functionalization step. Otherwise, the interfaces between the protein subunits might be also functionalized with APTS groups; thus, it will result in nonnative assembly (monomeric) in the NPs.
2. The purification of APTS-Protein processes should carried out under acidic conditions to prevent hydrolysis of APTS groups and the aggregation of APTS-Proteins. For the same reason, the purified APTS-Protein should be used immediately for the next step.
3. To make sure that protein molecules are encapsulated one by one, the purity of the purified APTS-Protein could be checked by Fast Protein Liquid Chromatography on a HiTrap Q HP at pH 8.0 to make sure that there is no free protein (the nonfunctionalized free protein will not be encapsulated in NPs) or aggregation of APTS-Proteins (which will result in multiple proteins encapsulated in a single NP).
4. The size of the protein-encapsulated silica NPs depends on two factors, the size of the protein and the thickness of the silica shell.
5. The shell thickness can be adjusted by the total amount of TEOS added. It is critical to add TEOS solution slowly to prevent the formation of new nucleation seeds and solid silica NPs without protein encapsulation. Give enough time for the previous batch of TEOS to complete the hydrolysis before adding another round.

3.1.2 Application Examples

3.1.2.1 Individually Coating NIRFP in Silica NPs for Bioimaging

NP-based fluorescent probes, including carbon dots (Yang et al., 2009), QDs (Cassette et al., 2010; Chan & Nie, 1998; Gu, Cui, Zhang, Xie, &

Pang, 2012), and inorganic/organic dye-encapsulated dots (Muddana et al., 2009; Zhou, Liu, & Li, 2012), have been developed very fast for fluorescence imaging in biomedical fields recently. Comparing to the conventional fluorescence dyes, NP-based fluorescence probes have many advantages, such as improved optical properties, enhanced stabilities, and multiple functions (Ma, Zhao, & Liang, 2011). However, most of the current nanoprobes have certain level of toxicity, limiting their long-term in vivo applications in human body. Thus, developing an ideal in vivo nanoprobe with low toxicity and high biosafety is a dream for chemists.

Though free FPs are not stable in the complex biosystems, FPs armord in NPs shows greater stability and better optical properties that are suitable for bioimaging. As demonstrated in Section 1, FPs encapsulated in silica NPs could be used as a robust fluorescent nanoprobe. However, the previous "batch-by-batch" methods usually result in large NPs (>40 nm) that are relatively easy to aggregate and adsorb proteins in vivo. Besides, the reverse emulsion methods may leave some surfactants on the surface of silica NPs, which is a hazard for in vivo applications. To improve this FP@silica nanoprobe, we developed the above "one-by-one" lysine-catalyzed Stöber method to individually coat the near-infrared FPs (NIRFPs) (eqFP650) with very thin silica nanoshell (Yang, Xiang, et al., 2013).

The overall routine of this lysine-catalyzed method to individually coat NIRFP protein is shown in Scheme 3. First, APTS was covalently linked to the ASP/GLU residues on the surface of NIRFPs by the traditional EDC/NHS approach, producing plenty of APTS groups on the surface of NIRFP. The APTS-NIRFP was purified by desalting. The APTS of the APTS-NIRFP were then hydrolyzed in aqueous solution along with a low

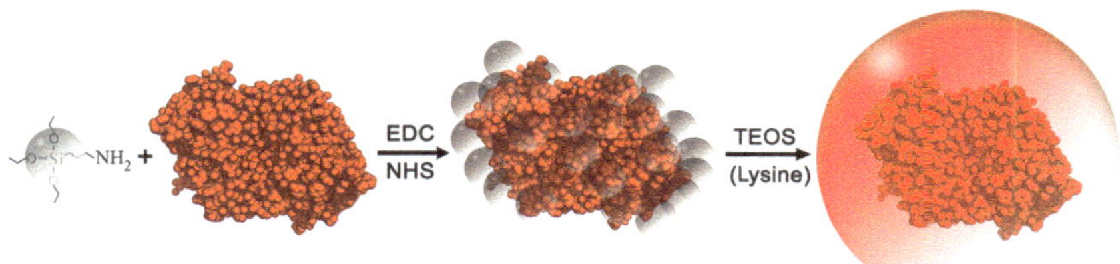

Scheme 3 Procedure for individually coating NIRFP with a thin silica shell. *Reproduced with permission: Yang, Y., Xiang, K., Yang, Y.-X., Wang, Y.-W., Zhang, X., Cui, Y., … & Cao, A. (2013). Individually coated near-infrared fluorescent protein as a safe and robust nanoprobe.* Nanoscale, *5, 10345–10352, Copyright 2013, The Royal Society of Chemistry.*

concentration of TEOS in the presence of catalyst lysine. In this process, a very thin silica shell coated every single NIRFP molecule. Finally, the produced NIRFP@silica NPs can be purified and concentrated by ultracentrifugation.

During the synthesis process, the high "local concentration" of $-Si(O-)_3$ groups, which are produced by the hydrolysis of APTS groups on the surface of each NIRFP molecule, serve as nucleation sites for the growth of silica shell from the hydrolyzation of TEOS. It is very important to keep the concentration of TEOS very low during the whole process to prevent the formation of pure silica particles.

The TEM imaging of the synthesized NIRFP@silica showed that they are monodispersed and there is no aggregation. The average particle size is only about 6 nm (Fig. 5), very close to the size of the free NIRFP, which is a dimer with a size about 5 nm (Shcherbo et al., 2010). This result demonstrates that only one NIRFP dimer is entrapped in each NIRFP@silica NP. The size of the NIRFP@silica NP can be increased by adding more TEOS stepwise. The NIRFP@silica NP showed excellent dispensability in various media, such as phosphate buffer, cell culture medium, and serum.

After coating by silica shell, the NIRFP shows greater stability against denaturants and protease K. For example, the free NIRFP is quickly denatured in 6 *M* Gd·HCl in 10 min, while NIRFP@silica maintains more than 60% of its original fluorescence intensity in 10–30 min. In addition, we further examined the biostability of NIRFP@silica through performing the standard metabolic assay in a mouse liver S9 fraction. The results showed that no fluorescence changes could be detected for the NIRFP@silica after

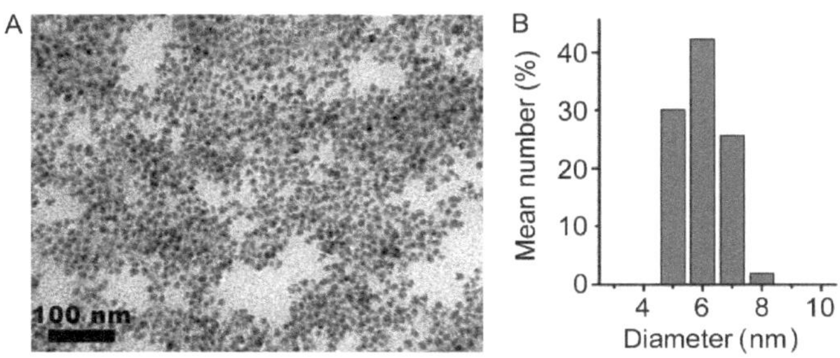

Fig. 5 TEM image (A) and size distribution (B) of NIRFP@silica NPs prepared according to the "one-by-one" encapsulation protocol. *Reproduced with permission: Yang, Y., Xiang, K., Yang, Y.-X., Wang, Y.-W., Zhang, X., Cui, Y., ... & Cao, A. (2013). Individually coated near-infrared fluorescent protein as a safe and robust nanoprobe.* Nanoscale, 5, *10345–10352, Copyright 2013, The Royal Society of Chemistry.*

3 h of incubation, while the free NIRFP was digested almost completely. The fluorescence quantum yield of the NIRFP@silica increased from 0.24 for the free NIRFP to 0.33, which can be explained by the silica confinement effect. The photostability of NIRFP@silica is also significantly improved when compared to the free NIRFP.

The cell viability assays were performed on three cell lines to evaluate cytotoxicity of the NIRFP@silica using a CCK-8 kit. The results showed no cell viability loss for all the cell lines after incubating with 200 μ*M* NIRFP@silica for 24 h, indicating that the NIRFP@silica is no-toxic at the cellular level. At animal level, no obvious abnormality was observed in mice after injecting the NIRFP@silica into nude mice via the tail vein at a high dose up to 30 mg/kg body weight.

The whole body in vivo imaging experiments were performed on nude mice after intravenous injection of NIRFP@silica with doses ranging from 1 to 30 mg/kg body weight. The results showed the bright red color of NIRFP@silica spread the whole body of the mouse without enrichment at any organ except for the bladder. Interestingly, there is no accumulation of NPs in lungs or liver (both are usually the main accumulation organs for NPs), showing the good biocompatibility of the synthesized NIRFP@silica.

Fig. 6 showed the fluorescence changes in mouse along with the time postinjection of 10 mg/kg NIRFP@silica via the tail vein to nude mouse. Just after 10 min postinjection, bright fluorescence spread all over the mouse. As time passed, the fluorescence intensity decreased gradually and almost disappeared after 24 h. The urine of mice showed high fluorescence, indicating that the NIRFP@silica are cleared out through urine. In addition, ex vivo imaging experiments were performed to detect the accurate distribution of NIRFP@silica NPs in mouse organs, including liver, kidneys, livers, lungs, spleen, and hearts. No accumulation of NIRFP@silica was observed in lungs. All these data show that individually coated NIRFP@silica NP is a robust and safe fluorescent nanoprobe for in vivo applications.

3.2 Individually Coating Protein Molecules in CS NFs

3.2.1 Protocol

3.2.1.1 Materials and Chemicals

Ethanol (GR) and glutaraldehyde (AR) was purchased from Sinopharm Chemical Reagent Co., Ltd. (Shanghai, China). Chitosan oligosaccharide (COS, deacetylation degree of 90% and MW = 1 kDa) was purchased from Golden-Shell Biochemical Co., Ltd. (Zhejiang, China).

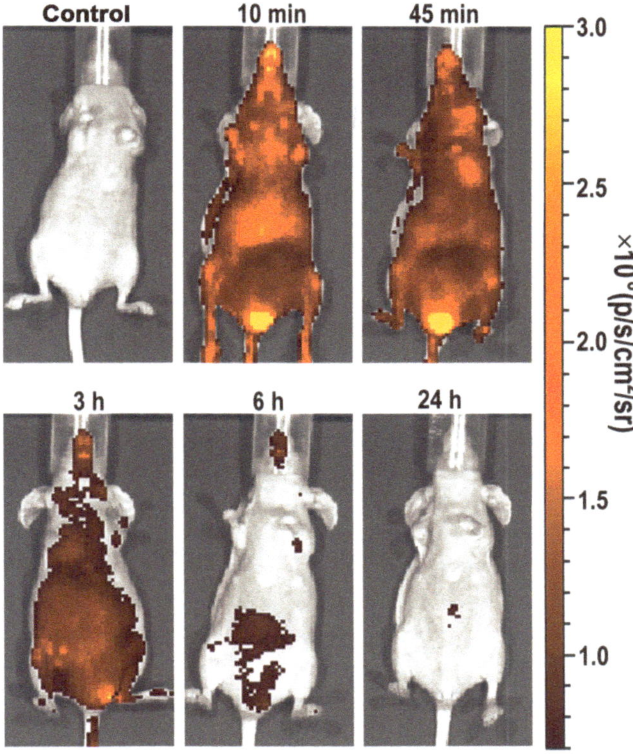

Fig. 6 The fluorescence imaging of a nude mouse after injection of 10 mg/kg of NIRFP@silica via the tail vein. *Modified with permission: Yang, Y., Xiang, K., Yang, Y.-X., Wang, Y.-W., Zhang, X., Cui, Y., … & Cao, A. (2013). Individually coated near-infrared fluorescent protein as a safe and robust nanoprobe.* Nanoscale, 5, *10345–10352, Copyright 2013, The Royal Society of Chemistry.*

3.2.1.1.1 Equipment U3010 UV/Vis spectrophotometer (Hitachi, Japan), F-7000 fluorescence spectrophotometer (Hitachi, Japan), BS224S electronic scale (Sartorius, Germany), Forma Orbital Shaker (Thermo, USA), HITACH CR21G II high speed freezing centrifuge (Hitachi, Japan), JEM 200CX microscope (JEOL, Japan), JEM-2010F microscope (JEOL, Japan), nanosizer (DLS, Malvern Nano ZS90, Malvern, UK), FV1000 confocal fluorescence microscope (Olympus, Japan), circular dichroism spectra (CD, Jasco J-815, Japan), and ÄKTA purifier 10 (GE, USA).

3.2.1.2 Procedures

1. The COS (5 g) is protonated in 25 mL of 1 *M* HCl.
2. Then, 75–125 mL of absolute ethanol is added into the above solution to precipitate protonated COS (COS^+). COS^+ powder is obtained by centrifugation and lyophilization.
3. The protein to be encapsulated (10 mg) was dissolved in 20 mL of phosphate buffer (20 m*M*, pH 8.0). Then COS^+ (2000 amino glucose units of COS^+ for 1 protein molecule) added and stirred at room temperature for 30 min.
4. Then glutaraldehyde (Glu, 2.5%) was added as the cross-linker. The molar ratio of protein:amino glucose units of COS^+:Glu is 1:2000:700.
5. Let the polymerization reaction continue for 12 h at room temperature.
6. Finally, Protein@CS NPs are purified by dialysis or ultracentrifugation (Millipore MWCO 100 kDa ultrafilter) to remove unreacted species.

3.2.1.2.1 Tips

1. Low MW COS is preferred for two reasons. First, low MW COS is more biocompatible than the higher ones. Second and more importantly, lower MW COS causes less cross-linking problems.
2. The pH value in step 3 is critical for the success of encapsulation. The pH 8 is optimized for FPs. Other proteins (Proteins with $pI < 7$ are preferred, because they will be negatively charged to bind with positively charged COS) could be encapsulated by this procedure too, but the pH value should be optimized case by case. The rule of thumb to choose optimum pH is that the pH condition will make the charges of COS and protein different, so that COS could bind on the surface of the protein.
3. The concentration of protein should be kept low to reduce protein–protein cross-linking.
4. The size of the protein-encapsulated CS NPs depends on two factors, the size of the protein, and the thickness of the CS shell. Thickness of the CS shell depends on the molar ratio of protein:amino glucose units of COS^+.
5. Protein@CS NPs can be stored as solid powder (obtained by lyophilization), which is readily suspended in weak acidic buffers.

3.2.2 Application Example

3.2.2.1 CS-Coated Red Fluorescent Protein as a Dual-Functional siRNA Carrier

Small interfering RNA (siRNA), with length in 20–25 bp, is a short double-stranded RNA molecule. siRNA has been recognized as a promising therapeutic for antiviral, anticancer, or other genetic diseases

(Li & Huang, 2006; Xue, Liu, & Wong, 2014; Zhao, Mi, & Feng, 2013). Currently, biomedical application of siRNA is still limited due to the high cost and risk of mutations and oncogenic effects, as well as that siRNA cannot enter cells by itself. In addition, most current siRNA nanovectors can only deliver siRNA into the cytoplasm, which serve at posttranscriptional level (Verdel et al., 2004). Rare vectors can carry siRNA into cell nucleus to act at the transcriptional level (Langlois et al., 2005) with a better siRNA silencing efficiency.

CS is believed to be an ideal biopolymer for the siRNA carrier, due to its high biocompatibility and good biodegradability (Mao, Sun, & Kissel, 2010). Many reports show that CS is much safer than the widely used commercial siRNA vector, Lipofectamine 2000 (He, Yin, Tang, & Yin, 2013; Otake et al., 2006; Wei et al., 2013). For in vivo applications, the optimum size for siRNA vectors should be around 20 nm (Li et al., 2007). Because very small size (<5 nm) NPs can be quickly cleared out from the body, while the big size (>100 nm) NPs usually have a short half-life in blood.

Herein, we developed an efficient approach to synthesize an optimal-sized dual-functional siRNA vector, RFP@CS, by individually coating the red fluorescent protein (RFP) with CS (Liu et al., 2015). The siRNA can efficiently bind on the surface of RFP@CS NPs and then be delivered into the cells, even into the nucleus to silence its target genes.

Scheme 4 shows the synthesis procedure of this RFP@CS. The main idea of this approach is to use the negatively charged RFP molecules as the nuclei to attract the positively charged COS (MW $\sim$ 1 kDa) in order to form COS coating shells. Glutaraldehyde was used to cross-link and to fix the COS shells, forming a positively charged nanovectors to carry siRNA. The prepared RFP@CS can be easily purified by dialysis or ultracentrifugation. This nanovector is very stable in both aqueous suspension and as a solid powder. The RFP@CS in powder form, which is obtained

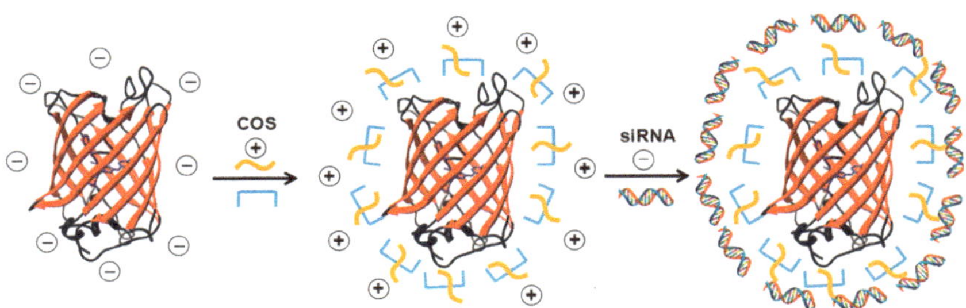

Scheme 4 Procedure for the synthesis of chitosan-coated red fluorescent protein to deliver siRNA.

by lyophilization, is very convenient for storage, and it can be readily suspended in acidic solution with full performance identical to the freshly synthesized NPs.

TEM images showed that RFP@CS NPs have an average size ~13.5 nm, within the optimum size range for delivering siRNA (Liu et al., 2015). By coating with the CS shell, the stability of RFP was greatly enhanced. In addition, RFP@CS shows excellent fluorescence stability against photobleaching and quenching regents, such as positively charged Cs^+ and negatively charged I^-. The improved property indicates that RFP@CS can be potentially used as a fluorescent probe to monitor the delivery of siRNA into cells.

The synthesized RFP@CS binds siRNA on its surface tightly and efficiently with a loading capacity of 30%. The bound siRNA shows greater stability against RNase A, and binding on the surface of RFP@CS might stiffen the chain of siRNA and/or prevent siRNA reaching into the active site of RNase. This advantage is important for the RFP@CS as a good siRNA nanovector.

The delivery of siRNA was monitored by confocal fluoresce microscopy, with carboxyfluorescein (FAM)-labeled siRNA loaded on RFP@CS. After loading FAM-labeled siRNA, RFP@CS was incubated with HeLa cells. Both fluorescence of the FAM and RFP can be observed inside cells, and their locations are well overlapped, demonstrating that RFP@CS could deliver siRNA into the cells (Fig. 7). Further test showed that the RFP@CS, similar to the other known nanovectors entered cells through endosome and resided in lysosomes. In addition, the confocal images (Fig. 7) showed that some of the RFP and siRNA could escape from lysosome and enter cell nuclei.

After siRNA transfection, the siRNA gene-silencing efficacy was monitored at mRNA level by RT-PCR, as well as at protein expression level by western blotting. At both levels, the gene silencing of siRNA delivered by RFP@CS is highly effective, much better than the commercial vector Lipofectamine 2000. In addition, no obvious cytotoxicity of the RFP@CS vector was observed at both cellular and animal levels. Therefore, this safe dual-functional siRNA-delivering vector, Protein@CS, has the potential for in vivo applications.

4. SUMMARY AND OUTLOOK

In this chapter, we have described several methods developed in our group to armor proteins (including enzymes and peptides) in NPs for various

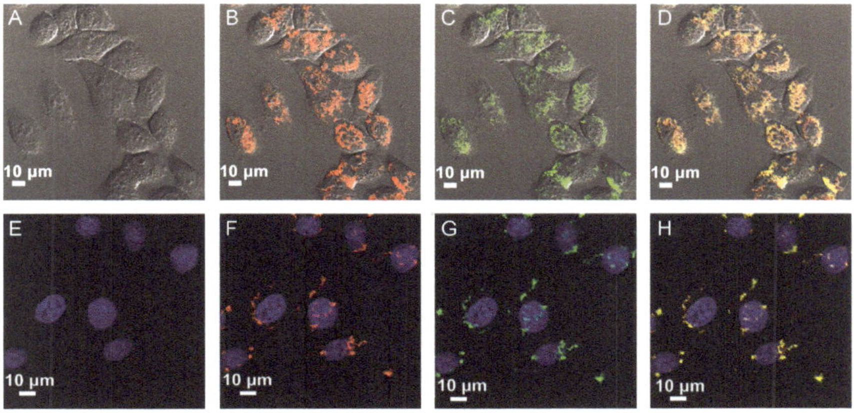

Fig. 7 The confocal fluorescence images of RFP@CS delivering FAM-siRNA into HeLa cells after 6 h treatment. *Upper panel* (A–D), cell nuclei are not stained; *bottom panel* (E–H), cell nuclei were stained with purple dye (DAPI). (D) The overlap of (A–C). (H) The overlap of (E–G). *Red color* represents RFP@CS; *green color* represents FAM-siRNA; *purple color* represents the nuclei; and *yellow color* indicates the colocalization of FAM-siRNA and RFP@CS. *Reproduced with permission: Liu, Y., Song, Z.-M., Deng, X., Cui, Y., Yang, Y., Han, K., … & Cao, A. (2015). Chitosan-coated red fluorescent protein nanoparticle as a potential dual-functional siRNA carrier.* Nanomedicine (London, England), 10, *2005–2016, Copyright 2015, Future Medicine Ltd.*

applications. These methods can be divided into two categories, the one that encapsulated multiple molecules in one NP, and the other that encapsulated a single molecule in one NP. Each method has its own advantages and disadvantages, as well as suitable applications. Table 1 summarizes some of the merits/demerits and their suitable applications of these methods.

The main purpose of protein encapsulation in NPs is to improve the stability of proteins, making it possible to use the fascinating functions of natural proteins for various applications. In this aspect, either batch by batch or one by one, covalently, or noncovalently, all our methods could significantly improve the stability of the encapsulated proteins against denaturants, long-term storage, and protease digestion. Notably, there seems to be a certain rule on the stability improvement. Our results show that the stability improvement depends on the number of anchor points of each protein molecule to the coating nanoshell. For example, the covalently encapsulated FPs (Cai et al., 2011; Yang, Liu, et al., 2013; Yang, Xiang, et al., 2013) are much more stable than the noncovalently encapsulated FPs (Cao et al., 2010). The mechanism is likely due to the fact that each covalently bound FP molecule has dozens of covalent bonds linked to the silica shell, while each noncovalently bound FP had only one His-tag to link to the Ca^{2+} ions on the silica shell. Those bonds

Table 1 Comparison of Different Protein Encapsulation Methods

	Batch-by-Batch Encapsulation		One-by-One Encapsulation	
	Noncovalently	**Covalently**	**Silica**	**Chitosan**
Merits	Simple procedure; Less impact on protein structure; Large particles that easy to recover	Highly stable; Large particles that easy to recover	Highly stable; Small particles; Excellent dispersibility; Less protein absorption; Good biocompatibility; Fast kinetics for encapsulated enzymes; Good for in vivo application	Highly stable; Small particles; Excellent dispersibility; Less protein absorption; Good biocompatibility; Good for in vivo application
Demerits	Large particles; Easy to aggregate; No so good for in vivo applications; slow kinetics for encapsulated enzymes; Easy to leak	Large particles; easy to aggregate; no so good for in vivo applications; slow kinetics for encapsulated enzymes	Small particles that are difficult to recover	Small particles that are difficult to recover
Applications	Protein and peptide drugs delivery; other short-term applications	Enzymes encapsulated for long-term applications in tough conditions	Encapsulated fluorescent proteins for bioimaging; pseudo-homogeneous catalysis of encapsulated enzymes	Nucleic acid drug delivery; encapsulated fluorescent proteins for bioimaging

may help to maintain the functional structure of the encapsulated proteins against denaturation, or keep proteins in refoldable conformations under strong denaturing conditions. Another interesting feature is that the soft polymer shell, i.e., CS, has better protection effects on encapsulation proteins than the rigid silica shell FPs (Liu et al., 2015).

A critical issue for a successful protein encapsulation is to maintain the unique functions of proteins. Unfortunately, many encapsulation methods may have deleterious effects on the encapsulated proteins. Those deleterious effects include changing the structure/conformation of encapsulated proteins, causing aggregation of proteins inside NPs ("batch-by-batch" methods), blocking the active pockets of enzymes, and obstructing/retarding the diffusing of enzyme substrates/products. In principle, noncovalent encapsulation methods would have less deleterious effects on the structure of proteins and are the primary choice for drug delivery applications. Our His-tag based noncovalent method (Cao et al., 2010) provides an easy general way to armor proteins and peptides, no matter what kind of charge they have. More importantly, as our current study shows (unpublished data), this His-tag based noncovalent encapsulation method also provides a novel controllable release mechanism for protein and peptide drugs.

Due to the permanent immobilization inside NPs and the blocking of the NP shells, covalently encapsulated proteins are usually not suitable for the functions that involve large interfaces such as protein–protein interactions. Fortunately, some functions may not be affected by the NP shells. For example, the fluorescence is not reduced because the silica shell is transparent to the fluorescence; actually, nanoconfinement can enhance fluorescence. As we demonstrated, silica encapsulated FPs are excellent nanoprobes for bioimaging (Cai et al., 2011; Cao et al., 2010; Yang, Liu, et al., 2013; Yang, Xiang, et al., 2013). Comparing to the organic or inorganic fluorophores, these encapsulated FPs fluorescent probes have many advantages, such as less toxicity and high biocompatibility. In addition, there are numerous FPs and their engineered variants with distinct optical properties, which can be encapsulated similarly and used for bioimaging with different demands. It is worthy to point out that these methods could also be used to armor fluorescent dye-labeled proteins, which may be cheaply obtained in large quantity, to perform similar functions.

Many efforts have been made to armor enzymes for catalytic applications. For catalytic reactions, kinetics is an important parameter, which is often affected by nanoencapsulation. Most current methods encapsulated a large number of enzymes in a large NP, which not only causes diffusion problem but also makes the crowded enzyme molecules inefficient.

Our "one-by-one" encapsulation method (Yang, Xiang, et al., 2013) might be a good solution to deal with these problems, with each protein molecule being wrapped by a thin (<1 nm) silica shell. The individually encapsulated proteins have rapid mass transfer and large specific surface area, and the resulted small NPs are so nicely dispersed in solution that they could behave almost homogeneously like single molecules. But it has to be pointed out that it also makes the recovery of small NPs difficult. Currently, we are developing methods to solve the recovery problem of small NPs.

Numerous studies have demonstrated how nanotechnology could make an impact on proteins, making the impossible possible. On the other hand, proteins could also be used to make better NPs. For example, proteins could also serve as good templates to synthesize NPs. As demonstrated earlier, RFP encapsulated RFP@CS NPs could serve as excellent dual-functional siRNA nanovectors (Liu et al., 2015). In fact, other cheap commercial proteins could be coated by CS similarly to deliver siRNA or small molecular drugs. Therefore, the research on protein encapsulation benefits both protein science/technology and nanoscience/technology. Past decades have witnessed tremendous progresses on the interaction between biotechnology and nanotechnology, and the emerging of nanotechnology. The field of protein armoring is an excellent example, with more fascinating applications yet to come.

ACKNOWLEDGMENTS

This work was supported by the China Natural Science Foundation (Nos. 21371117 and 31571024), the National Basic Research Program of China (No. 2016YFA0201602), and the Shanghai Pujiang Program (No. 16PJ1403400).

REFERENCES

Aebi, H. (1984). Catalase in vitro. *Methods in Enzymology*, *105*, 121–126.

Barbosa, O., Torres, R., Ortiz, C., Berenguer-Murcia, A., Rodrigues, R. C., & Fernandez-Lafuente, R. (2013). Heterofunctional supports in enzyme immobilization: From traditional immobilization protocols to opportunities in tuning enzyme properties. *Biomacromolecules*, *14*, 2433–2462.

Benkovic, S. J., & Hammes-Schiffer, S. (2003). A perspective on enzyme catalysis. *Science*, *301*, 1196–1202.

Cai, Z., Ye, Z., Yang, X., Chang, Y., Wang, H., Liu, Y., & Cao, A. (2011). Armord enhanced green fluorescence protein in silica nanoparticle for cellular imaging. *Nanoscale*, *3*, 1974–1976.

Cao, A., Wang, G., Tang, Y., & Lai, L. (2002). Linear correlation between thermal stability and folding kinetics of lysozyme. *Biochemical and Biophysical Research Communications*, *291*, 795–797.

Cao, A., Ye, Z., Cai, Z., Dong, E., Yang, X., Wang, H., … Liu, Y. (2010). A facile method to armor proteins in silica nanoparticles: Armord green fluorescent protein as a robust fluorescence probe. *Angewandte Chemie International Edition, 49*, 3022–3025.
Cassette, E., Pons, T., Bouet, C., Helle, M., Bezdetnaya, L., Marchal, F., & Dubertret, B. (2010). Synthesis and characterization of near-infrared Cu-In-Se/ZnS core/shell quantum dots for in vivo imaging. *Chemistry of Materials, 22*, 6117–6124.
Chalfie, M. (1994). Green fluorescent protein as a marker for gene expression. *Trends in Genetics, 10*, 151.
Chan, W. C., & Nie, S. (1998). Quantum dot bioconjugates for ultrasensitive nonisotopic detection. *Science, 281*, 2016–2018.
Chudakov, D. M., Lukyanov, S., & Lukyanov, K. A. (2005). Fluorescent proteins as a toolkit for in vivo imaging. *Trends in Biotechnology, 23*, 605–613.
De, M., Ghosh, P. S., & Rotello, V. M. (2008). Applications of nanoparticles in biology. *Advanced Materials, 20*, 4225–4241.
Eslamian, M., & Shekarriz, M. (2009). Recent advances in nanoparticle preparation by spray and microemulsion methods. *Recent Patents on Nanotechnology, 3*, 99–115.
Fernandez-Lafuente, R. (2009). Stabilization of multimeric enzymes: Strategies to prevent subunit dissociation. *Enzyme and Microbial Technology, 45*, 405–418.
Garcia-Galan, C., Berenguer-Murcia, A., Fernandez-Lafuente, R., & Rodrigues, R. C. (2011). Potential of different enzyme immobilization strategies to improve enzyme performance. *Advanced Synthesis and Catalysis, 353*, 2885–2904.
Goth, L. (1991). A simple method for determination of serum catalase activity and revision of reference range. *Clinica Chimica Acta, 196*, 143–151.
Gu, Y. P., Cui, R., Zhang, Z. L., Xie, Z. X., & Pang, D. W. (2012). Ultrasmall near-infrared Ag_2Se quantum dots with tunable fluorescence for in vivo imaging. *Journal of the American Chemical Society, 134*, 79–82.
He, X., Chen, J., Wang, K., Qin, D., & Tan, W. (2007). Preparation of luminescent Cy5 doped core-shell SFNPs and its application as a near-infrared fluorescent marker. *Talanta, 72*, 1519–1526.
He, J., Li, X. F., Evans, D. G., Duan, X., & Li, C. Y. (2000). A new support for the immobilization of penicillin acylase. *Journal of Molecular Catalysis B: Enzymatic, 11*, 45–53.
He, C. B., Yin, L. C., Tang, C., & Yin, C. H. (2013). Multifunctional polymeric nanoparticles for oral delivery of TNF-α siRNA to macrophages. *Biomaterials, 34*, 2843–2854.
Kester, M., Heakal, Y., Fox, T., Sharma, A., Robertson, G. P., Morgan, T. T., … Ruiz-Velasco, V. (2008). Calcium phosphate nanocomposite particles for in vitro imaging and armord chemotherapeutic drug delivery to cancer cells. *Nano Letters, 8*, 4116–4121.
Kim, J., Grate, J. W., & Wang, P. (2008). Nanobiocatalysis and its potential applications. *Trends in Biotechnology, 26*, 639–646.
Kornberg, A., & Baker, T. A. (1992). *DNA replication: Vol. 3*. New York: W.H. Freeman.
Korzeniowska, B., Nooney, R., Wencel, D., & McDonagh, C. (2013). Silica nanoparticles for cell imaging and intracellular sensing. *Nanotechnology, 24*, 442002.
Kumar, R., Roy, I., Ohulchanskyy, T. Y., Goswami, L. N., Bonoiu, A. C., Bergey, E. J., … Prasad, P. N. (2008). Covalently dye-linked, surface-controlled, and bioconjugated organically modified silica nanoparticles as targeted probes for optical imaging. *ACS Nano, 2*, 449–456.
Lai, B., Cao, A., & Lai, L. (2000). Organic cosolvents and hen egg white lysozyme folding. *Biochimica et Biophysica Acta—Protein Structure and Molecular Enzymology, 1543*, 115–122.
Langlois, M. A., Boniface, C., Wang, G., Alluin, J., Salvaterra, P. M., Puymirat, J., … Lee, N. S. (2005). Cytoplasmic and nuclear retained DMPK mRNAs are targets for

RNA interference in myotonic dystrophy cells. *Journal of Biological Chemistry, 280*, 16949–16954.
Lee, J., Kim, J., Kim, J., Jia, H., Kim, M. I., Kwak, J. H., … Wang, P. (2005). Simple synthesis of hierarchically ordered mesocellular mesoporous silica materials hosting crosslinked enzyme aggregates. *Small, 1*, 744–753.
Li, H. Y., Guo, S. C., Roll, R., Li, J., Diao, Z. J., Shao, N., … Guo, P. X. (2007). Phi29 pRNA vector for efficient escort of hammerhead ribozyme targeting survivin in multiple cancer cells. *Cancer Biology & Therapy, 6*, 697–704.
Li, S. D., & Huang, L. (2006). Gene therapy progress and prospects: Non-viral gene therapy by systemic delivery. *Gene Therapy, 13*, 1313–1319.
Liu, J. K. (2014). The history of monoclonal antibody development—Progress, remaining challenges and future innovations. *Annals of Medicine and Surgery, 3*, 113–116.
Liu, Y., Song, Z.-M., Deng, X., Cui, Y., Yang, Y., Han, K., … Cao, A. (2015). Chitosan-coated red fluorescent protein nanoparticle as a potential dual-functional siRNA carrier. *Nanomedicine (London, England), 10*, 2005–2016.
Lu, R., Cao, A., Lai, L., & Xiao, J. (2006). Interactions of beta-lactoglobulin with sodium decylsulfonate, decyltriethylammonium bromide, and their mixtures. *Journal of Colloid and Interface Science, 299*, 617–625.
Lu, R., Cao, A., Lai, L., Zhu, B., Zhao, G., & Xiao, J. (2005). Interaction between bovine serum albumin and equimolarly mixed cationic-anionic surfactants decyltriethylammonium bromide-sodium decyl sulfonate. *Colloids and Surfaces B: Biointerfaces, 41*, 139–143.
Lu, R., Xiao, J., Cao, A., Lai, L., Zhu, B., & Zhao, G. (2005). Surfactant-induced refolding of lysozyme. *Biochimica et Biophysica Acta—General Subjects, 1722*, 271–281.
Ma, X., Zhao, Y., & Liang, X. (2011). Theranostic nanoparticles engineered for clinic and pharmaceutics. *Accounts of Chemical Research, 44*, 1114–1122.
Mao, S. R., Sun, W., & Kissel, T. (2010). Chitosan-based formulations for delivery of DNA and siRNA. *Advanced Drug Delivery Reviews, 62*, 12–27.
Mateo, C., Palomo, J. M., Fernandez-Lorente, G., Guisan, J. M., & Fernandez-Lafuente, R. (2007). Improvement of enzyme activity, stability and selectivity via immobilization techniques. *Enzyme and Microbial Technology, 40*, 1451–1463.
Min, K., & Yoo, Y. J. (2014). Recent progress in nanobiocatalysis for enzyme immobilization and its application. *Biotechnology and Bioprocess Engineering, 19*(4), 553–567.
Muddana, H. S., Morgan, T. T., Adair, J. H., & Butler, P. J. (2009). Photophysics of Cy3-armord calcium phosphate nanoparticles. *Nano Letters, 9*, 1559–1566.
Otake, K., Shimomura, T., Goto, T., Imura, T., Furuya, T., Yoda, S., … Abe, M. (2006). One-step preparation of chitosan-coated cationic liposomes by an improved supercritical reverse-phase evaporation method. *Langmuir, 22*, 4054–4059.
Qi, L. M. (2010). Colloidal chemical approaches to inorganic micro- and nanostructures with controlled morphologies and patterns. *Coordination Chemistry Reviews, 254*, 1054–1071.
Rodrigues, R. C., Ortiz, C., Berenguer-Murcia, A., Torres, R., & Fernández-Lafuente, R. (2013). Modifying enzyme activity and selectivity by immobilization. *Chemical Society Reviews, 42*, 6290–6307.
Rothman, J. E. (1994). Mechanisms of intracellular protein transport. *Nature, 372*, 55–63.
Secundo, F. (2013). Conformational changes of enzymes upon immobilisation. *Chemical Society Reviews, 42*, 6250–6261.
Shcherbo, D., Shemiakina, I. I., Ryabova, A. V., Luker, K. E., Schmidt, B. T., Souslova, E. A., … Zaraisky, A. G. (2010). Near-infrared fluorescent proteins. *Nature Methods, 7*, 827–829.
Tsien, R. Y. (1998). The green fluorescent protein. *Annual Review of Biochemistry, 67*, 509–544.

Verdel, A., Jia, S., Gerber, S., Sugiyama, T., Gygi, S., Grewal, S. I., & Moazed, D. (2004). RNAi-mediated targeting of heterochromatin by the RITS complex. *Science*, *303*, 672–676.
Wang, P., Dai, S., Waezsada, S. D., Tsao, A. Y., & Davison, B. H. (2001). Enzyme stabilization by covalent binding in nanoporous sol–gel glass for nonaqueous biocatalysis. *Biotechnology and Bioengineering*, *74*, 249–255.
Wang, F., Tan, W., Zhang, Y., Fan, X., & Wang, M. (2006). Topical review: Luminescent nanomaterials for biological labelling. *Nanotechnology*, *17*, 1–13.
Wei, W., Lv, P. P., Chen, X. M., Yue, Z. G., Fu, Q., Liu, S. Y., … Ma, G. H. (2013). Codelivery of mTERT siRNA and paclitaxel by chitosan-based nanoparticles promoted synergistic tumor suppression. *Biomaterials*, *34*, 3912–3923.
Xue, H. Y., Liu, S., & Wong, H. L. (2014). Nanotoxicity: A key obstacle to clinical translation of siRNA-based nanomedicine. *Nanomedicine*, *9*, 295–312.
Yang, X., Cai, Z., Ye, Z., Chen, S., Yang, Y., Wang, H., … Cao, A. (2012). In situ synthesis of porous silica nanoparticles for covalent immobilization of enzymes. *Nanoscale*, *4*, 414–416.
Yang, S.-T., Cao, L., Luo, P. G., Lu, F., Wang, X., Wang, H., … Sun, Y.-P. (2009). Carbon dots for optical imaging in vivo. *Journal of the American Chemical Society*, *131*, 11308–11309.
Yang, S.-T., Liu, Y., Wang, Y.-W., & Cao, A. (2013). Biosafety and bioapplication of nanomaterials by designing protein nanoparticle interactions. *Small*, *9*, 1635–1653.
Yang, Y., Xiang, K., Yang, Y.-X., Wang, Y.-W., Zhang, X., Cui, Y., … Cao, A. (2013). Individually coated near-infrared fluorescent protein as a safe and robust nanoprobe. *Nanoscale*, *5*, 10345–10352.
Yokoyama, S. (2003). Protein expression systems for structural genomics and proteomics. *Current Opinion in Chemical Biology*, *7*, 39–43.
Zhao, J., Mi, Y., & Feng, S. S. (2013). siRNA-based nanomedicine. *Nanomedicine*, *8*, 859–862.
Zhou, J., Liu, Z., & Li, F. Y. (2012). Upconversion nanophosphors for small-animal imaging. *Chemical Society Reviews*, *41*, 1323–1349.

CHAPTER TWO

Enzyme Adsorption on Nanoparticle Surface Probed by Highly Sensitive Second Harmonic Light Scattering

Anindita Das*, Abhijit Chakrabarti†, Puspendu K. Das*,[1]
*Indian Institute of Science, Bangalore, India
†Saha Institute of Nuclear Physics, HBNI, Kolkata, India
[1]Corresponding author: e-mail address: pkdas@ipc.iisc.ernet.in

Contents

Abstract

Recent developments in second harmonic light scattering technique and the associated theoretical models have provided a deeper insight of molecular interactions on micro- and nanoparticle surfaces. This technique is extended to probe the thermodynamics of protein adsorption on nanoparticle surface which is crucial for understanding the fate of nanoparticle-based formulations in biomedical applications. A modified Langmuir adsorption model has been applied to extract the thermodynamic parameters from the experimental data. The general applicability of the technique is established by extracting free energy change, association constant, and binding stoichiometry of adsorption of a moderate size protein, alcohol dehydrogenase, and a small size protein, insulin, on gold nanoparticles. The free energy change for the adsorption is found to be

Methods in Enzymology, Volume 590
ISSN 0076-6879
http://dx.doi.org/10.1016/bs.mie.2017.01.020
© 2017 Elsevier Inc.
All rights reserved.

of the order of −55 kJ/mol, which indicates that the interaction of proteins with the nanoparticle surface involves weak forces. On the other hand, the low value of the free energy change makes the detachment of the protein from the particle surface easier and guarantees reversibility of the binding process. In addition, one gets the binding stoichiometry of the proteins with the nanoparticle surface which opens up the possibility of controlling the payload of the protein- or peptide-based therapeutics in future biomedical applications.

1. INTRODUCTION

Nanoparticles offer many new applications in biomedicine, diagnostics, therapeutics, and targeted drug delivery. Among various nanoparticles, gold nanoparticles (GNPs) are in the forefront and have been studied most aggressively because of their tunable surface, optical, and electronic properties. These properties are rapidly changing many areas of medicine and medical technology. GNPs are inert, less toxic, and small enough to enter the cells and organelles of all parts of human body enabling newer and more sensitive devices for medical applications. GNPs of different shapes and sizes have been used in photothermolysis of cancer cells and tumors (Abadeer & Murphy, 2016; Mocan et al., 2016; Nair, Nazeer, Jayasree, & Ajayaghosh, 2015; Oh, Yu, Kim, & Nam, 2015), genomics and bioimaging of cells and tissues (Fan, Senapati, Singh, & Ray, 2013; Jaque et al., 2016; Sinha et al., 2016), biosensors (Paul et al., 2015; Ray, 2010), drug carriers (Niikura, Iyo, Matsuo, Mitomo, & Ijiro, 2013; Swiech et al., 2016), and nanophotonic devices (Gür, Schwarz, Ye, Diez, & Schmidt, 2016; Liang et al., 2017). A large volume of work has already been done in the area of analytical diagnostics (Neely et al., 2009), delivery of drugs (Bhumkar, Joshi, Sastry, & Pokharkar, 2007; Dhar, Daniel, Giljohann, Mirkin, & Lippard, 2009; Pelaz et al., 2012), and on-site delivery of therapeutics for *in vitro* as well as *in vivo* application of GNPs (Dreaden, Alkilany, Huang, Murphy, & El-Sayed, 2012; Dykman & Khlebtsov, 2012). Greater understanding at the molecular level on the interactions of GNPs with biofluids containing enzymes and proteins e.g., blood, plasma, cerebrospinal fluid, and urine is thus needed when nanoparticle-based cocktails are directed to infected cells or tumor cells (Cedervall et al., 2007; Lacerda et al., 2010; Li, Mu, Zhang, & Yan, 2010; Saptarshi, Duschl, & Lopata, 2013).

Interaction of nanoparticles with proteins marks the basis of their bioreactivity. The association or dissociation depends on the rates, the binding affinities, and stoichiometries of protein–nanoparticle conjugate formation. It is thus important to understand this interaction at the fundamental level (Aldeek, Safi, Zhan, Palui, & Mattoussi, 2013). The kinetics of interaction of enzymes and proteins with gold and other nanoparticles have been probed by various linear and nonlinear spectroscopic techniques, and the readers may consult excellent reviews on this subject (Dinkel, Braunschweig, & Peukert, 2016; Gan, Xu, & Dai, 2011) which will not be discussed in this chapter. Our focus in this chapter is to provide a comprehensive description of binding affinities, stoichiometries, and thermodynamics of GNP–enzyme interactions and the advancements made in the measurement techniques and data analysis (Cedervall et al., 2007; Das, Chakrabarti, & Das, 2015, 2016; Lacerda et al., 2010).

In a given biological fluid, proteins associate with nanoparticles, and the amount and presentation of the proteins on the surface of the particles may lead to certain cellular responses. Proteins compete for the nanoparticle surface, leading to the formation of an adsorbed protein layer that largely defines the identity of the nanoparticle (Aggarwal, Hall, McLeland, Dobrovolskaia, & McNeil, 2009; Cedervall et al., 2007; Palchetti et al., 2016; Radauer-Preiml et al., 2016; Ritz et al., 2015). Even after adsorption, the structure, stability, and function of the adsorbed proteins could be altered giving rise to adverse health effects (Brun & Sicard-Roselli, 2014; Saptarshi et al., 2013). For example, adsorption of cytochrome *c* onto silica nanoparticles results in disruption of protein structure and significant changes in the heme microenvironment (Shang et al., 2009). Knowledge of thermodynamic parameters, such as changes in the free energy, enthalpy, and entropy, may help understand the molecular basis of the interaction between a macromolecule and ligand molecules (Klebe, 2015; Núñez, Venhorst, & Kruse, 2012; Yang, Catalano, & Maluf, 2015). Taken together, for controlled and safe application of nanomaterials in healthcare, it is important to understand the dose–response relationship and the loading strength of proteins/enzymes on nanoparticles. These, in turn, affect the extent and ease of desorption as well as the exchange of the adsorbed ligand with other enzymes or ligands. Thus, the study of thermodynamics of enzyme adsorption on nanoparticles, in detail, is not only necessary but also critical for enzyme-based drug formulation, storage, and delivery.

2. TECHNIQUES FOR PROBING NANOPARTICLE–PROTEIN INTERACTION

Several techniques are available for probing nanoparticle–protein adsorbates in solution. Since GNPs are strongly colored due to surface plasmon resonance absorption around 520 nm for 10–100 nm size GNPs, colorimetric sol-particle immunoassay (SPIA) was initially developed (Leuvering, Thal, van der Waart, & Schuurs, 1980) and later extended for DNA detection (Mirkin, Letsinger, Mucic, & Storhoff, 1996). Binding affinity and mole ratios of nanoparticle–protein interactions have also been probed by fluorescence (Li et al., 2010) and size exclusion chromatography (Cedervall et al., 2007). Surface plasmon resonance spectroscopy is capable of detecting adsorbed enzymes in submonolayer quantities on metal surfaces (Bakhtiar, 2013; Nguyen, Park, Kang, & Kim, 2015; Patching, 2014; Schasfoort & Tudos, 2008; Stenberg, Persson, Roos, & Urbaniczky, 1991). It is suitable for ultrasensitive detection of biomolecular conjugate, and it is portable for field applications (Guo et al., 2016; Liu et al., 2015; Puiu & Bala, 2016; Spoto & Minunni, 2012). This technique has been used for probing thermodynamics as well as kinetics of adsorption of enzymes on metal surfaces (Baek, Wark, & Lee, 2014; Breault-Turcot, Chaurand, & Masson, 2014; Mrksich, Sigal, & Whitesides, 1995). But the difficulty associated with this technique is in the proper functionalization of the metal surface and anchoring the protein on it (Baek et al., 2014; Cedervall et al., 2007).

Another widely used technique is isothermal titration calorimetry (ITC) (Huang & Lau, 2016; Jelesarov & Bosshard, 1999; Wiseman, Williston, Brandts, & Lin, 1989), and the strength of the interaction is represented by Wiseman C parameter (Wiseman et al., 1989), defined as

$$C = K_a[M]_{\mathrm{Tot}} \tag{1}$$

where K_a is the binding constant and $[M]_{\mathrm{Tot}}$ is the concentration of nanoparticles. In order to obtain sufficient signal-to-noise ratio, the C parameter should be ≥ 10. In order to extract thermodynamic parameters of an interacting system with $C \geq 10$, the GNP concentration should be high and binding stoichiometry should be 1:1 (Turnbull & Daranas, 2003), which may not be always achievable. This limits the scope of ITC measurements particularly at low concentrations of nanoparticles and proteins (Gourishankar, Shukla, Ganesh, & Sastry, 2004; Joshi, Shirude, Bansal, Ganesh, & Sastry, 2004).

When proteins bind to nanoparticles, the size of the latter can be analyzed by dynamic light scattering (Ipe et al., 2006). However, this method is qualitative since the size of the conjugate is subject to change due to solvation, counter ion binding, etc. Mass spectrometry techniques are also widely used in studying enzyme adsorption on nanoparticles by identifying nanoparticle–protein conjugates. This technique is also limited by the qualitative nature of the outcome (Saptarshi et al., 2013). Thus, a quantitative and sensitive technique was needed to overcome the limitations of conventional techniques.

The second harmonic light scattering (SHLS) signal from goat antihuman IgG bound to GNP at 1064 nm was enhanced greatly due to the aggregation of GNPs (Zhang, Zhang, Wang, Tang, & Lu, 2003), and the measurement has been carried out at much lower concentrations than by colorimetric methods. This strategy of aggregation induced enhancement of SHLS signal was used for the ultrasensitive detection of biomarkers (Neely et al., 2009; Russier-Antoine et al., 2008). The specific interaction of prostate cancer-specific antigen (PSA) with anti-PSA aptamer functionalized GNP breaks internanoparticle coupling and results in reduced SHLS signal (Sinha et al., 2015). These significant findings suggested the potentiality of SHLS for probing nanoparticle–enzyme interactions in a quantitative manner. This chapter aims to establish the versatility of SHLS in exploring the thermodynamics of enzyme adsorption on nanoparticles, especially for the weakly interacting enzyme–nanoparticle systems.

3. SECOND HARMONIC LIGHT SCATTERING

3.1 Background

Radiation incident on a molecule or material induces polarization by changing the electron charge distribution within the molecule or the material. The electric field associated with the radiation interacts with electrons in the molecule or medium and reorganizes the charge distribution. Since the electric field, $E(t)$, is a function of time, the total polarization $P(t)$ can be expressed as a function of the incident optical field in a Taylor series as (Boyd, 2003; Chemla & Zyss, 1987; Verbiest, Clays, & Rodriguez, 2009)

$$P(t) = \chi^{(1)} E(t) + \chi^{(2)} E^2(t) + \chi^{(3)} E^3(t) + \cdots \quad (2)$$

where $\chi^{(1)}$ is the linear susceptibility, $\chi^{(2)}$ is the second-order susceptibility, and $\chi^{(3)}$ the third-order susceptibility of the medium and so on. A similar expression can be written at the molecular level where the susceptibilities are replaced by polarizabilities of various orders as

$$p(t) = p_0 + \alpha E(t) + \beta E^2(t) + \gamma E^3(t) + \cdots \tag{3}$$

where $p(t)$ is the total polarization and p_0 is the polarization in the absence of the radiation field which originates from the permanent dipole moment of a molecule, and α, β, and γ are the linear, second-order, and third-order molecular polarizabilities, respectively.

The macroscopic bulk susceptibility and the microscopic molecular second-order polarizability have been related to each other by Zyss and Oudar (1982) using a oriented gas model by the expression

$$\chi^{(2)} E^2(t) = Nf\beta E^2(t) \tag{4}$$

where N is the number of molecules in the macroscopic medium exposed to radiation, and f is the local field factor which couples the externally applied field to the field actually experienced by the molecules at the incident and second harmonic frequencies.

The second-order nonlinear phenomena, being an even order process, are forbidden in isotropic materials within the electric dipole approximation. At the molecular level in order to have nonzero second-order nonlinearity, the molecules or crystals should not possess a center of symmetry. A nonzero β or $\chi^{(2)}$ molecules or materials can combine two light waves and produce new waves at sum and difference frequencies. A special case of sum frequency generation is second harmonic generation, where both the light waves which are incident on the molecules are degenerate and produce a light wave with twice the frequency of the incident wave. When second harmonic light is scattered from molecules in solution, the light is incoherently scattered due to random orientation of the molecules in the region of interaction of light with molecules. The bulk centrosymmetry of the medium is broken due to instantaneous orientational fluctuations of molecules in the timescale of the interaction. The light is scattered in all directions in the scattering plane and for a finite size detector, the scattered second harmonic (SH) signal is weak.

3.2 SHLS From GNPs

At the surface/interface such as metallic nanoparticles in solution, the symmetry is broken due to nonuniformity of the surface, structural discontinuity, field discontinuity, defects, etc. As a result of all or some of these, a net surface nonlinear polarization develops which acts as a source for SHLS. Incoherent SHLS signal from spherical GNPs has been reported as early as 1998 (Vance, Lemon, & Hupp, 1998). There are a large number of applications of coherent second harmonic generation from surfaces and interfaces, as documented earlier (Gonella & Dai, 2014).

The incoherent second harmonic scattering signal or hyper Rayleigh scattering (HRS) signal observed from metal nanoparticles was a factor of 10^4–10^5 higher than expected, mainly due to a large surface dipole created from the collective oscillations of electrons known as surface plasmons and also due to surface plasmon resonance enhancement at excitation and/or second harmonic wavelengths, unless the measurement wavelengths are carefully chosen to avoid two-photon resonance. With all these cautions in place, SHLS provides high surface sensitivity and a label-free method for investigating processes happening at the surface.

3.3 Apparatus

A typical SHLS (or HRS) setup for measuring the SHLS from solution vary from laboratory to laboratory, but it is still more or less the same as the original setup described by the inventor of the technique (Terhune, Maker, & Savage, 1965). Clays and Persoons who revived the technique in the early 1990s used a set-up similar to ours (Clays & Persoons, 1992). The setup to measure SHLS is shown in Fig. 1. A Q-switched Nd:YAG laser that delivers 1064 nm radiation at a repetition rate of 10 Hz with a pulse width of 10 ns was used as the incident light source. The beam was passed through a long pass colored glass filter (RG 830) to allow IR light to pass but block ambient visible light and focused on the sample by a plano-convex lens of focal length 20 cm to a point 2–3 cm after the sample cell, and the fine adjustment of focusing was done by a translation stage. If the focusing is too close to the window of the cell, the focused beam would drill a hole in the cell. The incident power was varied in the range of 3–12 mJ per pulse (mJ/P) by a rotating polarizer. The sample solutions were taken in a cylindrical glass cell of 2.5 cm diameter and ~20 cc volume. The second harmonic scattered

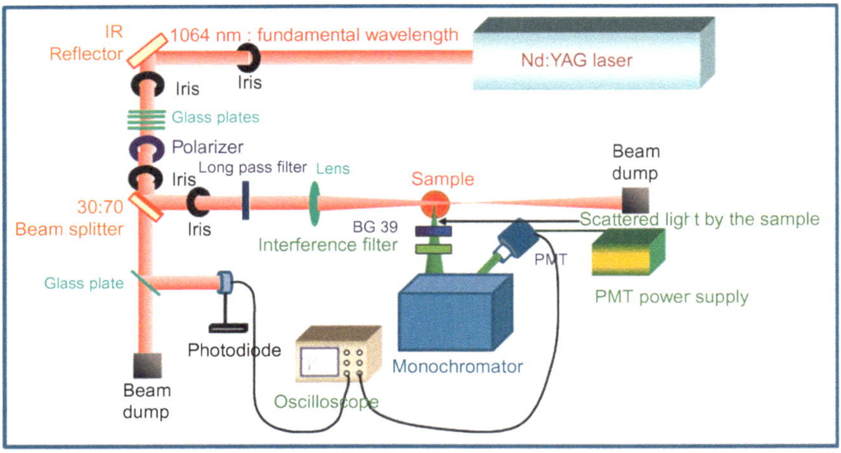

Fig. 1 Schematic diagram for the second harmonic light scattering experimental setup.

signal is known to increase quadratically with the input intensity, and this is checked before collecting the data from unknown samples. At high laser powers with focusing other nonlinear optical processes such as stimulated Raman scattering, stimulated Brillouin scattering (Liu et al., 2009; Shen & Bloembergen, 1965) dielectric breakdown (Sacchi, 1991), etc., may start to contribute. In order to avoid such processes which compete with SHLS the incident laser power is kept as low as possible. Wang, Zhang, Cui, Duan, and Liu, (2005), while studying the aggregation effect of ~10 nm silver nanoparticles in water on HRS signal, have controlled the energy of their picosecond laser pulse at 1064 nm below 10 mJ/P and focused the beam 15 cm after the sample cell to avoid the unwanted nonlinear processes. In our setup, as we are using a nanosecond laser beam which has less peak power than a picosecond beam, the input energy was maintained below 15 mJ/P, and above 25 mJ/P input energy, dielectric breakdown has been noted. The scattered photons from the sample in our experiment were collected at 90 degree with respect to the incident beam. The SH signal is generally accompanied by photons generated by other processes such as linear Rayleigh scattering, Raman scattering, and one or multiphoton fluorescence and thus, required to be separated properly before detection.

Russier-Antoine et al. have reported huge background signal at the SH wavelength from GNPs (∼11 nm) when irradiated with a Ti:Sapphire laser at 1.19 eV (Russier-Antoine, Jonin, Nappa, Bénichou, & Brevet, 2004). They assigned the background as intense photoluminescence from 11 nm GNPs.

Dinkel et al. (2016), while irradiating GNPs with 810 nm laser, have observed a broad background signal at the wings of the sharp SH scattered signal peaking at 405 nm. The background to the red side of the SH signal from 410 to 625 nm was ascribed to originate from two-photon fluorescence while the fluorescence between 350 and 400 nm was assigned to three-photon fluorescence and higher order nonlinear processes. In our setup the SHS signal passed through a broad band IR filter (BG-39) followed by a narrow-band interference filter (CVI-Melles Griot, USA, 532 ± 4 nm) before being dispersed through a two-stage monochromator (JobinYvon Triax 550, Japan) and then collected on a UV–vis photomultiplier tube (Philips Photonics, France, XP2262B).

The signal was averaged typically over 512 shots in a 500-MHz digital storage oscilloscope (Tektronix, USA, TDS 3520B) and stored in a PC. A part of the incident beam was guided to a photodiode (Thorlabs) for the measurement of the power incident on the sample solution during the experiment. The laser power was kept below 5 mJ/P during the SH titrations of GNPs with proteins since high laser power can damage protein samples, although alcohol dehydrogenase (ADH) and insulin are completely transparent at 1064 nm. The scattered signal at 532 nm which is detected, in principle, can be absorbed by the protein. The molar absorption coefficients of ADH and insulin used in this study at 532 nm are 0, and therefore, this possibility is ruled out in our experiments. However, in case of an inner filter effect affecting the signal intensity, a correction would be necessary (Ray & Das, 1995). A schematic diagram of the experimental setup is shown in Fig. 1.

The dependence of the intensity of second harmonic scattered light on input light intensity should be quadratic in nature. The quadratic dependence can be described by the following equation:

$$I_{2\omega} = A\,(I_\omega)^2 \tag{5}$$

where $I_{2\omega}$ is the intensity of the SHS light, I_ω is the intensity of the incident light beam, and A is a proportionality constant which contains the hyperpolarizability and other terms shown in Eq. (4). In the logarithmic form Eq. (5) is written as

$$\log I_{2\omega} = \log A + 2\log I_\omega \tag{6}$$

In general a nonlinear process of order n can be represented by the following equation:

$$\log I_{n\omega} = \log A + n\log I_\omega \tag{7}$$

The scattered light intensity in the range of two-photon wavelength of the incident light was recorded at different incident light energy measured by a power meter (Scientech Vector H410). The scattered signal for a representative sample of GNPs and its dependence on the incident laser intensity are shown in Fig. 2. From Fig. 2A it is evident that the scattered photons are centered at 532 nm which is the second harmonic of the 1064 nm incident light. The fitting of either of the area or the height to the scattered signal with Eq. (7) was found to produce *n* close to 2 which ensures the quadratic nature of the process generating the SH signal (Fig. 2B and C).

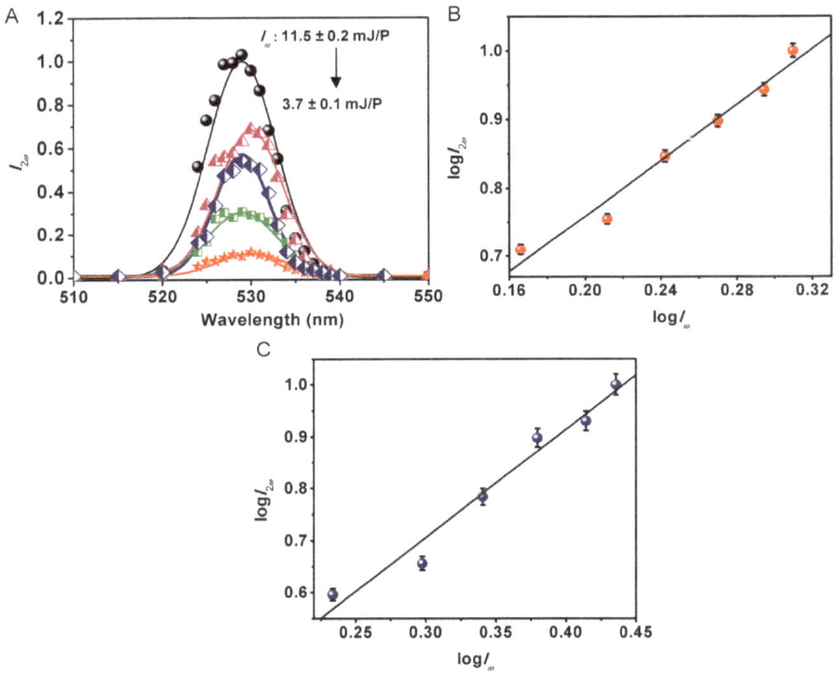

Fig. 2 (A) Monochromator scan of signal (*symbols*) plotted against the incident light intensity at 1064 nm for 0.013 n*M* 45 nm GNPs in 10 m*M* pH 7 phosphate buffer. The smooth curve is a Gaussian fit through the experimental data points, which shows that the peak maximum is not exactly at 532 nm although that is what it should be and the monochromator dial has a backlash of −0.4 nm. Also the signal at different laser powers appears to peak at slightly different wavelengths due the fitting of the data to a Gaussian. (B) Log–log plot of the signal obtained from the area under the Gaussian fit of the experimental data against the incident laser power in mJ/pulse. The straight line is a fit through the experimental points with $\chi^2 = 0.99$ and the slope of the straight line (*n*) is 1.98; (C) the same plot as (B) except the height of the Gaussian is plotted along the *X*-axis. The slope of the straight line fit (*n*) is 2.08 with $\chi^2 = 0.99$.

4. SAMPLE PREPARATION

4.1 Gold Nanoparticles

Turkevich et al. reported the first structural assessment of GNPs by varying the reaction conditions (Turkevich, Stevenson, & Hillier, 1951) and Fren's perfected Turkevich's citrate-mediated growth method and produced narrowly dispersed 16–150 nm GNPs (Frens, 1973). Several strategies for the synthesis of GNPs using chemical reduction route are since available, and they include the reduction of gold chloride using sodium borohydride (Jana, Gearheart, & Murphy, 2001a,2001b), trisodium citrate (TSC) (Frens, 1973), ascorbic acid (AA) (Ziegler & Eychmüller, 2011), hydroquinone (Perrault & Chan, 2009), 2-mercaptosuccinic acid (Niu, Zhu, & Liu, 2007), hydroxylamine (Brown, Walter, & Natan, 2000), etc. The reduction of acid salt of gold chloride by TSC or AA is most popular among the reduction methods. For the synthesis of larger nanoparticles multistep seeded growth procedure has been proven to be advantageous for controlling particle size distribution and shape compared to one-step synthesis routes (Jana et al., 2001a,2001b; Kumar, Mutreja, & Sykes, 2016; Niu et al., 2007; Oh, Susumu, Goswami, & Mattoussi, 2010; Ziegler & Eychmüller, 2011).

A seeded growth procedure was followed by us for the synthesis of 30–60 nm size GNPs starting from 15 nm GNPs (Das et al., 2015) following the method described by Ziegler and Eychmüller (2011) with certain modifications. At first 2.5 mL of $HAuCl_4{\cdot}3H_2O$ solution (0.2%, w/v) was diluted to 50 mL of water taken in a 250-mL two-necked round bottom flask. The solution was then heated to boiling followed by addition of 2 mL of TSC solution (1%, w/v containing 0.05%, w/v citric acid) to the boiling solution under vigorous stirring. After the addition the color of the solution slowly turned red from yellow. The solution was further kept boiling for 10 min and then allowed to cool to room temperature. Following this method, we have synthesized 15 nm GNPs, which were used subsequently as seeds for the next step growth.

For the synthesis of 30 nm GNPs, 6 mL of the 15 nm GNPs solution was diluted to a volume of 40 mL in a 250-mL round bottom flask. To this, a 20-mL aliquot of the precursor solution A containing $HAuCl_4{\cdot}3H_2O$, and a 20-mL of solution B containing TSC as a capping agent and AA as a reducing agent, were added separately at room temperature over an hour under vigorous stirring. The precursor solution A was prepared by dilution of 0.8 mL of 0.2% (w/v) $HAuCl_4 \cdot 3H_2O$ solution to 20 mL. The precursor solution B was prepared by dilution of 0.2 mL 1% (w/v) AA and 0.1 mL 1% (w/v) TSC solutions to 20 mL. After the addition of solutions A and B onto the 15 nm GNPs solution, the resulting mixture was heated to boiling. The solution was boiled for about 30 min and then allowed to cool to room temperature.

For the synthesis of 45 nm GNPs, a similar procedure was adopted starting with 15 nm GNPs as seeds. Only the precursor solution A was prepared by dilution of 4 mL of 0.2% (w/v) $HAuCl_4 \cdot 3H_2O$ solution to 20 mL, and precursor solution B was prepared by dilution of 1 mL 1% (w/v) AA and 0.5 mL 1% (w/v) TSC solutions to 20 mL. This provided a higher amount of $HAuCl_4 \cdot 3H_2O$ to meet the requirement for making larger size particles.

For the preparation of 60 nm GNPs, 45 nm seeds were used with precursor solution A being prepared by dilution of 1.2 mL of 0.2% (w/v) $HAuCl_4 \cdot 3H_2O$ solution to 20 mL, and precursor solution B being prepared by dilution of 0.3 mL 1% (w/v) AA and 0.15 mL 1% (w/v) TSC solutions to 20 mL. The amount of gold precursor ($HAuCl_4 \cdot 3H_2O$) required for the growth of seed particles to achieve a certain diameter has been calculated by knowing the diameter and concentration of seed particle and following the procedure described by Liu, Atwater, Wang, and Huo (2007) The approximations based on which the calculation was done are complete transformation of seed particles to growth particles and formation of uniform spherical particles. The size of the nanoparticles was determined from the statistical distribution of ~500 particle's bright field transmission electron microscope images (Das et al., 2015).

The GNPs were purified from the unreacted reducing and capping agents by centrifugation of the as-synthesized GNPs solution in microcentrifuge tubes. The centrifugation allowed nanoparticles to settle down at the bottom of the tube and the unreacted reagents in the supernatant. The centrifugate (GNPs) was dispersed in milli-Q water. The process was repeated twice to ensure ~100% purification. The centrifugation speed and time employed for purifying different size GNPs are as follows: 15,000 rpm for 30 min (15 nm GNPs), 9000 rpm for 15 min (30 nm GNPs), 6000 rpm for 15 min (45 nm GNPs), 3000 rpm for 15 min (60 nm GNPs). The concentration of the as-prepared GNPs was determined from the amount of gold precursor used in the synthesis and diameter of the particles obtained from the TEM analysis following a method suggested by Liu et al. (2007). The molar extinction coefficients of GNPs were then determined from the absorbance vs concentration plot which were used for the determination of the concentration of the purified GNPs in solution throughout. The molar extinction coefficient of GNPs at their SPR band maxima is found as following: $\varepsilon_{520\,\mathrm{nm}}^{15\,\mathrm{nm}} = 4.80 \times 10^8 M^{-1}\mathrm{cm}^{-1}$, $\varepsilon_{522\,\mathrm{nm}}^{30\,\mathrm{nm}} = 1.96 \times 10^9 M^{-1}\mathrm{cm}^{-1}$, $\varepsilon_{524\,\mathrm{nm}}^{45\,\mathrm{nm}} = 9.91 \times 10^9 M^{-1}\mathrm{cm}^{-1}$, $\varepsilon_{529\,\mathrm{nm}}^{60\,\mathrm{nm}} = 2.70 \times 10^{10} M^{-1}\mathrm{cm}^{-1}$. The nanoparticles were further dispersed in a volume of 15 mL, where the concentration and pH were adjusted to 10 m*M* and 7, respectively, by 50 m*M* potassium phosphate buffer of pH < 7 (to adjust the pH to 7 during dilution). The concentration of GNPs in 10 m*M* buffer was maintained to 0.100 n*M* (15 nm GNPs), 0.032 n*M* (30 nm GNPs), 0.013 n*M* (45 nm GNPs), 0.008 n*M* (60 nm GNPs).

The concentration of GNPs in solution in molarity was calculated as

$$C = \frac{N_{\mathrm{Total}}}{NVN_{\mathrm{A}}}, \tag{8}$$

where C is concentration, N_{Total} is the total number of gold atoms in the starting material $HAuCl_4 \cdot 3H_2O$, N is the average number of gold atoms in a nanoparticle, V is the final volume of the solution in L, and N_{A} is the Avogadro's number.

4.2 Proteins

The stock solutions of ADH and insulin were prepared in 10 m*M* phosphate buffer of pH 7.0, and the concentrations were determined using the extinction coefficients of $\varepsilon_{280\,\mathrm{nm}} = 3.53 \times 10^4 M^{-1}\mathrm{cm}^{-1}$ and $\varepsilon_{276\,\mathrm{nm}} = 6100\, M^{-1}\mathrm{cm}^{-1}$, respectively (Dunn & Hutchison, 1973; Gammeltoft, 1981). The stock solutions were diluted to produce different concentrations in the phosphate buffer for SHLS experiments (Das et al., 2015, 2016).

5. INTERACTION OF ADH WITH GNPs

The change in the SHLS intensity of GNPs upon addition of small aliquots of ADH was monitored at 298 K, and the data are shown in Fig. 3. The HRS light intensity decreases as a function of the protein concentration. A fixed amount of GNPs is dispersed in 15 mL buffer and aliquots of small amounts of protein in 400 μL of buffer is added in a step-wise fashion into the GNP solution. Since the volume change upon addition of protein solution is negligible compared to the total volume, no volume correction was done. With each addition of protein

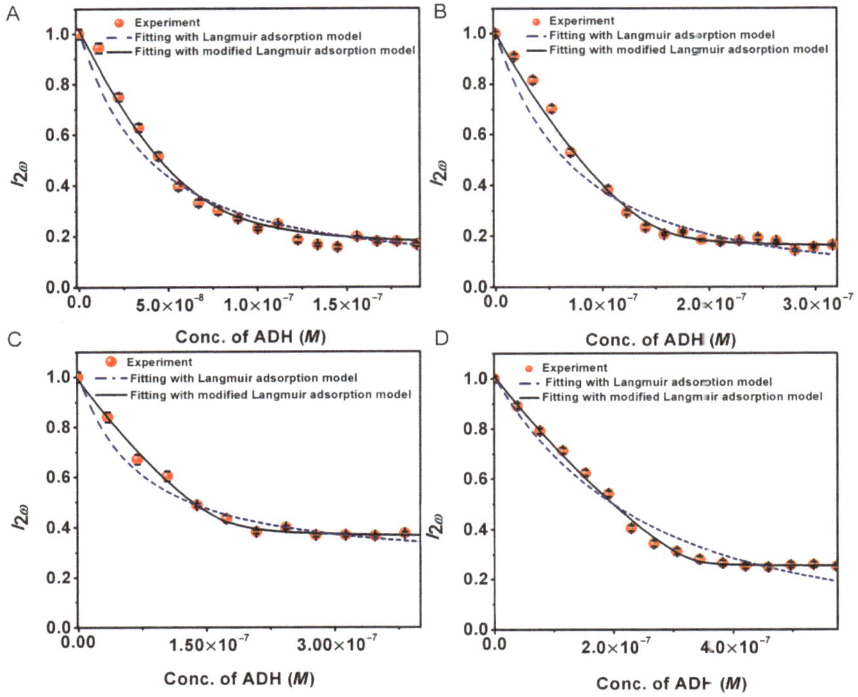

Fig. 3 Normalized SH intensity at 532 nm from (A) 0.1 nM 15 nm GNPs, (B) 0.032 nM 30 nm GNPs, (C) 0.013 nM 45 nm GNPs, (D) 0.008 nM 60 nm GNPs at 10 mM pH 7 phosphate buffer as a function of addition of ADH. Each data points were recorded after ~5 min of ADH addition. The SH intensity was normalized with respect to the SH intensity from pure GNP solution before addition of ADH.

solution the decrease in the normalized SH intensity is significant (~5%–10%) until it reaches a point when no further protein molecule binds to the surface and the signal reaches steady state and does not change any further. This reduction in SH signal is due to the adsorption of ADH as it binds to the surface of the GNPs. At neutral pH in 10 mM phosphate buffer the GNPs are evenly dispersed in solution. At higher concentrations of the buffer, the citrate capped GNPs aggregate in solution, which is detrimental to their application at neutral pH (Das et al., 2016). The citrate-capped GNPs and the ADH molecules at pH 7.0 are negatively charged which were ascertained from the zeta potential measurements (Das et al., 2015).

The kinetics of adsorption of protein molecule on GNP surface needs to be mentioned here. In the absence of external stirring the binding kinetics of protein is expected to be guided purely by diffusion. The timescale of adsorption of thiol containing molecules has been probed by several groups in the past (Dinkel et al., 2016; Gan, Gonella, Zhang, Dai, 2011; Gan, Xu, et al., 2011). For thiol adsorption on citrate-capped GNPs, the SHS signal typically reaches its steady-state value within minutes or so. Since we are interested in equilibrium adsorption and desorption of protein molecules on the GNP surface, we have incubated the GNP protein solution for a few minutes before taking the SHS signal scan through the monochromator.

The changes in the SHLS intensity as shown in Fig. 3 can be quantitatively related to the ADH coverage on the GNP surface and the binding constant of the protein to the GNP surface using a Langmuir-type adsorption model. In the Langmuir model of adsorption, the surface is treated as a lattice of identical noninteracting sites, and the adsorption is described as a binding equilibrium between the enzyme and the surface sites. This has been demonstrated for adsorption of small organic dyes on the micrometer-sized polystyrene beads (Wang, Yan, Liu, & Eisenthal, 1998). The SHS signal measured as a function of added concentration, C of the protein in the bulk solution can be easily converted to the molecular number density, N of protein molecules adsorbed on the surface where $N_{\max}$ is the number density at maximum surface coverage θ, where $\theta = N/N_{\max}$. If the depletion of the bulk protein concentration due to adsorption is assumed to be small compared to the bulk concentration, we obtain the usual expression for the Langmuir adsorption

$$\frac{dN}{dt} = k_1 \frac{(C-N)}{55.5}(N_{\max} - N) - k_{-1}N, \tag{9}$$

where N and $N_{\max}$ and C pertain to ADH and k_1 and k_{-1} are the rate constants for adsorption and desorption, respectively. After a while when equilibrium is established, the rates of adsorption and desorption become equal, and the equilibrium constant replaces the rate constants as

$$\frac{N}{N_{\max}} = \frac{1}{1 + \dfrac{55.5}{KC}} \tag{10}$$

where θ influences the SH scattered signal, $I_{2\omega}$ from GNPs in solution. The implicit assumption in the Langmuir adsorption model is that adsorbed molecules do not interact with each other, and the free energy change in the process is independent of the surface coverage which in any case cannot exceed a monolayer of substance on the surface. In writing Eq. (10) we have assumed that the number of adsorbed ADH molecules is very small compared to the bulk concentration of ADH and thus $(C-N) \approx C$.

Adsorption of the protein molecules on the GNP surface causes the SH signal to decrease and the decrease is proportional to θ. As we have detected incoherent second harmonic light in a fixed perpendicular detection geometry, the normalized SHS light intensity can be written as (Gan, Gonella, et al., 2011)

$$I_{\mathrm{SH}} = |1 - Q'\theta|^2 \tag{11}$$

where Q' is the overall portion which can be quenched. In other words, if the product of Q' and θ is 1, the SH signal from the GNPs will be fully quenched by addition of protein. Now combining the Langmuir adsorption model and Eq. (11) we can write

$$I_{\mathrm{SH}} = \left|1 - Q'\left(\frac{1}{1 + \frac{55.5}{KC}}\right)\right|^2 \tag{12}$$

The fit to the experimental data points in Fig. 3 by Eq. (12) using a nonlinear least square fit with two parameters Q' and K is shown by blue dotted lines. It is apparent that the fit is reasonable but not perfect. A careful look at the data plotted in Fig. 3 shows that the SHS signal reaches a steady value after about 80% quenching and remains unchanged thereafter. According to Eq. (12) this is not possible and thus in the Langmuir fit the lower part of the decay curve is not reproduced well. In fact, the fit to Eq. (12) shown by the blue dotted line gives large standard deviation indicating a poor fit. Instead, the data can be fitted better to a modified Langmuir adsorption model, where the assumption that most of the ADH molecules remain in the bulk is done away with. In the modified Langmuir model, the SH intensity is related to protein concentration as

$$\mathrm{I}_{\mathrm{SH}} = \left|1 - Q'\left(\frac{\left(C + N_{\max} + \frac{55.5}{K}\right) - \sqrt{\left(C + N_{\max} + \frac{55.5}{K}\right)^2 - 4CN_{\max}}}{2N_{\max}}\right)\right|^2 \tag{13}$$

where N_{max} refers to the maximum number of protein molecules that can adsorb (number of available binding sites) on the surface. When Eq. (13) is fitted to Fig. 3, the fitted curve is shown in black solid, and it reproduces the titration data very well. The values of K, N_{max} for the adsorption of ADH on the surface of GNPs obtained after fitting the data with a modified Langmuir adsorption model have been listed in Table 1. The number of protein molecules adsorbed per nanoparticle at saturation ($n_{ads} = N_{max}/N_{GNP}$) has been calculated from the analysis and is shown in Table 1 for each nanoparticle size.

The Gibb's free energy change for the binding of the protein on the surface sites has also been calculated from the equilibrium constant using the thermodynamic relation ($\Delta G^0 = -RT \ln K$). The values of the Gibb's free energy change for the adsorption of protein on the nanoparticle surface are listed in Table 1. The results indicate that the number of ADH molecules at saturation of adsorption per nanoparticle (n_{ads}) increases with size of the nanoparticles. This is quite expected as the surface area of a nanoparticle increases with size and a larger surface area can accommodate a larger number of protein units.

The free energy of adsorption at room temperature (25°C) was calculated from the equilibrium constant, and the value is −55 kJ/mol which is much lower than the free energy required for chemisorption (Humpola, Odetti, Fertitta, & Vincente, 2013). The low value of free energy change indicates that the interaction is weak, and ADH is held on the surface by weak forces leading to physisorption. The weak binding makes the detachment of ADH from the particle surface easier. In an earlier study our group has shown that (Das et al., 2016) the adsorbed ADH can be quantitatively recovered by simple centrifugation of the protein–nanoparticle conjugate.

Table 1 Thermodynamic Parameters for the Adsorption of ADH on GNPs Surface

Size of GNPs (nm)	N_{max} for ADH ($\times 10^{-8}$ M)	n_{ads}	K ($\times 10^9$ M^{-1})	$\Delta G°$ (kJ/mol)
15	9.25	925	2.98	−54.0
30	15.13	4,727	10.97	−57.3
45	16.89	12,798	17.72	−58.5
60	33.29	41,611	57.73	−61.4

6. INTERACTION OF INSULIN WITH GNPs

Our investigation was extended to a different protein to establish the utility of SHLS to probe the thermodynamics of protein–nanoparticle interaction and generalize such observation. Insulin was chosen for which the time taken to attain the equilibrium adsorption on GNP surface was found to be shorter than ADH adsorption. The titration data for insulin are shown in Fig. 4 and have been summarized in Table 2. The data are similar to what was observed for ADH and thus the SHLS method for studying NP–Protein interaction is general and do not depend on the nature of the protein or the NP.

7. STOICHIOMETRY OF THE PROTEIN–GNP CONJUGATE

Two important quantities are obtained from the SHS studies of the GNP–enzyme conjugate. The binding constant or the free energy of adsorption of the enzyme on the surface of GNP and the maximum number of protein molecules that are adsorbed (n_{ads}) per GNP have been measured. The first quantity gives us the conventional binding constant assuming a model that is built on the basis of 1:1 binding. This is certainly not applicable to enzyme adsorption on GNP since we would expect one particle of gold to bind to many molecules of protein. Protein corona is formed on the surface. How many proteins are there on a nanoparticle surface? The second quantity

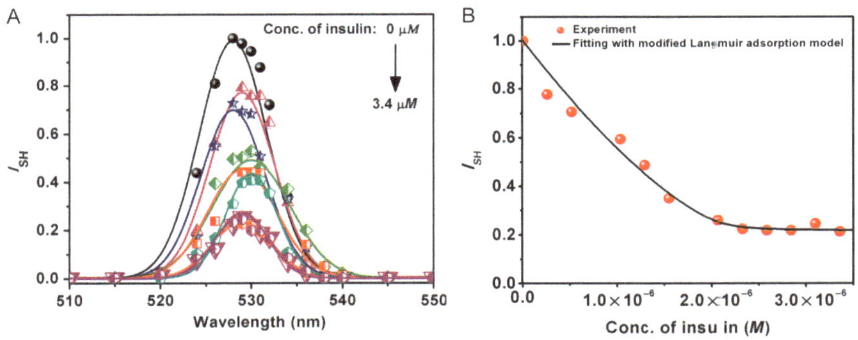

Fig. 4 (A) Normalized scattered light intensity in the range of 510–550 nm from 0.013 n*M* 45 nm GNPs in phosphate buffer as a function of insulin concentration. Each data points were recorded after ~2 min of insulin addition. (B) Normalized area under Gaussian curves as a function of addition of insulin. The intensity was normalized with respect to the highest intensity from pure GNPs solution before addition of insulin.

Table 2 Thermodynamic Parameters for the Adsorption of Insulin on GNPs

Size of GNPs (nm)	N_{max} for insulin ($\times 10^{-6}$ M)	n_{ads}	K ($\times 10^9$ M^{-1})	$\Delta G°$ (kJ/mol)
15	1.32	13,200	1.10	−51.6
30	1.73	54,100	2.84	−53.9
45	2.29	176,100	10.33	−57.1
60	2.43	303,700	25.29	−59.3

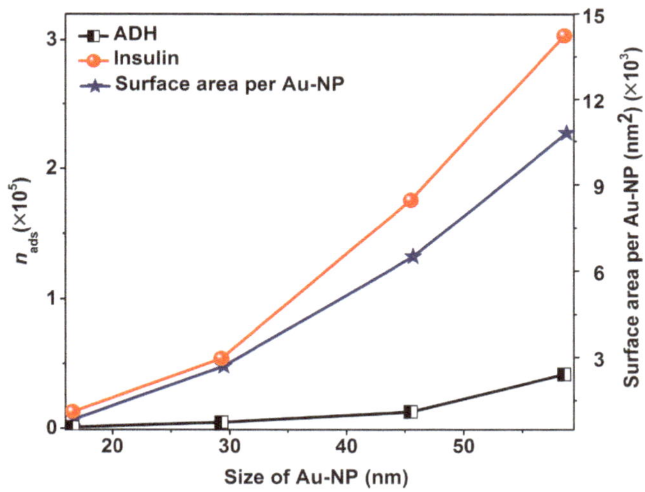

Fig. 5 The variation of n_{ads} as a function of size of the gold nanoparticle and its surface area.

that is the stoichiometry of the binding interaction, i.e., n_{ads} increases with increasing size of the GNP and correlates well with increasing surface area of the GNPs with size (Fig. 5). The data shown in Fig. 5 point out clearly two things: (1) a larger surface area adsorbs more number of proteins and small protein molecules bind on the surface in more numbers, and (2) the number of proteins bound to the surface increases in a nonlinear fashion with the size of the particle but it correlates well with the surface area. This observation also raises some important questions: how are the proteins arranged physically on the surface or why is there a nonlinear size effect of protein binding to nanoparticles? In the literature there are reports on protein corona on nanoparticles. In fact, several investigators have used DLS to infer that the thickness of the adsorbed protein layer increases with increasing size or surface area of the particle. By indiscriminate binding study of citrate-capped GNPs with blood serum proteins, Dobrovolskaia et al. estimated that 14.4 μg total protein is bound to 3534×10^{12} nm^2 surface of 30 nm GNPs (Dobrovolskaia et al., 2009). This number turns out to be slightly lower than what is found out by us. On the other hand, Tenzer et al. after investigating human blood plasma corona on silica nanoparticles in a proteomic analysis concluded that electrostatic effects alone are not the major driving force regulating such interaction and small change in particle size can drastically change the nanoparticle corona. In fact, the size of the particle is a crucial determinant of the nature and stoichiometry of the adsorbed protein layer on the nanoparticle surface (Tenzer et al., 2011).

8. SUMMARY AND FUTURE OUTLOOK

Protein adsorption on the surface of nanomaterials has been a subject of immense interest in recent past and of particular interest in designing of devices in biomedical diagnostics. The first step toward understanding the nature of nanoparticle interaction with proteins in model systems as well as in biofluids is characterization and analysis of different levels of adsorbed proteins on the surface of nanoparticles of varying chemical composition, size, shape, and polarity. The nanoparticle bound proteins may also enable differential release, facilitate translocation across cellular membranes, and allow targeting (clearance or accumulation) in vital organs of human body. Systematic analysis of the nature of binding of nanoparticles of varying size, shape, and other properties with proteins having different structural and functional features will substantiate our existing knowledge of nanoparticle–protein interactions. Out of many different spectroscopic and calorimetric techniques used in these studies, we have discussed the usefulness and potential of the highly sensitive, SHLS technique in this chapter. Nanoparticle-based drugs display better solubility, biodistribution, pharmacokinetics and are easier to administer with less toxicity and side effects. Particle size regulates the decoration of the surface with the number of protein molecules. The potency of a protein- or peptide-based drug may be altered upon binding with nanoparticles and based on the local concentration of the drug in the corona, the effectiveness may go up. The conjugation process between the drug and the nanoparticle may lead to higher efficacy at low concentrations which will, in turn, reduce the side effects of such formulations. However, many of these possibilities remain to be investigated and explored and brought to fruition.

Our studies, we hope, will not only bring clarity in the synthesis of nanoparticle–protein bioconjugates and lead to better understanding on the behavior of the nanoconjugates *in vivo* but also initiate more research advancements in protein-based therapeutics and diagnostics.

ACKNOWLEDGMENT

P.K.D. thanks the Ministry of Communication and Information Technology for the center of excellence in Nanoelectronics, Phase II grant for generous funding of the research described here.

REFERENCES

Abadeer, N. S., & Murphy, C. J. (2016). Recent progress in cancer thermal therapy using gold nanoparticles. *The Journal of the Physical Chemistry C*, *120*, 4691–4716.

Aggarwal, P., Hall, J. B., McLeland, C. B., Dobrovolskaia, M. A., & McNeil, S. E. (2009). Nanoparticle interaction with plasma proteins as it relates to particle biodistribution, biocompatibility and therapeutic efficacy. *Advanced Drug Delivery Reviews*, *61*, 428–437.

Aldeek, F., Safi, M., Zhan, N., Palui, G., & Mattoussi, H. (2013). Understanding the self-assembly of proteins onto gold nanoparticles and quantum dots driven by metal-histidine coordination. *ACS Nano*, 7, 10197–10210.

Baek, S. H., Wark, A. W., & Lee, H. J. (2014). Dual nanoparticle amplified surface plasmon resonance detection of thrombin at subattomolar concentrations. *Analytical Chemistry*, *86*, 9824–9829.

Bakhtiar, R. (2013). Surface plasmon resonance spectroscopy: A versatile technique in a biochemist's toolbox. *Journal of Chemical Education*, *90*, 203–209.

Bhumkar, D. R., Joshi, H. M., Sastry, M., & Pokharkar, V. B. (2007). Chitosan reduced gold nanoparticles as novel carriers for transmucosal delivery of insulin. *Pharmaceutical Research*, *24*, 1415–1426.

Boyd, R. W. (2003). *Nonlinear optics* (2nd ed.). California, CA: Academic press.

Breault-Turcot, J., Chaurand, P., & Masson, J.-F. (2014). Unravelling nonspecific adsorption of complex protein mixture on surfaces with SPR and MS. *Analytical Chemistry*, *86*, 9612–9619.

Brown, K. R., Walter, D. G., & Natan, M. J. (2000). Seeding of colloidal Au nanoparticle solutions. 2. Improved control of particle size and shape. *Chemistry of Materials*, *12*, 306–313.

Brun, E., & Sicard-Roselli, C. (2014). Could nanoparticle corona characterization help for biological consequence prediction? *Cancer Nanotechnology*, *5*, 7.

Cedervall, T., Lynch, I., Lindman, S., Berggård, T., Thulin, E., Nilsson, H., et al. (2007). Understanding the nanoparticle–protein corona using methods to quantify exchange rates and affinities of proteins for nanoparticles. *PNAS*, *104*, 2050–2055.

Chemla, D. S., & Zyss, J. (Eds.), (1987). *Nonlinear optical properties of organic molecules and crystals: Vol. 1*. Florida: Academic press, Inc.

Clays, K., & Persoons, A. (1992). Hyper-Rayleigh scattering in solution. *Review of Scientific Instruments*, *63*, 3285–3289.

Das, A., Chakrabarti, A., & Das, P. K. (2015). Suppression of protein aggregation by gold nanoparticles: A new way to store and transport proteins. *RSC Advances*, *5*, 38558–38570.

Das, A., Chakrabarti, A., & Das, P. K. (2016). Probing protein adsorption on a nanoparticle surface using second harmonic light scattering. *Physical Chemistry Chemical Physics*, *18*, 24325–24331.

Dhar, S., Daniel, W. L., Giljohann, D. A., Mirkin, C. A., & Lippard, S. J. (2009). Polyvalent oligonucleotide gold nanoparticle conjugates as delivery vehicles for platinum (IV) warheads. *Journal of the American Chemical Society*, *131*, 14652–14653.

Dinkel, R., Braunschweig, B., & Peukert, W. (2016). Fast and slow ligand exchange at the surface of colloidal gold nanoparticles. *The Journal of the Physical Chemistry C*, *120*, 1673–1682.

Dobrovolskaia, M. A., Patri, A. K., Zheng, J., Clogston, J. D., Ayub, N., Aggarwal, P., et al. (2009). Interaction of colloidal gold nanoparticles with human blood: Effects on particle size and analysis of plasma protein binding profiles. *Nanomedicine*, *5*, 106–117.

Dreaden, E. C., Alkilany, A. M., Huang, X., Murphy, C. J., & El-Sayed, M. A. (2012). The golden age: Gold nanoparticles for biomedicine. *Chemical Society Reviews*, *41*, 2740–2779.

Dunn, M. F., & Hutchison, J. S. (1973). Roles of zinc ion and reduced coenzyme in the formation of a transient chemical intermediate during the equine liver alcohol dehydrogenase catalyzed reduction of an aromatic aldehyde. *Biochemistry*, *12*, 4882–4892.

Dykman, L., & Khlebtsov, N. (2012). Gold nanoparticles in biomedical applications: Recent advances and perspectives. *Chemical Society Reviews*, *41*, 2256–2282.

Fan, Z., Senapati, D., Singh, A. K., & Ray, P. C. (2013). Theranostic magnetic core–plasmonic shell star shape nanoparticle for the isolation of targeted rare tumor cells from whole blood, fluorescence imaging, and photothermal destruction of cancer. *Molecular Pharmaceutics*, *10*, 857–866.

Frens, G. (1973). Controlled nucleation for the regulation of the particle size in monodisperse gold suspensions. *Nature Physical Science*, *241*, 20–22.

Gammeltoft, S. (1981). Receptor binding of biosynthetic human insulin on isolated pig hepatocytes. *Diabetes Care*, *4*, 235–237.

Gan, W., Gonella, G., Zhang, M., & Dai, H.-L. (2011). Communication: Reactions and adsorption at the surface of silver nanoparticles probed by second harmonic generation. *The Journal of Chemical Physics*, *134*, 041104.

Gan, W., Xu, B., & Dai, H.-L. (2011). Activation of thiols at a silver nanoparticle surface. *Angewandte Chemie, International Edition*, *50*, 6622–6625.

Gonella, G., & Dai, H.-L. (2014). Second harmonic light scattering from the surface of colloidal objects: Theory and applications. *Langmuir*, *30*, 2588–2599.

Gourishankar, A., Shukla, S., Ganesh, K. N., & Sastry, M. (2004). Isothermal titration calorimetry studies on the binding of DNA bases and PNA base monomers to gold nanoparticles. *Journal of the American Chemical Society*, *126*, 13186–13187.

Guo, T., Liu, F., Liang, X., Qiu, X., Huang, Y., Xie, C., et al. (2016). Highly sensitive detection of urinary protein variations using tilted fiber grating sensors with plasmonic nanocoatings. *Biosensors and Bioelectronics*, *78*, 221–228.

Gür, F. N., Schwarz, F. W., Ye, J., Diez, S., & Schmidt, T. L. (2016). Toward self-assembled plasmonic devices: High-yield arrangement of gold nanoparticles on DNA origami templates. *ACS Nano*, *10*, 5374–5382.

Huang, R., & Lau, B. L. T. (2016). Biomolecule–nanoparticle interactions: Elucidation of the thermodynamics by isothermal titration calorimetry. *Biochimica et Biophysica Acta*, *1860*, 945–956.

Humpola, P. D., Odetti, H. S., Fertitta, A. E., & Vincente, J. L. (2013). Thermodynamic analysis of adsorption models of phenol in liquid phase on different activated carbons. *Journal of the Chilean Chemical Society*, *58*, 1541–1544.

Ipe, B. I., Shukla, A., Lu, H., Zou, B., Rehage, H., & Niemeyer, C. M. (2006). Dynamic light-scattering analysis of the electrostatic interaction of hexahistidine tagged cytochrome P450 enzyme with semiconductor quantum dots. *Chemphyschem: A European Journal of Chemical Physics and Physical Chemistry*, 7, 1112–1118.

Jana, N. R., Gearheart, L., & Murphy, C. J. (2001a). Seed-mediated growth approach for shape-controlled synthesis of spheroidal and rod-like gold nanoparticles using a surfactant template. *Advanced Materials*, *13*, 1389–1393.

Jana, N. R., Gearheart, L., & Murphy, C. J. (2001b). Seeding growth for size control of 5-40 nm diameter gold nanoparticles. *Langmuir*, *17*, 6782–6786.

Jaque, D., Richard, C., Viana, B., Soga, K., Liu, X., & Solé, J. G. (2016). Inorganic nanoparticles for optical bioimaging. *Advances in Optics and Photonics*, *8*, 1–103.

Jelesarov, I., & Bosshard, H. R. (1999). Isothermal titration calorimetry and differential scanning calorimetry as complementary tools to investigate the energetics of biomolecular recognition. *Journal of Molecular Recognition*, *12*, 3–18.

Joshi, H., Shirude, P. S., Bansal, V., Ganesh, K. N., & Sastry, M. (2004). Isothermal titration calorimetry studies on the binding of amino acids to gold nanoparticles. *The Journal of Physical Chemistry B*, *108*, 11535–11540.

Klebe, G. (2015). Applying thermodynamic profiling in lead finding and optimization. *Nature Reviews Drug Discovery*, *14*, 95–110.

Kumar, D., Mutreja, I., & Sykes, P. (2016). Seed mediated synthesis of highly monodispersed gold nanoparticles in the presence of hydroquinone. *Nanotechnology*, *27*, 355601.

Lacerda, S. H. D. P., Park, J. J., Meuse, C., Pristinski, D., Becker, M. L., Karim, A., et al. (2010). Interaction of gold nanoparticles with common human blood proteins. *ACS Nano*, *4*, 365–379.

Leuvering, J. H., Thal, P. J., van der Waart, M., & Schuurs, A. H. (1980). Sol particle immunoassay (SPIA). *Journal of Immunoassay*, *1*, 77–91.

Li, L., Mu, Q., Zhang, B., & Yan, B. (2010). Analytical strategies for detecting nanoparticle-protein interactions. *Analyst*, *135*, 1519–1530.

Liang, F.-X., Ge, C.-W., Zhang, T.-F., Xie, W.-J., Zhang, D.-Y., Zou, Y.-F., et al. (2017). Plasmonic hollow gold nanoparticles induced high-performance Bi_2S_3 nanoribbon photodetector. *Nanophotonics*, *6*, 494–501.

Liu, X., Atwater, M., Wang, J., & Huo, Q. (2007). Extinction coefficient of gold nanoparticles with different sizes and different capping ligands. *Colloids and Surfaces B, Biointerfaces*, *58*, 3–7.

Liu, Y., Liu, Q., Chen, S., Cheng, F., Wang, H., & Peng, W. (2015). Surface plasmon resonance biosensor based on smart phone platforms. *Scientific Reports*, *5*, 12864.

Liu, D., Shi, J., Ouyang, M., Chen, X., Liu, J., & He, X. (2009). Pumping effect of stimulated Brillouin scattering on stimulated Raman scattering in water. *Physical Review A*, *80*, 033808.

Mirkin, C. A., Letsinger, R. L., Mucic, R. C., & Storhoff, J. J. (1996). A DNA-based method for rationally assembling nanoparticles into macroscopic materials. *Nature*, *382*, 607–609.

Mocan, L., Matea, C. T., Bartos, D., Mosteanu, O., Pop, T., Mocan, T., et al. (2016). Advances in cancer research using gold nanoparticles mediated photothermal ablation. *Clujul Medical*, *89*, 199–202.

Mrksich, M., Sigal, G. B., & Whitesides, G. M. (1995). Surface plasmon resonance permits in situ measurement of protein adsorption on self-assembled monolayers of alkanethiolates on gold. *Langmuir*, *11*, 4383–4385.

Nair, L. V., Nazeer, S. S., Jayasree, R. S., & Ajayaghosh, A. (2015). Fluorescence imaging assisted photodynamic therapy using photosensitizer-linked gold quantum clusters. *ACS Nano*, *9*, 5825–5832.

Neely, A., Perry, C., Varisli, B., Singh, A. K., Arbneshi, T., Senapati, D., et al. (2009). Ultrasensitive and highly selective detection of Alzheimer's disease biomarker using two-photon Rayleigh scattering properties of gold nanoparticle. *ACS Nano*, *3*, 2834–2840.

Nguyen, H. H., Park, J., Kang, S., & Kim, M. (2015). Surface plasmon resonance: A versatile technique for biosensor applications. *Sensors*, *15*, 10481–10510.

Niikura, K., Iyo, N., Matsuo, Y., Mitomo, H., & Ijiro, K. (2013). Sub-100 nm gold nanoparticle vesicles as a drug delivery carrier enabling rapid drug release upon light irradiation. *ACS Applied Materials & Interfaces*, *5*, 3900–3907.

Niu, J., Zhu, T., & Liu, Z. (2007). One-step seed-mediated growth of 30–150 nm quasispherical gold nanoparticles with 2-mercaptosuccinic acid as a new reducing agent. *Nanotechnology*, *18*, 325607.

Núñez, S., Venhorst, J., & Kruse, C. G. (2012). Target–drug interactions: First principles and their application to drug discovery. *Drug Discovery Today*, *17*, 10–22.

Oh, E., Susumu, K., Goswami, R., & Mattoussi, H. (2010). One-phase synthesis of water-soluble gold nanoparticles with control over size and surface functionalities. *Langmuir*, *26*, 7604–7613.

Oh, M. H., Yu, J. H., Kim, I., & Nam, Y. S. (2015). Genetically programmed clusters of gold nanoparticles for cancer cell-targeted photothermal therapy. *ACS Applied Materials & Interfaces*, *7*, 22578–22586.

Palchetti, S., Digiacomo, L., Pozzi, D., Peruzzi, G., Micarelli, E., Mahmoudi, M., et al. (2016). Nanoparticles-cell association predicted by protein corona fingerprints. *Nanoscale*, *8*, 12755–12763.

Patching, S. G. (2014). Surface plasmon resonance spectroscopy for characterisation of membrane protein–ligand interactions and its potential for drug discovery. *Biochimica et Biophysica Acta*, *1838*, 43–55.

Paul, A. M., Fan, Z., Sinha, S. S., Shi, Y., Le, L., Bai, F., et al. (2015) Bioconjugated gold nanoparticle based SERS probe for ultrasensitive identification of mosquito-borne viruses using Raman fingerprinting. *Journal of Physical Chemistry C*, *119*, 23669–23675.

Pelaz, B., Jaber, S., de Aberasturi, D. J., Wulf, V., Aida, T., de la Fuente, J. M., et al. (2012). The state of nanoparticle-based nanoscience and biotechnology: Progress, promises, and challenges. *ACS Nano*, *6*, 8468–8483.

Perrault, S. D., & Chan, W. C. W. (2009). Synthesis and surface modification of highly monodispersed, spherical gold nanoparticles of 50-200 nm. *Journal of the American Chemical Society*, *131*, 17042–17043.

Puiu, M., & Bala, C. (2016). SPR and SPR imaging: Recent trends in developing nanodevices for detection and real-time monitoring of biomolecular events. *Sensors*, *16*, 870.

Radauer-Preiml, I., Andosch, A., Hawranek, T., Luetz-Meindl, U., Wiederstein, M., Horejs-Hoeck, J., et al. (2016). Nanoparticle-allergen interactions mediate human allergic responses: Protein corona characterization and cellular responses. *Particle and Fibre Toxicology*, *13*, 3.

Ray, P. C. (2010). Size and shape dependent second order nonlinear optical properties of nanomaterials and their application in biological and chemical sensing. *Chemical Reviews*, *110*, 5332–5365.

Ray, P. C., & Das, P. K. (1995). Dissociation constants of weak organic acids in protic solvents obtained from their first hyperpolarizabilities in solution. *The Journal of Physical Chemistry*, *99*, 17891–17895.

Ritz, S., Schöttler, S., Kotman, N., Baier, G., Musyanovych, A., Kuharev, J., et al. (2015). Protein corona of nanoparticles: Distinct proteins regulate the cellular uptake. *Biomacromolecules*, *16*, 1311–1321.

Russier-Antoine, I., Huang, J., Benichou, E., Bachelier, G., Jonin, C., & Brevet, P.-F. (2008). Hyper Rayleigh scattering of protein-mediated gold nanoparticles aggregates. *Chemical Physics Letters*, *450*, 345–349.

Russier-Antoine, I., Jonin, C., Nappa, J., Bénichou, E., & Brevet, P.-F. (2004). Wavelength dependence of the hyper Rayleigh scattering response from gold nanoparticles. *The Journal of Chemical Physics*, *120*, 10748–10752.

Sacchi, C. A. (1991). Laser-induced electric breakdown in water. *Journal of the Optical Society of America B*, *8*, 337–345.
Saptarshi, S. R., Duschl, A., & Lopata, A. L. (2013). Interaction of nanoparticles with proteins: Relation to bio-reactivity of the nanoparticle. *Journal of Nanobiotechnology*, *11*, 26.
Schasfoort, R. B. M., & Tudos, A. J. (Eds.), (2008). *Handbook of surface plasmon resonance*. Cambridge: RSC Publishing
Shang, W., Nuffer, J. H., Muñiz-Papandrea, V. A., Colón, W., Siegel, R. W., & Dordick, J. S. (2009). Cytochrome c on silica nanoparticles: Influence of nanoparticle size on protein structure, stability, and activity. *Small*, *5*, 470–476.
Shen, Y. R., & Bloembergen, N. (1965). Theory of stimulated Brillouin and Raman scattering. *Physical Review*, *137*, A1787–A1805.
Sinha, S. S., Jones, S., Demeritte, T., Chavva, S. R., Shi, Y., Burrell, J., et al. (2016). Multimodal nonlinear optical imaging of live cells using plasmon-coupled DNA-mediated gold nanoprism assembly. *Journal of Physical Chemistry C*, *120*, 4546–4555.
Sinha, S. S., Paul, D. K., Kanchanapally, R., Pramanik, A., Chavva, S. R., Nellore, B. P. V., et al. (2015). Long-range two-photon scattering spectroscopy ruler for screening prostate cancer cells. *Chemical Science*, *6*, 2411–2418.
Spoto, G., & Minunni, M. (2012). Surface plasmon resonance imaging: What next? *Journal of Physical Chemistry Letters*, *3*, 2682–2691.
Stenberg, E., Persson, B., Roos, H., & Urbaniczky, C. (1991). Quantitative determination of surface concentration of protein with surface plasmon resonance using radiolabeled proteins. *Journal of Colloid and Interface Science*, *143*, 513–526.
Swiech, O. A., Opuchlik, L. J., Wojciuk, G., Stepkowski, T. M., Kruszewski, M., & Bilewicz, R. (2016). Doxorubicin carriers based on Au nanoparticles—Effect of shape and gold-drug linker on the carrier toxicity and therapeutic performance. *RSC Advances*, *6*, 31960–31967.
Tenzer, S., Docter, D., Rosfa, S., Wlodarski, A., Kuharev, J., Rekik, A., et al. (2011). Nanoparticle size is a critical physicochemical determinant of the human blood plasma corona: A comprehensive quantitative proteomic analysis. *ACS Nano*, *5*, 7155–7167.
Terhune, R. W., Maker, P. D., & Savage, C. M. (1965). Measurements of nonlinear light scattering. *Physical Review Letters*, *14*, 681–684.
Turkevich, J., Stevenson, P. C., & Hillier, J. (1951). A study of the nucleation and growth processes in the synthesis of colloidal gold. *Discussions of the Faraday Society*, *11*, 55–75.
Turnbull, W. B., & Daranas, A. H. (2003). On the value of c: Can low affinity systems be studied by isothermal titration calorimetry? *Journal of the American Chemical Society*, *125*, 14859–14866.
Vance, F. W., Lemon, B. I., & Hupp, J. T. (1998). Enormous hyper-Rayleigh scattering from nanocrystalline gold particle suspensions. *The Journal of Physical Chemistry B*, *102*, 10091–10093.
Verbiest, T., Clays, K., & Rodriguez, V. (Eds.), (2009). *Second-order nonlinear optical characterization techniques: An introduction*. Florida, FL: CRC Press
Wang, H., Yan, E. C. Y., Liu, L., & Eisenthal, K. B. (1998). Energetics and population of molecules at microscopic liquid and solid surfaces. *The Journal of Physical Chemistry B*, *102*, 4446–4450.
Wang, G., Zhang, Y., Cui, Y., Duan, M., & Liu, M. (2005). Study on the behavior of hyper-Rayleigh scattering for silver nanoparticles with aggregation effects. *The Journal of Physical Chemistry B*, *109*, 1067–1071.
Wiseman, T., Williston, S., Brandts, J. F., & Lin, L. N. (1989). Rapid measurement of binding constants and heats of binding using a new titration calorimeter. *Analytical Biochemistry*, *179*, 131–137.
Yang, T.-C., Catalano, C. E., & Maluf, N. K. (2015). Analytical ultracentrifugation as a tool to study nonspecific protein–DNA interactions. *Methods in Enzymology*, *562*, 305–330.

Zhang, C. X., Zhang, Y., Wang, X., Tang, Z. M., & Lu, Z. H. (2003). Hyper-Rayleigh scattering of protein-modified gold nanoparticles. *Analytical Biochemistry, 320*, 136–140.

Ziegler, C., & Eychmüller, A. (2011). Seeded growth synthesis of uniform gold nanoparticles with diameters of 15-300 nm. *Journal of Physical Chemistry C, 115*, 4502–4506.

Zyss, J., & Oudar, J. L. (1982). Relations between microscopic and macroscopic lowest-order optical nonlinearities of molecular crystals with one- or two-dimensional units. *Physical Review A, 26*, 2028–2048.

CHAPTER THREE

Armoring Enzymes by Metal–Organic Frameworks by the Coprecipitation Method

Miao Hou, Jun Ge[1]
Key Lab for Industrial Biocatalysis, Ministry of Education, Tsinghua University, Beijing, China
[1]Corresponding author: e-mail address: junge@mail.tsinghua.edu.cn

Contents

Methods in Enzymology, Volume 590
ISSN 0076-6879
http://dx.doi.org/10.1016/bs.mie.2016.12.002
© 2017 Elsevier Inc.
All rights reserved.

Abstract

Enzymatic processes have great potential for applications in industrial biocatalysis, biosensing, and biomedical engineering. However, poor stability and difficulty in the reuse of enzymes still create challenges for their use in catalysis and other applications. In recent years, enzyme-incorporated metal–organic frameworks (enzyme@MOF) have emerged as a new type of armored enzymes and are thought to be an appealing solution to the above challenges. To date, the preparation of enzyme-MOF composites can be divided into three categories, including physical adsorption, covalent conjugation, and coprecipitation. The coprecipitation method integrates the synthesis of MOFs and encapsulation of enzymes into one step, allowing common MOFs with regular pore sizes to be used as carriers for enzyme binding. In this chapter, we present a protocol for the synthesis of enzyme@MOF composites using the coprecipitation approach with particular details.

1. INTRODUCTION

Enzymes are naturally evolved and highly efficient catalysts with high chemo-, region-, and stereoselectivity, which make them a viable option for achieving more environmentally friendly industries (Benkovic & Hammes-Schiffer, 2003). The exploration of using enzyme catalysts instead of chemical catalysts has been extensively investigated by academics as well as industry (Bornscheuer et al., 2012; Kirk, Borchert, & Fuglsang, 2002).

However, the industrial application of enzymes is still challenged by their low operational stability and poor reusability. Enzyme immobilization has been proven as an effective way to enhance enzyme stability (DiCosimo, McAuliffe, Poulose, & Bohlmann, 2013). Multiple covalent attachments with the carrier matrix are generally shown to improve the stabilities of monomeric and multimeric enzymes. The improved stability and reusability can significantly reduce the cost of enzyme use and thus, make their industrial application more economically feasible.

Furthermore, immobilization may also improve other enzyme properties such as activity or selectivity (Ansari & Husain, 2012; Mateo, Palomo, Fernandez-Lorente, Fernandez-Lafuente, & Guisan, 2007). The application of nanotechnology in the preparation of immobilized enzymes has been demonstrated as a novel and effective strategy. Nanomaterials with high-specific surface area, low mass transfer limitation, and abundant functionality are recognized as excellent carriers for enzyme encapsulation. Enzymes encapsulated in polymer nanoparticles (Ge, Yang, Zhu, Lu, & Liu, 2012; Wang et al., 2013), polymer nanogels (Ge, Lei, & Zare, 2012), and flower-like inorganic crystals (Ge, Lei, et al., 2012) have been proven to exhibit both high enzyme activity and stability.

In recent years, metal–organic frameworks (MOFs), which are an emerging type of porous nanomaterials with a tunable pore size (Hayashi, Côté, Furukawa, O'Keeffe, & Yaghi, 2007) and built in functional groups (Kong et al., 2013), have attracted quite a lot of attention. MOFs also known as coordination polymers, which are made of metal ions or metal-containing clusters with organic ligands through self-assembly. MOFs feature structural diversity and amenability to be designed with specific functionality. Their adjustable but uniform pore sizes and functional pore walls allow them to accommodate a variety of guest species. These features allow for great promise for a variety of applications including gas storage (Mason et al., 2013; Xie et al., 2012), separation (Centrone, Santiso, & Hatton, 2011; Hasan & Jhung, 2015), catalysis (Kong et al., 2013), sensing (Greathouse et al., 2010), fuel cells (Yoon, Suh, Natarajan, & Kim, 2013), as well as hosting many guest species including quantum dots, nanoparticles, and proteins (Gao et al., 2014; Mao et al., 2014).

Very recently, MOFs have shown their great promise in the preparation of encapsulated enzymes. Once encapsulated in MOFs, it can significantly improve the stability and preserve the activity due to the confinement in the nanoporous frameworks of MOFs (Lyu, Zhang, Zare, Ge, & Liu, 2014; Wu, Ge, Yang, Hou, & Liu, 2015). In addition to the high enzyme activity and stability, postencapsulation, the intrinsic micropores of MOFs can enhance size selectivity and impart new functions in enzymatic catalysis (Shieh et al., 2015).

The preparation of enzyme-MOF composites can be divided into three categories, including physical adsorption, covalent conjugation, and coprecipitation. Physical adsorption of proteins into the presynthesized MOFs is the most straightforward method. Ma and coworkers reported the adsorption of microperoxidase-11(MP-11) on a mesoporous MOF, Tb-TATB for enzymatic catalysis. The MP-11@Tb-TATB composite was obtained by simply incubating the Tb-TATB in HEPES (4-(2-hydroxyethyl)-1-piperazineethanesulfonic acid) buffer containing the enzyme. The bound MP-11 showed a 48.7% substrate conversion in the oxidation reaction of 3,5-di-tert-butylcatechol (DTBC) (Lykourinou et al., 2011). However, due to the relatively large size of protein molecules when compared to the small pore size of common MOFs, this method requires the utilization of special MOFs with large pore sizes, which can accommodate protein molecules in the MOF matrix.

A more flexible way for armoring enzymes with MOFs is the bioconjugation of enzymes with MOFs by covalent bonds. Some typical organic linkers of MOFs containing carboxylate groups or amine groups can display the functional groups on the surface of MOF crystals after synthesis, which can be further activated and used for bioconjugation. Lin and coworkers synthesized the trypsin-immobilized MOFs by using MIL-101 (Cr), MIL-88B (Cr), and MIL-88B-NH2 (Cr) as the carriers (Shih et al., 2012). The enzyme-MOF conjugates showed good reusability owing to the stable covalent bonding between protein and MOF carriers.

Markedly different from the above two methods, a coprecipitation method has been proposed by us and other groups (Lyu et al., 2014; Mao et al., 2014; Shieh et al., 2015; Wu et al., 2015). The coprecipitation method integrates the synthesis of MOFs and encapsulation of enzymes in one step, allowing MOFs with pore sizes smaller than the size of enzymes to be used as carriers for encapsulation. Very recently, this method has also been used in the encapsulation of cells and viral particles in MOFs (Li et al., 2016; Liang et al., 2016).

The enzyme@MOF composites prepared by the coprecipitation method have relatively uniform size and display similar structure and morphology as the pure MOF crystals. Different types of enzymes including horseradish peroxidase (HRP), cytochrome *c* (Cyt *c*), glucose oxidase (GOx), *Candida antarctica* lipase B (CALB) have been used to synthesize the armored biocatalysts, giving the loading amount around 5 wt%–10 wt%, which can be determined by methods including thermal gravimetric

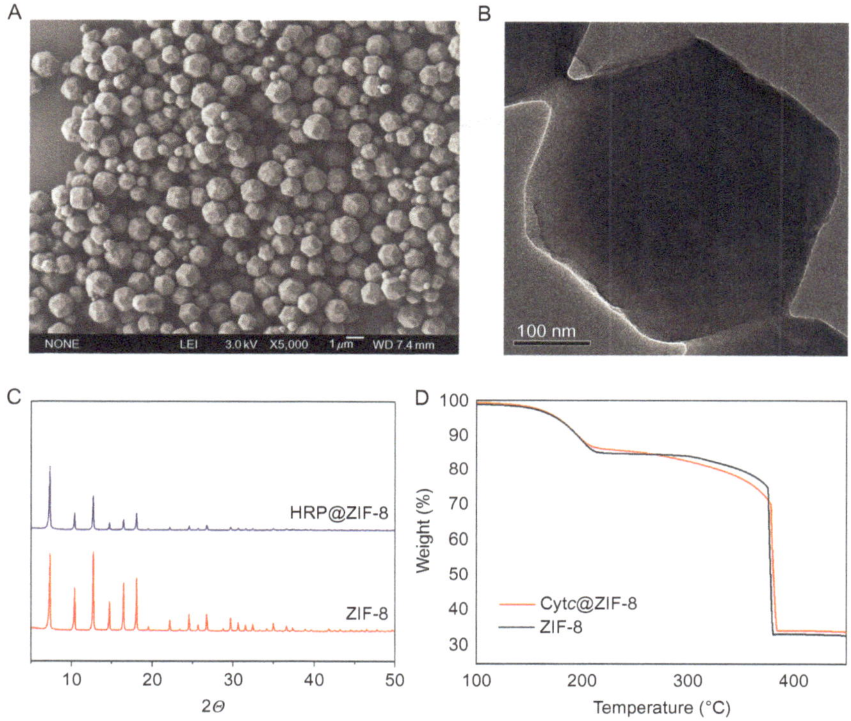

Fig. 1 (A) SEM image of the Cyt *c*/ZIF-8, 1 μm size bar. (B) TEM image of the Cyt *c*/ZIF-8. (C) XRD patterns of ZIF-8 and the HRP/ZIF-8. (D) TGA curves of ZIF-8, the Cyt *c*/ZIF-8.

analysis (TGA), size exclusion chromatography (SEC), and inductively coupled plasma-mass spectrometry (ICP-MS) (Fig. 1; Lyu et al., 2014; Mao et al., 2014; Shieh et al., 2015; Wu et al., 2015). The as-prepared composites exhibited high catalytic selectivity and stability. The rigid structure of the MOF would increase the structural stability of the embedded enzymes, thus improving the enzymes' stabilities, making it promising for various applications. Wu et al. (2015) reported the synthesis of GOx&HRP@ZIF-8 composite, which after incubation in phosphate buffer at room temperature for 7 days retained ~80% of the initial activity. The stability of the composite against protease and ethylene diaminetetraacetic acid (EDTA) was also obviously increased when compared to that of the free enzymes. After being digested by trypsin for 30 min at 37°C, the GOx&HRP@ZIF-8 composite retained almost the same activity (Fig. 2). Shieh et al. (2015) reported the synthesis of catalase@ZIF-90 composite with ~5 wt% protein loading. The composites showed catalytic activity in the degradation of hydrogen peroxide even in the presence of protease, proteinase K.

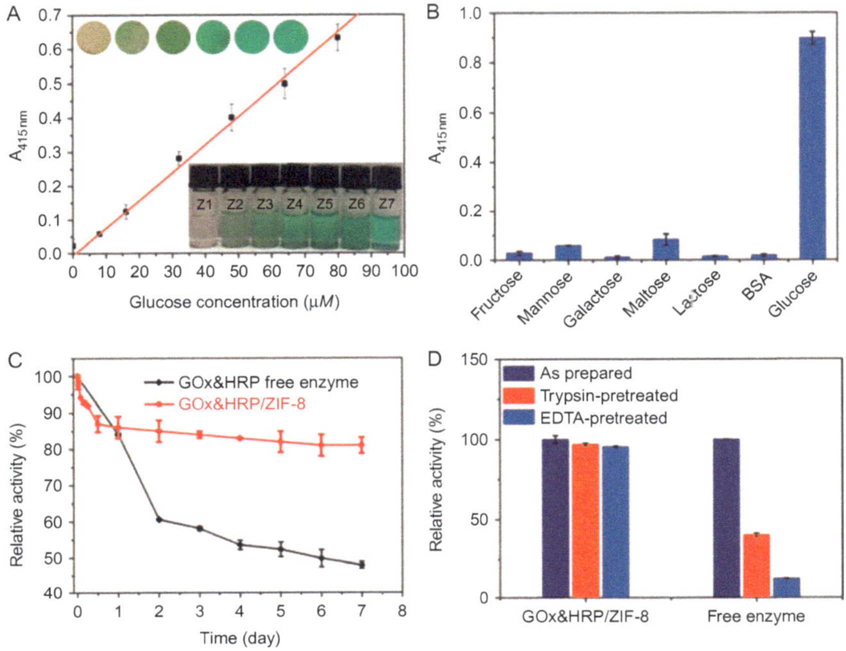

Fig. 2 (A) Detection of glucose at concentrations of 0–100 m*M*. (B) Determination of the selectivity of the GOx&HRP/ZIF-8. (C) Long-term stability of the GOx&HRP/ZIF-8 when compared those of the corresponding free enzymes incubated in pH 7.4 phosphate buffered saline at room temperature. (D) Stability of the GOx&HRP/ZIF-8 composite compared with free enzymes incubated in pH 7.4 phosphate buffered saline, containing 1 mg/mL of trypsin or 1 wt% of EDTA (Wu et al., 2015).

In this protocol, we describe the coprecipitation method for the synthesis of enzyme@MOF composites in detail, using HRP as the model enzyme and zeolitic imidazolate framework (ZIF-8) as the model MOF. The HRP@ZIF-8 composite was formed within 30 min at 25°C by simply mixing aqueous solutions containing zinc ions, 2-methylimidazole, and enzymes. The use of water as the solvent would preserve the enzyme activity, and the rigid structure of MOF would increase the structural stability of the embedded enzyme, therefore, improving the overall enzyme stability.

2. EQUIPMENT

Beaker, electronic balance, sonicator bath, magnetic stirring apparatus, magnetic stir bar, centrifuge, 50-mL centrifuge tubes, measuring cylinder, pipette, crucible, 1-mL quartz cuvette, UV–Vis spectrophotometer, drying cabinet, Zetasizer Nano ZS (Malvern Instruments), laser scanning confocal microscope, scanning electron microscope (SEM), transmission electron microscope (TEM), X-ray diffraction (XRD) instrument, Fourier transform infrared spectrometer (FT-IR), muffle, high-performance liquid chromatography (HPLC), TSK-GEL G2000SWxL column, thermal gravimetric analysis (TGA) instrument, and ICP-MS.

3. MATERIALS

Peroxidase from Horseradish (HRP) (e.g., Sigma-Aldrich, CAT #77332),
Zinc nitrate hexahydrate (99.998%) (e.g., Sigma-Aldrich, CAT #230006),
2-Methylimidazole (e.g., Sigma-Aldrich, CAT #M50650),
2,2′-Azino-bis (3-ethylbenzothiazoline-6-sulfonic acid) diammonium salt (ABTS) (e.g., Sigma-Aldrich, CAT #1888888),
Fluorescein isothiocyanate (FITC) (e.g., Sigma-Aldrich, CAT #F7250),
Rhodamine B isothiocyanate (RhB) (e.g., Sigma-Aldrich, CAT #283924),
H_2O_2, Na_2SO_4, NaN_3, ethanol, methyl, alcohol, acetic acid, deionized (DI) water, sodium dihydrogen phosphate, disodium hydrogen phosphate, sodium carbonate, sodium bicarbonate.

4. PROTOCOL

4.1 Duration

12 h for synthesis, about 2 days for characterization.

4.2 Preparation

A schematic for the synthesis of enzyme@MOF composites is given in Fig. 3. This protocol may be scaled up or down, depending on the desired amount. Additional microcentrifuge tubes can be added for greater number of samples.

5. SAFETY

When using $Zn(NO_3)_2$ and 2-methylimidazole, avoid the formation of dust and aerosols. Provide appropriate exhaust ventilation at places where dust is formed. Normal measures for preventive fire protection are necessary. In addition, abide by the operational rules when using the instruments.

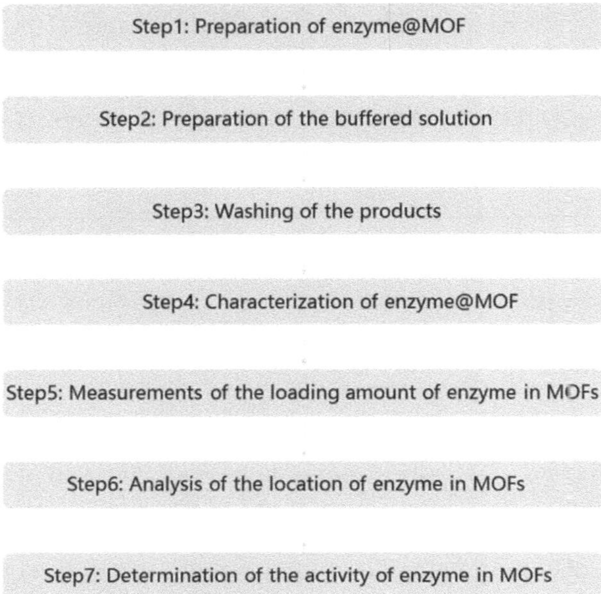

Fig. 3 Flowchart of the complete protocol.

6. STEP1: PREPARATION OF ENZYME@MOF

6.1 Overview

The HRP@ZIF-8 composite is chosen as the model to describe the detailed protocol. There are three major steps for the synthesis: (a) Prepare precursor solutions (enzyme solution, metal ion solution, and organic ligand solution), (b) mix three precursor solutions and incubate the mixture at room temperature, and (c) purify by washing off residual reactants. A schematic of synthesis is given in Fig. 4.

6.2 Methods

60 min

1.1 Dissolve 30 mg HRP in 4 mL DI water, 0.372 g $Zn(NO_3)_2 \cdot 6H_2O$ in 4 mL DI water and 4.1 g 2-methylimidazole in 40 mL DI water.

1.2 Degas the zinc nitrate solution and 2-methylimidazole solution in an ultrasonic bath for 5 min to obtain two homogeneous mixtures.

1.3 Mix the zinc nitrate solution with the 2-methylimidazole solution in a glass beaker while stirring and immediately add the enzyme solution.

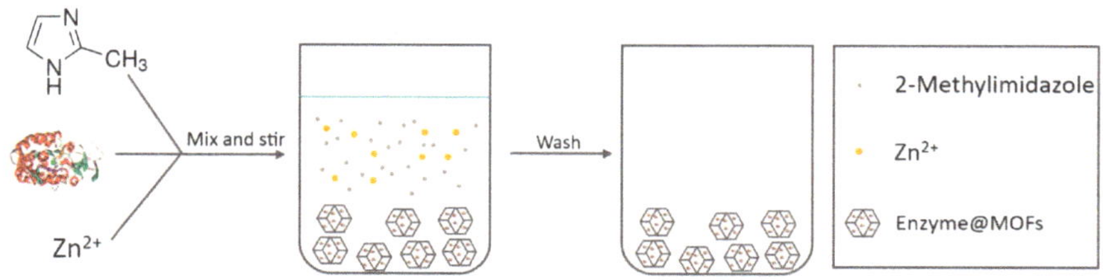

Fig. 4 The preparation of HRP@ZIF-8 composite.

1.4 Seal the glass beaker using sealing film.

1.5 Stir the mixture for 30 min at room temperature with a rotation speed of 500 rpm.

6.3 Tip

The choice of beaker and stirrer with moderate size is necessary to improve the mixing effect. In this experiment, the reaction volume is 44 mL, 50 mL beaker and 20 mm stirrer are used. In addition, a moderate stirring rate is also required, because a high stirring rate will destroy the crystal structure of MOFs, while a low stirring rate will reduce the mixing uniformity.

6.4 Tip

The enzyme solution needs to be poured into the mixed solution containing metal ions and organic ligands as soon as possible, because the growth of ZIF-8 crystals is very fast. It is also inappropriate to mix the enzyme solution with either metal ion solution or organic ligand solution in advance, because the interaction between metal ions and enzyme and the strong acid condition of organic ligand solution may deactivate the enzyme.

6.5 Tip

The procedure can be scaled up or scaled down by keeping the molar ratio of precursors the same and adjusting the volume of DI water to maintain the appropriate concentrations. In this experiment, the procedure can be easily scaled up to ~1 g composite/batch, and down to ~10 mg composite/batch in laboratory.

7. STEP2: PREPARATION OF BUFFERS

7.1 Overview

There are three kinds of buffer solutions used in this process including phosphate buffer (PB) solution (50 m*M*, pH 7.4), PB solution (50 m*M*, pH 7.0), and carbonate buffer solution (100 m*M*, pH 9.5). These buffer solutions are used in the washing of enzyme@MOF composite, determination of enzyme activity, and for fluorescent labeling of enzyme.

7.2 Methods

60 min

2.1 For the PB buffer solution (50 m*M*, pH 7.0), using a 200-mL volumetric flask, dissolve all components at desired concentrations in 100 mL DI water, and then adjust the pH to 7.0. See Table 1.

2.2 For the PB buffer solution (50 m*M*, pH 7.4), using a 200-mL volumetric flask, dissolve all components at desired concentrations in 100 mL DI water, and then adjust pH to 7.4. See Table 1.

2.3 For the carbonate buffer solution (100 m*M*, pH 9.5), using a 200-mL volumetric flask, dissolve all components at desired concentrations in 100 mL DI water, and then adjust pH to 9.5. See Table 2.

7.3 Tip

Use PB buffer solution instead of DI water when determining the activity of the HRP@MOF and free HRP. The PB buffer solution can provide an optimal pH condition for the enzyme to display high catalytic activity.

7.4 Tip

The buffer solutions need to be freshly prepared prior to use, and the pH value of the buffer solutions should be measured and adjusted before each experiment.

Table 1 Formulation of PB Buffer Solution

Component	Sodium Dihydrogen Phosphate (mg)	Sodium Hydrogen Phosphate (mg)	Water (mL)
pH 7.4	114	1383	100
pH 7.0	234	1041	100

Table 2 Carbonate Buffer

Component	Sodium Carbonate (mg)	Sodium Bicarbonate (mg)	Water (mL)
pH 9.5	424	504	100

8. STEP3: PURIFICATION OF THE PRODUCT

8.1 Overview

The composites are washed with PB buffer solution after coprecipitation. This washing step is used to remove unreacted reagents and free enzymes. After washing, the composites are freeze-dried and stored at 4°C.

8.2 Methods

60 min

3.1 Centrifuge the mixture after coprecipitation for 10 min at 6000 rpm at 25°C.

3.2 Remove the supernatant from each centrifuge tube and store in a proper container.

3.3 Add PB buffer solution to each centrifuge tube and vortex.

3.4 Centrifuge the mixture for 10 min at 6000 rpm.

3.5 Repeat the washing and centrifugation steps for three times.

3.6 After washing, the precipitates are freeze-dried for 6 h to obtain the products.

3.7 Store the products at 4°C.

8.3 Tip

Be careful not to accidently discard the products when washing and removing the supernatant.

8.4 Tip

Ensure the centrifuge tubes are spun well. All the products need to be well suspended in solution before continuing to the next washing step.

9. STEP4: CHARACTERIZATION OF ENZYME@MOF

9.1 Overview

After synthesis of the enzyme@MOF composites, they are subjected to a series of characterizations including the characterization of morphology, particle size, chemical structure, and crystal structure.

9.2 Methods

4.1 Characterize the morphology of composite using scanning electron microscopy (SEM) and transmission electron microscopy (TEM).

4.2 For the SEM analysis of enzyme@MOF composites, samples are prepared by first suspending the composites in methanol and then drop 1–10 μL of the sample solution onto a silica wafer. After all methanol is evaporated, attach the silica wafer to a carbon paste and then sputter-coat it with a thin layer of conductive gold to improve the electrical conductivity.

4.3 For the TEM analysis of enzyme@MOF composites, samples are prepared by adding a drop of methanol suspension containing the synthesized enzyme@MOF composites on a carbon grid and drying at room temperature.

4.4 Characterize the crystal structure using powder X-ray diffraction (XRD). The XRD patterns are collected between $2\theta = 5$ degree and 50 degree at a scan rate of 2 degree/min.

4.5 Characterize the chemical structure of composite using FT-IR.

9.3 Tip

For the powder XRD, the samples need to be carefully grinded using an agate mortar. The crystal structure of the composite can also be validated by comparing the result with the standard crystal structure of ZIF-8.

9.4 Tip

For FT-IR spectra, the samples need to be fully dried to remove the residual water.

10. STEP5: MEASUREMENT OF ENZYME LOADING IN MOFS

10.1 Overview

The loading amount of enzyme in the composite can be determined by methods including thermogravimetric analysis (TGA), SEC, and ICP-MS.

10.2 Methods

5.1 Determine the loading amount of enzyme in composite by TGA by heating the samples from room temperature to 600°C at a rate of 20°C/min under air atmosphere. A weight loss may occur during the second-stage when compared with the pure ZIF-8, which is attributed to protein decomposition. Enzyme loading is calculated using the weight loss before 400°C.

5.2 Determine the concentration of unencapsulated HRP in the supernatant after synthesis of HRP@ZIF-8 by SEC. Compare the amount of unencapsulated HRP with the total amount of free HRP used in synthesis to calculate the loading.

5.3 For the SEC, 20 μL of sample solution is injected and eluted using phosphate buffer (0.1 *M*) containing 0.1 *M* Na_2SO_4, 0.1% NaN_3 at an elution speed of 1 mL/min. Protein detection is carried out by UV–Vis absorbance at 280 nm on a SHIMADZU HPLC system equipped with a TSK-GEL G2000SWxL column.

5.4 Determine the content of Fe in the composite by ICP-MS as there is one equivalent of Fe per HRP.

10.3 Tip

There is a gradual weight loss before 200°C, corresponding to the removal of guest molecules (mainly be H_2O) from the cavities of MOFs and unreacted species. Please take this weight loss into account when calculating the accurate loading of enzyme.

10.4 Tip

ICP-MS can only be applied to determine the loading amount of metalloenzymes.

10.5 Tip

Conventional protein concentration assays such as Coomassie Brilliant Blue method and BCA method are frequently used to determine the protein concentration in the supernatant after immobilization to calculate loading. However, these methods should not be applied in this system, because the unreacted Zn^{2+} and dimethylimidazole in the supernatant may cause some interference in the assays.

11. STEP6: ANALYSIS OF THE LOCATION OF ENZYMES IN MOFS

11.1 Overview

Confocal laser scanning microscopy (CLSM) can be used to prove the presence of HRP molecules in ZIF-8. In addition, SEM and TEM analysis after the calcination of the HRP@MOF can be used to investigate the location of HRP molecules in ZIF-8.

11.2 Methods

6.1 Remove protein molecules from the composite by a calcination process at 325°C for 2 h. The temperature is chosen based on thermogravimetric analysis data.

6.2 The TEM and SEM images of HRP@ZIF-8 composite after calcination show the existence of small cavities with size ranging from 5 to 20 nm where protein molecules were presumably encapsulated.

6.3 For the CLSM of the composite, observe the fluorescently labeled HRP@ZIF-8 composite under laser scanning confocal microscope. The detection wavelengths are 488 and 543 nm for FITC and RhB, respectively.

11.3 Tip

For the protein staining, dissolve 8 mg of FITC or rhodamine B isothiocyanate (RhB) in dimethyl sulfoxide (DMSO) with a final concentration of 2 mg/mL. Slowly add the solution into 1 mL of HRP solution (5 mg/mL of enzyme in 0.5 *M*, pH 9.5 carbonate buffer). Shake the mixture for 6 h at 300 rpm at room temperature in dark. Remove free FITC or RhB via dialysis against DI water. The fluorescent molecule-labeled enzymes are freeze-dried. Dissolve the fluorescent molecule-labeled enzymes in water for the subsequent synthesis of fluorescently labeled HRP@ZIF-8 composite.

11.4 Tip

The fluorescently labeled enzyme and enzyme@MOF composite need to be stored in the dark and prepared right before use.

11.5 Tip

Notably, during the dialysis process, water or buffer solution needs to be changed multiple times and the dialysis time must be long enough to thoroughly remove the unreacted fluorescent molecules. The dialysis procedure generally lasts for about 48 h which depends on the amount of sample, until the fluorescence of the dialysis solution is not detectable.

12. STEP7: DETERMINATION OF ENZYME ACTIVITIES

12.1 Overview

The catalytic activity and stability are two important properties of enzyme@MOF composites. For HRP, the standard method is used to determine the enzyme activity. The 2,2′-azino-bis (3-ethylbenzothiazoline-6-sulfonic acid)-diammonium salt (ABTS) and H_2O_2 are used as the substrates.

12.2 Methods

7.1 Prepare HRP and HRP@ZIF-8 composite solutions with an enzyme concentration of 10 μg/mL and use H_2O_2 with 0.3% mass fraction and 0.5 m*M* ABTS in PB buffer (50 m*M*, pH 7.0) as the substrates.

7.2 Add 50 μL HRP and 50 μL H_2O_2 solution to 900 μL of the substrate solution (ABTS in 50 m*M* phosphate buffer, pH 7.0) in the quartz colorimetric cell.

7.3 Record the increase of absorbance at 415 nm for 1 min on the UV–Vis spectrophotometer to calculate the enzyme activity.

12.3 Tip

Make sure that all the enzymatic assays are carried out at the same protein concentration for the comparison of the activities of free enzyme with that of the HRP@ZIF-8.

13. CONCLUSIONS

A protocol for the aqueous synthesis of enzyme-incorporated ZIF-8 nanocrystals using HRP as the model enzyme has been presented here. The synthesis, characterization of enzyme@MOF composites, and some tips are given. The as-prepared enzyme@MOFs have an average size of around 1 μm and display similar structure and morphology as the corresponding pure MOF crystals. The high enzyme activity and high stability of enzyme@MOF under harsh conditions make this biocatalyst promising for various applications. In addition, the coprecipitation approach shows the advantages of facile procedure, applicable to various enzymes, and common MOFs, efficient in preventing enzyme leaching and improving enzyme stability. This new method for the preparation of enzyme@MOF composites opens avenues for easily combining together the properties and functionalities of various enzymes with those of the MOF materials, and fabricating various types of MOF-based building blocks containing particular biological components. This combination of functional biological components with inorganic crystals could enable new applications in biotechnology, industrial biocatalysis, biosensing, and biomedical engineering.

REFERENCES

Ansari, S. A., & Husain, Q. (2012). Potential applications of enzymes immobilized on/in nano materials: A review. *Biotechnology Advances*, *30*, 512–523.

Benkovic, S. J., & Hammes-Schiffer, S. (2003). A perspective on enzyme catalysis. *Science*, *301*, 1196–1202.

Bornscheuer, U., Huisman, G., Kazlauskas, R., Lutz, S., Moore, J., & Robins, K. (2012). Engineering the third wave of biocatalysis. *Nature*, *485*, 185–194.

Centrone, A., Santiso, E. E., & Hatton, T. A. (2011). Separation of chemical reaction intermediates by metal-organic frameworks. *Small*, *7*, 2356–2364.

DiCosimo, R., McAuliffe, J., Poulose, A. J., & Bohlmann, G. (2013). Industrial use of immobilized enzymes. *Chemical Society Reviews*, *42*, 6437–6474.

Gao, Y., Wu, J., Zhang, W., Tan, Y., Zhao, J., & Tang, B. (2014). The electrochemical performance of sno 2, quantum dots@zeolitic imidazolate frameworks-8 (zif-8) composite material for supercapacitors. *Materials Letters*, *128*, 208–211.

Ge, J., Lei, D., & Zare, R. N. (2012). Protein-inorganic hybrid nanoflowers. *Nature Nanotechnology*, *7*, 428–432.

Ge, J., Yang, C., Zhu, J., Lu, D., & Liu, Z. (2012). Nanobiocatalysis in organic media: Opportunities for enzymes in nanostructures. *Topics in Catalysis*, *55*, 1070.

Greathouse, J. A., Ockwig, N. W., Criscenti, L. J., Guilinger, T. R., Pohl, P., & Allendorf, M. D. (2010). Computational screening of metal-organic frameworks for large-molecule chemical sensing. *Physical Chemistry Chemical Physics*, *12*, 12621–12629.

Hasan, Z., & Jhung, S. H. (2015). Removal of hazardous organics from water using metal-organic frameworks (MOFs): Plausible mechanisms for selective adsorptions. *Journal of Hazardous Materials*, *283*, 329–339.

Hayashi, H., Côté, A. P., Furukawa, H., O'Keeffe, M., & Yaghi, O. M. (2007). Zeolite A imidazolate frameworks. *Nature Materials*, *6*, 501–506.

Kirk, O., Borchert, T. V., & Fuglsang, C. C. (2002). Industrial enzyme applications. *Current Opinion in Biotechnology*, *13*, 345–351.

Kong, X., Deng, H., Yan, F., Kim, J., Swisher, J. A., Smit, B., et al. (2013). Mapping of functional groups in metal-organic frameworks. *Science*, *341*, 882–885.

Li, S., Dharmarwardana, M., Welch, R. P., Ren, Y., Thompson, C. M., Smaldone, R. A., et al. (2016). Template-directed synthesis of porous and protective core–shell bionanoparticles. *Angewandte Chemie, International Edition*, *55*, 10849–10854.

Liang, K., Richardson, J. J., Cui, J., Caruso, F., Doonan, C. J., & Falcaro, P. (2016). Biomimetics: Metal-organic framework coatings as cytoprotective exoskeletons for living cells. *Advanced Materials*, *28*, 8066.

Lykourinou, V., Chen, Y., Wang, X. S., Meng, L., Hoang, T., Ming, L. J., et al. (2011). Immobilization of MP-11 into a mesoporous metal-organic framework, MP-11@mesomof: A new platform for enzymatic catalysis. *Journal of the American Chemical Society*, *133*, 10382–10385.

Lyu, F., Zhang, Y., Zare, R. N., Ge, J., & Liu, Z. (2014). One-pot synthesis of protein-embedded metal-organic frameworks with enhanced biological activities. *Nano Letters*, *14*, 5761–5765.

Mao, Y., Li, J., Cao, W., Ying, Y., Hu, P., Liu, Y., et al. (2014). General incorporation of diverse components inside metal-organic framework thin films at room temperature. *Nature Communications*, *5*, 5532.

Mason, J. A., McDonald, T. M., Bae, T. H., Bachman, J. E., Sumida, K., Dutton, J. J., et al. (2013). Evaluating metal–organic frameworks for natural gas storage. *Chemical Science*, *5*, 32–51.

Mateo, C., Palomo, J. M., Fernandez-Lorente, G., Fernandez-Lafuente, R., & Guisan, J. M. (2007). Immobilization of enzymes on heterofunctional epoxy supports. *Enzyme and Microbial Technology*, *40*, 1451–1463.

Shieh, F. K., Wang, S. C., Yen, C. I., Wu, C. C., Dutta, S., Chou, L. Y., et al. (2015). Imparting functionality to biocatalysts via embedding enzymes into nanoporous materials by a de novo approach: Size-selective sheltering of catalase in metal–organic framework microcrystals. *Journal of the American Chemical Society*, *137*, 4276–4279.

Shih, Y. H., Lo, S. H., Yang, N. S., Singco, B., Cheng, Y. J., Wu, C. Y., et al. (2012). Trypsin-immobilized metal–organic framework as a biocatalyst in proteomics analysis. *Chempluschem*, *77*, 982–986.

Wang, R., Zhang, Y., Lu, D., Ge, J., Liu, Z., & Zare, R. N. (2013). Functional protein–organic/inorganic hybrid nanomaterials. *Wiley Interdisciplinary Reviews-Nanomedicine and Nanobiotechnology*, *5*, 320–328.

Wu, X., Ge, J., Yang, C., Hou, M., & Liu, Z. (2015). Facile synthesis of multiple enzyme-containing metal-organic frameworks in a biomolecule-friendly environment. *Chemical Communications*, *51*, 13408–13411.

Xie, Y., Yang, H., Wang, Z. U., Liu, Y., Zhou, H. C., & Li, J. R. (2012). Recent advances in carbon dioxide capture with metal-organic frameworks. *Greenhouse Gases Science & Technology*, *2*, 239–259.

Yoon, M., Suh, K., Natarajan, S., & Kim, K. (2013). Proton conduction in metal–organic frameworks and related modularly built porous solids. *Angewandte Chemie, International Edition*, *52*, 2688–2700.

CHAPTER FOUR

Enzyme Armoring by an Organosilica Layer: Synthesis and Characterization of Hybrid Organic/Inorganic Nanobiocatalysts

M. Rita Correro, Sabine Sykora, Philippe F.-X. Corvini, Patrick Shahgaldian[1]

School of Life Sciences, University of Applied Sciences and Arts Northwestern Switzerland, Muttenz, Switzerland

[1]Corresponding author: e-mail address: patrick.shahgaldian@fhnw.ch

Contents

Abstract

The availability of highly stable and reusable enzymes is one of the main challenges in bio-based industrial processes. Enzyme immobilization and encapsulation represent promising strategies to reach this goal. In this chapter, the synthetic strategy to produce hybrid organic/inorganic nanobiocatalysts (NBC) is reported. This strategy is based on the sequential immobilization of an enzyme on the surface of silica nanoparticles followed by the growth, at the surface of the nanoparticles, of a shielding layer which

Methods in Enzymology, Volume 590
ISSN 0076-6879
http://dx.doi.org/10.1016/bs.mie.2016.12.006
© 2017 Elsevier Inc.
All rights reserved.

serves as an armor to protect the enzyme against denaturation/degradation. This armor is produced through a thickness-controlled organosilane poly-condensation onto the nanoparticle surface around the enzyme to form a protective organosilica layer. The armored nanobiocatalysts present enhanced catalytic activity and improved stability against heat, pH, chaotropic agents, proteases, and ultrasound. The method is versatile in that it can be successfully adapted to a number of different enzymes.

1. INTRODUCTION

Continuous processing is not a new concept in industry (Plouffe, Macchi, & Roberge, 2014). Chemical and food industries have been progressively switching manufacturing operations to continuous processes over the last 2 decades, principally for reducing production costs and improving final product quality. In contrast, the pharmaceutical industry has been slower to move from batch to continuous manufacturing processes (Subramanian, 2014). Nevertheless, as pressure increases to reduce costs and improve product quality, the interest in continuous manufacturing processes in pharmaceutical production is increasing considerably. Enzymes are at the center of this technology leap as they allow for efficient biocatalysis with high conversion rates and excellent (enantio)-selectivity (Wells & Meyer, 2014). Enzymes are potent catalysts that can operate under mild conditions with high chemo-, regio-, and stereo-selectivities, elevated turnover rates and good substrate selectivity.

Notwithstanding the foremost importance of enzymes in continuous processes, their application is often hindered by their instability under non-physiological conditions and the impossibility to use native enzyme, which are inherently water soluble, in continuous processes (DiCosimo, McAuliffe, Poulose, & Bohlmann, 2013; Franssen, Steunenberg, Scott, Zuilhof, & Sanders, 2013; Polizzi, Bommarius, Broering, & Chaparro-Riggers, 2007).

Although protein engineering allows improving enzyme catalysis by enhancing, for example, the substrate selectivity or the enzyme thermal stability, significant limitations including time-consuming procedures and high costs remain unsolved. Additionally, the molecular modification of an enzyme is not a general strategy, but rather, a case-to-case method that depends on the structural and conformational properties of the enzyme (Bornscheuer et al., 2012). Enzyme immobilization/encapsulation on solid carriers or matrices represents an effective alternative strategy to protein engineering (Hanefeld, Gardossi, & Magner, 2009; Zhou & Hartmann, 2013). Besides allowing catalyst recycling, enzyme immobilization enables the use of biocatalysts in continuous operations. Enzyme anchoring on carrier materials of different nature such as (bio-)polymers, metals, and metalloid oxides has been achieved through numerous methods including physical adsorption and covalent binding. The covalent binding of an enzyme onto a support prevents enzyme leaching, which is usually observed when the enzyme is physically adsorbed onto a carrier (Rodrigues, Ortiz, Berenguer-Murcia, Torres, & Fernandez-Lafuente, 2013). Nevertheless, linking of a protein onto a solid surface can alter the three-dimensional conformation of the enzyme and thus, reduce its catalytic efficiency (Karshikoff, Nilsson, & Ladenstein, 2015). Among different methods of enzyme encapsulation, the silica sol–gel entrapment is a well-established procedure (Avnir, Coradin, Lev, & Livage, 2006; Ciriminna et al., 2013; Sanchez, Belleville, Popall, & Nicole, 2011). However, a general drawback is that the encapsulating matrices could prevent rapid diffusion of the substrate toward the catalytic site of the enzyme or retard the subsequent release of the product from the enzyme into the bulk phase.

To overcome some of these issues, we have recently demonstrated that viruses (Cumbo, Lorber, Corvini, Meier, & Shahgaldian, 2013), virus-like particles (Sykora et al., 2015), natural (Correro et al., 2016), and artificial enzymes (Hestericova et al., 2016) can be entrapped in designer organosilica shells of controlled composition and thickness (Fig. 1) with retention of high degree of specific activities.

Applied to enzymes, our strategy is based on the crosslinking of an enzyme onto the surface of carrier silica nanoparticles followed by building a protective armor with an organosilica shell of controlled thickness (Correro et al., 2016). The choice of silica as carrier is justified by several advantages. First, silicon being tetravalent possesses a bonding geometry

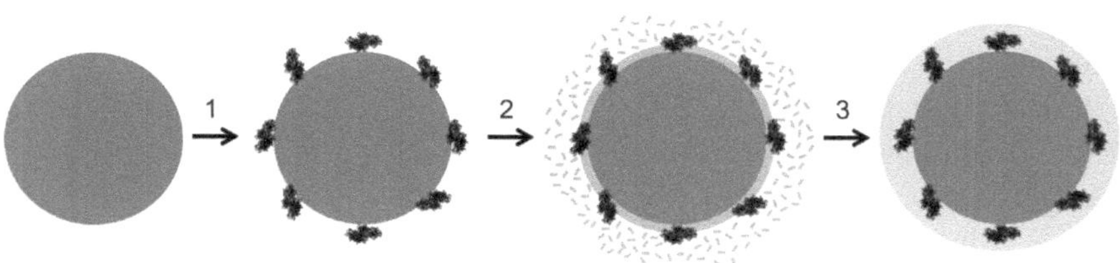

Fig. 1 Schematic representation of the synthetic strategy. (Step 1) The enzyme is covalently bound at the surface of silica nanoparticles. (Step 2) The immobilized enzyme is incubated with tetraethyl orthosilicate and (3-aminopropyl)triethoxysilane. (Step 3) Silane self-assembly and poly-condensation.

similar to that of carbon and is prone to form large covalent networks. Moreover, the chemical reactivity of silicates is distinct from that of organic molecules and chemical modifications of silica are straightforward.

In the present work, a β-galactosidase from *Kluyveromyces lactis* is used as model enzyme; this tetrameric enzyme catalyzes the hydrolysis of lactose into galactose and glucose and finds a variety of applications in the dairy industry for the production of lactose-free milk and derivatives.

We have demonstrated that the soft protective silica shell allows the enzyme to maintain a sufficient conformational mobility without the loss of its catalytic activity. Moreover, we have shown that the protective shell confers a remarkable stability against stress conditions (i.e., pH, temperature, chaotropes, ultrasound, and proteases). The thickness of the armor can be easily controlled through the reaction conditions and can be optimized to the size of the enzyme. This strategy of enzyme armoring herein reported is versatile and has already been applied to a number of natural and artificial enzymes (Correro et al., 2016; Hestericova et al., 2016).

2. EQUIPMENTS AND CHEMICALS

Equipments

1. Round-bottom flask
2. 20-mL glass vials
3. 50-mL centrifuge tubes
4. Pipettes and Pasteur glass pipettes
5. Mica sheets
6. Magnetic stirrer
7. Water bath
8. pH meter
9. Centrifuge
10. Ultrapure water purification system
11. Thermomixer
12. Refrigerated circulator
13. Spectrophotometer
14. Precision balance, sputter coater for gold deposition
15. Ultrasonic bath
16. Freeze dryer
17. Field-emission scanning electron microscope
18. Olympus® analysis software package

Reagents

1. Tetraethyl orthosilicate (TEOS, ≥99%) (Sigma-Aldrich, Switzerland)
2. (3-Aminopropyl)triethoxysilane (APTES, ≥98%) (Sigma-Aldrich, Switzerland)
3. Ammonium hydroxide (ACS grade, 28–30%) (Sigma-Aldrich, Switzerland)
4. Ethanol (ACS grade, anhydrous) (Sigma-Aldrich, Switzerland)
5. Disodium phosphate (Sigma-Aldrich, Switzerland)
6. Glutaraldehyde (Grade I, 25% in water) (Sigma-Aldrich, Switzerland)
7. Liquid nitrogen
8. β-Galactosidase from *K. lactis* (Sigma-Aldrich, Switzerland)
9. *o*-Nitrophenyl-β-D-galactopyranoside (ONPG) (Sigma-Aldrich, Switzerland)
10. 2-(N-morpholino)ethanesulfonic acid (MES) (Sigma-Aldrich, Switzerland)
11. $MgCl_2$ (Sigma-Aldrich, Switzerland)

3. SAFETY GUIDELINES

Always wear protective gloves, safety goggles, and a lab coat. Wear protective masks when handling TEOS, APTES, glutaraldehyde, and ammonium hydroxide. Use cryogenic gloves to protect your hands and arms when handling liquid nitrogen. Dispose all chemicals and in particular, environmentally hazardous chemicals (e.g., ammonium hydroxide and glutaraldehyde) in appropriate waste containers.

4. PROCEDURES

4.1 General Synthetic Strategy at a Glance

Hybrid organic/inorganic nanobiocatalysts are produced following the sequence: (i) silica nanoparticle synthesis; (ii) enzyme immobilization onto the silica nanoparticles and (iii) enzyme armoring in an organosilica layer.

4.2 Silica Nanoparticles Synthesis

Silica nanoparticles are prepared in a single reaction following the conventional procedure (Stober, Fink, & Bohn, 1968). This method allows producing monodispersed spherical silica nanoparticles with a diameter of around 200 nm by the hydrolysis in alcohol and the consecutive condensation of tetraethyl orthosilicate (TEOS) molecules (Fig. 2). In a fume hood, place a water bath on a magnetic stirrer; place a 1-L round-bottom flask in the water bath (Fig. 3). Set bath temperature at 20°C.

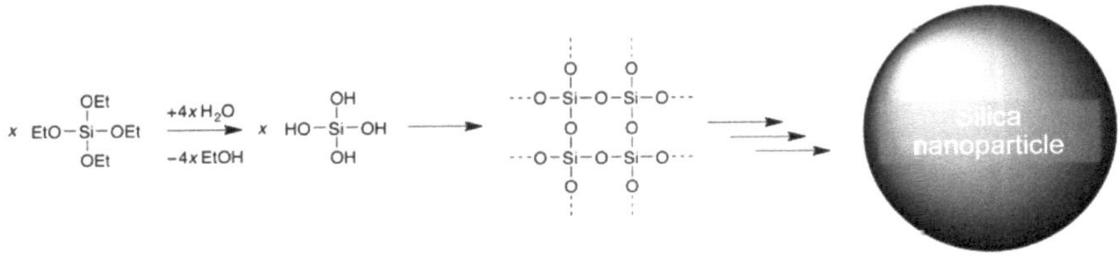

Fig. 2 Schematic representation of the Stöber reaction of TEOS poly-condensation.

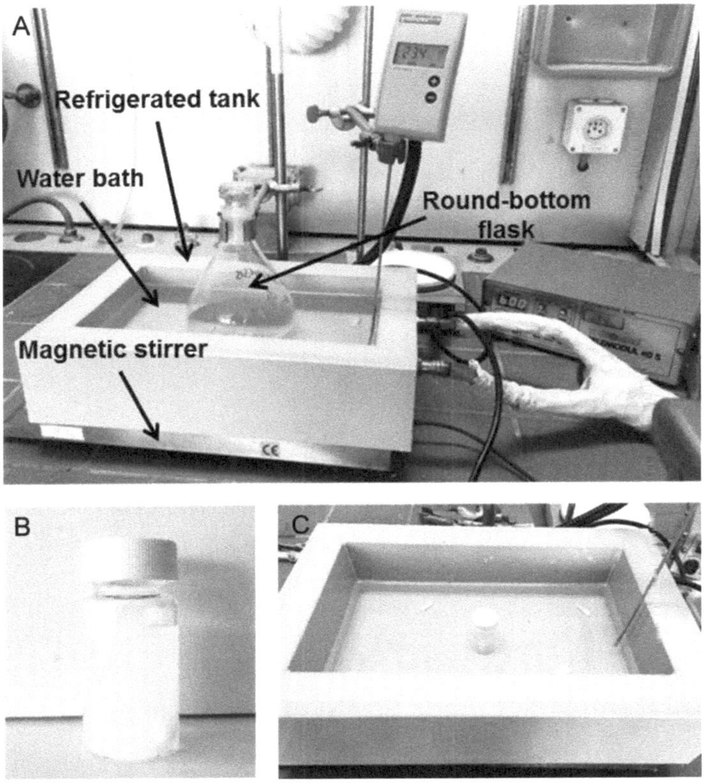

Fig. 3 Experimental set-up. (A) Set-up for the silica nanoparticles synthesis. A 1-L round-bottom flask is placed in a water bath, which is connected to a refrigerator/heater circulator. The water bath is placed on a magnetic stirrer. (B) 20-mL glass vial containing a silica nanoparticles suspension. (C) For enzyme immobilization/shielding, the 20-mL glass vial is placed in the same apparatus as in (A).

4.2.1 Silica Nanoparticle Synthesis Protocol

1. Pour 345 mL ethanol and 39 mL ammonia (25%) in a round-bottom flask and equilibrate at 20°C for 1 h under stirring (600 rpm).
2. Add 15.3 mL of TEOS.

3. Incubate for 24 h at 20°C.
4. Divide the particle suspension into 50-mL tubes and centrifuge for 15 min at 3220 × *g*.
5. Discard the supernatant and resuspend the pellet in 25 mL ethanol in the ultrasonic bath.
6. Centrifuge for 15 min at 3220 × *g*.
7. Discard the supernatant and resuspend the pellet in 25 mL nanopure water in the ultrasonic bath.
8. Repeat Steps 6 and 7.
9. Centrifuge for 15 min at 3220 × *g*.
10. Discard the supernatant and after snap freezing the particles pellet in liquid nitrogen, freeze dry for 12 h.
11. Weigh the particles.
12. Resuspend the pellet in 50 mL nanopure water in the ultrasonic bath.

4.2.2 Silica Nanoparticles Characterization

(1) Dilute the particle suspension 1:100 in water to a final volume of 100 μL and sonicate for 5 min.
(2) Spread 2 μL of particle suspension on a freshly cleaved mica sheet.
(3) The sample previously dried in a dry chamber is coated for 15 s at 10 mA with a gold–platinum alloy by using a sputter coater.
(4) Image the silica nanoparticles by scanning electron microscopy using the secondary electron InLens mode with an accelerating voltage of 20 kV.
(5) Acquire micrographs at a magnification of 150,000X.
(6) Acquire a number of micrographs so that at least 100 particles can be measured.
(7) Use an image analysis software to measure the diameter of the particles.
(8) Report in an excel file the measured values and calculate mean, standard deviation, and standard error.
(9) Plot a frequency histogram (Fig. 4).

4.3 Tips on Silica Nanoparticle Synthesis

The quality and monodispersity of the silica nanoparticles depends on: (i) a homogenous stirring speed and (ii) a constant temperature during the synthesis. Make sure that these parameters are checked before starting. After 30 min incubation of the ethanol–ammonia–TEOS mixture, the solution becomes milky indicating the formation of silica nanoparticles.

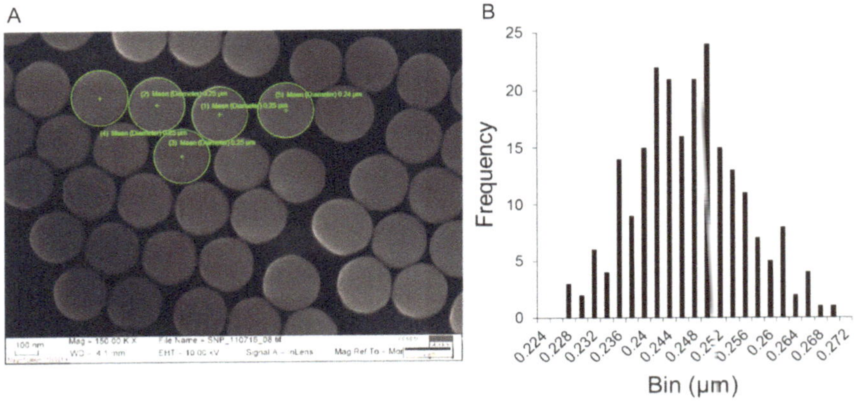

Fig. 4 Silica nanoparticles characterization. (A) Scanning electron micrograph of silica nanoparticles produced through the Stöber reaction and (B) histogram representing the distribution of the measured particles.

4.4 Enzyme Immobilization and Shielding

4.4.1 General Procedure and Experimental Set-Up

Enzyme immobilization is performed after functionalization of the silica nanoparticles' surface. First, the surface of the silica carrier is amino modified by incubating the particle suspension with 3-aminopropyl triethoxysilane (APTES). Afterward, in order to provide anchoring residues for the enzyme, the amino-modified particles are chemically modified with the homo-bifunctional crosslinker glutaraldehyde. As model enzyme, β-galactosidase from *K. lactis* is chosen. The enzyme shielding is performed by incubating the immobilized β-galactosidase with a mixture of tetraethyl orthosilicate and (3-aminopropyl)triethoxysilane at 20°C (Fig. 5).

The experiment is performed in glass vials, at 400 rpm under magnetic stirring at 20°C (Fig. 3). A β-galactosidase stock solution (0.63 mg mL^{-1}) in buffer (100 m*M* phosphate, 5 m*M* $MgCl_2$, pH 6.5) is prepared and kept on ice prior to use.

4.4.2 Silica Nanoparticle Functionalization

(1) Pour 18 mL of silica nanoparticles at a concentration of 3.2 mg mL^{-1} in a 20-mL glass vial.

(2) Add 11 μL of 3-aminopropyl triethoxysilane (APTES) and incubate for 30 min at 400 rpm stirring.

(3) Transfer the particle suspension to a 50-mL tube and centrifuge for 15 min at 3220 × *g*.

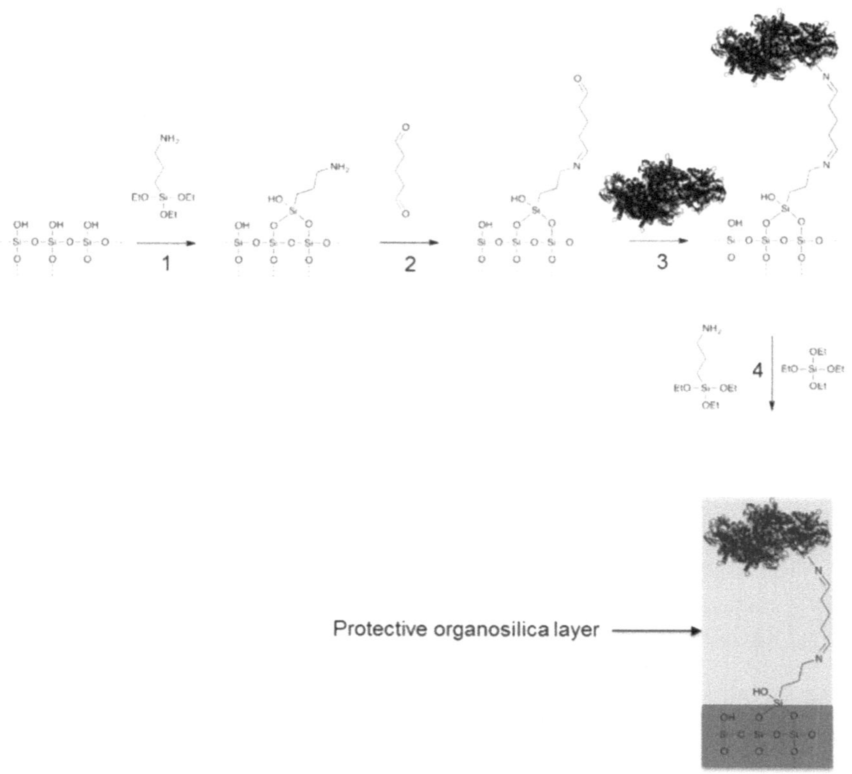

Fig. 5 Schematic representation of the strategy developed for enzyme immobilization and shielding. The silica nanoparticles suspension is sequentially incubated with (3-aminopropyl)triethoxysilane (APTES) (step 1), glutaraldehyde (step 2), enzyme (step 3), and with APTES and tetraethyl orthosilicate (step 4) for the synthesis of the protective organosilica layer. Details of the chemical reactions occurring at the surface of the silica nanoparticles during the reagents addition are showed.

(4) Discard the supernatant and resuspend the pellet in 20 mL nanopure water in the ultrasonic bath and centrifuge for 15 min at 3220 × *g*.
(5) Discard the supernatant and resuspend the pellet in 18 mL 1% (v/v) aqueous glutaraldehyde solution.
(6) Incubate for 30 min at 400 rpm stirring.
(7) Transfer the particle suspension in a 50-mL tube and centrifuge for 15 min at 3220 × *g*.
(8) Discard the supernatant and resuspend the pellet in 20 mL nanopure water in the ultrasonic bath and centrifuge for 15 min at 3220 × *g*.

4.4.3 Enzyme Crosslinking and Shielding

(1) Transfer the particle suspension in a 50-mL tube and centrifuge for 15 min at 3220 × *g*.
(2) Discard the supernatant and resuspend the pellet in 17 mL buffer (10 m*M* MES, 5 m*M* $MgCl_2$, pH 6.2) in the ultrasonic bath.
(3) Transfer the particle suspension into a 20-mL glass vial.
(4) Add 1 mL enzyme at a concentration of 0.63 mg protein mL^{-1} into the particle suspension and incubate at 20°C for 1 h at 400 rpm stirring.
(5) Add 86 μL of TEOS and incubate for 1 h at 400 rpm stirring.
(6) Add 21 μL APTES and incubate for 15 h at 400 rpm stirring.
(7) Transfer the suspension in a 50-mL tube and centrifuge the particle suspension for 15 min at 3220 × *g*.
(8) Discard supernatant and resuspend the pellet in 20 mL nanopure water and centrifuge for 15 min at 3220 × *g*.
(9) Discard supernatant and resuspend the pellet in 18 mL buffer (100 m*M* phosphate, 5 m*M* $MgCl_2$, pH 6.5).
(10) Incubate the particles at room temperature (20–22°C) for 20 h.
(11) Store the particles at 4°C.

4.4.4 Nanobiocatalyst Characterization

The enzyme immobilization yield is typically correlated to the chemical properties of the protein and in particular to the isoelectric point (Section 4.8). Another important parameter to consider for an optimal immobilization yield is the available SNPs surface (Section 4.7). When 0.035 mg mL^{-1} of β-galactosidase are incubated with 1 mL of SNPs at a concentration of 3.2 mg mL^{-1}, a 99% immobilization yield is obtained (cf. Section 4.5). After immobilization, 60% of the catalytic activity is retained. This catalytic activity decreases after the protective layer growth to a value of 28% but recovered (68%) after curing. NBCs obtained can be repetitively used (10 ×) by cycles of centrifugation/resuspension in fresh buffer without significant loss of activity. These are evaluated for their resistance to physicochemical stresses (i.e., temperature, ultrasounds, pH, and chaotropes) and compared to those of the free enzyme (FE) and to the immobilized unprotected reference (IUR).

4.4.4.1 Temperature Stress

When the activity assay is performed at 45, 50, 55, and 60°C, NBCs displayed enzymatic activities of 118%, 123%, 114%, and 88% of the activity measured at 40°C, respectively. FE and IUR gradually lost activity reaching complete inactivity at 60°C. When incubated at 50°C for 60 min, NBCs preserved more than 25% of initial activity, while FE and IUR were totally inactive.

The resistance to five freezing–thawing cycles was tested. Whereas FE and IUR lost, after the first cycle, as much as 45% and 25% of their activity, respectively, the loss for NBCs was only 5%. The biocatalytic activity of NBC remained relatively unchanged at 95% after the fifth cycle.

4.4.4.2 Ultrasound Stress

As additional physical stress test, FE, IUR, and the NBC were subjected to ultrasound treatment. Although the biocatalytic activity of both FE and IUR decayed rapidly over time, the loss of activity of NBC was moderate. It kept as much as 88% of activity after 40 min (18% activity was preserved for FE and 5% for IUR).

4.4.4.3 pH Stress

FE, IUR, and NBC were incubated for 15 min at increasing pH values. For values lower than the optimum pH of 6.5, IUR and FE showed a similar trend, with a loss of more than 40% activity at pH 5 and 100% at pH 4. NBCs lost only 29% and 71%, respectively. IUR lost more than 45% and 75% of activity at pH 7 and 8, respectively, whereas NBC lost only 10% and 15% activity. At these pH values, the most stable system is FE, which retained 100% activity. At pH 9 and 10, IUR and FE both lost more than 95% of activity, whereas NBC lost only 19% and 64%, respectively. At higher pH values (9, 10, and 11), the activity of IUR and FE was reduced to 0%, whereas NBC retained 7% of initial activity.

4.4.4.4 Chaotropic Stresses

The effect of the organosilica layer against chaotropic stress by using urea (6 M) and sodium dodecylsulfate (1%) was studied. In the presence of urea, neither FE nor IUR exhibited any biocatalytic activity; however, 12% activity was measured for the NBC. In the case of SDS treatment, FE was completely inactive. The activity of the IUR was as low as 7%, while NBCs maintained 45% of their initial activity.

4.4.4.5 Proteinase K Digestion

After incubation with proteinase K for 60 min, no activity was measured for FE and a 30% loss of activity was measured for IUR while NBCs preserved 100% of their initial activity.

Table 1 Immobilization and Shielding of Different Enzymes

Enzyme	Class	Immobilization Yield (%)	Shell (nm)	Shielded Enzyme Activity (U mg^{-1})
β-Galactosidase	EC 3.2.1.23	99	17	0.05
Acid phosphatase	EC 3.1.3.2	98	12	0.023
Laccase	EC 1.10.3.2	54	19	0.011
Alcohol dehydrogenase	EC 1.1.1.1	84	13	0.0015
Aspartate transaminase	EC 2.6.1.1	60	18	0.075

4.4.5 Shielding of Different Enzymes

The general strategy of NBCs synthesis was applied for shielding different enzymes such as acid phosphatase, laccase, alcohol dehydrogenase, and aspartate transaminase (Table 1). For all tested enzymes, the strategy was successful and functional NBCs were obtained.

4.5 Tips on the Immobilization Yield and on the Activity of the Immobilized Enzyme

The immobilization yield is obtained by performing a protein quantification assay (Bradford assay, Bradford, 1976) in the supernatant collected after enzyme binding. The immobilization yield (IY) is defined as:

$$IY = \frac{[Enzyme]_{total} - [Enzyme]_{unbound}}{[Enzyme]_{total}}$$

The catalytic activity of the immobilized enzyme was measured by using the artificial substrate *ortho*-nitrophenyl-β-galactoside (ONPG), which is hydrolyzed into a spectrophotometrically detectable product (*ortho*-nitrophenolate formed by the deprotonation of *ortho*-nitropenol in basic conditions). Activity assays were performed on supernatants collected after enzyme binding and on the nanoparticles.

4.6 Tips on the Curing of the Shielded Enzymes

The duration of incubation at room temperature of the shielded enzyme, called "curing time," is of particular importance to obtain a stable and soft organosilica layer. During this time, a rearrangement of the chemical interactions between the silanes of the shell and the protein surface occurs. This molecular rearrangement allows the enzyme to recover a favorable three-dimensional conformation, thus better retention of catalytic activity is possible.

4.7 Tips on the Calculation of the Carrier Surface

The available surface of the carrier material depends on the particles size. Once determined, the average value of the particles diameter, the available surface can be estimated as follows:

(i) Calculation of the volume of one particle.

(ii) Calculation of the number of particles per mg of silica (density of silica: c. 2.65 g cm^{-3}).

(iii) Calculation of the surface of one particle.

(iv) Calculation of the total surface of all particles in the sample.

Similarly, the shape and size of the enzyme (from the X-ray crystal structure) allow estimating the surface occupied onto silica. For example, the three-dimensional structure of β-galactosidase used in this set of experiments can be approximated as a tri-axial ellipsoid with dimensions of $15.9 \times 9.3 \times 5.3$ nm^3.

4.8 Tips for a Successful Enzyme Crosslinking and Shielding

The isoelectric point of the enzyme is of importance. During the enzyme immobilization on the carrier, the buffer should have a pH value favoring electrostatic interactions with the carrier surface. In more detail, since the silica nanoparticles are negatively charged, the enzyme should be positively charged in the coupling pH conditions. A negative charge at the enzyme surface would cause a repulsion of the particles and prevent the enzyme binding.

The poly-condensation of organosilanes is correlated to the following parameters: temperature, pH, and time. Additionally, the shell formation rate is dependent on the organosilane composition and ratio. APTES acts as a catalyst of the poly-condensation reaction.

The poly-condensation should be monitored overtime in order to reach a thickness adequate to the immobilized enzyme.

4.9 Tips on Particle Loss

Perform carefully the steps of centrifugation/resuspension of the particles pellet. A loss of particles would change the amount of enzymatic activity between replicates.

5. CONCLUSIONS

The synthesis of a novel nanobiocatalyst has been described. The developed system consists of an enzyme immobilized onto silica nanoparticles and armored with a protective layer. The process of armoring is based on the self-assembly and poly-condensation of organosilane building blocks at the surface of silica nanoparticles and around the immobilized enzyme. The organosilanes sorting onto the available surface depends not only on the chemical functionality of the building blocks but also on the chemical complementarity with the amino acid residues exposed at the surface of the protein. The nanobiocatalysts produced here have enhanced catalytic activity and improved stability against stresses of different nature (heat, pH, chaotropic agents, proteases, and ultrasound). The method herein reported is versatile in that it can be adapted to a number of different enzymes successfully and the method has a wide scope for enzyme armoring approaches.

ACKNOWLEDGMENTS

Financial supports from the Swiss Nanoscience Institute (SNI) through the NanoArgovia program (NanoZyme project), the Swiss commission for technology and innovation (grant agreement 16437.1 PFEN-NM), the EU-Eurostar program (grant agreement E!6894), and the Swiss State Secretariat for Education, Research, and Innovation (EU-H2020 INMARE project) are gratefully acknowledged.

REFERENCES

Avnir, D., Coradin, T., Lev, O., & Livage, J. (2006). Recent bio-applications of sol-gel materials. *Journal of Materials Chemistry*, *16*, 1013–1030.

Bornscheuer, U. T., Huisman, G. W., Kazlauskas, R. J., Lutz, S., Moore, J. C., & Robins, K. (2012). Engineering the third wave of biocatalysis. *Nature*, *485*, 185–194.

Bradford, M. M. (1976). A rapid and sensitive method for the quantitation of microgram quantities of protein utilizing the principle of protein-dye binding. *Analytical Biochemistry*, *72*, 248–254.

Ciriminna, R., Fidalgo, A., Pandarus, V., Beland, F., Ilharco, L. M., & Pagliaro, M. (2013). The sol-gel route to advanced silica-based materials and recent applications. *Chemical Reviews*, *113*, 6592–6620.

Correro, M. R., Moridi, N., Schutzinger, H., Sykora, S., Ammann, E. M., Peters, E. H., et al. (2016). Enzyme shielding in an enzyme-thin and soft organosilica layer. *Angewandte Chemie, International Edition*, *55*, 6285–6289.

Cumbo, A., Lorber, B., Corvini, P. F. X., Meier, W., & Shahgaldian, P. (2013). A synthetic nanomaterial for virus recognition produced by surface imprinting. *Nature Communications*, *4*, 1503–1509.

DiCosimo, R., McAuliffe, J., Poulose, A. J., & Bohlmann, G. (2013). Industrial use of immobilized enzymes. *Chemical Society Reviews*, *42*, 6437–6474.

Franssen, M. C., Steunenberg, P., Scott, E. L., Zuilhof, H., & Sanders, J. P. (2013). Immobilised enzymes in biorenewables production. *Chemical Society Reviews*, *42*, 6491–6533.

Hanefeld, U., Gardossi, L., & Magner, E. (2009). Understanding enzyme immobilisation. *Chemical Society Reviews*, *38*, 453–468.

Hestericova, M., Correro, M. R., Lenz, M., Corvini, P. F., Shahgaldian, P., & Ward, T. R. (2016). Immobilization of an artificial imine reductase within silica nanoparticles improves its performance. *Chemical Communications*, *52*, 9462–9465.

Karshikoff, A., Nilsson, L., & Ladenstein, R. (2015). Rigidity versus flexibility: The dilemma of understanding protein thermal stability. *FEBS Journal*, *282*, 3899–3917.

Plouffe, P., Macchi, A., & Roberge, D. M. (2014). From batch to continuous chemical synthesis—A toolbox approach. *Organic Process Research & Development*, *18*, 1286–1294.

Polizzi, K. M., Bommarius, A. S., Broering, J. M., & Chaparro-Riggers, J. F. (2007). Stability of biocatalysts. *Current Opinion in Chemical Biology*, *11*, 220–225.

Rodrigues, R. C., Ortiz, C., Berenguer-Murcia, A., Torres, R., & Fernandez-Lafuente, R. (2013). Modifying enzyme activity and selectivity by immobilization. *Chemical Society Reviews*, *42*, 6290–6307.

Sanchez, C., Belleville, P., Popall, M., & Nicole, L. (2011). Applications of advanced hybrid organic-inorganic nanomaterials: From laboratory to market. *Chemical Society Reviews*, *40*, 696–753.

Stober, W., Fink, A., & Bohn, E. (1968). Controlled growth of monodisperse silica spheres in micron size range. *Journal of Colloid and Interface Science*, *26*, 62–69.

Subramanian, G. (Ed.), (2014). *Continuous processing in pharmaceutical manufacturing*. Weinheim: Wiley-VCH.

Sykora, S., Cumbo, A., Belliot, G., Pothier, P., Arnal, C., Dudal, Y., et al. (2015). Virus-like particles as virus substitutes to design artificial virus-recognition nanomaterials. *Chemical Communications*, *51*, 2256–2258.

Wells, A., & Meyer, H.-P. (2014). Biocatalysis as a strategic green technology for the chemical industry. *ChemCatChem*, *6*, 918–920.

Zhou, Z., & Hartmann, M. (2013). Progress in enzyme immobilization in ordered mesoporous materials and related applications. *Chemical Society Reviews*, *42*, 3894–3912.

CHAPTER FIVE

Strategies for Biophysical Characterization of Protein–Polymer Conjugates

Cameron Williams, Melissa L. Dougherty, Katherine Makaroff, Jacob Stapleton, Dominik Konkolewicz[1], Jason A. Berberich[1], Richard C. Page[1]

Miami University, Oxford, OH, United States

[1]Corresponding authors: e-mail address: konkold@miamioh.edu; berberj@miamioh.edu; pagerc@miamioh.edu

Contents

Abstract

Protein–polymer conjugates are increasingly viewed as promising avenues to producing industrial enzymes with high activity capable of withstanding potentially harsh reaction conditions, or to designing novel therapeutics with triggered release, controlled masking, or increased resistance to proteolytic degradation. Common among these applications are the desire to improve the stability of protein–polymer conjugates to unfolding by exposure to chemicals or thermal stress. Thus, assays that allow researchers to robustly and easily characterize protein–polymer conjugates by obtaining thermodynamic parameters for folding stand to play an important role in the development of improved protein–polymer conjugates. Herein, we describe two techniques, differential scanning fluorimetry and intrinsic tryptophan fluorescence, used in our laboratories to obtain thermodynamic parameters of unfolding that allow for direct comparison of protein–polymer conjugates and the myriad effects of variations in attachment site, polymer identity, and polymer length. These two experiments, which are easily amenable to parallelization, are presented as high-throughput replacements for more traditionally employed circular dichroism experiments and as complements to functional chemical stability or functional thermal stability experiments. Each assay is presented in a parallelized format that allows for rapid scaling and high-throughput analysis of

Methods in Enzymology, Volume 590
ISSN 0076-6879
http://dx.doi.org/10.1016/bs.mie.2016.11.008

© 2017 Elsevier Inc.
All rights reserved.

protein–polymer conjugate libraries. Descriptions of the assays include a discussion of advantages and disadvantages alongside protocol details and approaches to data analysis.

1. INTRODUCTION

Though often viewed through a purely biological lens, proteins are first and foremost linear polymers assembled from a library of 20 distinct monomer units. While this library of 20 amino acid monomers enables incredible diversity of structure and function (Das, Dawson, & Orengo, 2015; Goncearenco & Berezovsky, 2015; Sawada & Honda, 2006), nature has employed posttranslational modifications as a mechanism to further modulate function (Das et al., 2015). In the laboratory, protein function can be modulated by modifying proteins with organic polymers constructed via various methods including biocompatible reversible deactivation radical polymerization methods such as atom transfer radical polymerization and reversible addition-fragmentation chain transfer polymerization (Averick et al., 2011; Boyer et al., 2007; Cobo, Li, Sumerlin, & Perrier, 2015; De, Li, Gondi, & Sumerlin, 2008; Lele, Murata, Matyjaszewski, & Russell, 2005; Li, Bapat, Li, & Sumerlin, 2011; Li, Li, Yu, Bapat, & Sumerlin, 2011; Mancini, Lee, & Maynard, 2012; Sumerlin, 2011). The modification of proteins with polymers offers the opportunity to further expand functional space by increasing enzyme function in nonnative solvents (Brogan & Hallett, 2016), increasing stability at high temperatures (Mancini et al., 2012; Mudhivarthi et al., 2012), conferring stability in the presence of denaturants (Lucius et al., 2016), or decreasing the rate of proteolytic digestion (Falatach, Li, et al., 2015; Riccardi et al., 2014). Additionally, there is evidence to suggest that modification of proteins can reduce the entropy of the unfolded state, thereby lessening the entropic gain from unfolding and favoring the folded state in comparison to the unmodified protein (Zhou, 2004).

In addition to questions that arise by viewing protein–polymer conjugates from the polymer chemistry field (Hoffman, 1995; Murata, Cummings, Koepsel, & Russell, 2013), a view from a protein chemistry vantage point raises additional questions regarding the properties, interactions, and behaviors of the polymer-modified protein. A primary set of questions center on the effect of polymer modification upon protein structure, dynamics, function, and stability. Studies of protein–polymer conjugates often characterize protein structure via low-resolution techniques such as circular dichroism (CD) (Brogan & Hallett, 2016; Magnusson, Bersani, Salmaso, Alexander, & Caliceti, 2010; tenCate, Rettig, Bernhardt, & Börner, 2005), intrinsic fluorescence, or absorbance of chromophores or cofactors (Brogan & Hallett, 2016). Due to the difficulty in obtaining high-resolution structures of protein–polymer conjugates, functional characterizations are often used as a proxy for determining the effect of conjugated polymers on protein structure (Boyer et al., 2007; Danielson et al., 2016; Falatach, Li, et al., 2015; Lucius et al., 2016; Mancini et al., 2012) since retention of activity is highly correlated with retention of structure (Masso, Lu, & Vaisman, 2006). Similar to structural characterization, determination of the effect of conjugated polymers on the dynamics of proteins is difficult; however, molecular dynamics simulations (Brogan, Sessions, Perriman, & Mann, 2014; Lucius et al., 2016) and neutron scattering (Russo et al., 2015) have begun to provide a glimpse.

Similar to the importance of structure and dynamics, maintained access to enzyme active sites is often a prerequisite for producing functional protein–polymer conjugates. Our recent study of lysozyme–polymer conjugates discussed potential negative effects of placing large polymers near the enzyme active site (Falatach, McGlone, et al., 2015). Other studies have utilized interference in the active site by a conjugated polymer as a means to regulate enzymatic activity (Cummings, Murata, Koepsel, & Russell, 2014; Murata et al., 2013). Similarly, while polymers conjugated to antibodies offer avenues to producing antibody drug conjugates or increasing the circulation lifetime of antibody therapeutics (Srivastava, O'Connor, Pandit, & Gerard Wall, 2014), the masking of epitopes by polymers is a factor that can likely be tuned (Bianchi & Pessi, 2010). Underlying these effects are not only the importance of the locations of modification sites within a single protein but also the role of protein–polymer interactions in modulating stability and activity (Lucius et al., 2016). Recent studies of lysozyme–polymer conjugates suggest that protein–polymer interactions depend upon the identity of the polymer and the location of the attachment on the enzyme (Brogan et al., 2014; Lucius et al., 2016). Polymers capable of changing structure and interactions on the basis of pH or temperature have enabled the development of responsive protein–polymer conjugates (Brogan et al., 2014; Cummings et al., 2014; De et al., 2008; Murata et al., 2013). The combination of complex effects that govern modulation of activity, partitioning, and protein–polymer interactions underscores the importance of characterizing enzymatic activity and stability for protein–polymer conjugates. In comparison to activity assays, which are specific to the enzyme in use, stability assays are more widely applicable and provide an opportunity to compare the effect of a particular polymer upon stability for a range of proteins. The sheer number of possible modifications due to variations in attachment site, polymer identity, and polymer length makes it increasingly important that researchers have analysis techniques available that can rapidly assess the activity and stability of samples within large and diverse protein–polymer conjugate libraries.

Although much research has been dedicated to the study of how to best modify proteins with polymers, studies have taken different approaches to characterizing the function and stability of protein–polymer conjugates. While the function of protein–polymer conjugates can be appropriately studied by applying the relevant enzymatic activity assay, many of which are easily parallelized, the appropriate methods for determination of changes in stability are not as clearly developed. The definition of stability and the technique used are both important parameters for interpreting the results of protein–polymer conjugations. We have chosen to define thermal stability as the melting point, which can be determined by differential scanning fluorimetry (DSF) (Cimmperman & Matulis, 2011; Lucius et al., 2016; Matulis, Kranz, Salemme, & Todd, 2005; Pantoliano et al., 2001; Vollrath, Hawkins, Porter, Holland, & Boulet-Audet, 2014), rather than utilizing functional thermal stability (Mancini et al., 2012). We have also defined chemical stability in terms of resistance to unfolding by chemical additives, most commonly denaturants, as measured by fitting a two-state folding model to intrinsic tryptophan fluorescence (ITF) measurements (Lucius et al., 2016). With an eye to biophysics, stability assessments by DSF and ITF offer valuable opportunities to characterize the Gibbs free energy, enthalpy, and entropy of folding.

In comparison to DSF and ITF, CD is a commonly used method that assesses stability by correlating CD spectral features with secondary structure content (Greenfield, 1996; Kelly, Jess, & Price, 2005; Whitmore & Wallace, 2004; Yang, Wu, & Martinez, 1986). Tracking secondary structure content as a function of temperature (Joshi, Shivach, Yadav, & Rathore, 2014) or denaturant concentration (Rizzo et al., 2015) allows for determination of melting points (T_m) or equilibrium unfolding concentrations, similar to measures obtained by DSF or ITF. Previous studies have utilized CD to monitor the effects of polymer conjugation upon protein secondary structure (Isarov, Lee, & Pokorski, 2016; Thilakarathne, Briand, Zhou, Kasi, & Kumar, 2011; Xia, Tang, & Olsen, 2013). Similar to DSF and ITF, CD has a low barrier to entry in that it utilizes widely available commercial instrumentation. Unfortunately, high-throughput screening is a challenge with CD, even in experimental setups where a flow-cell apparatus is paired with a CD spectropolarimeter. For example, measurement of thermal unfolding by CD requires approximately 70 min of experimental time per sample (Joshi et al., 2014); thus, the data acquired in triplicate by a single DSF run in a 96-well plate allow for measurement of 32 samples simultaneously. Acquiring CD thermal unfolding data for 32 samples with an acquisition time of 70 min per sample, without replicates, would require approximately 37 h of experimental time by CD. The poor translation of CD to high-throughput analysis limits the breadth of variations in protein–polymer conjugates that can be studied. In contrast, both DSF and ITF make use of 96- or 384-well plates and are readily parallelized. Thus, while CD is a valuable technique that provides for an evaluation of secondary structure content, high-throughput DSF and ITF experiments enable far wider and more rapid exploration of protein–polymer conjugate libraries.

In contrast to CD, DSF, and ITF, functional stability measurements bridge stability and activity, as opposed to reporting on thermodynamic parameters of stability. Functional stability measurements take a variety of forms ranging from proteolytic stability (Cummings et al., 2014; Falatach, Li, et al., 2015; Murata et al., 2013) to temperature stability (Cummings et al., 2014; Mancini et al., 2012; Murata et al., 2013; Thilakarathne et al., 2011) or pH stability (Cummings et al., 2014; Murata et al., 2013). Recent studies have applied proteolytic stability measurements to chymotrypsin–polymer conjugates (Cummings et al., 2014; Falatach, Li, et al., 2015; Murata et al., 2013); however, circulation lifetime measurements alongside trypsin and chymotrypsin digestion assays have been used historically as a measure of resistance to proteolytic digestion (Abuchowski, McCoy, Palczuk, van Es, & Davis, 1977). Functional thermal stability measurements range throughout the literature from room temperature measurements to assess shelf-life (Thilakarathne et al., 2011) to high-temperature studies at 50°C (Cummings et al., 2014; Murata et al., 2013) or 90°C (Mancini et al., 2012). Functional chemical stability measurements have traditionally examined stability as a function of pH (Cummings et al., 2014; Murata et al., 2013); however, activity upon exposure to denaturants or solvents (Depp, Kaar, Russell, & Lele, 2008) is also possible provided the solvent or denaturant used does not interfere with the enzymatic assay. A tangible benefit of functional stability measurements is that parallelization of enzymatic assays is readily achievable and enables high-throughput screening of protein–polymer conjugate libraries. While functional stability measurements are widely used and should continue to be part of the analysis toolkit for researchers in the protein–polymer field, we do not present protocols for functional chemical or functional thermal stability measurements here as the myriad possible enzymes and accompanying activity protocols are too diverse for single manuscript.

Herein, we discuss two techniques used in our laboratories to biophysically characterize protein–polymer conjugates. Although presented with respect to our recent studies of lysozyme–polymer conjugates, these protocols are applicable to polymer conjugates of other proteins and, with the exception of requiring at least one tryptophan residue for ITF, there are no special requirements that preclude widespread application. The assays described consume small quantities of protein enabling the researcher to obtain thermodynamic parameters from small-scale conjugation reactions. In the parallelized formats presented here, DSF and ITF experiments can each be run for multiple protein–polymer conjugates simultaneously. This opens the door to more quickly and more fully exploring polymer functional group space and will enable labs to more readily identify promising polymer modifications that confer enhanced stability with respect to thermal and chemical denaturation.

2. BIOPHYSICAL METHODS

2.1 Differential Scanning Fluorimetry

2.1.1 Summary

DSF, often colloquially referred to by the trademarked name ThermoFluor™ (Roche, Indianapolis, IN), is a robust method for quickly and simply identifying the melting temperature (T_m) for a protein sample (Niesen, Berglund, & Vedadi, 2007; Vollrath et al., 2014). The method relies upon a solvatofluorochromic dye, typically the merocyanine dye SYPRO Orange (ThermoFisher, Molecular Probes, Waltham, MA), for which a substantial increase in fluorescence quantum yield is observed upon partitioning into a hydrophobic environment with a lower dielectric constant than aqueous solutions (Hawe, Sutter, & Jiskoot, 2008). At low temperatures, the fluorescence of SYPRO Orange is low due to poor quantum yield of the fluorophore in aqueous solution (Fig. 1A, yellow). As the temperature increases, the protein begins to unfold, exposing hydrophobic amino acids that were previously packed within the protein core. Unfolding

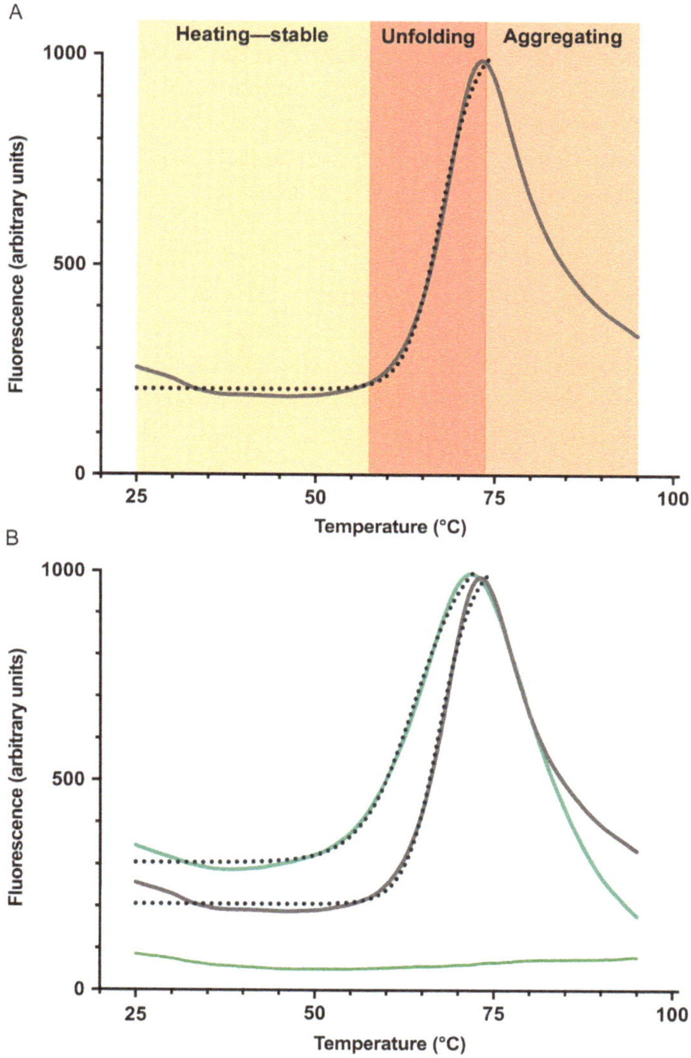

Fig. 1 Differential scanning fluorimetry for analyzing thermal stability. (A) Example DSF data for hen egg white lysozyme (HEWL). Stages of the trace including heating with no stability changes (*yellow*), heating with unfolding (*red*), and heating with aggregation (*orange*) are highlighted. The DSF trace is shown as a *gray line*, the nonlinear Boltzmann fits is shown as *black dots*. (B) DSF traces for native HEWL (*gray*), a HEWL-DP50-acrylamide conjugate (*cyan*), and a DP50 acrylamide polymer alone (*green*). A nonlinear Boltzmann fit to the native HEWL and HEWL-DP50-acrylamide traces are shown as *black dots*. No fit is shown to the DP50 acrylamide polymer alone as no detectable unfolding transition was present. *Adapted with permission from Lucius, M., Falatach, R., McGlone, C., Makaroff, K., Danielson, A. P., Williams, C., et al. (2016). Investigating the impact of polymer functional groups on the stability and activity of lysozyme-polymer conjugates.* Biomacromolecules, 17*(3), 1743–1134. Copyright (2016) American Chemical Society.*

of the protein allows interaction of SYPRO Orange with these hydrophobic amino acids and a concomitant increase in fluorescent signal (Fig. 1A, red). The T_m is found by fitting a Boltzmann function to the SYPRO Orange fluorescence vs temperature, or determining the inflection point of the fluorescence vs temperature, for the folded-to-unfolded transition as a function of time (Niesen et al., 2007) (Fig. 1A and B).

2.1.2 Advantages and Disadvantages

DSF is rapid and widely applicable to a range of proteins and protein–polymer conjugates. It does not require special or rarely encountered functional groups and, for soluble globular proteins is largely independent of amino acid sequence. Chief advantages of DSF are the low concentration requirements for protein samples, the capability to be parallelized, and ease of data collection. The overarching protein requirement is that hydrophobic residues residing within the protein core in the folded state become solvent exposed upon unfolding. A key disadvantage of DSF is the reliance upon a solvatofluorochromic dye which increases fluorescence upon partitioning into hydrophobic environments. Proteins with large hydrophobic surface patches, hydrophobic polymers, or suitably hydrophobic buffer components can produce a high background signal that obscures the folded-to-unfolded protein transition.

2.1.3 Sample Requirements and Preparation

DSF can be run on relatively small quantities of protein as a single well of 25–50 μL requires only 5 μ*M* protein. We have observed optimal performance with 5 μ*M* protein samples, though 1 μ*M* is often sufficient for proteins on the order of 25 kDa or larger that expose significant numbers of hydrophobic residues upon unfolding. The ability to use lower protein concentrations can be particularly helpful for samples that are difficult to obtain or produce. Samples should be prepared in a buffer in which the protein of interest is stable at room temperature. The use of denaturants, such as guanidine hydrochloride, should be avoided as these will destabilize the protein at low temperature, raising the background signal prior to the folded-to-unfolded transition. In our experience, guanidine hydrochloride can trigger increased fluorescence of the solvatofluorochromic dye due to hydrophobic interactions between the dye and the guanidine moiety (Godawat, Jamadagni, & Garde, 2010; Macdonald & Khajehpour, 2015).

2.1.4 Instrumentation and Consumables

2.1.4.1 Thermocycler

We utilize a BioRad CFX Connect quantitative real-time PCR (qPCR) system (Bio-Rad, Hercules, CA). In theory, any qPCR system can be used, provided that appropriate excitation (472 nm) and emission (570 nm) wavelengths are available (Niesen et al., 2007). In practicality, strong absorbance for SYPRO Orange is seen between 445 and 525 nm, and strong emission is observed from 570 to 610 nm, which allows for a broad range of excitation and emission wavelengths and to be utilized (Niesen et al., 2007). In our case, we utilize the preset "HEX" channel, typically utilized for hexachlorofluorescein, available on the CFX Connect. The use of a 96-well qPCR system offers additional benefits of fine control over temperature and parallelized data collection for 96 samples during a single run.

2.1.4.2 Plates

A wide range of plates and seals can be used; however, care must be taken to ensure the plastics can withstand temperatures used in the experiment. In particular, optically clear seals must be used that are compatible with heating to the lid temperature of 105°C, necessary in order to prevent condensation of samples near the top of each well. We utilize a 96-well FrameStar PCR plate (Phenix Research Products, Cat. No. MPC-96HS4-WW, Candler, NC) which features a plate skirt and white wells. The skirted plate facilitates facile sealing on the bench top, while the white wells ensure maximal reflection of light back to the fluorimeter. We pair the PCR plates with Excel Scientific ThermalSeal RT clear seals featuring a pressure sensitive encapsulated adhesive (Phenix Research Products, Cat. No. LMT-THER-RTQ, Candler, NC). The pressure-sensitive seal facilitates proper alignment of the seal onto the plate prior to activation of the adhesive by a manual pressure with a rubber brayer.

2.1.4.3 Solvatofluorochromic Dye

Although the merocynanine dye SYPRO Orange is most commonly used, other dyes including 8-anilino-1-naphthalenesulfonate (ANS) and 4,4′-dianilino-1,1′-binaphthyl-5,5′-disulfonate (Bis-ANS) can be used. We purchase SYPRO Orange as a 5000× concentrated stock (ThermoFisher, Cat. No. S6650, Waltham, MA) in DMSO and use the dye at 10× concentration in DSF assays. Ideally, the solvatofluorochromic dye will exhibit limited interactions with the protein in the folded state and will increase fluorescence when interacting with the unfolded protein. Undesirable dye–protein interactions may be manifested by high initial baselines, while insufficient interactions result in little or no detectable unfolding. In the case of high initial background or no visible folded-to-unfolded transition, additional dyes such as ANS or Bis-ANS should be tested. We find that using SYPRO Orange at 10× concentration yields excellent results for protein–polymer conjugates.

2.1.5 Protocol

1. Safety considerations: personal safety equipment such as safety glasses, nitrile gloves, and a laboratory coat should be worn at all times. SYPRO Orange is a mild irritant and is dissolved in DMSO, a flammable liquid. Volumes utilized are small; however, appropriate care should be taken with all chemicals used. Consult the MSDS for each buffer component to identify hazards.
2. Prepare enough protein for triplicates with 50 μL well volumes at 5 μ*M* protein. Ideally, the protein will be sufficiently concentrated to only require a 1- to 2-μL volume of protein to be added to each well.
3. Prepare enough buffer with 10× SYPRO Orange, typically we budget 50 μL per well to account for loss upon pipetting or transfer. We recommend 20 m*M* HEPES, pH 7.5, 10× SYPRO Orange, 150 m*M* NaCl; however, phosphate-buffered saline or any Good buffer (Good et al., 1966) is sufficient provided that the protein of interest is stable at the beginning of the assay.
4. Pipet 48 μL DSF buffer into each well. Add 2 μL protein to each well. We recommend setting up data in triplicate. Logical layout of replicates is discussed later in Section 2.1.6.
5. Seal the plate with an appropriate heat-tolerant clear seal. *Tip: make sure to get a good seal across all wells. Poorly sealed wells will result in rapid evaporation and poor quality or no data.*
6. Let the plate sit for 15 min to allow for mixing of the protein with the DSF buffer. This can be aided by briefly shaking or vortexing the plate. Quick centrifugation for 5 min at 3000 × *g* may be necessary to ensure all samples are at the bottom of each well prior to the start of the experiment. However, be aware that refrigeration may result in unwanted buildup of moisture on plastic seals above sample wells.
7. Begin the DSF data collection run. We recommend heating from 25°C to 95°C in 0.5°C steps with a 20-s equilibration time at each temperature

step and a lid temperature of 105°C. The preset "HEX" channel available on qPCR instruments is sufficient for excitation and emission with SYPRO Orange. Using these settings, a typical run is approximately 50 min long.

8. Export and analyze data as described in Section 2.1.6. Plates should be discarded after use.

2.1.6 Data Analysis and Typical Data

Data collection for DSF results in fairly large datasets, with each well providing 140 data points. Conducted in triplicate, facile data analysis is aided by logical setup of samples within the 96-well plate. Although data output formats may vary between qPCR manufacturers, the Bio-Rad CFX Connect qPCR produces data that can be exported to a .xlsx formatted file in which data for a single well are listed in a column, and columns are written in a sequential read format from right to left across the plate yielding data provided as A1–A12, followed by B1–B12, followed by C1–C12, proceeding through the plate and ending in H1–H12. Due to this format, replicates should be placed in the plate sequentially, such as triplicates for sample one in A1–A3, triplicates for sample two in A4–A6. Doing so enables rapid copying of data from Excel to Prism (GraphPad Software, Inc.), speeds up data analysis, and reduces the opportunity for introducing errors due to formatting data by hand.

A typical DSF melting curve (Fig. 1A) can be subdivided into three separate phases. The first phase is characterized by heating of the sample that results in no detectable unfolding and exhibits low fluorescence values (Fig. 1A, yellow). Variations between proteins or between different protein–polymer conjugates can increase fluorescence values in this region, often characterized by a high average value across this region, or by an initially high value that decreases to a minimum prior to reaching the second phase of the melting curve. Both behaviors are evidence of interaction between the SYPRO Orange and hydrophobic patches on the protein or polymer that are solvent exposed when the protein is in the folded state. Except in cases where high initial fluorescence obscures the second phase, these behaviors can be tolerated. In addition to interaction with hydrophobic regions of the protein, SYPRO Orange may interact with hydrophobic groups within the polymer and produce an observable fluorescence signal. To detect this behavior, a control experiment with polymer only is advised in order to identify whether the polymer causes changes in SYPRO Orange fluorescence across the experimental temperature range (Fig. 1B). A DSF curve for a polymer-only sample may be compared to a sample of SYPRO Orange in buffer to identify any increases in fluorescence due to interactions between SYPRO Orange and the polymer. The second phase of the melting curve (Fig. 1A, red) is characterized by a rapid increase in fluorescence up to a maximal value for the entire temperature range. This peak is caused by unfolding of the protein and exposure of hydrophobic residues to the aqueous solution. The melting point, halfway between the minimum and maximum fluorescence values, also approximated by the inflection point for the unfolding curve, is found in this region of the curve. The third phase of the melting curve (Fig. 1A, orange) is characterized by a decrease in fluorescence values, often to a global minimum. This sharp decrease in fluorescence results from aggregation of proteins which nonspecifically cluster via hydrophobic groups. This aggregation removes hydrophobic groups from solution resulting in an environment with minimal groups capable of interacting with the solvatofluorochromic dye and results in decreased fluorescence.

Data reduction, such as background subtraction, can be done with ease in Excel prior to export to Prism. In our experience, background subtraction is rarely required and often does not result in significant changes to the T_m value obtained from Boltzmann fitting. The use of Prism (GraphPad Software, Inc., San Diego, CA) or a similar graphical analysis and curve fitting program is desirable as these programs allow for rapid analysis of samples using nonlinear fitting. Nonlinear fitting of DSF data is traditionally performed by either finding the value at which the second derivative equals zero (inflection point) or by applying a Boltzmann fit (Niesen et al., 2007) (Eq. 1), where F_{min} and F_{max} are the minimum and maximum fluorescence values in the heating range prior to aggregation, m is a measure the slope of the unfolding curve with units °C, T is the temperature in units of °C, and T_m, expressed in units of °C, is the melting point to be fit using the least squares method to minimize differences between calculated fluorescence (F_{calc}) and the observed DSF data. The use of a Boltzmann fit allows for reliable fitting of the data from the first data point through the maximum of the unfolding peak and the only intervention required in Prism is to exclude data points after the peak (Fig. 1A, orange). In contrast, we have observed that the use of inflection point tends to be noisier and requires users to exclude data both before and after the unfolding portion of the heating curve In our experience, Boltzmann fits provide an easy to use, rapid approach to determining T_m values from DSF data. Combining T_m values with the definition that at the T_m 50% of the protein is folded, we define that at the T_m the ΔG of folding is zero, therefore, at the T_m, $\Delta H = -T_m \Delta S$ and $T_m = -{}^{\Delta H}/_{\Delta S}$. Thus, measurement of the T_m by DSF provides a convenient route to estimating the ratio of enthalpy and entropy ($\Delta H/\Delta S$) of folding for protein–polymer conjugates

$$F_{calc} = F_{min} + \frac{F_{max} - F_{min}}{1 + e^{\left(\frac{T_m - T}{m}\right)}}. \qquad (1)$$

2.2 Intrinsic Tryptophan Fluorescence

2.2.1 Summary

The use of extrinsic fluorophores, such as SYPRO Orange, to detect hydrophobic groups exposed upon unfolding is typically not compatible with the use of denaturants due to interactions between the denaturants and the hydrophobic extrinsic fluorophores (Godawat et al., 2010; Macdonald & Khajehpour, 2015). In contrast, tryptophan can be used as an intrinsic fluorophore that reports upon the folding state of the protein. Unfolding typically results in a red shift for λ_{max} that can be tracked as a function of denaturant concentration. Fitting the λ_{max} data to a sigmoidal two-state folding curve allows for determination of denaturant concentration required to unfold 50% of the sample, the D_{50} value. Comparing the D_{50} value across protein–polymer conjugate samples allows researchers to quantify increased resistance to denaturation and enhancements in protein stability conferred by polymer conjugation.

2.2.2 Advantages and Disadvantages

By utilizing intrinsic fluorescence of tryptophan residues native to the protein of interest, no additional modifications are required, nor do exogenous fluorophores need to be added. Despite this advantage, the absence of tryptophan residues, while uncommon (Carugo, 2008), does occur naturally, such as in the small protein ubiquitin (Ciechanover, Elias, Heller, Ferber, & Hershko, 1980). A further disadvantage is that the method does not provide conclusive information on the folding state of the protein. An inherent assumption is that the protein is folded in the absence of denaturant and unfolded once the λ_{max} value reaches maximum. While this assumption is likely valid for most proteins, deviations from this assumption will introduce errors in fitting observed behavior to a two-state unfolding model. An additional assumption is the reversibility of folding. To confirm reversibility, we recommend measuring the λ_{max} at multiple denaturant concentrations, followed by the dilution of the enzyme/denaturant solution into fresh buffer and a second measurement of λ_{max}. For systems with reversible unfolding, the second λ_{max} measurement should be identical to the λ_{max} value in the absence of denaturant. For systems found to be nonreversible, the protocol described later can be utilized to measure relative stability; however, full quantitation of the free energy of folding should be avoided. Although these disadvantages preclude use as an exact description of residual structure, the experiment does provide for quantitative measure of relative unfolding and can be reliably used to assess chemical stability across a family of protein–polymer conjugates. Since exogenous fluorophores are not needed, the experiment can be used not only with denaturants but also with other chemical additives that may normally destabilize the protein or interact with an exogenous fluorophore. To confirm that changes in observed fluorescence do not arise from fluorescence of the polymer or other sample components, we recommend a control experiment in which ITF data for a protein-free sample are collected to check for fluorescence of the polymer in the 320–350 nm region. This control experiment allows for detection of any fluorescence arising from the polymer, buffer, denaturants, additives, or other components within the sample.

2.2.3 Sample Requirements and Preparation

Our assay is designed for 40 μL samples containing 20 μ*M* protein or protein–polymer conjugate. Although an appropriate protein concentration should be optimized for each protein using native protein, we have found that 20 μ*M* is good starting point that serves to balance the need for producing a detectable fluorescence signal while conserving valuable samples. Ideally, proteins and protein–polymer conjugates will be sufficiently concentrated to allow for addition of only 1–2 μL per well. Lower concentration samples will decrease the maximum denaturant concentration achievable. Whenever possible, protein and protein–polymer conjugate samples should be dialyzed into the same buffer to be used in the assay (see Section 2.2.4.3). All protein samples should be centrifuged at 10,000 × *g* prior to use in the assay to remove aggregates or insoluble material.

2.2.4 Instrumentation and Consumables

2.2.4.1 Fluorimeter

We utilize a BioTek Synergy HT (BioTek Inc., Winooski, VT) plate reader and select excitation and emission wavelengths using monochromators.

The use of monochromators allows for a fixed excitation wavelength of 280 nm paired with a fluorescence spectral scan with 1 nm resolution between 310 and 400 nm.

2.2.4.2 Plates

We carry out our experiments in 384-well Nunc black plates with optical (clear) bottom wells (Fisher, Cat. No. 12-566-2, Waltham, MA). The clear bottom plates perform double-duty in our lab, allowing use both for the tryptophan fluorescence assay and visible wavelength absorbance enzyme activity assays in a "bottom read mode." Paired with a plate reader in "top read mode" the 384-well plates with black side walls lower background by minimizing cross talk between wells. We utilize optically clear seals (Greiner Bio-One, Cat. No. 676070, Monroe, NC) to minimize evaporation during data collection. However, plate seals are not required while acquiring data. If high quality, optically clear seals are not available, data can be acquired without a plate seal. In this scenario, 3-in. wide clear packing tape can be used as an inexpensive and quick method of sealing plates before and after data collection to minimize sample evaporation.

2.2.4.3 Buffers and Denaturant

A range ofbuffers can be used provided that the buffer does not absorb or fluoresce in the region utilized for the assay. Although phosphate or Tris buffers can be used, we recommend HEPES, or another Good buffer (Good et al., 1966). Our typical assay utilizes 20 m*M* HEPES, pH 7.5, 150 m*M* NaCl, and this buffer should be used with solid guanidine hydrochloride (Fisher, Cat. No. BP178 500) to prepare the 6 *M* guanidine hydrochloride stock solution to minimize changes in pH across the assay. Other denaturants or chemicals, such as urea, can be used in place of guanidine hydrochloride; however, care should be taken to ensure that the denaturant does not interfere with fluorescence in the 300–400 nm region and that it is fully miscible with water. All buffers should be filtered and degassed prior to use.

2.2.5 Protocol

1. Safety considerations: personal safety equipment such as safety glasses, nitrile gloves, and a laboratory coat should be worn at all times. Normal safety protocols used for handling protein samples should be used for ITF samples. Although volumes utilized are small, appropriate care should be taken with all chemicals used within the protein sample. We recommend consulting the MSDS for each buffer component to identify hazards.

2. Prepare 40 μL of 20 μ*M* sample in each well using triplicates for each concentration of denaturant. Allow samples to equilibrate for 20 min. *Tip: Pipetting samples into a 384-well plate is aided by use of multichannel pipettes. Pipetting by hand is made easier by placing the plate at an angle on a white or light colored surface.*
3. Centrifuge the plate to ensure all samples are at the bottom of each well and to remove any bubbles introduced upon mixing.
4. Place the plate into a fluorescence-capable plate reader. Plate seals are not necessary, though if desired the plate seal must be transparent to light in the 280–400 nm window.
5. Optimization runs benefit from a fluorescence spectral scan with a 310–400 nm window. Once wavelengths for folded and unfolded protein are found, a spectral scan within the optimal range ±10 nm, such as a spectral scan between 320 and 350 nm, can significantly speed data acquisition times.
6. Export to Excel format and analyze as described in Section 2.2.6. Plates can be reused until all wells have been used. Once all wells are used, plates should be discarded.

2.2.6 Data Analysis and Typical Data

ITF data, particularly in 384-well plates can yield very large datasets as a full plate with spectral scanning from 320 to 350 nm produces 11,520 data points. While the BioTek plate reader utilized here provides data in 1 nm increments, additional resolution could be achieved using an instrument with more finely incremented read intervals. Similar to output from qPCR instruments for DSF data, the output format for the plate reader used for the assay should be used to guide logical layout of replicates. For sequential read data output in A1–A12 left-to-right layout, replicates should be placed sequentially as detailed earlier for DSF data. Background subtraction and λ_{max} determination can then quickly follow using functions built into modern spreadsheet software. We perform background subtraction for each denaturant concentration by averaging the fluorescence values from a set of triplicate wells without protein. This average value, at each wavelength, is subtracted from the data in wells containing protein or protein–polymer conjugates. Background subtracted data are then used to search for λ_{max} values, a search that can be simplified by confining candidate wells to wavelengths between 330 and 345 nm. Maximal wavelength values are then plotted against denaturant concentration and fit to a two-state folding model in Prism using Eq. (2) to extract D_{50}, the concentration at which 50% of the protein or protein–polymer conjugate is unfolded. Eq. (2) fits the minimum and maximum fluorescence values, F_{min} and F_{max}, the denaturant concentration $[D]$ at each data point, and the D_{50} value, by minimizing differences between calculated fluorescence (F_{calc}) and the observed ITF data using the least squares method. Within Eq. (2), $[D]$ and the D_{50} value should be used as unitless values. The application of nonlinear fitting to ITF data provides for a rapid and easy method to quantitate differences in stability using data with limited resolution. Typical data compatible with fitting to a two-state model of unfolding, such as that in Fig. 2, begin with λ_{max} values near 330 nm that redshift upon unfolding. When compared to native, unmodified protein, shifting of the fits to higher or lower D_{50} values is valuable and rapid indicators for increased or decreased stability, respectively.

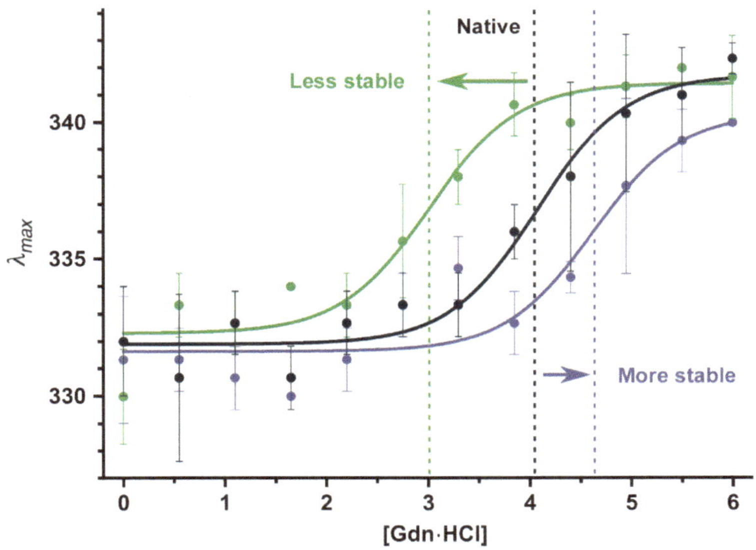

Fig. 2 Intrinsic tryptophan fluorescence data for analyzing chemical stability. The wavelength exhibiting maximal fluorescence for each sample is plotted against guanidine hydrochloride concentration, and nonlinear regression using a two-state folding model is shown as a *solid line* for HEWL (*black*), a HEWL-DP5-acrylamide conjugate (*green*), and a HEWL-DP200-acrylamide/acrylic acid conjugate (*blue*). *Dotted lines* indicate the denaturant concentration required to reach 50% unfolding (D_{50}), as determined by nonlinear regression fitting. A rightward shift in D_{50} suggests stabilization of the protein, while a leftward shift in D_{50} value indicates destabilization. *Adapted with permission from Lucius, M., Falatach, R., McGlone, C., Makaroff, K., Danielson, A. P., Williams, C., et al. (2016). Investigating the impact of polymer functional groups on the stability and activity of lysozyme-polymer conjugates.* Biomacromolecules, 17*(3), 1743–1134. Copyright (2016) American Chemical Society.*

Using the D_{50} value as an equilibrium constant for unfolding, and the equation $\Delta G = -RT \ln D_{50}$, ITF data can be used to provide an estimate for the Gibbs free energy for protein folding (ΔG) at each concentration of denaturant. Linear regression analysis of ΔG vs denaturant concentration provides a best fit line enabling the ΔG for protein folding in absence of denaturant to be estimated by extrapolating back to zero denaturant concentration (Lucius et al., 2016). Paired with the T_m value from DSF, a complete suite of thermodynamic parameters of folding can be obtained for a protein–polymer conjugate (Lucius et al., 2016).

$$F_{calc} = F_{min} + \frac{F_{max} - F_{min}}{1 + 10^{(D_{50} - [D])}}. \quad (2)$$

3. CONCLUSIONS

Polymer conjugation of proteins holds the promise of enabling a new generation of industrial enzymes and therapeutic proteins with behaviors tuned by customizing the polymer identity, length, and locations of attachment. Rapid and robust biophysical analyses are essential to allowing researchers to home in on the most promising protein–polymer conjugates. As the protein–polymer conjugate field continues to grow, these analyses may also enable future comparisons of polymer performance across a range of proteins, perhaps identifying the polymers that most often increase stability while maintaining activity. The two techniques presented here in parallelized formats, using 96- or 384-well plates, allow for high-throughput biophysical characterizations of protein–polymer conjugates and determination of Gibbs free energy (ΔG), enthalpy (ΔH), and entropy (ΔS) of folding. These techniques will enable laboratories with standard biochemical instrumentation to quickly incorporate biophysical analyses into protein–polymer characterization workflows and offer researchers valuable insights into reliably comparing the stability of protein–polymer conjugates across a diverse library of polymer modifications.

ACKNOWLEDGMENTS

The authors acknowledge institutional support from Miami University (to R.C.P., J.A.B., and D.K.), and the Robert H. and Nancy J. Blayney Professorship (to R.C.P.).

REFERENCES

Abuchowski, A., McCoy, J. R., Palczuk, N. C., van Es, T., & Davis, F. F. (1977). Effect of covalent attachment of polyethylene glycol on immunogenicity and circulating life of bovine liver catalase. *The Journal of Biological Chemistry*, *252*, 3582–3586.

Averick, S., Simakova, A., Park, S., Konkolewicz, D., Magenau, A. J. D., Mehl, R. A., et al. (2011). ATRP under biologically relevant conditions: Grafting from a protein. *ACS Macro Letters*, *1*, 6–10. http://dx.doi.org/10.1021/mz200020c.

Bianchi, E., Pessi, A., 2010. Method for shielding functional sites or epitopes on proteins.

Boyer, C., Bulmus, V., Liu, J., Davis, T. P., Stenzel, M. H., & Barner-Kowollik, C. (2007). Well-defined protein-polymer conjugates via in situ RAFT polymerization. *Journal of the American Chemical Society*, *129*, 7145–7154. http://dx.doi.org/10.1021/ja070956a.

Brogan, A. P. S., & Hallett, J. P. (2016). Solubilizing and stabilizing proteins in anhydrous ionic liquids through formation of protein-polymer surfactant nanoconstructs. *Journal of the American Chemical Society*, *138*, 4494–4501. http://dx.doi.org/10.1021/jacs.5b13425.

Brogan, A. P. S., Sessions, R. B., Perriman, A. W., & Mann, S. (2014). Molecular dynamics simulations reveal a dielectric-responsive coronal structure in protein-polymer surfactant hybrid nanoconstructs. *Journal of the American Chemical Society*, *136*, 16824–16831. http://dx.doi.org/10.1021/ja507592b.

Carugo, O. (2008). Amino acid composition and protein dimension. *Protein Science*, *17*, 2187–2191. http://dx.doi.org/10.1110/ps.037762.108.

Ciechanover, A., Elias, S., Heller, H., Ferber, S., & Hershko, A. (1980). Characterization of the heat-stable polypeptide of the ATP-dependent proteolytic system from reticulocytes. *The Journal of Biological Chemistry*, *255*, 7525–7528.

Cimmperman, P., & Matulis, D. (2011). Protein thermal denaturation measurements via a fluorescent dye. In A. Podjarny, A. Dejaegere, & B. Kieffer (Eds.), *RSC biomolecular sciences. Biophysical approaches determining ligand binding to biomolecular targets: Vol. 22.* (pp. 247–274). Cambridge: Royal Society of Chemistry. chapter 8.

Cobo, I., Li, M., Sumerlin, B. S., & Perrier, S. (2015). Smart hybrid materials by conjugation of responsive polymers to biomacromolecules. *Nature Materials*, *14*, 143–159. http://dx.doi.org/10.1038/nmat4106.

Cummings, C., Murata, H., Koepsel, R., & Russell, A. J. (2014). Dramatically increased pH and temperature stability of chymotrypsin using dual block polymer-based protein engineering. *Biomacromolecules*, *15*, 763–771. http://dx.doi.org/10.1021/bm401575k.

Danielson, A. P., Van Kuren, D. B., Lucius, M. E., Makaroff, K., Williams, C., Page, R. C., et al. (2016). Well-defined macromolecules using horseradish peroxidase as a RAFT initiase. *Macromolecular Rapid Communications*, *37*, 362–367. http://dx.doi.org/10.1002/marc.201500633.

Das, S., Dawson, N. L., & Orengo, C. A. (2015). Diversity in protein domain superfamilies. *Current Opinion in Genetics & Development*, *35*, 40–49. http://dx.doi.org/10.1016/j.gde.2015.09.005.

De, P., Li, M., Gondi, S. R., & Sumerlin, B. S. (2008). Temperature-regulated activity of responsive polymer-protein conjugates prepared by grafting-from via RAFT polymerization. *Journal of the American Chemical Society*, *130*, 11288–11289. http://dx.doi.org/10.1021/ja804495v.

Depp, V., Kaar, J. L., Russell, A. J., & Lele, B. S. (2008). Enzyme sheathing enables nanoscale solubilization of biocatalyst and dramatically increases activity in organic solvent. *Biomacromolecules*, *9*, 1348–1351. http://dx.doi.org/10.1021/bm701281x.

Falatach, R., Li, S., Sloane, S., McGlone, C., Berberich, J. A., Page, R. C., et al. (2015). Why synthesize protein–polymer conjugates? The stability and activity of chymotrypsin-polymer bioconjugates synthesized by RAFT. *Polymer*, *72*, 382–386. http://dx.doi.org/10.1016/j.polymer.2015.04.010.

Falatach, R., McGlone, C., Al-Abdul-Wahid, M. S., Averick, S., Page, R. C., Berberich, J. A., et al. (2015). The best of both worlds: Active enzymes by grafting-to followed by grafting-from a protein. *Chemical Communications*, *51*, 5343–5346. http://dx.doi.org/10.1039/c4cc09287b.

Godawat, R., Jamadagni, S. N., & Garde, S. (2010). Unfolding of hydrophobic polymers in guanidinium chloride solutions. *The Journal of Physical Chemistry B*, *114*, 2246–2254. http://dx.doi.org/10.1021/jp906976q.

Goncearenco, A., & Berezovsky, I. N. (2015). Protein function from its emergence to diversity in contemporary proteins. *Physical Biology*, *12*, 045002. http://dx.doi.org/10.1088/1478-3975/12/4/045002.

Good, N. E., Winget, G. D., Winter, W., Connolly, T. N., Izawa, S., & Singh, R. M. (1966). Hydrogen ion buffers for biological research. *Biochemistry*, *5*, 467–477. http://dx.doi.org/10.1021/bi00866a011.

Greenfield, N. J. (1996). Methods to estimate the conformation of proteins and polypeptides from circular dichroism data. *Analytical Biochemistry*, *235*, 1–10. http://dx.doi.org/10.1006/abio.1996.0084.

Hawe, A., Sutter, M., & Jiskoot, W. (2008). Extrinsic fluorescent dyes as tools for protein characterization. *Pharmaceutical Research*, *25*, 1487–1499. http://dx.doi.org/10.1007/s11095-007-9516-9.

Hoffman, A. S. (1995). "Intelligent" polymers in medicine and biotechnology. *Macromolecular Symposia*, *98*, 645–664. http://dx.doi.org/10.1002/masy.19950980156.

Isarov, S. A., Lee, P. W., & Pokorski, J. K. (2016). "Graft-to" protein/polymer conjugates using polynorbornene block copolymers. *Biomacromolecules*, *17*, 641–648. http://dx.doi.org/10.1021/acs.biomac.5b01582.

Joshi, V., Shivach, T., Yadav, N., & Rathore, A. S. (2014). Circular dichroism spectroscopy as a tool for monitoring aggregation in monoclonal antibody therapeutics. *Analytical Chemistry*, *86*, 11606–11613. http://dx.doi.org/10.1021/ac503140j.

Kelly, S. M., Jess, T. J., & Price, N. C. (2005). How to study proteins by circular dichroism. *Biochimica et Biophysica Acta*, *1751*, 119–139. http://dx.doi.org/10.1016/j.bbapap.2005.06.005.

Lele, B. S., Murata, H., Matyjaszewski, K., & Russell, A. J. (2005). Synthesis of uniform protein–polymer conjugates. *Biomacromolecules*, *6*, 3380–3387. http://dx.doi.org/10.1021/bm050428w.

Li, H., Bapat, A. P., Li, M., & Sumerlin, B. S. (2011). Protein conjugation of thermoresponsive amine-reactive polymers prepared by RAFT. *Polymer Chemistry*, *2*, 323–327. http://dx.doi.org/10.1039/C0PY00178C.

Li, H., Li, M., Yu, X., Bapat, A. P., & Sumerlin, B. S. (2011). Block copolymer conjugates prepared by sequentially grafting from proteins via RAFT. *Polymer Chemistry*, *2*, 1531–1535. http://dx.doi.org/10.1039/C1PY00031D.

Lucius, M., Falatach, R., McGlone, C., Makaroff, K., Danielson, A. P., Williams, C., et al. (2016). Investigating the impact of polymer functional groups on the stability and activity of lysozyme-polymer conjugates. *Biomacromolecules*, *17*, 1123–1134. http://dx.doi.org/10.1021/acs.biomac.5b01743.

Macdonald, R. D., & Khajehpour, M. (2015). Effects of the protein denaturant guanidinium chloride on aqueous hydrophobic contact-pair interactions. *Biophysical Chemistry*, *196*, 25–32. http://dx.doi.org/10.1016/j.bpc.2014.08.006.

Magnusson, J. P., Bersani, S., Salmaso, S., Alexander, C., & Caliceti, P. (2010). In situ growth of side-chain PEG polymers from functionalized human growth hormone—A new technique for preparation of enhanced protein-polymer conjugates. *Bioconjugate Chemistry*, *21*, 671–678. http://dx.doi.org/10.1021/bc900468v.

Mancini, R. J., Lee, J., & Maynard, H. D. (2012). Trehalose glycopolymers for stabilization of protein conjugates to environmental stressors. *Journal of the American Chemical Society, 134*, 8474–8479. http://dx.doi.org/10.1021/ja2120234.

Masso, M., Lu, Z., & Vaisman, I. I. (2006). Computational mutagenesis studies of protein structure–function correlations. *Proteins, 64*, 234–245. http://dx.doi.org/10.1002/prot.20968.

Matulis, D., Kranz, J. K., Salemme, F. R., & Todd, M. J. (2005). Thermodynamic stability of carbonic anhydrase: Measurements of binding affinity and stoichiometry using ThermoFluor. *Biochemistry, 44*, 5258–5266. http://dx.doi.org/10.1021/bi048135v.

Mudhivarthi, V. K., Cole, K. S., Novak, M. J., Kipphut, W., Deshapriya, I. K., Zhou, Y., et al. (2012). Ultra-stable hemoglobin–poly(acrylic acid) conjugates. *Journal of Materials Chemistry, 22*, 20423–20433. http://dx.doi.org/10.1039/C2JM34434C.

Murata, H., Cummings, C. S., Koepsel, R. R., & Russell, A. J. (2013). Polymer-based protein engineering can rationally tune enzyme activity, pH-dependence, and stability. *Biomacromolecules, 14*, 1919–1926. http://dx.doi.org/10.1021/bm4002816.

Niesen, F. H., Berglund, H., & Vedadi, M. (2007). The use of differential scanning fluorimetry to detect ligand interactions that promote protein stability. *Nature Protocols, 2*, 2212–2221. http://dx.doi.org/10.1038/nprot.2007.321.

Pantoliano, M. W., Petrella, E. C., Kwasnoski, J. D., Lobanov, V. S., Myslik, J., Graf, E., et al. (2001). High-density miniaturized thermal shift assays as a general strategy for drug discovery. *Journal of Biomolecular Screening, 6*, 429–440. http://dx.doi.org/10.1089/108705701753364922.

Riccardi, C. M., Cole, K. S., Benson, K. R., Ward, J. R., Bassett, K. M., Zhang, Y., et al. (2014). Toward "stable-on-the-table" enzymes: Improving key properties of catalase by covalent conjugation with poly(acrylic acid). *Bioconjugate Chemistry, 25*, 1501–1510. http://dx.doi.org/10.1021/bc500233u.

Rizzo, J. M., Shi, S., Li, Y., Semple, A., Esposito, J. J., Yu, S., et al. (2015). Application of a high-throughput relative chemical stability assay to screen therapeutic protein formulations by assessment of conformational stability and correlation to aggregation propensity. *Journal of Pharmaceutical Sciences, 104*, 1632–1640. http://dx.doi.org/10.1002/jps.24408.

Russo, D., Plazanet, M., Teixeira, J., Moulin, M., Härtlein, M., Wurm, F. R., et al. (2015). Investigation into the relaxation dynamics of polymer–protein conjugates reveals surprising role of polymer solvation on inherent protein flexibility. *Biomacromolecules, 17*, 141–147. http://dx.doi.org/10.1021/acs.biomac.5b01269.

Sawada, Y., & Honda, S. (2006). Structural diversity of protein segments follows a power-law distribution. *Biophysical Journal, 91*, 1213–1223. http://dx.doi.org/10.1529/biophysj.105.076661.

Srivastava, A., O'Connor, I. B., Pandit, A., & Gerard Wall, J. (2014). Polymer-antibody fragment conjugates for biomedical applications. *Progress in Polymer Science, 39*, 308–329.

Sumerlin, B. S. (2011). Proteins as initiators of controlled radical polymerization: Grafting-from via ATRP and RAFT. *ACS Macro Letters, 1*, 141–145. http://dx.doi.org/10.1021/mz200176g.

tenCate, M. G. J., Rettig, H., Bernhardt, K., & Börner, H. G. (2005). Sequence-defined polypeptide–polymer conjugates utilizing reversible addition fragmentation transfer radical polymerization. *Macromolecules, 38*, 10643–10649. http://dx.doi.org/10.1021/ma0519415. American Chemical Society.

Thilakarathne, V., Briand, V. A., Zhou, Y., Kasi, R. M., & Kumar, C. V. (2011). Protein polymer conjugates: Improving the stability of hemoglobin with poly(acrylic acid). *Langmuir, 27*, 7663–7671. http://dx.doi.org/10.1021/la2015034.

Vollrath, F., Hawkins, N., Porter, D., Holland, C., & Boulet-Audet, M. (2014). Differential scanning fluorimetry provides high throughput data on silk protein transitions. *Scientific Reports, 4*, 5625. http://dx.doi.org/10.1038/srep05625.

Whitmore, L., & Wallace, B. A. (2004). DICHROWEB, an online server for protein secondary structure analyses from circular dichroism spectroscopic data. *Nucleic Acids Research, 32*, W668–W673. http://dx.doi.org/10.1093/nar/gkh371.

Xia, Y., Tang, S., & Olsen, B. D. (2013). Site-specific conjugation of RAFT polymers to proteins via expressed protein ligation. *Chemical Communications, 49*, 2566–2568. http://dx.doi.org/10.1039/C3CC38976F.

Yang, J. T., Wu, C.-S. C., & Martinez, H. M. (1986). [11] Calculation of protein conformation from circular dichroism. *Methods in Enzymology, 130*, 208–269.

Zhou, H.-X. (2004). Loops, linkages, rings, catenanes, cages, and crowders: Entropy-based strategies for stabilizing proteins. *Accounts of Chemical Research, 37*, 123–130. http://dx.doi.org/10.1021/ar0302282.

CHAPTER SIX

Guide to the Preparation of Molecularly Imprinted Polymer Nanoparticles for Protein Recognition by Solid-Phase Synthesis

Jingjing Xu, Paulina X. Medina-Rangel, Karsten Haupt[1], Bernadette Tse Sum Bui[1]
Sorbonne Universités, Université de Technologie de Compiègne, CNRS Enzyme and Cell Engineering Laboratory, Rue Roger Couttolenc, Compiègne Cedex, France
[1]Corresponding authors: e-mail address: karsten.haupt@utc.fr; jeanne.tse-sum-bui@utc.fr

Contents

Abstract

Molecularly imprinted polymers (MIPs) are synthetic antibody mimics possessing specific cavities designed for a target molecule. Nowadays, molecular imprinting of proteins still remains a challenge as the generation of selective imprinted cavities is extremely difficult, due to their flexible structure and the presence of a multitude of functional sites. To overcome this difficulty, we propose a solid-phase synthesis strategy to prepare MIPs specific for any protein that can be immobilized in an oriented way on a solid support. Trypsin and kallikrein were used as model proteins. The solid-phase support consists of glass beads functionalized with two affinity ligands of the enzymes, the competitive inhibitor *p*-aminobenzamidine to orient the enzymes via their active site, or a Cu^{2+}

Methods in Enzymology, Volume 590
ISSN 0076-6879
http://dx.doi.org/10.1016/bs.mie.2017.02.004
© 2017 Elsevier Inc.
All rights reserved.

chelate to orient via the surface histidine residues of the enzyme. Thermoresponsive molecularly imprinted polymer nanoparticles (MIP-NPs) are then synthesized around the immobilized enzyme. The MIP-NPs are released by a simple temperature change, resulting in protein-free polymers endowed with improved binding site homogeneity since all binding sites have the same orientation. The MIP-NPs exhibit apparent dissociation constants between 0.02 and 2 n*M* toward their target proteins, which is comparable to those of natural antibodies. Moreover, these water-compatible polymers, targeting different domains of the enzyme, can also function as protective agents (armor), hence preventing the target proteins from denaturation by heat or pH.

1. INTRODUCTION

Synthetic receptors for specific recognition of proteins are attractive in various fields including sensors, immunoassays, proteomics, and clinical diagnostics. One class of such receptors are molecularly imprinted polymers (MIPs) which are tailor-made materials able to recognize and bind target molecules with affinities and specificities comparable to those of natural receptors such as antibodies. MIPs are produced by a templating process at the molecular level, by copolymerization of functional and cross-linking monomers, followed by the subsequent removal of the template after polymerization (Fig. 1; Beyazit, Tse Sum Bui, Haupt, & Gonzato, 2016; Haupt, Linares, Bompart, & Tse Sum Bui, 2012; Ye, 2015). The MIP can now specifically bind the target protein, and in principle, also protect the bound enzyme against denaturation, and thus could serve as nanoarmor, theme of this volume.

Advantageously, MIPs possess a higher chemical, physical, and thermal stability than biological receptors, are easily obtained and cost-effective, and can be engineered and adapted to standard industrial fabrication processes. However, the generation of MIPs for proteins still remains a challenge, mainly due to the complex structure of the protein in its native conformation, that has to be preserved during the polymerization process (Li, Cao, Whitcombe, & Piletsky, 2014; Takeuchi & Hishiya, 2008). Furthermore, the generation of selective imprinted cavities for proteins is difficult due to the multitude of functional sites, and the large size of the protein hampers its extraction from the polymer, thus permanent physical entrapment of the template protein is often encountered. However, as previously demonstrated with trypsin, if the polymerization of the MIP around the protein is orientated with an anchoring monomer derived from its competitive inhibitor *p*-aminobenzamidine (PAB), then selective MIPs can be obtained (Cutivet, Schembri, Kovensky, & Haupt, 2009). Indeed, the resulting MIP showed higher affinity for trypsin when compared to other proteins of similar size and isoelectric points. Still, the problem of the removal of the bulky

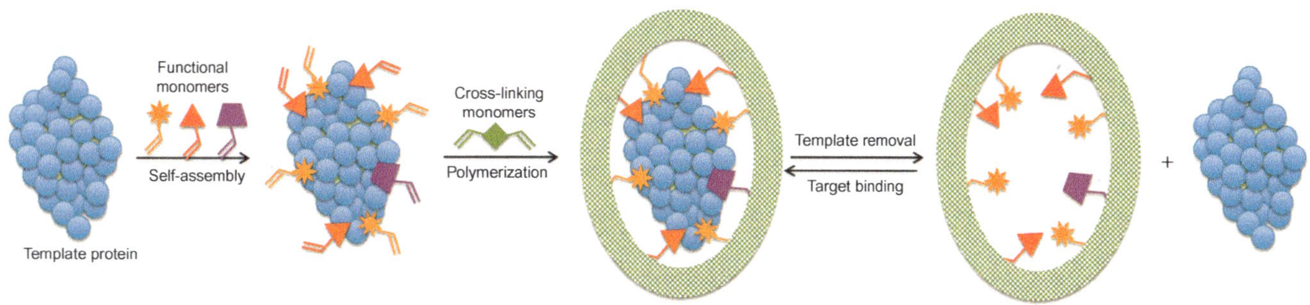

Fig. 1 Synthesis of a molecularly imprinted polymer (MIP) for a protein.

template from the MIP after polymerization remains (Verheyen et al., 2011). One solution to this problem is to employ a solid-phase synthesis approach where the protein is attached to the solid in a preferred orientation before imprinting (Ambrosini, Beyazit, Haupt, & Tse Sum Bui, 2013; Canfarotta, Poma, Guerreiro, & Piletsky, 2016; Xu et al., 2016). Preferentially, the protein should be immobilized through an affinity ligand, instead of direct chemical attachment to the support (Poma, Guerreiro, Caygill, Moczko, & Piletsky, 2014), which would enable an oriented immobilization on the solid surface.

Herein, we describe this method with two examples of serine proteases, trypsin and kallikrein. Two different affinity ligands are used to immobilize the proteins in an oriented manner: PAB, a competitive inhibitor of the enzymes, and a Cu^{2+} chelate [Cu^{2+}-IDA (iminodiacetic acid)], which has affinity for histidines present at the surface of the protein (Fig. 2). Molecularly imprinted polymer nanoparticles (MIP-NPs) are then synthesized around the immobilized enzyme. *N*-isopropylacrylamide (NIPAM) is used as the major component in the polymerization mixture to obtain thermoresponsive MIPs. pNIPAM is well known to undergo a volume-phase transition at its lower critical solution temperature at 32°C (Marchyk, Maximilien, Beyazit, Haupt, & Tse Sum Bui, 2014; Schild, 1992). The polymer is synthesized in the collapsed state (at 37°C), and upon cooling the polymer to below 32°C, MIP swells causing its release from the protein, resulting an easy and complete template removal (Fig. 2). Using this strategy, all binding sites should have the same orientation and be located at the surface of the particle, thus improving binding site homogeneity and accessibility for binding the intended protein to the MIP. Indeed, our results show that the resulting MIPs are water soluble, have a uniform size in the nanometer range (Fig. 5), and exhibit high affinity and specificity for their corresponding targets with apparent dissociation constants in the n*M* range comparable to antigen–antibody interactions (Fig. 6; Xu et al., 2016).

Moreover, the different orientations of the enzymes on the solid phase give rise to MIP-NPs that bind different locations on the enzymes. From Fig. 2A, we can see that the resulting MIPs immobilized via PAB, referred to as MIP1, should wrap but not inhibit their target enzymes. MIP1 has a binding site orientation such that when it binds the target enzyme, access of the substrate to the enzyme's active site is not blocked. On the other hand, the MIPs obtained using immobilization via the immobilized metal chelate, hereafter referred to as MIP2, should wrap the enzymes but could also perturb the enzyme's active site (Fig. 2B).

Due to this oriented selective binding of MIP1 and MIP2, we further tested them as stabilizing agents for protecting the enzymes from thermal

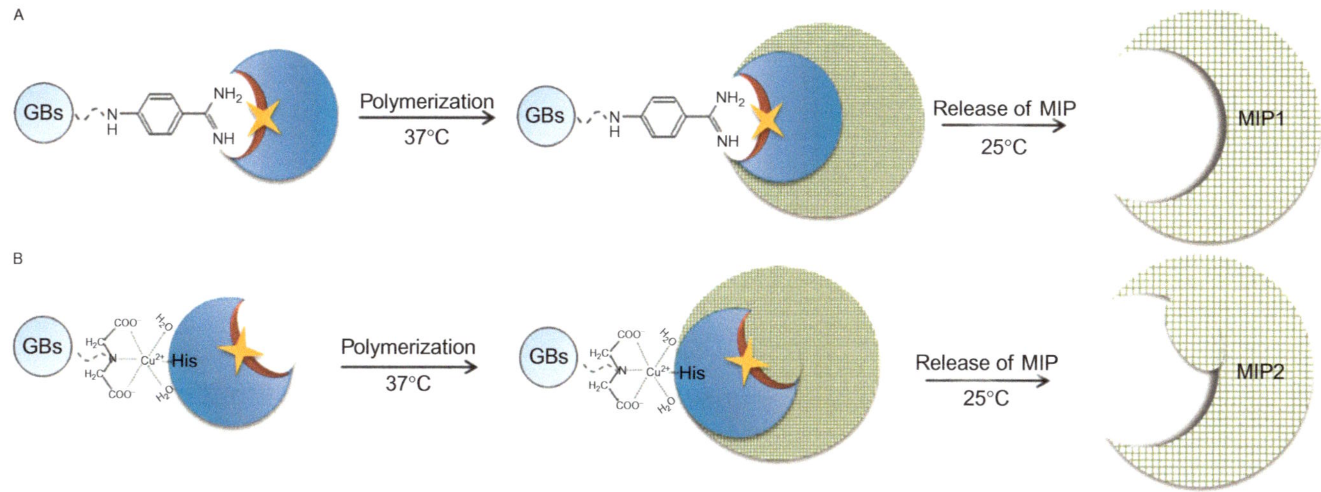

Fig. 2 Preparation of MIP-NPs wrapping different domains of the enzyme. (A) *p*-aminobenzamidine is immobilized on glass beads (GBs) to attract the enzyme via its active site, represented by a star. (B) Cu^{2+} chelate is immobilized on GBs to attract the enzyme via its surface histidine, leaving its active site free. This orientated conformation of the enzyme will in turn yield MIP-NPs interacting with different locations of the enzyme.

and pH denaturation. Trypsin is the most frequently used protease in proteomics experiments and also in numerous biotechnological processes, where high temperature processing is often required (Li et al., 2012; Vandermarliere, Mueller, & Martens, 2013), but it is not stable at high temperatures and also rapid autolysis occurs at basic pH (Šebela et al., 2006). Therefore, maintaining the stability of this enzyme would be useful in several domains of research. Our results show that $MIP1_{Trypsin}$ is able to protect trypsin from thermal denaturation at 45°C and pH 7.8 but not $MIP2_{Trypsin}$, which tends to inactivate the enzyme (Fig. 7A), in accord with our hypothesis.

In contrast to trypsin, kallikrein is thermally stable up to 60°C and even retains considerable activity (70%) at 80°C, but it is denatured at low pHs (Chao et al., 2010; Raspi, 1996; Zuber & Sache, 1972). Kallikrein MIPs are herein tested as protective agents against low pH denaturation. Our results show that $MIP1_{Kallikrein}$ slows down the denaturation of kallikrein at pH 4, almost suppressing it, whereas $MIP2_{Kallikrein}$ is less efficient (Fig. 7B). The development of such polymeric stabilizers (nanoarmor) for kallikrein would be useful, for instance to protect kallikrein from low pH environment like in the stomach, when kallikrein is administered as an oral drug for treating hypertension or other diseases (Chao et al., 2010; Overlack, Stumpe, Kolloch, Ressel, & Krueck, 1981).

This solid-phase synthesis strategy can certainly be generalized to prepare MIP-NPs of other proteins. These soluble artificial antibodies can be very attractive and useful for a broad range of applications in the field of biotechnological and biomedical processes where enzyme stabilization is required (Guerreiro et al., 2014; Liu, Zhai, Dong, & Zhao, 2015) or in medicine as therapeutic agents (Hoshino, Kodama, Okahata, & Shea, 2008).

2. PROTOCOLS

The solid-phase synthesis of MIP-NPs is composed of three main parts:

- Activation of the glass beads (GBs)
- Functionalization with the affinity ligands (PAB or IDA-Cu^{2+})
- Immobilization of the enzyme and synthesis of the MIP-NPs

Safety and precautions

Wear personal protective equipment such as laboratory coat, safety glasses, disposable gloves and perform experiments under the fume cupboard when manipulating toxic and hazardous chemicals. Dispose of the resulting waste (solid and liquid) in the dedicated containers.

General equipment

- Milli-Q purification water system (Millipore, France)
- Laboratory heating/drying oven (Memmert GmbH + Co. KG, Germany)
- Advanced Hot Plate Stirrer (VWR, France)
- Stovall Belly Dancer Laboratory Shaker (Stovall Life Science Inc., USA)
- Analytical electronic balance EX324, Explorer® (OHAUS Europe GmbH, Switzerland)
- Quanta FEG 250 scanning electron microscope (FEI Europe, Netherlands)
- Cary 60 UV–vis spectrophotometer (Agilent Technologies, France)
- Zeta-sizer NanoZS (Malvern Instruments Ltd., UK) for dynamic light scattering (DLS) analysis
- JEM-2100F transmission electron microscope (JEOL Ltd., Japan) and 300-mesh carbon-coated copper grid from AGAR Scientific (Stansted, UK)
- Allegra 64R centrifuge (Beckman Coulter, France)
- Freeze dryer Alpha 1-4 LD plus (SciQuip LTD, UK)
- Magnetic agitator (ROTH Sochiel E.U.R.L., France)
- pH meter (Mettler-Toledo SAS, France)
- FluoroLog-3 spectrofluorimeter (Horiba Jobin Yvon, France)

2.1 Activation of GBs

Equipment

- 1-L Erlenmeyer flask (VWR, 89085-018)
- Silicone oil TECHNICAL for oil baths (VWR, France, 24610.363)
- Glass funnel (VWR, LENZ5.3180.33)
- Filter paper (VWR, grade 413, 516-0820)
- Tea strainer (local supermarket)
- pH paper (ROTH Sochiel E.U.R.L., C728.1)

Reagents

- GBs, diameter 0.1 mm (ROTH Sochiel E.U.R.L., N029.1)
- Milli-Q water
- Sodium hydroxide (VWR, 28245.298)

  Prepare 4 M sodium hydroxide in water
- Acetone AR (Biosolve Chimie, France, 01030505)

Method

1. Activate the GBs by boiling in 4 *M* NaOH (typically 100 g GBs in 100 mL NaOH) for 10 min in a 1-L Erlenmeyer flask placed in an oil bath thermostated at 100°C, so as to introduce –OH groups on their surface.

2. Decant the activated GBs on a funnel fitted with a filter paper and wash with 8 × 250 mL water to ensure that all the NaOH is removed (pH of water 6.5), followed by 250 mL acetone.
3. Dry in an oven at 50°C for 6 h. Sieve with a tea strainer to remove aggregated GBs.
   The activated GBs can be stored at 8°C for 6 months without any alteration.

2.2 Functionalization of GBs

Equipment

- Glass Petri dish, diameter of 200 mm (VWR, 391-0477)
- 500-mL round-bottom flask (VWR, 201-1383)
- Reflux condenser (VWR, 89053-824)
- Oil bath

Reagents

- Activated GBs as prepared in Section 2.1
- Toluene anhydrous (99.8%) (Sigma-Aldrich, France, 244511)
- (3-Aminopropyl)triethoxysilane (APTES) (Sigma-Aldrich, A3648)
  Prepare 2% (v/v) of APTES in anhydrous toluene
- Sodium phosphate dibasic heptahydrate, ACS (VWR, BDH9296)
- Sodium phosphate monobasic monohydrate, ACS (VWR, BDH9298)
- Sodium chloride, ACS (VWR, BDH9286)
  Prepare 100 mM sodium phosphate buffer + 150 mM NaCl, pH 7.4 (buffer A)
- Glutaraldehyde (50% in water, Sigma-Aldrich, 340855)
  Prepare 5% (v/v) glutaraldehyde solution in buffer A
- 4-Aminobenzamidine dihydrochloride (Sigma-Aldrich, 06880)
  Prepare 100 mM PAB in buffer A
- Sodium borohydride (Sigma-Aldrich, 452882)
  Prepare 100 mg $NaBH_4$ in 100 mL buffer A
- Ethanolamine 99.5% (Sigma-Aldrich, 411000)
  Prepare 41 mM ethanolamine solution in buffer A (0.25 mL ethanolamine 99.5% in 100 mL buffer A)
- Ethanol (VWR, 1.00983.1000)
  Prepare 1 M NaCl in water:ethanol (4:1)
- (3-Glycidyloxypropyl)-trimethoxysilane (Glymo) (Sigma-Aldrich, 440167)
- Sodium carbonate (VWR, 27771.29)
- Iminodiacetic acid (IDA) (Sigma-Aldrich, 220000)
  Prepare 0.75 M IDA, 0.34 M NaCl, and 2 M Na_2CO_3 in water. The final pH is ~11

Method

(A) Functionalization with PAB

The functionalization with PAB is performed as previously described (Ambrosini et al., 2013; Xu et al., 2016). A schematic representation of the procedure is shown in Fig. 3A.

1. Place 100 g of activated GBs in a Petri dish and add 100 mL APTES (2%) in toluene, with gentle shaking on a Stovall Belly Dancer Laboratory Shaker overnight at room temperature.
2. After silanization, wash the GBs on a funnel fitted with a filter paper with 2 × 250 mL toluene, 2 × 250 mL acetone and dry in an oven at 50°C for 3 h. Sieve with a tea strainer to remove aggregated GBs.
3. Place the GBs on a Petri dish and add 100 mL of buffer A containing glutaraldehyde 5% (v/v), with shaking overnight on the Belly Dancer Shaker, at 30°C. Wash on filter paper with 7 × 250 mL water and 1 × 250 mL buffer A.

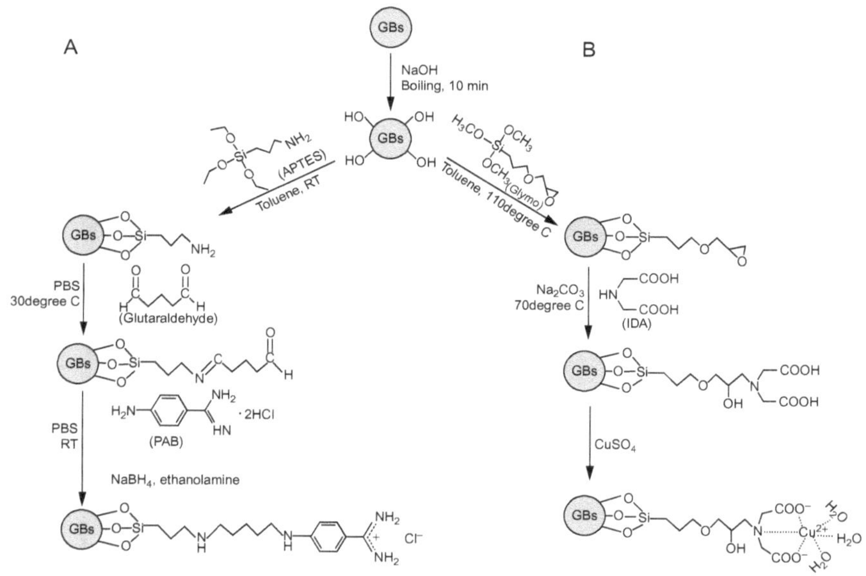

Fig. 3 Immobilization of (A) *p*-aminobenzamidine (PAB) and (B) IDA-Cu^{2+}, on GBs. *Reprinted with permission from Xu, J., Ambrosini, S., Tamahkar, E., Rossi, C., Haupt, K., & Tse Sum Bui, B. (2016). Toward a universal method for preparing molecularly imprinted polymer nanoparticles with antibody-like affinity for proteins.* Biomacromolecules, *17(1), 345–353; Copyright 2016. American Chemical Society.*

4. Transfer the glutaraldehyde-functionalized GBs to a Petri dish and add 100 mL of buffer A containing 100 m*M* PAB; incubate with shaking on the Belly Dancer Shaker at room temperature for 5 h. Wash the PAB-functionalized GBs on filter paper with 1 L buffer A.
5. Place the GBs in a Petri dish and add 100 mL buffer A containing 100 mg $NaBH_4$ (1 mg/mL) with shaking on the Belly Dancer Shaker for 30 min at room temperature, to reduce the Schiff base. Wash the GBs on a filter paper with 1 L buffer A.
6. Transfer the GBs to a Petri dish and add 100 mL of buffer A containing 41 m*M* ethanolamine with shaking on the Belly Dancer Shaker for 30 min so as to block the unreacted carbonyl groups.
7. Wash the GBs on filter paper with 2 × 250 mL buffer A, followed by 4 × 250 mL water (*see note A*, for the determination of the amount of PAB bound to the GBs). Use immediately or store at 8°C in a mixture of water:ethanol (4:1) containing 1 *M* NaCl.

The GBs can be kept as such for 6 months without alteration.

(B) Functionalization with IDA-Cu^{2+}

Fig. 3B shows the procedure of derivatization of GBs with IDA-Cu^{2+}. Activated GBs are silanized with Glymo, as previously described (Lei et al., 2013), with some modifications.

1. Incubate 50 g of activated GBs with a mixture of 200 mL of Glymo:toluene (1:1) in a 500-mL round-bottom flask equipped with a condenser and heat under reflux at 110°C for 17 h.
2. After silanization, wash the GBs on a funnel fitted with a filter paper with 2 × 250 mL toluene, 2 × 250 mL acetone and dry in an oven at 50°C for 3 h. Sieve with a tea strainer to remove aggregated GBs.
3. Transfer the GBs to an Erlenmeyer flask of 1 L and add 500 mL of a mixture of 0.75 *M* IDA, 0.34 *M* NaCl, and 2.0 *M* Na_2CO_3 with gentle stirring using a magnetic agitator at 70°C for 17 h. Wash on a filter paper with 8 × 250 mL water and dry the GBs on a Petri dish in an oven at 37°C overnight. Sieve with a tea strainer to remove aggregated GBs.

The IDA-functionalized GBs can be stored at 8°C for 4 months without alteration.

Before proceeding further, the integrity of the GBs after the different treatments was verified by scanning electron microscopy (SEM) (Fig. 4).

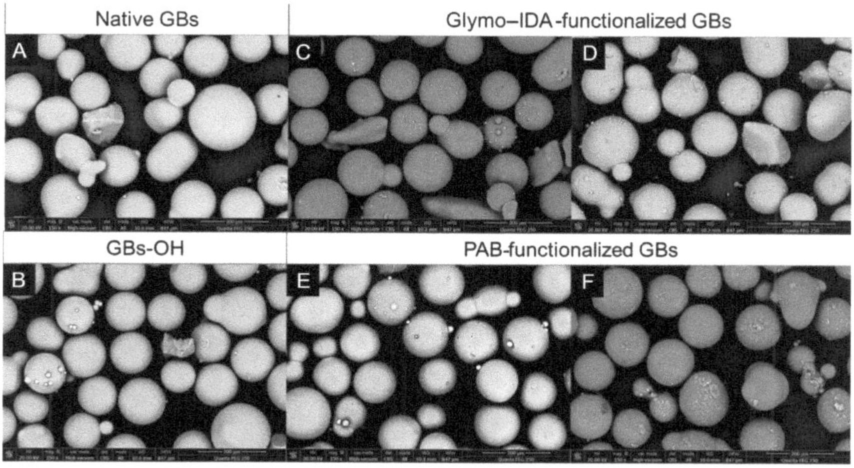

Fig. 4 SEM images of glass beads in their (A) native state, after (B) boiling in 4 *M* NaOH, (C) refluxing with Glymo in toluene at 110°C, (D) stirring with IDA at 70°C, (E) shaking with APTES in toluene, and (F) shaking with glutaraldehyde and PAB. Scale bar: 200 μm.

2.3 Enzyme Immobilization and Synthesis of MIP-NPs

Equipment

- Glass column with a thermostated jacket (XK26/40, GE Healthcare, France) with adapters (XK26, GE Healthcare, France)
- Circulating thermostated water bath (Bioblock Scientific Polystat 5, Fischer Scientific, France)
- Peristaltic pump (ISMATEC Rotary Piston Pumps, REGLO-CPF Analog, Germany)
- Nitrogen gas (Air Liquide, France)
- Glass beakers (VWR, 10754-948)
- Quartz cuvettes (Hellma Analytics, 100-5-20)

Reagents

- 25 m*M* sodium phosphate buffer pH 7 (buffer B)
- NIPAM (Sigma-Aldrich, 415324)
- *N*,*N*′-ethylenebis(acrylamide) (EbAm) (Sigma-Aldrich, 358878)
- Potassium persulfate (KPS) (Sigma-Aldrich, 379824)
- *N*,*N*,*N*′,*N*′-tetramethylethylenediamine (TEMED) (Sigma-Aldrich, T9281)
- Trypsin from bovine pancreas (Sigma-Aldrich, T8003)
  Prepare 0.5 or 1 mg/mL trypsin in buffer B

- Kallikrein from porcine pancreas (Sigma-Aldrich, K3627)
  Prepare 0.5 or 1 mg/mL kallikrein in buffer B
- Bradford reagent (Sigma-Aldrich, B6916)
- Copper (II) sulfate (Sigma-Aldrich, 451657)
  Prepare 50 mM $CuSO_4$ in H_2O
- Ethylenediaminetetraacetic acid (EDTA) (Sigma-Aldrich, 431788)
  Prepare 100 mM EDTA in H_2O

Method

A glass column with a thermostated jacket whose bed height can be adjusted with adapters was used. The temperature of the column was regulated by circulating water from a thermostated water bath. The solvents were pumped into the column using a peristaltic pump at a flow rate of 2.5 mL/min, unless stated otherwise.

A. Immobilization on PAB-functionalized GBs

1. Suspend 95 g of functionalized GBs in 20 mL water and pour into the glass column closed at the bottom with an adapter. Add more water to pack the column; this corresponds to a bed height of 10 cm. Insert the top adapter to close the column.
2. Connect the column to the peristaltic pump and pump 100 mL water at a high flow rate of 3.7 mL/min so as to expel the air bubbles trapped in between the GBs.
3. Equilibrate the GBs with 200 mL buffer B.
4. Inject 25 mL of a trypsin or kallikrein solution (0.5 mg/mL) at a flow rate of 1.2 mL/min; the flow-through is collected and passed through the column again for 1 h.
5. Wash the unbound protein by passing 300 mL buffer B.
   The amount of bound trypsin and kallikrein is 1.9 ± 0.2 and 1.4 ± 0.3 nmol/g of GBs, respectively (Xu et al., 2016).
6. Meanwhile, prepare the polymerization mixture. Mix (246.2 mg, 2.18 mmol) NIPAM and (19.2 mg, 0.114 mmol) EbAm (95:5 molar ratio) in 53 mL buffer B. Purge the solution with a flow of nitrogen gas for 30 min. Add the initiator composed of KPS (18.5 mg in 500 μL buffer A) and TEMED (1.35 μL).
7. Pump the polymerization mixture through the column and switch on the circulating water bath to set and maintain the temperature of the column to 37°C for 15 h (overnight).
8. To remove the low-affinity polymers and unreacted monomers, wash the column by volumes of 50 mL buffer B (warmed beforehand to 37°C). Normally 5 volumes of 50 mL buffer B are needed.

Check by DLS that the last fraction does not contain any particles before proceeding to the next step.

9. Leave the column to cool to room temperature. Elute the MIP-NPs by passing 5 mL volumes of buffer B. Normally 10 fractions are collected. Analyze the fractions by DLS.
10. Perform enzymatic activity (*see note B*) and Bradford assay on all the eluted fractions so as to check for the eventual leakage of the enzymes.
11. To verify if the immobilized enzyme is not denatured during the polymerization procedure, at the end of the experiment, wash the column with a mixture of 10 m*M* HCl + 0.5 *M* NaCl (3 × 50 mL) to desorb the protein and determine its activity by enzymatic assay. The enzyme should be active (Ambrosini et al., 2013).

B. Immobilization on IDA-Cu^{2+}-functionalized GBs

1. Suspend 50 g of functionalized GBs in 20 mL water and pour into the glass column closed at the bottom with an adapter. Add more water to pack the column; this corresponds to a bed height of 5.5 cm. Insert the top adapter to close the column.
2. Connect the column to the peristaltic pump and pump 100 mL water at a high flow rate of 3.7 mL/min so as to expel the air bubbles trapped in between the GBs.
3. Saturate the GBs with Cu^{2+} by passing 250 mL of 50 m*M* $CuSO_4$ solution; excess Cu^{2+} is removed by passing 150 mL water. The GBs are tainted with a uniform blue color.
4. Equilibrate the column with 150 mL buffer B.
5. Inject 50 mL of trypsin or kallikrein (1 mg/mL) through the column at a flow rate of 1.2 mL/min; the flow-through is collected and passed through the column again for 1 h.
6. Wash away the unbound enzyme with 250 mL buffer B.

   The amount of bound trypsin and kallikrein is 15.9 ± 0.8 and 14.2 ± 1.2 nmol per g of GBs, respectively (Xu et al., 2016).
7. Synthesize and recover the MIP-NPs as described earlier for the PAB-functionalized GBs (steps 6–9).
8. Perform enzymatic activity (*see note B*) and Bradford assay on all the eluted fractions so as to check for the eventual leakage of the enzymes.
9. Determine any presence of Cu^{2+} in all the eluted fractions by measuring the absorbance at 733 nm of a complex Cu^{2+}-EDTA (Xu et al., 2016).

3. PHYSICOCHEMICAL CHARACTERIZATION OF MIP-NPs

Equipment

– Disposable plastic cuvettes (VWR, 97000-590)
– 50 mL Oak Ridge, Nalgene centrifuge tubes (VWR, 525-2244)
– Dialysis membrane of 6–8 kDa cutoff (ROTH Sochiel E.U.R.L., 4570.1)

Method

1. Perform DLS analysis on all eluted fractions; MIPs having similar sizes and dispersities are pooled to constitute the stock of MIP-NPs (Fig. 5; Table 1).

   The MIP-NPs are very stable at 8°C as no aggregation is observed for a period of 6 months (as verified by DLS analysis).

2. In order to determine the concentration of the stock solution, pipet 2 mL of MIP-NPs in a centrifuge tube and centrifuge at 40,000 × g for 1 h at 37°C. Discard the supernatant by inverting the tube. Resuspend the precipitate in 2 mL of water and transfer to a 5-mL Eppendorf tube weighed beforehand. Rinse twice successively with 1 mL water and transfer to the 5-mL tube for lyophilization overnight. Weigh the dry MIP-NPs.

   Tip: the centrifuge machine must be preheated to 37°C before centrifuging the MIP. The supernatant must be discarded rapidly so as to avoid loss of MIP caused by temperature change.

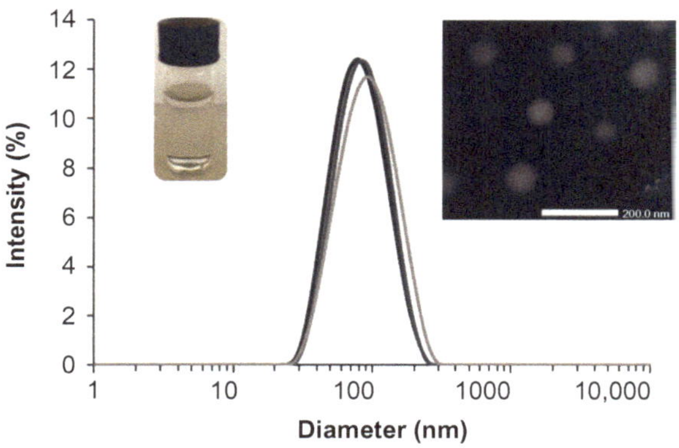

Fig. 5 Physicochemical characteristics of trypsin-imprinted MIP-NPs obtained on PAB-functionalized GBs. From *left* to *right*: 1 mg/mL MIP-NPs in water appear as a transparent solution at 25°C; size distribution as measured by DLS ($n = 3$); TEM image (scale bar: 200 nm).

Table 1 Physicochemical Characteristics of MIP-NPs

MIPs[a]	Diameter (nm) $n=3$	Polydispersity Index	Yield (mg/g GBs)
$MIP1_{Trypsin}$	119 ± 5	0.278	0.32 ± 0.02
$MIP1_{Kallikrein}$	97 ± 3	0.319	0.36 ± 0.04
$MIP2_{Trypsin}$	70 ± 6	0.284	0.45 ± 0.03
$MIP2_{Kallikrein}$	106 ± 5	0.395	0.65 ± 0.03

[a]MIP1 is obtained with the template enzymes immobilized on PAB, and MIP2 on IDA-Cu^{2+}.

Typically, concentrations of MIP-NPs for proteins immobilized via PAB ($MIP1_{Trypsin}$ and $MIP1_{Kallikrein}$) of ~0.6 mg/mL and via IDA-Cu^{2+} ($MIP2_{Trypsin}$ and $MIP2_{Kallikrein}$) of ~1.2 mg/mL are obtained. Knowing the mass of GBs and the total volume of stock MIP-NPs, the yield of MIP synthesis can be calculated, as milligrams of MIP-NPs per gram of GBs (Table 1).

3. Determine the morphology of the stock MIP-NPs by transmission electron microscopy (TEM) imaging. For this, dialyze 2 mL of the stock MIP-NPs overnight in 1 L of water using a dialysis membrane of 6–8 kDa cutoff.

4. EVALUATION OF THE BINDING CHARACTERISTICS OF MIP-NPs USING A QUARTZ CRYSTAL MICROBALANCE

Equipment

- Silicon dioxide-coated crystals (QSX 303) (Lot Quantum Design, France)
- UV/ozone cleaner (ProCleaner™ 220, BioForce Nanosciences, USA)
- Q-Sense E1 instrument and Q-sensor holder (Biolin Scientific, Sweden)
- Peristaltic pump (ISMATEC Rotary Piston Pumps, REGLO-CPF Analog, Germany)

Reagents

- Ammonium hydroxide, ACS, 28% (VWR, L13168.AU)
- Hydrogen peroxide 35% (Sigma-Aldrich, 1.08556.9025)
  Prepare NH_4OH (28%)/H_2O_2 (35%)/H_2O (1/1/6, v/v/v)
- Anhydrous toluene, APTES, acetone, 5% glutaraldehyde, 100 m*M* PAB, Glymo, IDA, $CuSO_4$, buffer A, buffer B, trypsin (2 mg trypsin in 2 mL buffer B), kallikrein (2 mg kallikrein in 2 mL buffer B) as described in the previous sections

- 0.6, 1.2, 2.4, 4.8, 9.6, and 50 μg/mL $MIP1_{Trypsin}$ prepared in buffer B, diluted from the stock (0.6 mg/mL)
- 0.6, 1.2, 2.4, 4.8, 9.6, and 50 μg/mL $MIP1_{Kallikrein}$ prepared in buffer B, diluted from the stock (0.6 mg/mL)
- 2.5, 10, 25, 50, 100, and 250 μg/mL $MIP2_{Trypsin}$ prepared in buffer B, 5 mL each, diluted from the stock (1.2 mg/mL)
- 2.5, 10, 25, 50, 100, and 250 μg/mL $MIP2_{Kallikrein}$ prepared in buffer B, 5 mL each, diluted from the stock (1.2 mg/mL)

Method

A quartz crystal microbalance (QCM) sensor was employed to evaluate the binding affinity and capacity of the MIP-NPs. All measurements were performed with a Q-Sense E1 in a flow-through mode using a peristaltic pump (flow rate: 50 μL/min). The functionalization of the QCM crystal (with IDA-Cu^{2+} or PAB and further immobilization of the protein) was carried out at 25°C, whereas the binding experiments with the MIP-NPs were performed at 37°C. All manipulations of the QCM crystals were done with tweezers.

A. Cleaning and activation of QCM crystals

1. Place four silicon dioxide-coated sensor crystals in a Q-sensor holder and immerse in a glass beaker containing 80 mL of NH_4OH (28%)/H_2O_2 (35%)/H_2O (1/1/6, v/v/v).
2. Put the beaker in a water bath at 70°C for 10 min.
3. Rinse each crystal with 5 × 5 mL water; dry under nitrogen gas.
4. Place the QCM crystals under UV/ozone irradiation (ProCleaner™ 220) for 10 min for further cleaning.
5. Rinse each crystal with 5 × 5 mL water; dry under nitrogen gas.

B. Functionalization with PAB, enzyme immobilization, and binding with $MIP1_{Trypsin}$ and $MIP1_{Kallikrein}$

1. Place the four QCM crystals in a 50-mL glass beaker containing 5 mL toluene and 5 mL APTES; incubate with gentle shaking on a Belly Dancer Shaker overnight at room temperature.
2. Rinse each crystal with 5 × 5 mL toluene followed by 5 × 5 mL acetone; dry them under nitrogen gas.
3. Place the QCM chips into 10 mL of buffer A containing 5% glutaraldehyde in the beaker; incubate them on a Belly Dancer Shaker overnight at room temperature.
4. Rinse each crystal with 10 × 5 mL buffer A; dry under nitrogen gas. *Glutaraldehyde-functionalized crystals can be stored at 8°C for 1 week without alteration.*

5. Place the crystal in the flow module of the Q-Sense instrument; pass buffer A until a stable baseline is reached.
   Here, start monitoring the change in frequency (ΔF).
6. Pass buffer A containing 100 m*M* PAB until saturation is observed.
7. After functionalization with PAB, wash the module with buffer B until the baseline is stable for more than 15 min.
8. Pump 2 mL of trypsin or kallikrein (1 mg/mL) through the sensor until saturation is observed, and then wash with buffer B for 15–30 min until the signal is stable.
9. Meanwhile, prepare 5 mL aliquots of $MIP1_{Trypsin}$ and $MIP1_{Kallikrein}$ at concentrations of 0.6, 1.2, 2.4, 4.8, 9.6, and 50 μg/mL and place in a water bath at 37°C.
10. Set the temperature control of the QCM apparatus to 37°C. When the temperature is stable, wash with buffer B at 37°C until the baseline is stable for more than 15 min.
11. Inject each concentration of $MIP1_{Trypsin}$ or $MIP1_{Kallikrein}$ at 37°C for 30 min until saturation is observed. Between each concentration, wash with buffer B for 15–30 min until the signal is stable.
12. Inject $MIP1_{Trypsin}$ or $MIP1_{Kallikrein}$ (0.6–50 μg/mL) sequentially on PAB-functionalized sensor chip alone so as to investigate the contribution of nonspecific interactions. Nonspecific interactions (Fig. 6A and B, light-colored lines) proved to be low.

    These results show that trypsin and kallikrein-imprinted MIP-NPs were specific as they bind their respective template proteins to a higher extent than to the sensor without immobilized proteins.

C. Functionalization with IDA-Cu^{2+}, enzyme immobilization, and binding with $MIP2_{Trypsin}$ and $MIP2_{Kallikrein}$
   1. Place the four QCM chips in a 50-mL glass beaker containing 5 mL toluene and 5 mL Glymo; incubate on a Belly Dancer Shaker overnight at room temperature.
   2. Rinse each crystal with 5 × 5 mL toluene followed by 5 × 5 mL acetone; dry them under nitrogen gas.
   3. Place the QCM chips in a glass beaker containing 10 mL of 0.75 *M* IDA, 0.34 *M* NaCl, and 2.0 *M* Na_2CO_3 in water; put the beaker in an oven maintained at 70°C for 4 h.
   4. Rinse each crystal with 10 × 5 mL water; dry under nitrogen gas.
      IDA-functionalized chips can be stored at 8°C for 2 weeks without alteration.
      Here, start monitoring the change in frequency (ΔF).

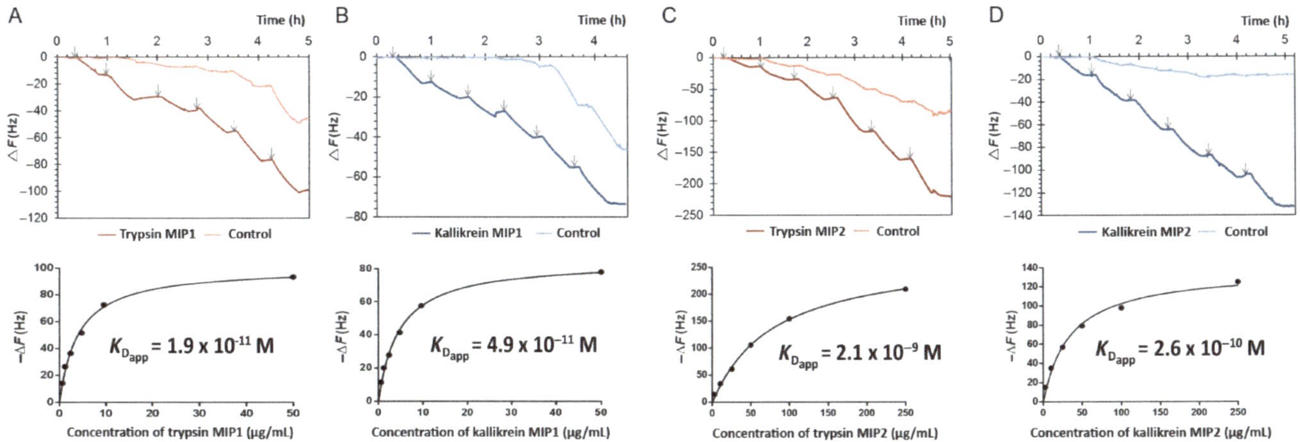

Fig. 6 Time course of the frequency change (ΔF), on PAB-derivatized QCM sensor of (A) immobilized trypsin and (B) immobilized kallikrein, following injection of $MIP1_{Trypsin}$ and $MIP1_{Kallikrein}$, respectively; on Cu^{2+} chelate-derivatized sensor of immobilized (C) trypsin with $MIP2_{Trypsin}$ and (D) kallikrein with $MIP2_{Kallikrein}$, in 25 m*M* sodium phosphate buffer pH 7, at 37°C. Injections at the time point are indicated by the *arrows*, and the concentrations were 0.6, 1.2, 2.4, 4.8, 9.6, and 50 μg/mL on PAB-derivatized QCM sensor and 2.5, 10, 25, 50, 100, and 250 μg/mL on Cu^{2+} chelate-derivatized sensor. Control means injection of the MIP on immobilized ligand only. Measurements were performed three times independently with two batches of MIPs. Corresponding binding isotherms of the MIPs for their respective targets are shown. *Adapted with permission from Xu, J., Ambrosini, S., Tamahkar, E., Rossi, C., Haupt, K., & Tse Sum Bui, B. (2016). Toward a universal method for preparing molecularly imprinted polymer nanoparticles with antibody-like affinity for proteins.* Biomacromolecules, 17*(1), 345–353; Copyright 2016. American Chemical Society.*

5. Place the crystal in the flow-module of the QCM apparatus, pass 50 mM $CuSO_4$ solution until saturation is observed, and then wash with water for ~30 min until the signal is stable.
6. After charging with Cu^{2+}, wash the system with buffer B until the baseline is stable for more than 15 min.
7. Pump 2 mL of trypsin or kallikrein (1 mg/mL) onto the IDA-Cu^{2+}-functionalized QCM sensor (saturation is observed) and then wash with buffer B for 15–30 min until the signal is stable.
8. Meanwhile, prepare 5 mL aliquots of $MIP2_{Trypsin}$ or $MIP2_{Kallikrein}$ at concentrations of 2.5, 10, 25, 50, 100, and 250 μg/mL and place in a water bath at 37°C.
9. Set the temperature control of the QCM apparatus to 37°C. When the temperature is stable, wash with buffer B at 37°C until the baseline is stable for more than 15 min.
10. Inject each concentration of $MIP2_{Trypsin}$ or $MIP2_{Kallikrein}$ at 37°C for 30 min until saturation is observed. Between each concentration, wash with buffer B for 15–30 min until the signal is stable.
11. Inject $MIP2_{Trypsin}$ or $MIP2_{Kallikrein}$ (2.5–250 μg/mL) sequentially on IDA-Cu^{2+}-functionalized sensor chip alone (Fig. 6C and D, pale-colored lines), so as to investigate the contribution of nonspecific interaction. In both cases, binding of MIPs to their respective templates was high, indicating the creation of imprinted sites.

   Overall, these results imply that the MIP-NPs synthesized by the solid-phase approach are specific toward their corresponding templates. Calculated apparent dissociation constants (K_{dapp}) for all MIP-NPs (*see note C*) were obtained by nonlinear fitting of the data to a single-site Langmuir-type binding isotherm and are shown in Fig. 6.

5. MIP-NPs PROTECT ENZYME FROM THERMAL AND pH DENATURATION

Equipment

- Gilson Pipetman P1000, P100, P20
- 50 mL Oak Ridge, Nalgene centrifuge tubes (VWR, 525-2244)
- 4 mL glass tubes (VWR, 548-0051)
- Water bath (Bioblock Scientific Polystat 5, Fischer Scientific, France)
- Digital thermometer (VWR, TD10 type T)

A. Protection of trypsin by $MIP1_{Trypsin}$ against thermal inactivation at 45°C

Reagents

- Hydrochloric acid, 37% (VWR, 20252.290)
- Calcium chloride (Sigma-Aldrich, C5670)
  Prepare a solution of 1 mM HCl + 10 mM $CaCl_2$ in water, the pH is ~5
- Trypsin
  Prepare 50 μM trypsin in 1 mM HCl + 10 mM $CaCl_2$
- Tris base (Sigma-Aldrich, T1378)
  Prepare 50 mM Tris–HCl pH 8. Use Tris base and adjust the pH with 1 M HCl
- N_α-*p*-tosyl-L-arginine methyl ester hydrochloride (TAME) (Sigma-Aldrich, T4626)
  Prepare 10 mM TAME in 50 mM Tris–HCl pH 8
- 50 m*M* sodium phosphate buffer pH 8
- $MIP1_{Trypsin}$ (11 mg/mL, obtained by centrifuging 10 mL of $MIP1_{Trypsin}$ in a centrifuge tube at 40,000 × *g* at 37°C and then dissolving the precipitate with the required volume of buffer B at room temperature)
- $MIP2_{Trypsin}$ (11 mg/mL in buffer B)

Method

Thermal inactivation of trypsin was done at 45°C. This temperature was chosen because it is the maximum temperature at which the MIP stays soluble. $MIP2_{Trypsin}$ and buffer B alone (no polymer) were used as controls.

1. To three separate glass tubes of 4 mL, pipet 900 μL of 50 m*M* sodium phosphate buffer pH 8 and place in a water bath at 45°C.
   Tip: Use a water bath instead of a dry block heater since the former produces a more homogeneous heating.
2. In three 4 mL glass tubes containing separately 90 μL of $MIP1_{Trypsin}$ (11 mg/mL) or $MIP2_{Trypsin}$ (11 mg/mL) (final concentration: 1 mg/mL, final volume: 1 mL) or buffer B alone, add 10 μL of a stock solution of 50 μ*M* trypsin (final concentration: 500 n*M*). Incubate in a water bath at 37°C for 30 min. Transfer the contents to tubes 1 (final pH 7.8, verified with a pH meter) and incubate at 45°C.
3. Withdraw 100 μL from each tube every hour, starting from $t = 0$ to $t = 6$ h; determine residual trypsin activity by assay with TAME at 25°C (*see note B*) (Fig. 7A).

B. Protection of kallikrein by $MIP1_{Kallikrein}$ against pH inactivation

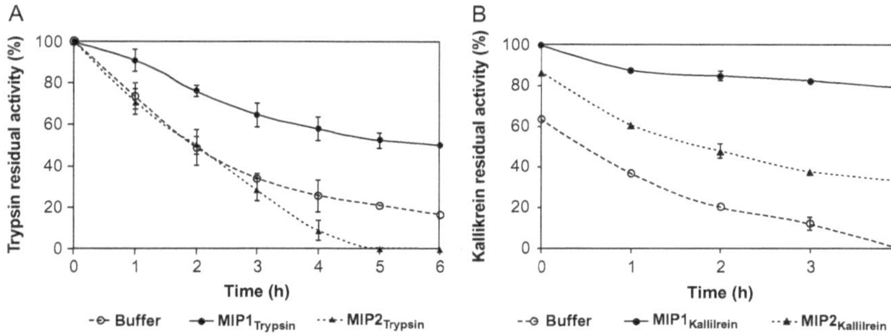

Fig. 7 Residual activity of (A) trypsin incubated without polymer (*dashed line*) or with 1 mg/mL MIP1$_{Trypsin}$ (*full line*) or MIP2$_{Trypsin}$ (*dotted line*) in 50 m*M* sodium phosphate buffer pH 7.8 at 45°C. (B) Kallikrein incubated without polymer (*dashed line*) or with 1 mg/mL MIP1$_{Kallikrein}$ (*full line*) or MIP2$_{Kallikrein}$ (*dotted line*) in 100 m*M* sodium acetate buffer pH 4 at 37°C. Data are means from three independent experiments with three different batches of polymers. The error bars represent standard deviations.

Reagents

- Stock solution of 50 μ*M* kallikrein in water
- MIP1$_{Kallikrein}$ (10 mg/mL in buffer B)
- MIP2$_{Kallikrein}$ (10 mg/mL in buffer B)
- Sodium acetate (Sigma-Aldrich, S2889)
- Acetic Acid, >99.7%, ACS (VWR, BDH3092-500MLP)

  Prepare 100 mM sodium acetate buffer pH 4
- N_α-benzoyl-L-arginine ethyl ester hydrochloride (BAEE) (Sigma-Aldrich, B4500)

  Prepare 10 mM BAEE in 50 mM Tris–HCl pH 8.7

Method

1. In three glass tubes of 4 mL containing separately 100 μL of MIP1$_{Kallikrein}$ (10 mg/mL) or MIP2$_{Kallikrein}$ (10 mg/mL) (final concentration: 1 mg/mL, final volume: 1 mL) or buffer B alone, add 10 μL of a stock solution of 50 μ*M* kallikrein (final concentration: 500 n*M*). Adjust the volume to 500 μL with buffer B. Incubate in a water bath at 37°C for 1 h. The temperature was additionally verified with a digital thermometer (accuracy: 0.1°C in the used temperature range).
2. Add 500 μL of 100 m*M* sodium acetate buffer pH 4 to the three tubes (the pH stays 4, as verified by pH paper).
3. Withdraw 100 μL from each tube every hour, starting from $t = 0$ to $t = 4$ h; determine residual kallikrein activity by assay with BAEE at 25°C (*see note B*) (Fig. 7B).

6. CONCLUDING REMARKS

The synthesis of soluble nanosized MIPs for proteins, based on a solid-phase approach, is described. GBs are functionalized with affinity ligands for specific recognition and binding of the target proteins. This configuration enables an oriented immobilization of the proteins, upon which thermo-responsive MIP-NPs are synthesized. The GBs play the role of both a reactor and a separation column since, after synthesis, polymer NPs without binding sites are washed away before recovery of the MIP-NPs. The MIP-NPs are then released from the support by a simple temperature change, resulting in protein-free polymers. This solid-phase synthesis approach is versatile and can be generalized to the synthesis of MIP-NPs for any serine protease inhibited by PAB or any enzyme-bearing surface histidine residues and potentially, engineered proteins bearing a His-tag. In the latter case, Ni^{2+} associated with another chelating group could be envisaged. More generally, this approach can be applied to obtain MIPs as synthetic antibody mimics specific for any protein that can be immobilized in an oriented way on a solid phase.

7. NOTES

A. Determination of PAB immobilized on glutaraldehyde-functionalized GBs

Reagents and materials

- Glutaraldehyde-functionalized GBs (Section 2.2, Method A)
- 100 m*M* PAB dissolved in buffer A
- Fluorescence cuvettes (Hellma Analytics, 111-061-40)

Method

PAB was quantified by measuring its fluorescence (λ_{ex}: 285 nm, λ_{em}: 365 nm, slit: 4 nm) at 25°C on a spectrofluorimeter. A calibration curve of PAB ranging from 1 to 10 μ*M* was constructed (Fig. 8). 1 g of glutaraldehyde-functionalized GBs was incubated with 1 mL of 100 m*M* PAB in buffer A. The amount of unbound PAB was determined on the supernatant, which was diluted with buffer A, so that the values fall within the calibration curve. Bound PAB was calculated by subtracting the amount of the unbound ligand from the total amount of ligand added to the GBs. The amount of PAB immobilized on the GBs was found to be 22.6 ± 2.3 nmol/g of GBs.

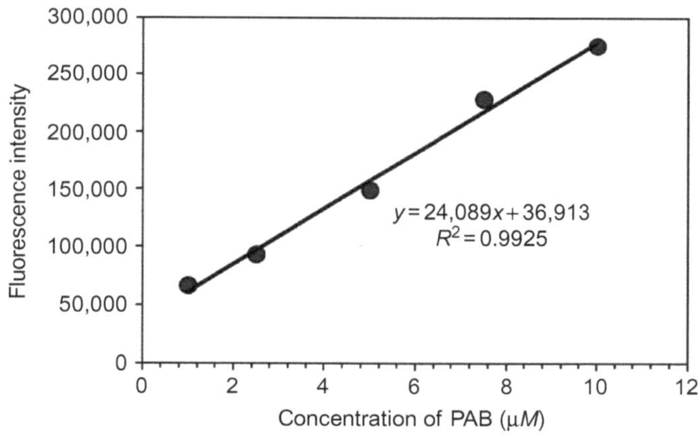

Fig. 8 Calibration curve of PAB in 100 m*M* sodium phosphate buffer + 150 m*M* NaCl pH 7.4 (λ_{ex}: 285 nm, λ_{em}: 365 nm, slit: 4 nm). Data represent the mean of two independent experiments.

B. Enzymatic activity assays

Reagents

- 1 μ*M* trypsin stock solution prepared in 1 m*M* HCl + 10 m*M* $CaCl_2$, pH ~ 5, kept in ice (diluted from 50 μ*M* trypsin prepared in 1 m*M* HCl + 10 m*M* $CaCl_2$, pH ~ 5)
- 50 m*M* Tris–HCl pH 8
- 10 m*M* TAME prepared in 50 m*M* Tris–HCl pH 8
- 2 μ*M* kallikrein stock solution prepared in H_2O (diluted from 50 μ*M* prepared in water)
- 50 m*M* Tris–HCl pH 8.7
- 10 m*M* BAEE in 50 m*M* Tris–HCl pH 8.7

Method

Enzymatic activity measurements were done spectrophotometrically on a CARY60 UV–vis spectrophotometer. Molecular weights of trypsin and kallikrein taken to prepare the stock solutions were 23.8 and 25.6 kDa, respectively. The calibration curves for trypsin (procedures 1–3) and kallikrein (procedures 4–6) were constructed as follows:

1. Pipet 50 μL of 10 m*M* TAME (final concentration: 500 μ*M*, final volume of assay: 1 mL) into a quartz cuvette containing 50 m*M* Tris–HCl, pH 8.
2. Add 10–100 μL of 1 μ*M* trypsin solution (final concentration: 10–100 n*M*) into the cuvette.

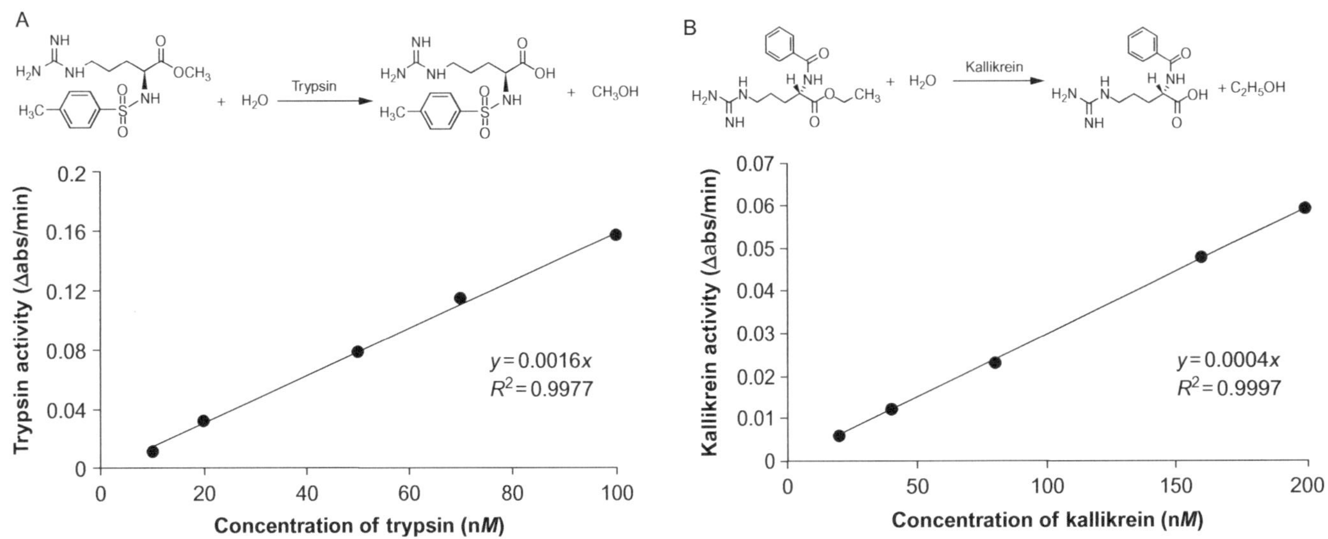

Fig. 9 Representative calibration curves of activity of (A) trypsin and (B) kallikrein, using TAME and BAEE as substrates, respectively.

3. Mix by inverting the cuvette upside-down rapidly and insert the cuvette immediately into the cell holder, and monitor the hydrolysis of TAME by measuring the change in absorbance at 247 nm for 1.5 min. The calibration curve for trypsin is shown in Fig. 9A.
4. Pipet 50 μL of 10 m*M* BAEE (final concentration: 500 μ*M*, final volume of assay: 1 mL) into a quartz cuvette containing 50 m*M* Tris–HCl, pH 8.7.
5. Add 10–100 μL of 2 μ*M* kallikrein (final concentration: 20–200 n*M*, final volume of assay: 1 mL).
6. Mix by inverting the cuvette upside-down rapidly and put the cuvette immediately into the cell holder, and monitor the hydrolysis of BAEE by measuring the change in absorbance at 253 nm for 5 min. The calibration curve for kallikrein is shown in Fig. 9B.

C. Determination of apparent dissociation constants (K_{dapp})

In order to determine the K_{dapp} (Fig. 6), first the apparent concentrations in (*M*) of the MIP-NPs must be determined using the following equation (Guerreiro et al., 2014; Hoshino et al., 2008):

$$[\mathrm{NPs}] = \left(6/\pi N_{\mathrm{A}} d^3 \rho\right) X$$

where N_{A} is the Avogadro's constant, d is the hydrodynamic diameter of the particles found by DLS (cm), ρ is the density of the particles (g/cm^3), and X is the polymer weight concentration (mg/mL). The polymer density was taken to be 0.36 ± 0.03 g/cm^3 (ρ values for NIPAM-based polymers in the collapsed state) (Meewes, Ricka, De Silva, Nyffenegger, & Binkert, 1991).

ACKNOWLEDGMENTS

J.X. thanks the China Scholarship Council for financial support. We thank the European Regional Development Fund and the Regional Council of Picardie (cofunding of equipment under CPER 2007–2013), as well as Frederic Nadaud and Caroline Boulnois for the SEM/TEM images.

REFERENCES

Ambrosini, S., Beyazit, S., Haupt, K., & Tse Sum Bui, B. (2013). Solid-phase synthesis of molecularly imprinted nanoparticles for protein recognition. *Chemical Communications*, *49*(60), 6746–6748.

Beyazit, B., Tse Sum Bui, B., Haupt, K., & Gonzato, C. (2016). Molecularly imprinted polymer nanomaterials and nanocomposites by controlled/living radical polymerization. *Progress in Polymer Science*, *62*, 1–21.

Canfarotta, F., Poma, A., Guerreiro, A., & Piletsky, S. (2016). Solid-phase synthesis of molecularly imprinted nanoparticles. *Nature Protocols*, *11*(3), 443–455.

Chao, J., Shen, B., Gao, L., Xia, C. F., Bledsoe, G., & Chao, L. (2010). Tissue kallikrein in cardiovascular, cerebrovascular and renal diseases and skin wound healing. *Biological Chemistry*, *391*(4), 345–355.

Cutivet, A., Schembri, C., Kovensky, J., & Haupt, K. (2009). Molecularly imprinted microgels as enzyme inhibitors. *Journal of the American Chemical Society*, *131*(41), 14699–14702.

Guerreiro, A., Poma, A., Karim, K., Moczko, E., Takarada, J., Vargas-Sansalvador, I. P., et al. (2014). Influence of surface-imprinted nanoparticles on trypsin activity. *Advanced Healthcare Materials*, *3*(9), 1426–1429.

Haupt, K., Linares, A. V., Bompart, M., & Tse Sum Bui, B. (2012). Molecularly imprinted polymers. *Topics in Current Chemistry*, *325*, 1–28.

Hoshino, Y., Kodama, T., Okahata, Y., & Shea, K. J. (2008). Peptide imprinted polymer nanoparticles: A plastic antibody. *Journal of the American Chemical Society*, *130*(46), 15242–15243.

Lei, W., Xue, M., Zhong, X., Meng, Z. H., Zhang, W. B., & Zhang, L. Y. (2013). Preparation of surface-imprinted silica using metal coordination for the separation of proteins. *Journal of Liquid Chromatography & Related Technologies*, *36*(15), 2196–2207.

Li, S., Cao, S., Whitcombe, M. J., & Piletsky, S. A. (2014). Size matters: Challenges in imprinting macromolecules. *Progress in Polymer Science*, *39*, 145–163.

Li, X., Xiong, H., Yang, K., Peng, D., Peng, H., & Zhao, Q. (2012). Optimization of the biological processing of rice dregs into nutritional peptides with the aid of trypsin. *Journal of Food Science and Technology*, *49*, 537–546.

Liu, Y., Zhai, J., Dong, J., & Zhao, M. (2015). Magnetic surface imprinted hydrogel nanoparticles for specific and reversible stabilization of proteins. *Molecular Imprinting*, *2*, 75–82.

Marchyk, N., Maximilien, J., Beyazit, S., Haupt, K., & Tse Sum Bui, B. (2014). One-pot synthesis of iniferter-bound polystyrene core nanoparticles for the controlled grafting of multilayer shells. *Nanoscale*, *6*(5), 2872–2878.

Meewes, M., Ricka, J., De Silva, M., Nyffenegger, R., & Binkert, T. (1991). Coil-globule transition of poly (N-isopropylacrylamide): A study of surfactant effects by light scattering. *Macromolecules*, *24*(21), 5811–5816.

Overlack, A., Stumpe, K. O., Kolloch, R., Ressel, C., & Krueck, F. (1981). Antihypertensive effect of orally administered glandular kallikrein in essential hypertension. Results of double blind study. *Hypertension*, *3*, 118–121.

Poma, A., Guerreiro, A., Caygill, S., Moczko, E., & Piletsky, S. (2014). Automatic reactor for solid-phase synthesis of molecularly imprinted polymeric nanoparticles (MIP NPs) in water. *RSC Advances*, *4*(8), 4203–4206.

Raspi, G. (1996). Kallikrein and kallikrein-like proteinases: Purification and determination by chromatographic and electrophoretic methods. *Journal of Chromatography. B, Biomedical Sciences and Applications*, *684*(1), 265–287.

Schild, H. G. (1992). Poly (N-isopropylacrylamide): Experiment, theory and application. *Progress in Polymer Science*, *17*(2), 163–249.

Šebela, M., Štosová, T. Á., Havliš, J., Wielsch, N., Thomas, H., Zdráhal, Z., et al. (2006). Thermostable trypsin conjugates for high-throughput proteomics: Synthesis and performance evaluation. *Proteomics*, *6*(10), 2959–2963.

Takeuchi, T., & Hishiya, T. (2008). Molecular imprinting of proteins emerging as a tool for protein recognition. *Organic & Biomolecular Chemistry*, *6*(14), 2459–2467.

Vandermarliere, E., Mueller, M., & Martens, L. (2013). Getting intimate with trypsin, the leading protease in proteomics. *Mass Spectrometry Reviews*, *32*, 453–465.

Verheyen, E., Schillemans, J. P., van Wijk, M., Demeniex, M. A., Hennink, W. E., & van Nostrum, C. F. (2011). Challenges for the effective molecular imprinting of proteins. *Biomaterials*, *32*(11), 3008–3020.

Xu, J., Ambrosini, S., Tamahkar, E., Rossi, C., Haupt, K., & Tse Sum Bui, B. (2016). Toward a universal method for preparing molecularly imprinted polymer nanoparticles with antibody-like affinity for proteins. *Biomacromolecules*, *17*(1), 345–353.

Ye, L. (2015). Synthetic strategies in molecular imprinting. *Advances in Biochemical Engineering Biotechnology*, *150*, 1–24.

Zuber, M., & Sache, E. (1972). Isolation and characterization of porcine pancreatic kallikrein. *Biochemistry*, *13*, 3098–3110.

CHAPTER SEVEN

Armored Urease: Enzyme-Bioconjugated Poly(acrylamide) Hydrogel as a Storage and Sensing Platform

Konda R. Kunduru*, S.N. Raju Kutcherlapati*, Dhamodaran Arunbabu†, Tushar Jana*,[1]

*School of Chemistry, University of Hyderabad, Hyderabad, India

†Madanapalle Institute of Technology and Science, Madanapalle, India

[1]Corresponding author: e-mail addresses: tusharjana@uohyd.ac.in; tjscuoh@gmail.com

Contents

Abstract

Jack bean urease is an important enzyme not only because of its numerous uses in medical and other fields but also because of its historical significance—the first enzyme to be crystallized and also the first nickel metalloenzyme. This enzyme hydrolyzes urea into

Methods in Enzymology, Volume 590
ISSN 0076-6879
http://dx.doi.org/10.1016/bs.mie.2017.02.008
© 2017 Elsevier Inc.
All rights reserved.

ammonia and carbon dioxide; however, the stability of this enzyme at ambient temperature is a bottleneck for its applicability. To improve urease stability, it was immobilized on different substrates, particularly on polymeric hydrogels. In this study, the enzyme was coupled covalently with poly(acrylamide) hydrogel with an yield of 18 μmol/cm^3. The hydrogel served as the nanoarmor and protected the enzyme against denaturation. The enzyme immobilized on the polymer hydrogel showed no loss in activity for more than 30 days at ambient temperature, whereas free enzyme lost its activity within a couple of hours. The Michaelis–Menten constant (K_m) for free and immobilized urease were 0.0256 and 0.2589 m*M*, respectively, on the first day of the study. The K_m of the immobilized enzyme was approximately 10 times higher than that of the free enzyme. The hydrogel technique was also used to prepare light diffracting polymerized colloidal crystal array in which urease enzyme was covalently immobilized. This system was applied for the detection of mercury (Hg^{2+}) with the lower limit as 1 ppb, which is below the maximum contaminant limit (2 ppb) for mercury ions in water. The experimental details of these studies are presented in this chapter.

1. INTRODUCTION

Enzymes are biomacromolecules that accelerate bioprocesses by catalysis. Enzyme biocatalysis offers solutions for the production of various chemicals that are very important to mankind and society by green and sustainable methodologies (Sheldon, 2000; Sheldon & van Rantwijk, 2004). Enzymatic reactions have proven to show various advantages. They work under mild reaction conditions (physiological pH and temperature) and provide high activities and yields of products with chemo-, regio-, and stereoselectivities. Furthermore, enzyme technology is often considered economical and green, when compared to traditional organic synthesis. Since the usage of enzyme may not require functional group protection, time- and cost-effective organic synthesis is possible. Plenty of industrial applications are possible either in native or in the immobilized form of enzymes (Sheldon, 2007).

Even though enzymes can function as catalysts in solution, they also function in bound or immobilized state. The term "immobilized" refers to the statement which reads as "enzymes physically confined or localized in a certain defined region of space with retention of their catalytic activities, and which can be used repeatedly and continuously" (Sheldon, 2007). Easier reactor operation, product separation, catalyst reuse, and wider choice of enzymes are some of the advantages of using immobilized enzymes. Enzyme immobilization can be classified into two different categories, physical and chemical. Covalent attachment of the enzyme to solid supports via specific cross-linking or coupling agents such as glutaraldehyde and 1-ethyl-3-(3-dimethylaminopropyl)carbodiimide hydrochloride (EDC)-mediated coupling has been reported extensively. Cross-linking with multifunctional agents, adsorption, gel entrapment, microencapsulation within solid or liquid membranes, containment of enzymes in membrane reactors, formation of Langmuir–Blodgett films, and layer-by-layer assembly or intercalation in layered solids are different techniques used to immobilize enzymes. Enzyme immobilization is important in the field of biotechnology for developing tools in medical diagnostics, bioaffinity chromatography, and biosensing (Sheldon, 2007).

Enzyme supports are mainly classified into two types, and these are organic and inorganic supports. Organic supports can be either natural polymers such as polysaccharides, dextrans, cellulose, agar, agarose, chitin, alginate, proteins, collagen, and albumin, or synthetic polymers such as polystyrene (PS), polyacrylamide (PAM), polymethylmethacrylate, polyamide (PA), vinyl, and allyl polymers. Desired properties of support include hydrophilicity, inertness toward enzymes, ease of derivatization, biocompatibility, resistance to microbial attack, and cost efficiency. Physical dimensions of the solid support such as particle dimension, mechanical and compression strength, swelling behavior, and absence of ionized groups play a vital role in the immobilization of enzymes. Also, binding capacity of the support plays an important role, which is a function of pore size and particle size of the supports. High surface of porous supports often leads to efficient loading of enzymes; hence porous supports are preferred (Brena, González-Pombo, & Batista-Viera, 2013).

A schematic representation of different immobilization methods is presented in Fig. 1. Covalent coupling of enzymes is considered as a stable immobilization platform. Covalent coupling of enzymes involves activating solid support either by addition of active group to the polymer or by modifying the polymer backbone with desired functionality. Enzymes are immobilized onto the support by amide, ether, thioether, or carbamate linkages. Also, various activation agents such as bisoxiranes (epoxides), epichlorohydrin, glutaraldehyde (Porath & Axén, 1976), cyanogen bromide (Axen, Porath, & Ernback, 1967), glycidol-glyoxyl (Guisán, 1988), and *N*-hydroxy-succinimidyl (Drobnik, Labský, Kudlvasrová, Saudek, & Švec, 1982; Miron & Wilchek, 1982) are used for the immobilization. In the entrapment-mediated immobilization, enzyme is embedded much like an occlusion within the polymer hydrogel matrix,

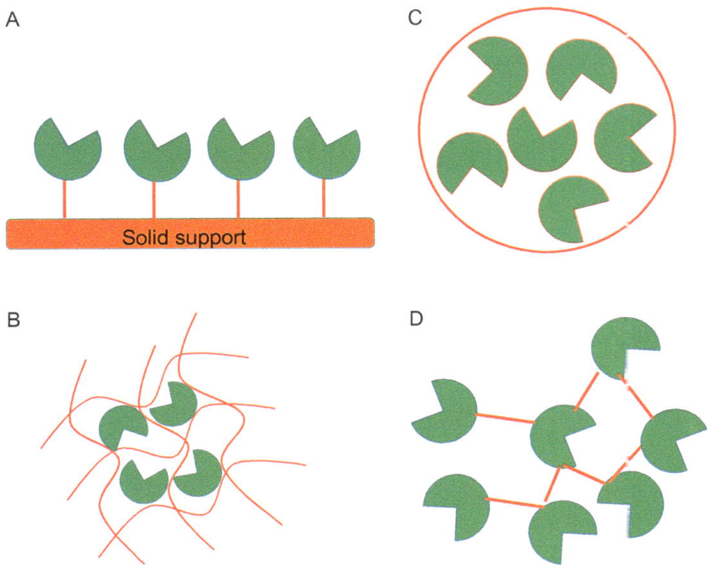

Fig. 1 Schematic representation of different methods for enzyme immobilization. (A) Covalent coupling, (B) Entrapment, (C) Encapsulation, and (D) Cross-linking.

which allows the substrate and the products pass through the medium without harming or leaving the enzyme intact. In this method, enzyme is not bound but occluded within the fibrous network. Entrapment within a gel matrix (Bernfeld & Wan, 1963), a fiber (Dinelli, Marconi, & Morisi, 1976), and an encapsulation (Wadiak & Carbonell, 1975) has been reported widely.

2. REVERSIBLE IMMOBILIZATION

Reversible immobilization enables retrieval of the enzyme under mild conditions, once the process is complete. The advantage of this method is the removal of deactivated enzyme and replacing it with the fresh enzyme, which makes the process a cost-efficient method. Several reversible immobilization techniques have been reported in the literature (Fig. 2). Among them nonspecific adsorption, ionic binding, hydrophobic adsorption, are affinity binding are the different adsorption phenomena, which are often used for reversible immobilization of enzymes onto various supports. Metal chelation and disulfide binding are also important reversible immobilization techniques (Brena & Batista-Viera, 2006).

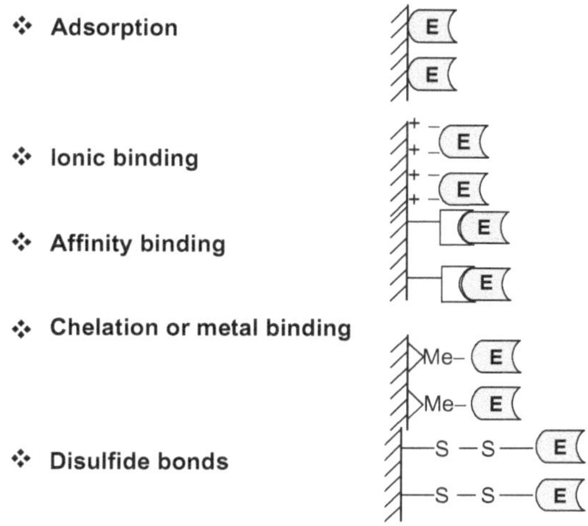

Fig. 2 Reversible immobilization techniques reported in the literature. *Adapted from Brena, B. M., & Batista-Viera, F. (2006). Immobilization of enzymes. In Guisan, J. M. (Ed.),* Immobilization of enzymes and cells *(2nd ed., pp. 15–30). Totawa, NJ: Human Press, with permission from Springer.*

3. JACK BEAN UREASE: ACTIVE SITE

Urease (urea amidohydrolase EC 3.3.1.5) was the first enzyme to be crystallized in 1926, and it is the first nickel metalloenzyme discovered. This has been crystallized in a hexagonal space group with a molecule in the asymmetric group. It forms a hammer or T-shaped molecule as shown in Fig. 3. (Balasubramanian & Ponnuraj, 2010; Zerner, 1991). It catalyzes urea hydrolysis yielding ammonia and carbonate, and the latter is subsequently converted to carbon dioxide at physiological pH as shown in Scheme 1.

The structure of this enzyme was established by both experimental and computational analysis. The activity of the enzyme was due to the presence of the binuclear Ni complex active site in the β-sheet structure and the mobile flap is located adjacent to the active site (Jabri & Karplus, 1996; Krajewska & Zaborska, 2007; Roberts, Miller, Roitberg, & Merz, 2012; Sharma, Mandani, & Sarma, 2013). The activity of this enzyme largely depends upon the opening and closing dynamics of this mobile flap. The wide opening of mobile flap could achieve higher availability of the active site for the substrate and can result in higher activity (Roberts et al., 2012). Due to this reason, the immobilization process has to be carried out in such a way that the wide opening of mobile flap of the enzyme must be ensured so

$$(NH_2)_2CO + 2H_2O \xrightarrow{\text{Urease}} CO_2 + 2NH_3 + H_2O$$

Scheme 1 Hydrolysis of urea by urease yields carbonate and ammonia.

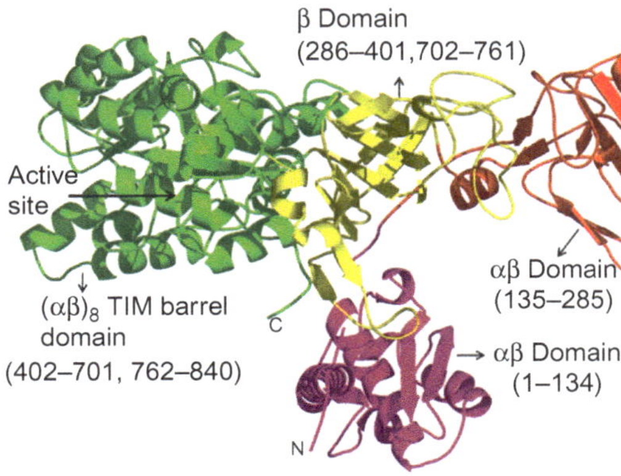

Fig. 3 Crystal structure of jack bean urease (JBU). A hammer or T-shaped JBU monomer consists of four domains: the N-terminal αβ domain (*magenta*, handle of the hammer), another αβ domain (*red*, one end of the hammer head), a β domain (*yellow*, the middle region of the hammer head), and C-terminal $(\alpha\beta)_8$ TIM barrel domain (*green*, other end of the hammer head). *Reproduced from Balasubramanian, A., & Ponnuraj, K. (2010). Crystal structure of the first plant urease from jack bean: 83 years of journey from its first crystal to molecular structure.* Journal of Molecular Biology, 400, *274–283, with permission from Elsevier.*

that the accessibility to the active site by the substrate becomes easier, and thereby a higher activity of the enzyme may be achieved.

4. IMMOBILIZATION OF JACK BEAN UREASE

Immobilization of enzyme may improve certain chemical, biochemical, mechanical, and kinetic properties of enzymes when compared to the free-state enzyme in buffer. Immobilization of enzymes on the polymer hydrogels network is a good alternative among the available immobilization techniques. Polymeric hydrogel is a cross-linked three-dimensional network, which retains high amount of water inside the matrices. The aqueous environment of the hydrogel will help the enzyme not to lose its activity, thereby enhancing the enzyme activity toward a specific function. Different polymeric hydrogel networks such as poly(ethyleneglycol), poly (2-hydroxylethylmethacryloate), poly(*N*-isopropyl acrylamide), and acrylamide copolymer-based hydrogels are commonly available for enzyme immobilization. Acrylamide-based copolymer hydrogels are very convenient for enzyme immobilization due to sufficient hydrophilic nature, resistance toward microbial or enzymatic attacks, and high chemical and mechanical stability.

Immobilized urease was used for the construction of artificial organs, biosensors (Bisht, Takashima, & Kaneto, 2005; Poźniak, Krajewska, & Trochimczuk, 1995), quantification of urea and Hg^{2+} (Arunbabu, Sannigrahi, & Jana, 2011; Kutcherlapati, Yeole, & Jana, 2016), dialysis regeneration systems in artificial kidney machines (Krajewska, Leszko, & Zaborska, 1989), removing urea from foods and beverages (Elçin, 1995), and many other biological applications. Immobilization of urease has been reported on different polymeric substrates (Bisht et al., 2005; Fahmy, Bagos, & Mohammed, 1998; Kuralay, Özyörük, & Yıldız, 2005; Laska, Włodarczyk, & Zaborska, 1999; Maia et al., 2007) for variety of uses. The enzyme was immobilized on the poly(*N*-3-aminopropyl pyrrole-*co*-pyrrole) film and electrochemically prepared onto an indium-tin-oxide-coated glass plate and this technique showed an improvement in the lifetime stability of the enzyme electrode (Bisht et al., 2005). In another work, urease was immobilized in a poly(vinylferrocenium) matrix, which was further coated on the platinum electrode and the resulting electrode was found to be stable for 24 days (Kuralay et al., 2005). Urease suffers from poor stability and deactivation at room temperature, and it is mainly because of the unfolding of the native protein structure. Hence, it is important to develop a method for the long-term stability of the urease enzyme at room temperature. Although several attempts were reported in the literature, they showed certain disadvantages in executing such methods in the real-time application. The noncovalent attachment might be the primary reason for such observations, in most of the reported cases.

We have recently developed a method where in the enzyme urease was immobilized in a polymer hydrogel matrix by covalent linking between enzyme and polymer backbone for improved storage stability at room temperature without affecting its activity. Urease–urea hydrolysis kinetics in aqueous media was used to study the stability of this enzyme at room temperature both in free urease form in solution and in the immobilized form in the hydrogel network. The physicochemical properties and the operational stability at room temperature are discussed in detail by comparing the stabilities of the free and boundurease (Kutcherlapati et al., 2016).

5. METHODOLOGY

5.1 Equipment and Reagents

5.1.1 General Equipment

1. *Safety*: Goggles, laboratory coat, latex or nitrile gloves, and ventilation hood.
2. Waste management: general aqueous waste receptacle
3. Deionized water
4. Screw cap glass vial (10 mL)
5. Volumetric micropipettes (20, 100, 200, and 1000 μL)
6. Beakers (100–1000 mL)
7. Stir/hot plate
8. Analytical balance
9. pH meter

5.1.2 Preparation of Hydrogel

1. Acrylamide (AA) (Sigma-Aldrich, Bengaluru, India)
2. *N,N′*-Methylenebisacrylamide (BAA) (Sigma-Aldrich, Bengaluru, India)
3. Diethoxyacetophenone (DEAP) (Sigma-Aldrich, Bengaluru, India)
4. *N,N,N′,N′*-Tetramethylethylenediamine (TEMED) (Sigma-Aldrich, Bengaluru, India)
5. Polymerization cell (size: 3 cm × 3 cm)
6. EDC (Sigma-Aldrich, Bengaluru, India)
7. Disposable syringe
8. Mercury lamp (Black Ray) (wavelength 365 nm)

5.1.3 Immobilization of Urease in the Hydrogel

1. NaOH (Merck)
2. 10% (v/v) *N,N,N′,N′*-tetramethylethylenediamine (TEMED) (Sigma-Aldrich, Bengaluru, India)
3. Urease (Jack bean, *Canavalia ensiformis*, MW 480 kDa) (Sigma-Aldrich, Bengaluru, India)
4. EDC
5. Phosphate-buffered saline (PBS) solution (pH 7.3)

5.1.4 Urease–Urea Assay

1. Disposable plastic cuvette
2. Reagent A: Phenol (5 g, 53 mmol) and sodium nitroprusside (25 mg, 0.084 mmol) in 500 mL of water
3. Reagent B: Dissolving NaOH (2.5 g, 0.0625 mmol) and NaOCl, which contains 5% active chlorine (4.2 mL) in 500 mL of water
4. UV–vis spectrophotometer
5. NaCl solution (150 m*M*)
6. Acetate buffer, phosphate buffer, and Tris–HCl buffer were used for pH range 4–5.5, 6–8, and 8.5–10, respectively

5.1.5 Characterization Instrumentation

1. Cary UV–vis spectrophotometer
2. Thermogravimetric and differential thermal analysis (TG-DTA) (Netzsch STA 409PC)
3. Differential scanning calorimetry (DSC) (Pyris Diamond DSC, Perkin Elmer)
4. Field emission scanning electron microscopy (FESEM) Carl Zeiss Ultra-55 microscope using EHT detector operating at 5 kV voltage
5. Circular dichroism spectrophotometer (JASCO J-815)

5.1.6 Preparation of Crystalline Colloidal Array

1. Styrene (Sisco, India) monomer
2. Divinylbenzene (DVB, Aldrich) cross-linker
3. Sodium salt of 2-propene sulfonic acid (2-PSA) as an ionic comonomer (Polyscience, USA) and sodium dodecyl sulfate (SDS) as surfactant (Merck, USA)
4. Sodium bicarbonate (Sisco, India)
5. Ammonium per sulfate (APS) (Merck, USA)
6. Dowex mixed-bed ion-exchange resin (Sigma-Aldrich, St. Louis, USA)

5.2 Preparation of Hydrogel

1. Acrylamide (0.100 g, 1.40 mmol), *N,N′*-methylenebisacrylamide (0.01 g, 0.03902 mmol), and 2.00 mL of water were mixed together using a vortex mixer in a 10 mL screw cap glass vial, and then 4–5 drops of 10% (v/v) DEAP (7.7 μL, 3.84 mol) in DMSO was added into the above reaction mixture and mixed thoroughly using vortex mixer.
2. This solution was nitrogen bubbled for 5 min to remove any dissolved oxygen. The polymerization mixture was then injected into a

polymerization cell (size: 3 cm × 3 cm) consisting of two quartz disks, separated by a 127-μm thick parafilm spacer with the help of disposable syringe.

3. The polymerization cell was placed under a mercury lamp (Black Ray) operating at a wavelength of 365 nm for 4 h.
4. After the completion of the reaction, the quartz cell was opened in HPLC grade water followed by thoroughly washing of obtained hydrogel with HPLC grade water for several times to remove the unreacted monomers and extra initiator.
5. Then the hydrogel was stored into the water for further use.

5.3 Immobilization of Urease in the Hydrogel

1. Polymer hydrogel (size: 3 cm × 3 cm) was hydrolyzed using 25 mL of aqueous 0.3N NaOH with 10% (v/v) *N,N,N′,N′*-tetramethylethylenediamine (TEMED) for 1.5 h.
2. After hydrolysis step hydrogel was thoroughly washed with deionized water.
3. 5 mg of urease and 5 mg of solid EDC were dissolved in 1 mL of PBS at pH 7.3. *TIP*: EDC is hygroscopic, so it should be weighed and used very quickly.
4. The urease–EDC solution was placed on the surface of the hydrolyzed hydrogel and kept at 25°C for 12 h.
5. The resulting urease-coupled hydrogel (UCG) was washed extensively with PBS solution.
6. The amount of urease attached to hydrogel was estimated by measuring the absorption at 278 nm of UCG against blank hydrogel without urease using UV–vis spectrophotometer.
7. In the end, UCG was cut into 1 cm × 1 cm-sized gels and stored in PBS solution in cold condition.

5.4 Urease–Urea Assay

1. This assay was carried out based on the Berthelot reaction.
2. This Berthelot assay was carried out for UCG and free urease (urease in solution).
3. A 1 cm × 1 cm-sized UCG or known amount of free urease was taken in a disposable plastic cuvette.
4. 1.5 mL of Reagent A, 0.5 mL urea solution of known concentration, and 1.5 mL of Reagent B were added to the plastic cuvette.

5. Immediately after the addition of Reagent B a blue-colored indophenol generation was started.
6. The absorption at 635 nm using a UV–vis spectrophotometer for indophenol blue was monitored for 1 h.
7. The absorption value at 635 nm is directly related to the dye concentration, which is equivalent to the ammonia concentration produced because of the urea–urease hydrolysis reaction.
8. The experiments were carried out with various urea concentrations to study the kinetics of UCG and free urease.
9. After every exposure, the UCG was washed thoroughly with 150 m*M* NaCl solution for 4–5 times and finally with the Reagent A
10. A fresh batch of enzyme was used after every exposure in case of free urease.
11. A gradual decrease in absorbance was observed with decreasing urea concentration.
12. The amount of urease enzyme is always constant for all the experiment both in UCG and in free state.
13. To study the assay as a function of days (time), both UCG in PBS and a known concentration of free urease stock solution in PBS were kept in the room temperature in the laboratory environment.
14. Then the assays as described earlier were carried out every 2 days interval up to 30 days for both UCG and free urease.

5.5 Analysis of Urease–Urea Assay With UCG and Free Urease

5.5.1 Michaelis–Menten Constant (K_m) Values

1. Lineweaver–Burk (L–B) plots, obtained from the urea–urease hydrolysis assay, were used to calculate Michaelis–Menten constant (K_m) values for free and immobilized ureases.
2. The concentrations of urea were in the range of 0.14–1.4 m*M* and 0.062–0.312 m*M* for UCG and free urease in PBS solution, respectively.
3. The concentrations of urease and UCG were 5.6 and 18 μ*M*, respectively.

5.5.2 Effect of pH, Thermal History, and Thermal Inactivation Studies

4. These concentrations of urease and urea were kept fixed for all the experiments carried out to see the effect of pH, thermal history, and thermal inactivation of urease enzyme.
5. Urea concentrations were 0.16 and 3.472 m*M* for free urease and UCG, respectively.

6. All the experiments were carried out in duplicate to check the reproducibility, and the data are well matched.
7. The activities in each case were represented as % of relative activity with respect to highest activity, which is considered as 100%.
8. Effect of pH on the activity of free urease and UCG was studied in different pHs in the range of pH 4–10. The pH of the assay solutions (Reagent A and Reagent B) was altered from 4 to 10 by using different buffers. Acetate buffer, phosphate buffer, and Tris–HCl buffer were used for pH range 4–5.5, 6–8, and 8.5–10, respectively.
9. The samples were incubated in different pH assay solutions for 1 h and then absorbance of indophenol dye was measured.
10. The time 1 h was chosen since by this time period urease–urea hydrolysis gets completed. For each pH, fresh piece of UCG and free urease were used.
11. Thermal history experiments of free urease and UCG were examined as follows: the free urease and UCG were kept in 100 m*M* PBS with NaCl (150 m*M*) at pH 7.3 for 15 min at each temperature from 20°C to 80°C.
12. After 15 min exposure to the temperature, the samples were incubated in assay solution for 1 h and then dye absorbance was measured. Here also for each temperature, fresh piece of UCG and fresh free urease were used. The data were collected for each 5°C interval.
13. Thermal inactivation studies of free urease and UCG were carried out as follows: the samples were incubated at 70°C in buffer for the indicated time period and assayed for residual activities as discussed earlier. Inactivation studies were carried out up to 250 min. Here, also for every data point fresh samples (UCG and free urease) were used.

5.6 Characterization Methods

1. Urease–urea hydrolysis kinetics was studied by UV–vis spectroscopic analysis in absorbance mode on a Cary 100 Bio UV–vis spectrophotometer.
2. TG-DTA: Thermal changes were studied starting from 25°C to 800°C with a scanning rate of 10°C/min in the presence of nitrogen flow.
3. DSC: All samples were kept at 25°C for 5 min and then heated from 25°C to 300°C (up to 150°C in case of urease) with a heating rate of 5°C/min.

4. FESEM: The instrument was operating at 5 kV voltage. A small amount of powdered sample was placed on a cleaned glass coverslip followed by gold coating and finally imaged in the FESEM machine.
5. Circular dichroism: 1 mg/mL of sample was prepared in 25 m*M* phosphate buffer and which was scanned from 300 to 190 nm using a scan rate of 50 nm/min and has an accumulation of 5 nm.

5.7 Methodology for Sensing Mercury

5.7.1 Preparation of Crystalline Colloidal Array

1. Cross-linked charged polystyrene colloidal particles were synthesized by an emulsion polymerization technique as reported by our group (Arunbabu, Sannigrahi, & Jana, 2008).
2. In a four-neck mercury-sealed round-bottom flask fitted with a reflux condenser, a teflon stirrer attached to a high-torque overhead mechanical stirrer, nitrogen, and reagent inlet was charged with 75 mL of triple-distilled water containing 0.1 g of sodium bicarbonate and the solution was deoxygenated by bubbling nitrogen for 40 min.
3. The temperature was maintained by placing the reaction vessel in a controlled temperature oil bath.
4. After thorough deoxygenation, the SDS (varied concentrations of SDS for different reactions, 0.348, 0.693, 1.734, 6.935, 13.870, 20.806, 27.741, and 69.353 m*M*) was dissolved in 10 mL of water and added, and the temperature of the reaction vessel was increased to 50°C.
5. Freshly deoxygenated styrene (33 g, 0.32 mol) and DVB (1.65 g, 0.013 mol) were injected slowly at a constant rate of 4 mL/min.
6. The 2-PSA sodium salt (1.98 g, 0.014 mol) was dissolved in 5 mL of water and injected for 10 min after the addition of styrene and DVB, and the temperature was increased to 70°C and the stirring speed was increased to 350 rpm.
7. After equilibration for 30 min, required amount of APS (17.54 m*M*) dissolved in 10 mL of water was injected into the reaction mixture and the reaction was refluxed for 3–4 h. A nitrogen blanket and the stirring rate of 350 rpm were maintained during refluxing.
8. A milky white colloidal solution was obtained upon the completion of the reaction and it was then allowed to cool and filtered through glass wool. The filtered solution was centrifuged for 40 min at 45,000 rpm at 15°C in an ultracentrifuge. The residue (solid white mass) was

thoroughly dispersed in triple-distilled water with the help of sonication and mixing in a vortex motor.

9. Centrifugation and dispersion process were repeated 3–4 times to remove all the impurities from the colloidal polystyrene particles. The particles showed bright iridescence after these purification steps. The colloidal solution was taken in a storage glass bottle and mixed-bed ion-exchange resin was added and placed on a vertical rotor for the thorough mixing of the particles with resin.

5.7.2 Preparation of Polymerized Crystalline Colloidal Array

1. Acrylamide (AA, 0.100 g, 1.40 mmol, Sigma-Aldrich), *N,N'*-methylenebisacrylamide (0.005 g, 0.01951 mmol), CCA (2.00 g, 8 wt%, 102 nm diameter cross-linked polystyrene particles prepared from the previously reported), and AG 501-X8 ion-exchange resin (0.10 g, 20–50 mesh, mixed bed, Bio-Rad) were mixed together in a 10-mL screw cap glass vial by shaking in the vortex mixer.
2. 10% (v/v) DEAP (7.7 μL, 3.84 mol) in dimethylsulfoxide was added in to the earlier reaction mixture and mixed with the help of vortex mixer. This reaction solution (polymerization mixture) is nitrogen bubbled thoroughly to remove the any dissolved oxygen.
3. The polymerization mixture was withdrawn from the vial using a thin needle to avoid ion-exchange resin and injected into a polymerization cell consisting of two quartz disks, separated by a 127-μm thick parafilm spacer. The polymerization cell was placed under mercury (Black Ray) lamp operating at 365 nm for 4 h. After the completion of the reaction the quartz cell was opened in Millipore water, and the obtained polymerized crystalline colloidal array (PCCA) was washed thoroughly with Millipore water several times.

5.7.3 Preparation of Ion Strength Responsive PCCA (Carboxylated PCCA)

The hydrolysis of PCCA makes the PCCA ions responsive. Partial hydrolysis of amide groups of PAM backbone of PCCA yields carboxylate groups. These covalently attached carboxylate groups carry counterions inside the hydrogel and the hydrogel (PCCA) swells owing to the osmotic pressure. Therefore, the PCCA becomes ionic strength responsive. The hydrolysis of PCCA was carried out as follows:

1. A small piece of PCCA was treated with 25 mL of aqueous hydrolysis solution containing 0.3 N NaOH (Merck) with 10% (v/v) *N,N,N′,N′*-tetramethylethylenediamine (TEMED, Sigma-Aldrich) for 1.5 h.
2. The hydrolyzed PCCA was washed thoroughly several times with Millipore water.

5.7.4 Preparation of Urease-Coupled PCCA Sensor for Mercury

1. 0.005 g of urease (Jack Bean, 0.005 g, MW 480 kDa SRL, India) and 0.005 g of solid ethyl-(3-dimethylaminopropyl)carbodiimidehydrochloride (EDC, 0.005 g, 0.026 mmol, Aldrich) were dissolved in 1 mL 0.1 m*M* PBS at pH 7.21.
2. The urease–EDC solution was placed on top of the hydrolyzed PCCA for 12 h. After 12 h the resulting urease-coupled PCCA (UPCCA) mercury sensor was washed extensively with Millipore water.

5.7.5 Measurement of UPCCA Mercury Sensor Response

All the diffraction measurements were carried out in reflection mode with an Ocean Optics (USB4000-UV-VIS-NIR) spectrophotometer. The sensing experiments of the UPCCA sensor were carried out as follows:

1. A small piece of UPCCA was fixed in a Petri dish (made of polystyrene, Hi-Glass, India) and the diffraction was recorded in water.
2. Then the UPCCA was exposed to the different concentrations of urea solution in water and their diffraction spectra were collected. A gradual blue shift was observed with increasing urea concentration.
3. UPCCA's initial diffraction in water was reverted by washing the UPCCA with water. Then the UPCCA was exposed to the 1 m*M* urea solution in water, which contains Hg^{2+} ion, and the diffraction was recorded. UPCCA was washed to retain its diffraction.
4. Then the UPCCA was exposed to the next higher Hg^{2+} concentration along with1 m*M* urea solution. After every exposure, the UPCCA was washed with water and obtained the original diffraction in water.

6. RESULTS AND DISCUSSION

The β-sheet structure of urease enzyme in the buffer solution (free urease) was confirmed by observing a peak at 215 nm in circular dichroism (CD) spectra (Sharma et al., 2013) and it is well known that the β-sheet conformation is responsible for the enzymatic activity of urease. The height of the 215 nm peak in the CD spectra, which reflects the β-sheet conformation and hence the enzymatic activity, decreases with storage time of free enzyme at 27–30°C, whereas the changes are negligible with the storage time for free urease at 4°C. The disruption of β-sheet structural conformation (deactivation) of the enzyme at 27–30°C is attributed as the reason behind such observation in CD spectra analysis. As the β-sheet structure gets disrupted, this would denature the urease when it was stored at room temperature for longer periods of time (Balasubramanian & Ponnuraj, 2010). Therefore, we have decided to improve the stability of the enzyme at room temperature for longer periods of time by immobilizing the urease in a polymer hydrogel as an armor, and we succeeded in improving the stability (Kutcherlapati et al., 2016).

Urease was immobilized on PAM–bisacrylamide hydrogel having different degrees of cross-linking. The synthetic protocol for immobilization of urease inside the hydrogel is shown in Scheme 2. The primary amide functionality of PAM hydrogel was treated with alkali for the partial hydrolysis, which resulted in the generation of carboxylate functionality without affecting the cross-linking in hydrogel matrix. In the following reaction, conjugation of urease was done using a chemical coupling protocol in buffer solution (pH 7.3). The UCG was stored in buffer and calculated the amount of enzyme coupled to the hydrogel system by measuring the absorbance of urease at 278 nm and was found to be 18 μmol/cm^3 (Kutcherlapati et al., 2016).

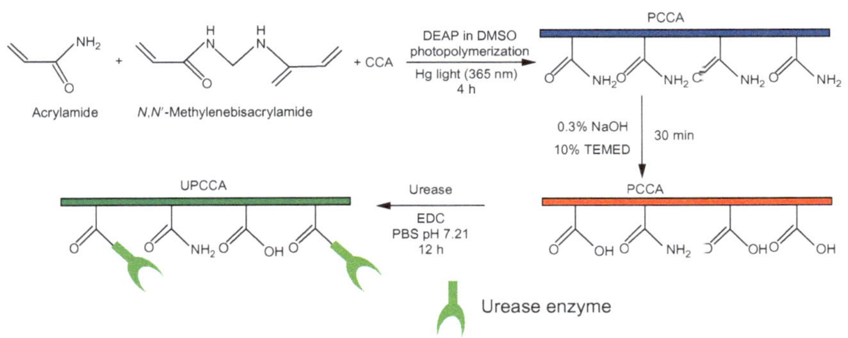

Scheme 2 Preparation of urease-coupled polymer hydrogel or can be called as immobilization of urease in the hydrogel matrices. *Reproduced from Arunbabu, D., Sannigrahi, A., & Jana, T. (2011). Photonic crystal hydrogel material for the sensing of toxic mercury ions (Hg^{2+}) in water.* Soft Matter, *7, 2592–2599, with permission from The Royal Society of Chemistry.*

6.1 Urease Stability at Room Temperature: Comparison Between Free Urease and Immobilized Urease

We have carried out a comparative study by setting two experimental conditions: (1) storage of free enzyme in buffer solution at room temperature and (2) storage of UCG (which is the immobilized urease in the hydrogel) at room temperature. The detailed kinetic studies for these two experimental conditions were done using urease–urea assay for different urea concentrations and also for several days (30 days). Very surprisingly, an unexpected behavior was observed from the kinetic plots under the studied conditions from this assay. The hydrolysis of urea by free urease decreases with increasing storage time, which is expected due to the denaturation of the enzyme by breaking the β-sheet structure of the enzyme over a period of storage time as observed from CD spectra. However, in the case of second experimental condition (immobilized state of enzyme; UCG state), with the increasing storage time at room temperature the production of ammonia increases, which indicates the retention of the activity of enzyme with storage time. To understand this unusual stability of the enzyme in hydrogel matrix at room temperature, we analyzed the L–B plots (Fig. 4) for both free urease and UCG hydrogel matrix and observed the increased activity of enzyme in case of UCG matrix at room temperature even after a month of storing time (Kutcherlapati et al., 2016).

The kinetic data such as maximum velocity (V_{max}), Michaelis–Menten constant (K_m), catalytic constant, or turnover number (K_{cat}), and catalytic efficiency (ε) were calculated from the slope and intercept of L–B plots (Fig. 4) for both free urease and UCG states and presented in Figs. 5 and 6.

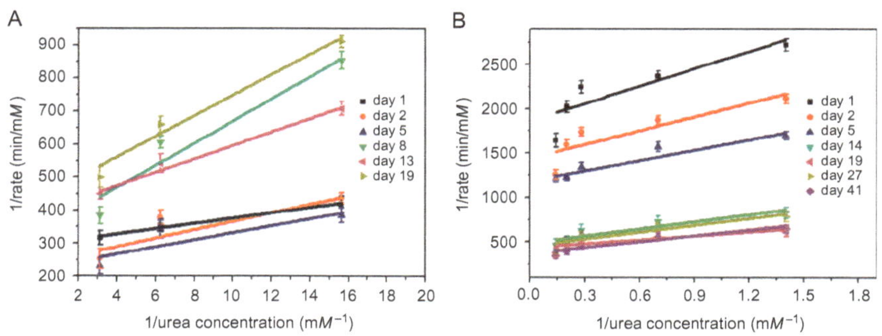

Fig. 4 L–B plots of urease–urea reaction (A) free urease and (B) UCG state. The kinetics was followed for several days after storing the sample at room temperature. *Reproduced from Kutcherlapati, S. R., Yeole, N., & Jana, T. (2016). Urease immobilized polymer hydrogel: Long-term stability and enhancement of enzymatic activity.* Journal of Colloid and Interface Science, 463, *164–172, with the permission from Elsevier.*

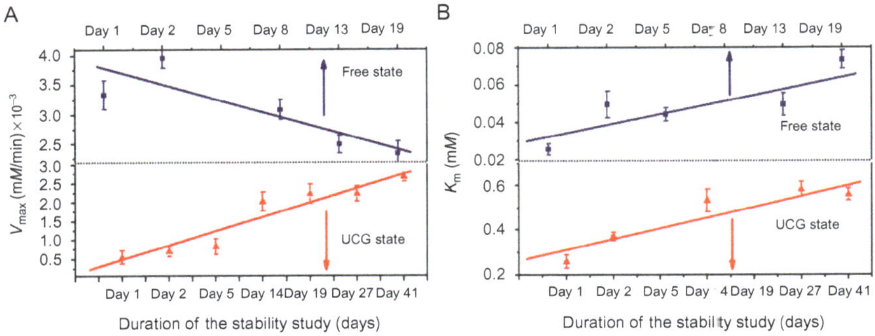

Fig. 5 Plots of V_{max} (A) and K_m (B) against the duration of the storage stability study for both free urease and UCG state. *Reproduced from Kutcherlapati, S. R., Yeole, N., & Jana, T. (2016). Urease immobilized polymer hydrogel: Long-term stability and enhancement of enzymatic activity.* Journal of Colloid and Interface Science, 463, *164–172, with permission from Elsevier.*

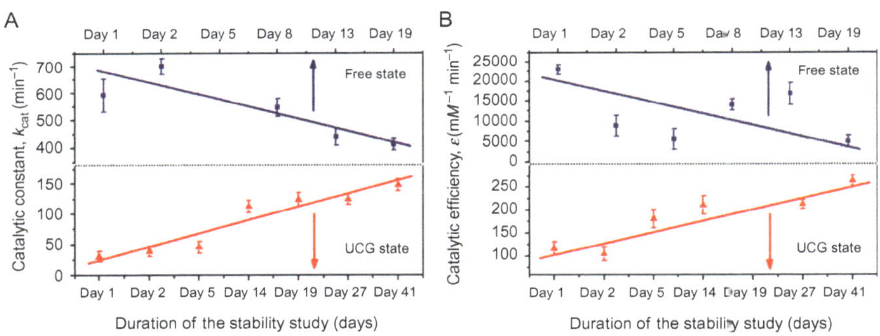

Fig. 6 Variation of catalytic constant or turnover number (K_{cat}) (A) and catalytic efficiency (ε) (B) for both free urease and UCG state as a function of storage stability duration at room temperature. *Reproduced from Kutcherlapati, S. R., Yeole, N., & Jana, T. (2016). Urease immobilized polymer hydrogel: Long-term stability and enhancement of enzymatic activity.* Journal of Colloid and Interface Science, 463, *164–172, with permission from Elsevier.*

The kinetic parameters (V_{max}, K_{cat}, and ε) were found to be higher for free urease when compared to that of UCG hydrogel matrix and the K_m value was reverse, i.e., free urease showed lower K_m when compared to that of the UCG state. The lower value of V_{max}, K_{cat}, and ε in UCG state compared to free urease may be due to the following reasons: (1) restricted free diffusion of the substrate and the product since urease–urea reaction has to take place inside the hydrogel matrix owing to the fact that the urease is covalently attached to PAM backbone, and thereby decreases the reaction speed, turnover of enzyme, and enzymatic catalytic efficiency; (2) another reason may be the steric crowding inside the hydrogel and most significantly; and (3) covalent bonding between enzyme and PAM chain may suppress the availability of active site of the urease to the substrate (Kutcherlapati et al., 2016). The lower affinity of the enzyme toward substrate is identified by the higher value of K_m. Although at the initial stage the affinity of the immobilized urease toward the substrate is 10 times lower than the free urease affinity, an increase in the K_m value was observed in the former case, with increase in the storage time. This could be due to multiple factors (Kutcherlapati et al., 2016). Similar observations were reported previously (Çevik, Senel, & Abasiyanik, 2011; Chen & Chiu, 2000; Poźniak et al., 1995).

6.2 Effect of pH and Temperature on the Enzyme-Coupled Hydrogel Matrix (UCG)

The pH stability of the urease has been improved by one pH unit after immobilization in the hydrogel matrix (Fig. 7), and this is probably due to the change in the microenvironment of the hydrogel after conjugation with enzyme. It has been reported by many researchers that the substrate used for the immobilization (Fahmy et al., 1998) can play a significant role in improving the pH stability of the enzyme. The one unit shift in pH as observed in PAM gel (Fig. 7) might be due to the negative charges of partially hydrolyzed hydrogel, which makes the immobilized enzyme less sensitive to pH variation (Kutcherlapati et al., 2016). The shift toward more basic pH upon immobilization might be due to retaining of higher concentration of ammonia in the microenvironment of enzyme inside the hydrogel

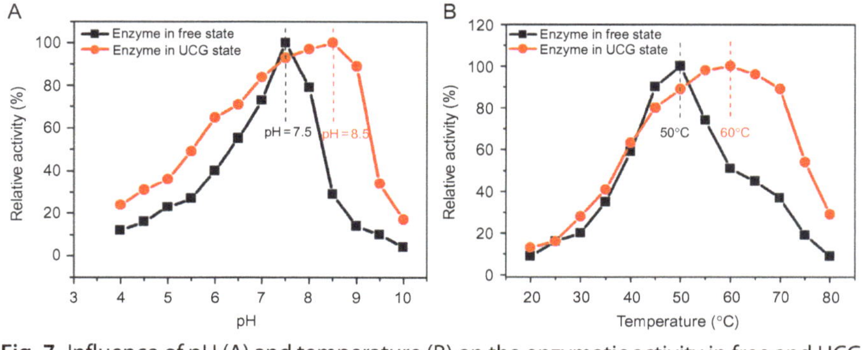

Fig. 7 Influence of pH (A) and temperature (B) on the enzymatic activity in free and UCG state. *Reproduced from Kutcherlapati, S. R., Yeole, N., & Jana, T. (2016). Urease immobilized polymer hydrogel: Long-term stability and enhancement of enzymatic activity.* Journal of Colloid and Interface Science, 463, *164–172, with permission from Elsevier.*

matrix when compared to bulk solution. This type of observations was previously reported where in the enzyme it was immobilized on EC-tri beads and also on chitosan beads (Çevik et al., 2011; Kayastha & Srivastava, 2001).

The optimum temperature was improved by 10°C for the enzyme-coupled hydrogel (UCG) when compared to that of the free urease enzyme (Fig. 7). The improved thermal stability might be attributed to the multi-point attachment of the enzyme in the hydrogel, which altered the physicochemical properties of the enzyme (Kutcherlapati et al., 2016).

6.3 Sensing of Mercury Pollutant in Water

One attractive application of the bound urease was the sensing of highly toxic mercury in water by using a new photonic crystal hydrogel (Fig. 8)

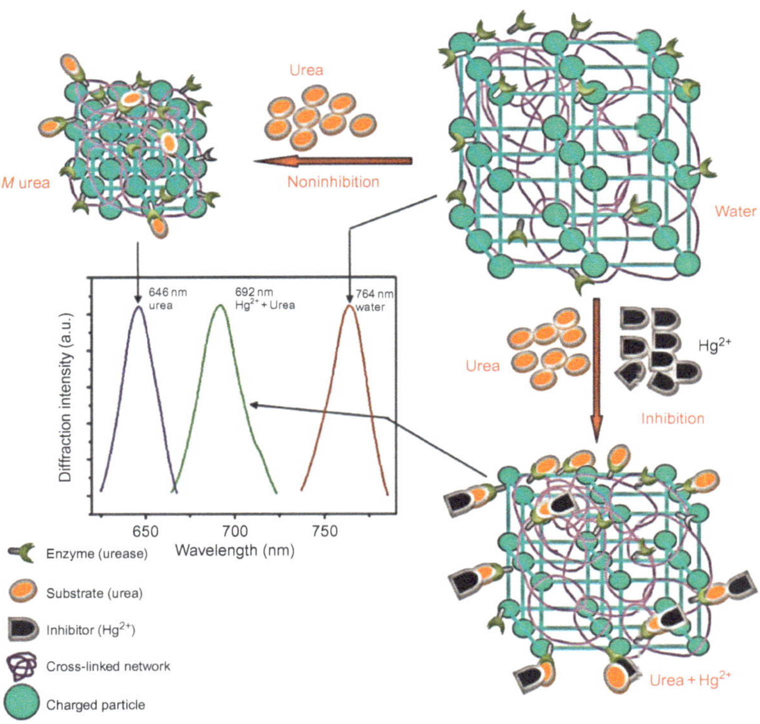

Fig. 8 Schematic representation of UPCCA sensor concept in which the sensing motif relies on a two-step-coupled spontaneous processes. *Reproduced from Arunbabu, D., Sannigrahi, A., & Jana, T. (2011). Photonic crystal hydrogel material for the sensing of toxic mercury ions (Hg^{2+}) in water.* Soft Matter, *7, 2592–2599, with permission from The Royal Society of Chemistry.*

(Arunbabu et al., 2011). The sensing principle was based on the principle of the PCCA technique (Arunbabu et al., 2011; Holtz & Asher, 1997; Sharma et al., 2004; Xu, Goponenko, & Asher, 2008). PCCA sensing technology has been previously reported for the sensing of different analytes like glucose, metal cations, creatinine, and others. The PCCA was composed of polymeric hydrogel and an array of colloidal particles, which is called crystalline colloidal array (CCA), and both CCA and PCCA diffract visible light. The principle of this method was that the wavelength of the diffracted light would be varied by altering the volume of the hydrogel, which altered the lattice of embedded CCA.

After the appropriate chemical or physical modification of the hydrogel, it can be tuned as a function of any corresponding analyte concentration if the modification of the hydrogel and the analyte respond to each other. Using this principle, mercury concentration in water was determined by monitoring light diffracted by the PCCA, which consists of urease and carboxylic functionalities. The heavy metal ions such as Hg^{2+}, Cu^{2+}, and Ag^{+} ions are known to inhibit urease activity (Zhylyak, Dzyadevich, Korpan, Soldatkin, & El'Skaya, 1995). These metal ions interfere with the hydrolysis of urea by urease and can suppress the formation of ammonium and bicarbonate ions (Arunbabu et al., 2011).

The urease immobilized PCCA (UPCCA) has been used for the determination of Hg^{2+} in aqueous media (Fig. 9). When aqueous urea was exposed to the UPCCA, urea was hydrolyzed to give ammonium and bicarbonate ions and which resulted in charge screening with the carboxylic acid groups of the ionic polyacrylic acid generated by the hydrolysis of PAM matrix. This resulted in the shrinkage of the hydrogel and the UPCCA exhibited the blue shift in the diffraction as a function of urea concentration. With the increasing concentration of urea in the exterior solution, the UPCCA diffraction shifts toward blue (Fig. 8) (Arunbabu et al., 2011). When the urea solution containing Hg^{2+} was exposed to UPCCA, mercury inhibited the hydrolysis of urea by urease and thereby suppressed the generation of ammonium and carbonate ions, which resulted in smaller blue shift of the UPCAA (Arunbabu et al., 2011). Hence, the diffraction wavelength of the UPCCA was monitored as a function of mercury concentration at a fixed urea concentration (Fig. 9). This UPCCA technology was successfully applied in the detection of Hg^{2+} at 1 ppb in contaminated water. The maximum contaminant level of mercury in water set by the environmental protection agency, USA, is 2 ppb. A detailed study on the urea–urease kinetics and their inhibition in the solution (free urease) and also as UPCCA state

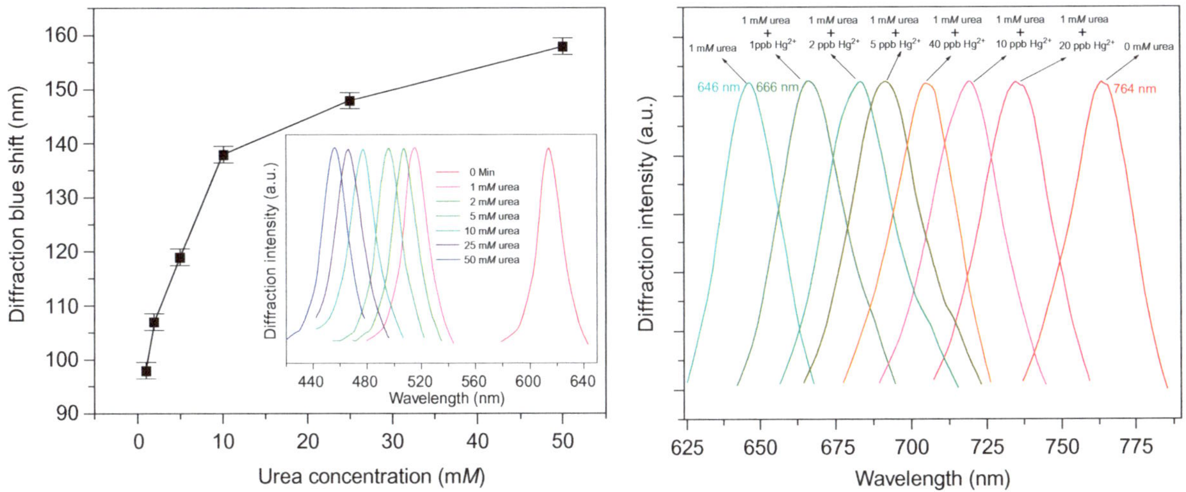

Fig. 9 (A) Dependence of UPCCA diffraction as a function of urea concentration in water. *Inset*: Diffraction spectra of UPCCA in response to various urea concentrations. (B) Diffraction spectra of UPCCA in response to various mercury ion (Hg^{2+}) concentrations in the presence of 1 m*M* urea in water. *Reproduced from Arunbabu, D., Sannigrahi, A., & Jana, T. (2011). Photonic crystal hydrogel material for the sensing of toxic mercury ions (Hg^{2+}) in water.* Soft Matter, *7, 2592–2599, with permission from The Royal Society of Chemistry.*

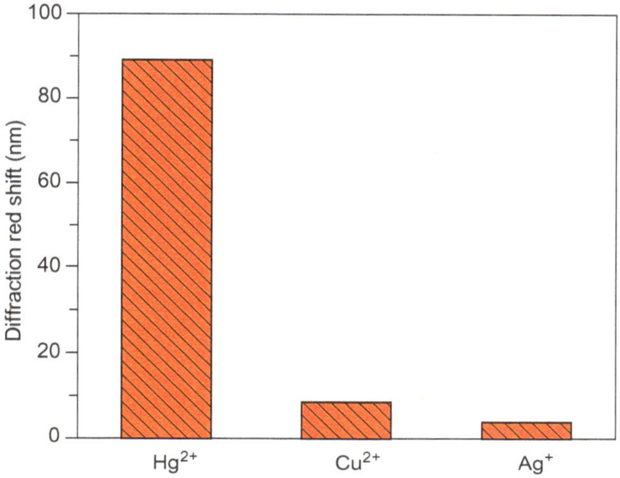

Fig. 10 UPCCA sensor selectivity toward all the three inhibitors in the presence of 1 m*M* urea. *Reproduced from Arunbabu, D., Sannigrahi, A., & Jana, T. (2011). Photonic crystal hydrogel material for the sensing of toxic mercury ions (Hg^{2+}) in water.* Soft Matter, *7, 2592–2599, with permission from The Royal Society of Chemistry*

(hydrogel) revealed the high specificity of the UPCCA for mercury when compared to copper and silver ions (Fig. 10) (Arunbabu et al., 2011).

7. CONCLUSIONS AND FUTURE PERSPECTIVES

Jack bean urease was immobilized covalently in the poly(acrylamide) hydrogel matrix by chemical coupling method and immobilized enzyme had significantly improved the enzymatic activity, under ambient conditions. The detailed study of the urea–urease assay and the kinetic data indicated that the urease in the hydrogel showed activity for more than a month at room temperature. The free urease enzyme in buffer solution has become inactive very quickly and lost its activity. The moisture retained in the hydrogel creates a conducive microenvironment for the enzyme to remain active for long times, during storage at room temperature.

The PCCA technology has been successfully implemented in detecting urea and mercury in aqueous medium with very high selectivity and sensitivity with a detection limit as low as 1 ppb of mercury ions.

In summary, this hydrogel process could be used for improving the storage of biomacromolecules at room temperature for medical and other uses. The UPCCA technology may be applied in the potable water systems for the detection of mercury, copper, and silver ions in the contaminated water.

ACKNOWLEDGMENTS

Dr. D.S. Kothari Fellowship (No. F.42/2006 (BSR)/CH/15-16/0157) of UGC, India is acknowledged for providing financial support to Dr. K.R.K. We thank the CSIR, Govt. of India for the funding through a research grant [02(0175)/14/EMR-II dated 07-05-2014].

REFERENCES

Arunbabu, D., Sannigrahi, A., & Jana, T. (2008). Tuning the particle size and charge density of the crosslinked polystyrene particles. *Journal of Applied Polymer Science*, *108*(4), 2718–2725.

Arunbabu, D., Sannigrahi, A., & Jana, T. (2011). Photonic crystal hydrogel material for the sensing of toxic mercury ions (Hg^{2+}) in water. *Soft Matter*, 7, 2592–2599.

Axen, R., Porath, J., & Ernback, S. (1967). Chemical coupling of peptides and proteins to polysaccharides by means of cyanogen halides. *Nature*, *214*, 1302–1304.

Balasubramanian, A., & Ponnuraj, K. (2010). Crystal structure of the first plant urease from jack bean: 83 years of journey from its first crystal to molecular structure. *Journal of Molecular Biology*, *400*, 274–283.

Bernfeld, P., & Wan, J. (1963). Antigens and enzymes made insoluble by entrapping them into lattices of synthetic polymers. *Science*, *142*, 678–679.

Bisht, V., Takashima, W., & Kaneto, K. (2005). An amperometric urea biosensor based on covalent immobilization of urease onto an electrochemically prepared copolymer poly (N-3-aminopropyl pyrrole-co-pyrrole) film. *Biomaterials*, *26*, 3683–3690.

Brena, B. M., & Batista-Viera, F. (2006). Immobilization of enzymes. In J. M. Guisan (Ed.), *Immobilization of enzymes and cells* (2nd ed., pp. 15–30). Totowa, NJ: Human Press.

Brena, B., González-Pombo, P., & Batista-Viera, F. (2013). Immobilization of enzymes: A literature survey. In J. M. Guisan (Ed.), *Immobilization of enzymes and cells* (3rd ed., pp. 15–31). New York: Springer Science + Business Media.

Çevik, E., Senel, M., & Abasiyanik, M. F. (2011). Immobilization of urease on copper chelated EC-tri beads and reversible adsorption. *African Journal of Biotechnology*, *10*, 6590–6597.

Chen, J. P., & Chiu, S. H. (2000). A poly (N-isopropylacrylamide-co-N-acryloxysuccinimide-co-2-hydroxyethyl methacrylate) composite hydrogel membrane for urease immobilization to enhance urea hydrolysis rate by temperature swing. *Enzyme and Microbial Technology*, *26*(5), 359–367.

Dinelli, D., Marconi, W., & Morisi, F. (1976). [18] Fiber-entrapped enzymes. *Methods in Enzymology*, *44*, 227–243.

Drobnik, J., Labský, J., Kudlvasrová, H., Saudek, V., & Švec, F. (1982). The activation of hydroxy groups of carriers with 4-nitrophenyl and N-hydroxysuccinimidyl chloroformates. *Biotechnology and Bioengineering*, *24*, 487–493.

Elçin, Y. M. (1995). Encapsulation of urease enzyme in xanthan-alginate spheres. *Biomaterials*, *16*, 1157–1161.

Fahmy, A. S., Bagos, V. B., & Mohammed, T. M. (1998). Immobilization of Citrullus vulgaris urease on cyanuric chloride DEAE-cellulose ether: Preparation and properties. *Bioresource Technology*, *64*, 121–129.

Guisán, J. (1988). Aldehyde-agarose gels as activated supports for immobilization-stabilization of enzymes. *Enzyme and Microbial Technology*, *10*, 375–382.

Holtz, J. H., & Asher, S. A. (1997). Polymerized colloidal crystal hydrogel films as intelligent chemical sensing materials. *Nature*, *389*, 829–832.

Jabri, E., & Karplus, P. A. (1996). Structures of the Klebsiella aerogenes urease apoenzyme and two active-site mutants. *Biochemistry*, *35*, 10616–10626.

Kayastha, A. M., & Srivastava, P. K. (2001). Pigeonpea (Cajanus cajan L.) urease immobilized on glutaraldehyde-activated chitosan beads and its analytical applications. *Applied Biochemistry and Biotechnology, 96*, 41–53.
Krajewska, B., Leszko, M., & Zaborska, W. (1989). Membrane immobilized urease for possible use in dialysate regeneration system of artificial kidney. *Environment Protection Engineering, 15*, 173–180.
Krajewska, B., & Zaborska, W. (2007). Jack bean urease: The effect of active-site binding inhibitors on the reactivity of enzyme thiol groups. *Bioorganic Chemistry, 35*, 355–365.
Kuralay, F., Özyörük, H., & Yıldız, A. (2005). Potentiometric enzyme electrode for urea determination using immobilized urease in poly (vinylferrocenium) film. *Sensors and Actuators B: Chemical, 109*, 194–199.
Kutcherlapati, S. R., Yeole, N., & Jana, T. (2016). Urease immobilized polymer hydrogel: Long-term stability and enhancement of enzymatic activity. *Journal of Colloid and Interface Science, 463*, 164–172.
Laska, J., Włodarczyk, J., & Zaborska, W. (1999). Polyaniline as a support for urease immobilization. *Journal of Molecular Catalysis B: Enzymatic, 6*, 549–553.
Maia, M. D. M. D., de Vasconcelos, E. A., de Mascena Diniz, P. F. C., da Costa Maciel, J., Cajueiro, K. R. R., da Silva, M. D. P. C., et al. (2007). Immobilization of urease on vapour phase stain etched porous silicon. *Process Biochemistry, 42*, 429–433.
Miron, T., & Wilchek, M. (1982). A spectrophotometric assay for soluble and immobilized N-hydroxysuccinimide esters. *Analytical Biochemistry, 126*, 433–435.
Porath, J., & Axén, R. (1976). [3] Immobilization of enzymes to agar, agarose, and Sephadex support. *Methods in Enzymology, 44*, 19–45.
Poźniak, G., Krajewska, B., & Trochimczuk, W. (1995). Urease immobilized on modified polysulphone membrane: Preparation and properties. *Biomaterials, 16*, 129–134.
Roberts, B. P., Miller, B. R., III, Roitberg, A. E., & Merz, K. M., Jr. (2012). Wide-open flaps are key to urease activity. *Journal of the American Chemical Society, 134*, 9934–9937.
Sharma, A. C., Jana, T., Kesavamoorthy, R., Shi, L., Virji, M. A., Finegold, D. N., et al. (2004). A general photonic crystal sensing motif: Creatinine in bodily fluids. *Journal of the American Chemical Society, 126*, 2971–2977.
Sharma, B., Mandani, S., & Sarma, T. K. (2013). Biogenic growth of alloys and core-shell nanostructures using urease as a nanoreactor at ambient conditions. *Scientific Reports, 3*, 2601.
Sheldon, R. A. (2000). Atom efficiency and catalysis in organic synthesis. *Pure and Applied Chemistry, 72*, 1233–1246.
Sheldon, R. A. (2007). Enzyme immobilization: The quest for optimum performance. *Advanced Synthesis & Catalysis, 349*, 1289–1307.
Sheldon, R. A., & van Rantwijk, F. (2004). Biocatalysis for sustainable organic synthesis. *Australian Journal of Chemistry, 57*, 281–289.
Wadiak, D., & Carbonell, R. (1975). Kinetic behavior of microencapsulated β-galactosidase. *Biotechnology and Bioengineering, 17*, 1157–1181.
Xu, X., Goponenko, A. V., & Asher, S. A. (2008). Polymerized polyHEMA photonic crystals: pH and ethanol sensor materials. *Journal of the American Chemical Society, 130*, 3113–3119.
Zerner, B. (1991). Recent advances in the chemistry of an old enzyme, urease. *Bioorganic Chemistry, 19*, 116–131.
Zhylyak, G. A., Dzyadevich, S. V., Korpan, Y. I., Soldatkin, A. P., & El'Skaya, A. V. (1995). Application of urease conductometric biosensor for heavy-metal ion determination. *Sensors and Actuators B: Chemical, 24*, 145–148.

CHAPTER EIGHT

Armored Enzyme–Nanohybrids and Their Catalytic Function Under Challenging Conditions

Omkar V. Zore*,†, Rajeswari M. Kasi*,†,1, Challa V. Kumar*,†
*University of Connecticut, Storrs, CT, United States
†Polymer Program, Institute of Materials Science, University of Connecticut, Storrs, CT, United States
[1]Corresponding author: e-mail address: rajeswari.kasi@uconn.edu

Contents

Abstract

Synthesis and characterization of highly stable and functional bienzyme–polymer triads assembled on layered graphene oxide (GO) are described here. Glucose oxidase (GOx) and horseradish peroxidase (HRP) were used as model enzymes and polyacrylic acid (PAA) as model polymer to armor the enzymes. PAA–armored GOx and HRP covalent conjugates were further protected from denaturation by adsorption onto GO nanosheets. Structure and morphology of this enzyme–polymer–nanosheet hybrid biocatalyst (GOx–HRP–PAA/GO) were confirmed by agarose gel electrophoresis, zeta

Methods in Enzymology, Volume 590
ISSN 0076-6879
http://dx.doi.org/10.1016/bs.mie.2017.02.007
© 2017 Elsevier Inc.
All rights reserved.

potential, circular dichroism, and transmission electron microscopy. The armored biocatalysts retained full enzymatic activities under challenging conditions of pH (2.5–7.4), warm temperatures (65°C), and presence of chemical denaturants, 4 m*M* sodium dodecyl sulfate, while GOx/HRP physical mixtures without the armor had very little activity under the same conditions. Therefore, this novel combination of two orthogonal approaches, enzyme conjugation with PAA and subsequent physical adsorption onto GO nanosheets, resulted in super stable hybrid biocatalysts that function under harsh conditions. Therefore, this general and powerful approach may be used to design environmentally friendly, green, biocompatible, and biodegradable biocatalysts for energy production in biofuel cell or biobattery applications.

1. INTRODUCTION

1.1 Need for Enzyme Armor

Enzymes are widely used in various biological processes as well as in detergent, food and beverage, animal feed, textile, pulp and paper, organic synthesis, leather, and personal care resulting in a $1.5 billion dollar market share (Kirk, Borchert, & Fuglsang, 2002). These applications are due to attractive catalytic properties of enzymes including high substrate specificity, high stereo/regioselectivity, increase in reaction rate due to a reduction in the activation energy, and the ability to perform catalysis under mild and ambient conditions (Fersht, 1999; Lele, Murata, Matyjaszewski, & Russel, 2005; Torchilin et al., 1978; Zore, Lenehan, Kumar, & Kasi, 2014). Often, enzymes are to be used under nonphysiological conditions such as high temperatures, extremes of pH, organic solvent, or in presence of denaturants. But most enzymes are denatured or deactivated under these challenging conditions. Therefore, enzyme stabilization under these extreme conditions is a necessity and this issue has not been fully addressed.

One approach to the problem of stabilizing enzymes under nonphysiological conditions is to protect the enzyme by encapsulating or wrapping its hierarchical 3D structure using synthetic polymers (Zore, Pattammattel, Gnanaguru, Kumar, & Kasi, 2015). Our hypothesis is that confinement of the enzyme by chemical conjugation with synthetic polymers will result in lowered conformational entropy of the enzyme, which may help stabilize the enzyme. One needs to bear in mind that small improvements in free energy of stabilization can have much larger improvements in the enzyme stability because of the logarithmic relation between the free energy change ($\Delta G°$) and the equilibrium constant for the denaturation (K), since $\Delta G^\circ = -RT\ln(K)$, where R is the gas constant and T is the absolute temperature. At the denaturation temperature, the equilibrium constant will have a value of 1, which implies equal molar concentrations of the native and the denatured states because ΔG° is zero. But, at room temperature $\Delta G^\circ > 0$ and denaturation is not spontaneous for most enzymes. Small increases in ΔG° will decrease K, thus contributing to improved enzyme stability (Scheme 1).

These gains in ΔG° are often obtained by decreasing ΔS° by restricting the conformational space of the enzyme, while keeping ΔH° unchanged. We assume that ΔH° is defined by the intrinsic primary sequence of the enzyme and, thus, kept constant when the enzyme is wrapped with the polymer. Thus, wrapping the enzyme in a polymer matrix will reduce ΔS° by restricting the motions of the denatured polypeptide chain, and hence raising the ΔG° when ΔH° is kept constant. The changes in ΔH° due to enzyme wrapping with a synthetic polymer are thought to be too small to be of considerable value, but they may play an important role as well. The above assumptions and arguments are yet to be fully demonstrated,

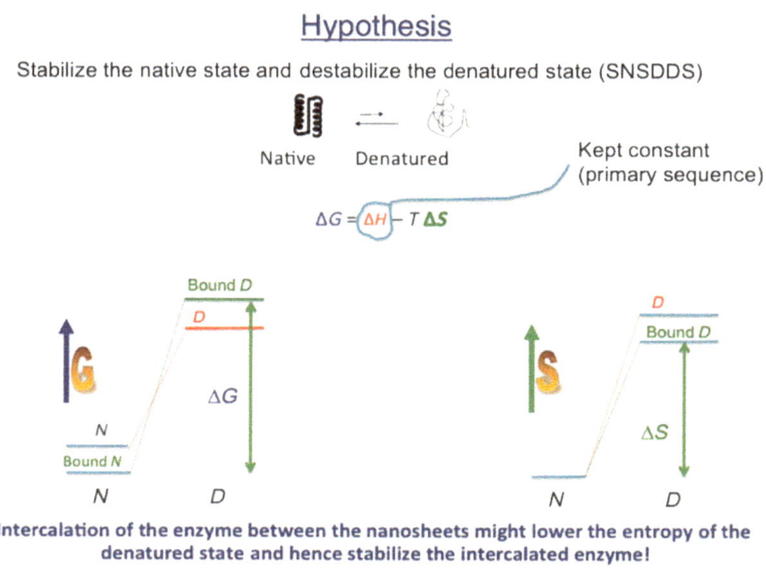

Scheme 1 Stabilization of the native state or destabilization of the denatured state, or both, could thermodynamically favor the native state of the enzyme. The Gibbs free energy and entropy changes for stabilized enzyme are shown.

but these fundamental thermodynamic principles provided inspirations to find rational methods of armoring enzymes with synthetic polymers (Lawrence et al., 2014, 2016).

The polymer–armored enzymes are thought to be suitable for applications in biofuel cells, photovoltaics, biocatalysis, and sensing. With these goals, a number of enzymes were conjugated or adsorbed to synthetic polymers, DNA or 2D materials such as graphene and graphene oxide (GO) or α-zirconium phosphate for stabilization, as the armor. In this chapter, we describe methods to prepare and characterize two enzymes conjugated simultaneously but randomly with polyacrylic acid (PAA) and the triad is physically assembled on GO nanosheets resulting in a polymer-conjugated hybrid nanocatalyst. These types of hybrid materials open up new possibilities for stabilizing enzymes for practical applications. We will first describe the synthesis and characterization of two enzymes (bienzyme) conjugated to PAA and then its assembly on GO followed by characterization of the bienzyme–PAA–GO hybrid conjugates. Biocatalytic activities of these hybrid conjugates and corresponding controls are also described.

1.2 Synthesis of Bienzyme–Polymer Conjugates

This chapter focuses on synthesis of bienzyme–polymer conjugates and hybrid materials for biocatalysis applications. For this purpose, glucose oxidase (GOx) and horseradish peroxidase (HRP) were chosen as model enzymes because they are widely used with applications in food and baking industries, in wastewater treatments, in biosensing, and in ELISA bioassays (Dalal & Gupta, 2007; Ghimire et al., 2015; Lavery et al., 2010; Munoz-Munoz et al., 2007; Naar et al., 2002; Song, Qu, Zhao, Ren, & Qu, 2010; Wong, Wong, & Chen, 2008). These two enzymes are often used together in assays for the detection of glucose or other substrates, and hence, armoring the two enzymes in one step would be of significant interest. Also, GOx and HRP are unstable at elevated temperatures (>50–70°C) and they have very narrow pH range (4.5–6) for activities (Gouda, Singh, Rao, Thakur, & Karanth, 2003; He et al., 2009; Jo, Lee, & Kim, 2009; Wong et al., 2008). Various methods, therefore, have been used to stabilize these enzymes, which include enzyme–polymer nanoparticle synthesis, use of various additives/stabilizers (Eremin, Budnikova, Sviridov, & Metelitsa, 2002; Paz-Alfaro, Ruiz-Granados, Uribe-Carvajal, & Sampedro, 2009), or intercalation in layered inorganic solids (Kumar & Chaudhari, 2000). Recent reports of enzyme–polymer conjugates from our lab demonstrated the enhanced stability of enzymes in the presence of proteases, inhibitors, high temperatures (85–90°C), and long-term storage (10 weeks) (Jeykumari & Narayanan, 2008; Riccardi et al., 2014; Rocha-Martín, Rivas, Muñoz, Guisán, & López-Gallego, 2012). However, for applications in biofuel cells, biobatteries, and bioassays, multiple enzymes are needed in a single reactor or in a cascade reaction where multiple reactions can occur within the same reactor and thus, stabilization of multiple enzymes in the same matrix becomes essential. Therefore, we describe the conjugation of the amine groups of these two enzymes with the carboxyl groups of PAA by using a water-soluble condensing agent such as a carbodiimide (Scheme 2), which is tested for enhanced stability while retaining full activity (Riccardi et al., 2014; Zore et al., 2014, 2015). Here, two enzymes are simultaneously conjugated to PAA in a random fashion such that each PAA chain may have one or more enzymes attached to it and then the conjugate is further stabilized by adsorption to GO. Specific procedures for the synthesis of these bienzyme–PAA conjugates are given under Section 3.2.

1.3 Adsorption of the Bienzyme–Polymer Onto GO

A lot of attention has been paid to polymer–GO composite materials due to their applicability in various fields including biobattery and biofuel cell applications (Gao et al., 2011; Potts, Dreyer, Bielawski, & Ruoff, 2011; Shim et al., 2014). GO interacts with different polymers by functional groups present on its basal plane (Vickery, Patil, & Mann, 2009). Due to these interactions, the composite materials exhibit superior mechanical, optical, and electronic properties when compared to those of the polymer or GO. The presence of functional groups (–OH, –COOH, etc.) on GO can also be used to link it to polymers by covalent bonds. These interactions can be further enhanced by π-stacking between hydrophobic regions on GO and the polymer motifs.

One step further, GO is used to enhance the properties of enzymes conjugated to polymers in biological or nonbiological environments. In one example, a protein–polymer–graphene composite was used. In this case, GO is covalently linked to PEG and then proteins are assembled and delivered to cells. In other instances, hybrid drug delivery systems were obtained, where the protein cargo is not released at low pH conditions in the stomach, but released at higher pH conditions in the intestines (Kavitha, Kang, & Park, 2013; Shen et al., 2012).

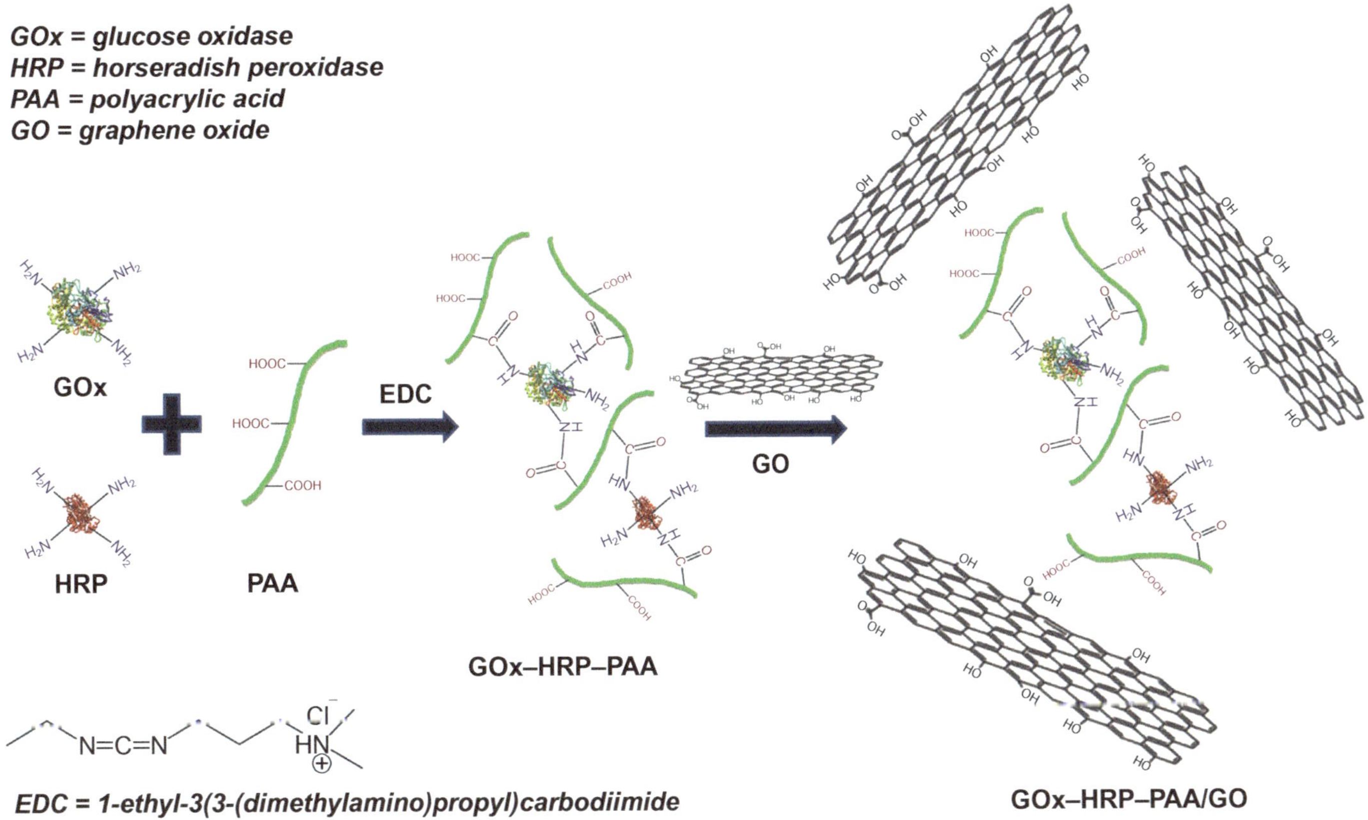

Scheme 2 Synthesis of GOx–HRP–PAA/GO conjugate using GOx and HRP bienzyme and PAA and GO. *Reprinted with permission from Zore, O. V., Pattammattel, A., Gnanaguru, S., Kumar, C. V., & Kasi, R. M. (2015). Bienzyme–polymer–graphene oxide quaternary hybrid biocatalysts: Efficient substrate channeling under chemically and thermally denaturing conditions.* ACS Catalysis, 5, *4979–4988. Copyright (2015)* American Chemical Society.

In this report, we describe how the enzyme–polymer conjugates were assembled onto GO sheets to further stabilize the enzymes against denaturing conditions while retaining full activity (Scheme 2). The enzyme is wrapped in the polymer and then loaded onto the GO nanosheets, protecting the encased enzyme while providing clear access to the active site of the enzyme by the substrates. The details of the synthesis and characterization are presented in the next few sections. Our hypothesis is that the combination of wrapping the enzyme with the polymer chain and then adsorption of the conjugate onto the layers of GO will super enhance the stability of the biocatalyst, which might be able to function even under harsh conditions when ordinary, unprotected enzyme denatures. The data presented here fully support this hypothesis, but further work is required to understand the relative contributions of the effectiveness of each of the components in armoring the enzyme.

2. METHODS: SYNTHESIS OF BIENZYME–POLYMER CONJUGATES

Conjugation of the two enzymes to PAA was carried out using GOx and HRP as enzymes and PAA as a model polymer. The bienzyme–polymer conjugates are named as GOx–HRP–PAA and were synthesized by covalent conjugation of the lysine amine groups of GOx and HRP with the carboxyl groups of PAA by using well-established carbodiimide chemistry. Single enzyme conjugates were synthesized as controls (GOx–PAA). A detailed synthesis protocol is given below.

2.1 Materials

2.1.1 Equipment

1. Safety equipment such as lab coat, safety goggles, and gloves
2. 15 mL glass vials
3. UV spectrophotometer
4. 100 μL and 1000 μL pipettes

2.1.2 Reagents

1. PAA (M_V = 450,000 g/mol) (Sigma-Aldrich)
2. GOx (Sigma-Aldrich)

3. 1-Ethyl-3-(3-dimethylaminopropyl)carbodiimide (EDC) Sigma-Aldrich (St. Louis, MO)
4. Agarose (Molecular Biology Hoefer Inc., Allison, MA)
5. HRP (Calzyme Laboratories Inc., San Luis Obispo, CA)
6. Gel electrophoresis apparatus
7. Electrophoresis running buffer (40 m*M* Tris acetate)

2.2 Synthesis of Bienzyme–Polymer Conjugates

Synthesis of bienzyme–polymer conjugates was carried out as described:

1. Prepare stock solution of PAA by dissolving 200 mg of PAA in 8 mL of DI water and dial pH to 7 using 6 *M* NaOH. The solution is made up to 10 mL resulting in 2% (w/v) solution of PAA.
2. Prepare stock solution of GOx by dissolving the enzyme in 10 m*M* phosphate buffer (PB) (pH 7.4) to make 15 mg/mL solution.
3. Dissolve HRP in PB to make 6 mg/mL solution.
4. Prepare 130 mg/mL EDC solution. This solution gets deactivated within half hour and should be used up as soon as possible.
5. Add 2.8 mL of 2% (w/v) PAA in PB to 1.7 mL EDC solution such that the molar ratio of carboxylic acid groups (–COOH) from PAA and EDC was maintained at 1:1.5.
6. To this solution add 800 μL of GOx stock solution followed by 540 μL of HRP stock solution in a dropwise manner.
7. The solution is stirred for 6 h.
8. This solution is dialyzed using 25 kDa dialysis membrane in pH 7.4 10 m*M* PB to remove any unreacted EDC and EDC–urea by-product.
9. Similar to bienzyme–polymer conjugate, single enzyme–polymer conjugates can also be synthesized. To prepare single enzyme–polymer conjugates, for example, GOx–PAA, HRP is excluded and the conjugate is made from PAA, GOx, and EDC stock solutions.

   Tip: Prepare all the solutions fresh at the time of synthesis.

The conjugation of enzyme(s) to PAA is achieved through the reaction of the lysine $-NH_2$ from enzyme and –COOH from PAA. Table 1 shows the number of lysines present in each of these enzymes, their molecular weights and the total number of residues. The synthesis of bienzyme and single enzyme polymer conjugates is confirmed by agarose gel, zeta potential, and transmission electron microscopy (TEM) studies. The characterization methodology is presented in the following sections.

Table 1 Key Properties of Glucose Oxidase (GOx) and Horseradish Peroxidase (HRP)

	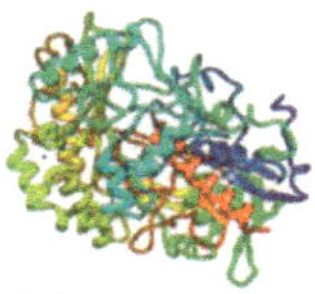	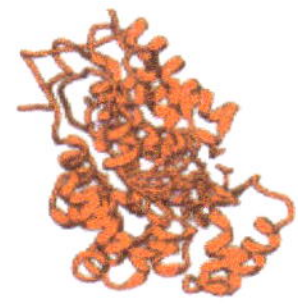
Enzyme	**GOx**	**HRP**
Molecular weight (kDa)	160	40
Total number of residues	1166	306
Number of lysine residues	30	6
p*I*	4.2 (Wong et al., 2008)	6.4 (Spadiut, Rossetti, Dietzsch, & Herwig, 2012)
Substrate	Glucose	Hydrogen peroxide
K_M (m*M*)	1.34 (Shin, Youn, Han, Kang, & Hah, 1993)	5 (Gilabert et al., 2004)
k_{cat} (s^{-1})	16×10^{-2} (Witt, Wohlfahrt, Schomburg, Hecht, & Kalisz, 2000)	90 (Gilabert et al., 2004)

3. METHODS: SYNTHESIS OF BIENZYME–POLYMER–GRAPHENE OXIDE (GOx–HRP–PAA/GO) HYBRID MATERIALS

3.1 Materials

3.1.1 Equipment

1. Safety equipment such as lab coat, safety goggles, and gloves
2. 15 mL glass vials
3. UV spectrophotometer
4. 100 μL and 1000 μL pipettes
5. Beakers, conical flasks, distillation apparatus

3.1.2 Reagents

1. Graphite (Sigma-Aldrich)
2. H_2SO_4 (Sigma-Aldrich)

3. $KMnO_4$ (Sigma-Aldrich)
4. 30% H_2O_2 (Sigma-Aldrich)
5. HCl (Sigma-Aldrich)

3.2 Synthesis of Bienzyme–Polymer GO Hybrid Materials

Synthesis of conjugate hybrid materials was carried out by first preparing GO solution from graphite. Synthesis of GO is described later:

1. Disperse 1 g of graphite in 30 mL of conc. H_2SO_4 and place this mixture in an ice bath.
2. Slowly add 3 g of $KMnO_4$ to the above mixture in small portions with stirring.
3. Heat the mixture to 50°C for 3 h, followed by addition of 70 mL of water to quench the reaction.
4. After 15 min, 300 mL of additional water is added along with 20 mL of 30% H_2O_2 and the solution turns bright yellow.
5. Wash this mixture with 10% HCl for three times followed by water until neutral pH is attained.
6. Sonicate the GO obtained in the previous step in 10 m*M* pH 7.4 PB for 45 min in bath sonicator.
7. Centrifuge the mixture at 4000 rpm for 1 h to remove any unexfoliated graphene.
8. The GO obtained was characterized using Raman spectroscopy, XRD, and zeta potential studies as reported (Pattammattel et al., 2013).

GO obtained from above is used to make hybrid bienzyme–polymer graphene oxide (bienzyme–PAA/GO) biocatalysts. Bienzyme–PAA conjugate made from Section 2.2 is used for the synthesis of bienzyme–PAA/GO composite. The synthetic protocol is as follows:

1. GO dispersions with concentration of 0.7 mg/mL is used here.
2. This dispersion of GO is added dropwise to GOx–HRP–PAA conjugate (refer to Section 2.2 for synthesis of conjugates) such that the weight ratio of total enzyme to GO is 1:1.5 and 1:2.
3. The conjugates after adsorption onto GO are named as GOx–HRP–PAA/GO (1:1.5) and GOx–HRP–PAA/GO (1:2).
4. Similarly, use single enzyme–polymer conjugate (GOx–PAA), which is physically adsorbed onto GO to make GOx–PAA/GO (1:1.5) and GOx–PAA/GO (1:2).
5. Mass percent and ratio of enzyme, polymer to GO used for synthesis of samples is presented in Table 3.

Tip: GO solution is to be added to protein–polymer conjugates dropwise.

Bienzyme and single enzyme–polymer conjugates and conjugate hybrid materials were characterized using agarose gel, zeta potential, and TEM studies. These data are presented later.

4. CHARACTERIZATION

4.1 Agarose Gel Electrophoresis

4.1.1 Equipment

1. Horizontal gel electrophoresis apparatus by Gibco Model 200

4.1.2 Reagents

1. Agarose (Sigma-Aldrich)
2. Glycerol (Sigma-Aldrich)
3. Bromophenol blue (Sigma-Aldrich)
4. Coomassie Blue (Sigma-Aldrich)

Conjugation of enzymes to polymer was confirmed using agarose gel electrophoresis. Following experimental protocol was used:

1. Horizontal gel electrophoresis apparatus by Gibco Model 200, Life Technologies Inc., Grand Island NY apparatus was used.
2. Make 0.5% (w/v) agarose in 40 m*M* pH 6 Tris acetate buffer.
3. Microwave the above solution and leave in the apparatus to gel.
4. Make samples for loading into the wells by mixing 20 μL each of sample and loading buffer.
5. Loading buffer constituted 50% (v/v) glycerol and 0.01% (m/v) bromophenol blue.
6. Load the samples in wells and carry out electrophoresis for 40 min at 100 mV.
7. Stain the gel using 10% (v/v) acetic acid and 0.02% (m/v) Coomassie Blue overnight.
8. Destain the gel using 10% (v/v) acetic acid overnight, photograph it, and report.
9. Fig. 1 shows the agarose gel electrophoresis data for the synthesized samples.

4.2 Zeta Potential

Zeta potential studies were carried out to determine electrophoretic mobility of the conjugates.

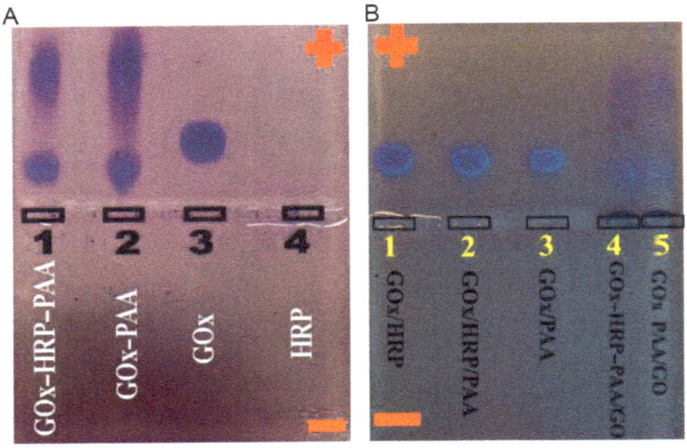

Fig. 1 (A) Agarose gel electrophoresis with GOx–HRP–PAA loaded in lane 1, GOx–PAA in lane 2, GOx in lane 3, and HRP in lane 4. HRP did not get stained well. (B) Agarose gel electrophoresis of GOx/HRP (lane 1), GOx/HRP/PAA (lane 2), GOx/PAA (lane 3), GOx–HRP–PAA/GO (1:2) (lane 4), and GOx–PAA/GO (1:2) (lane 5). GOx/HRP/PAA and GOx/PAA are physical mixtures, synthesized without EDC and lane 4 contains single and bienzyme conjugate hybrids. Agarose gel was done at pH 6 in 40 m*M* Tris acetate buffer. *Reprinted with permission from Zore, O. V., Pattammattel, A., Gnanaguru, S., Kumar, C. V., & Kasi, R. M. (2015). Bienzyme–polymer–graphene oxide quaternary hybrid biocatalysts: Efficient substrate channeling under chemically and thermally denaturing conditions.* ACS Catalysis, *5, 4979–4988. Copyright (2015) American Chemical Society.*

4.2.1 Equipment

1. Brookhaven zeta plus zeta potential analyzer

4.2.2 Materials

1. Polystyrene cuvettes (1 cm path length)

4.2.3 Procedure

1. Use Brookhaven zeta plus zeta potential analyzer (Brookhaven Instruments Corporation, Holtsville, NY) to estimate the zeta potentials of samples.
2. Take 1.5 mL of each sample, equivalent of 1.5 mL of GOx and HRP.
3. Fill the 3 mL of polystyrene cuvettes with 1.5 mL of the sample.
4. Use Smoluchowski fit by the software and matching electrophoretic mobility technique to calculate zeta potential in mV. Fig. 2 shows zeta potentials of samples.

4.3 Transmission Electron Microscopy

TEM studies were carried out to understand the morphology of the conjugates and hybrid materials.

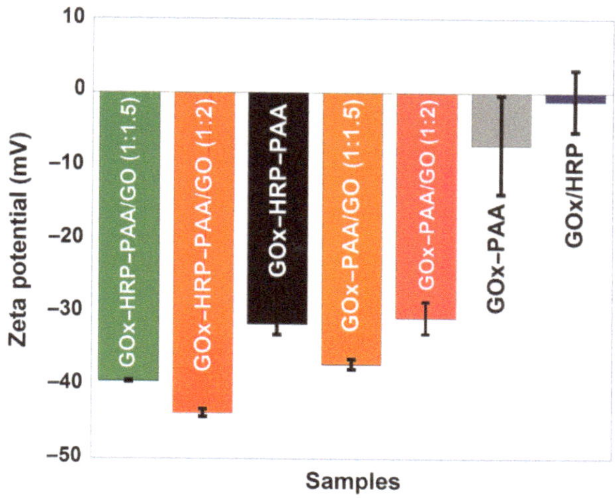

Fig. 2 Zeta potential of GOx–HRP–PAA/GO (1:1.5) (*green*), GOx–HRP–PAA/GO (1:2) (*red*), GOx–HRP–PAA (*black*), GOx–PAA/GO (1:1.5) (*brown*), GOx–PAA/GO (1:2) (*pink*), GOx–PAA (*gray*), and GOx/HRP (*blue*) were done at 25°C. All measurements were performed in 10 m*M* sodium phosphate buffer at pH 7.0. *Reprinted with permission from Zore, O. V., Pattammattel, A., Gnanaguru, S., Kumar, C. V., & Kasi, R. M. (2015). Bienzyme–polymer–graphene oxide quaternary hybrid biocatalysts: Efficient substrate channeling under chemically and thermally denaturing conditions.* ACS Catalysis, 5, *4979–4988. Copyright (2015) American Chemical Society.*

4.3.1 Equipment

1. Tecnai T12 TEM

4.3.2 Materials and Reagents

1. TEM copper grid (Ted Pella).
2. Uranyl acetate (Sigma-Aldrich).
3. Use Tecnai T12 TEM instrument.
4. Operate the Tecnai T12 instrument at accelerating voltage of 120 kV.
5. Prepare aqueous samples of GOx and HRP such that the concentration is ~1 n*M*.
6. Use a micropipette to dropcast 5 μL of the sample on the copper grid covered with a Formvar film.
7. Blot the excess solution using filter paper and air dry.
8. Stain the sample with 5 μL of 0.5 wt% of uranyl acetate.
9. Dry the sample and image using Tecnai T12 TEM.
10. TEM images are shown in Fig. 3.

Safety: Dispose of uranyl acetate according to EHS standards for disposal for radioactive waste.

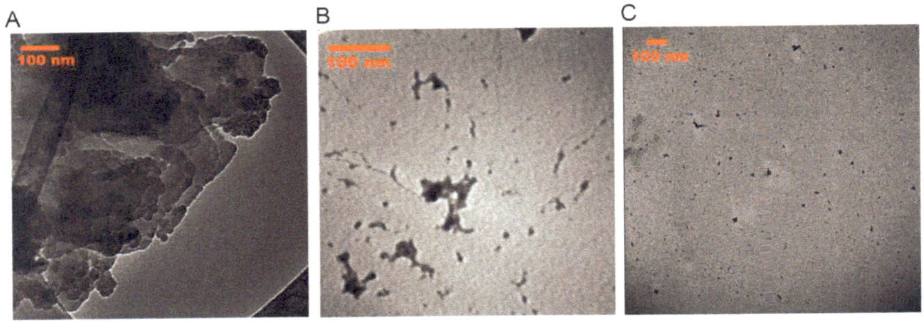

Fig. 3 TEM micrographs of (A) GOx–HRP–PAA/GO (1:2), (B) GOx–HRP–PAA, and (C) GOx/HRP. All samples were stained with 0.5% (w/w) uranyl acetate. GOx–HRP–PAA/GO (1:2) showed nanogels assembled on GO sheets, GOx–HRP–PAA showed nanogels, and GOx/HRP showed aggregated particles. *Reprinted with permission from Zore, O. V., Pattammattel, A., Gnanaguru, S., Kumar, C. V., & Kasi, R. M. (2015). Bienzyme–polymer–graphene oxide quaternary hybrid biocatalysts: Efficient substrate channeling under chemically and thermally denaturing conditions.* ACS Catalysis, 5, *4979–4988. Copyright (2015) American Chemical Society.*

4.4 Circular Dichroism Spectroscopy

Circular dichroism (CD) studies were carried out to determine enzyme structure retention before and after conjugation and after synthesis of hybrid materials. Following protocol was used.

4.4.1 Equipment

1. JASCO J710 spectropolarimeter

4.4.2 Materials

1. CD cuvettes (0.05 cm path length)

4.4.3 Procedure

1. Use JASCO J710 spectropolarimeter for measuring optical properties of polarized light when interacting with enzymes.
2. Dilute all samples to 1 n*M* of equivalent total enzyme concentration.
3. Load 0.5 mL sample into the quartz cuvette.
4. Record the CD spectra in the 195–250 nm wavelength.
5. Normalize the spectra to 1 cm path length and 1 μ*M* of total enzyme concentration.
6. Reference ellipticity at 222 nm for unmodified bienzyme at 100% and calculate the % ellipticity for all samples based on that.

4.4.4 Results

CD spectra and % ellipticity retention of all the samples are shown in Fig. 4. Successful synthesis and retention of secondary structure of enzyme–polymer conjugates and hybrid materials were tested and confirmed using agarose gel electrophoresis, zeta potential, TEM, and CD studies. The streaking observed in case of bienzyme and single enzyme conjugates and conjugate hybrids in the agarose gel electrophoresis were due to polydisperse nature of the samples (Fig. 1). Commercial PAA has a polydispersity index of 1.5–2.0. The conjugates are synthesized by randomly linking carboxylic acid groups in PAA with amino groups on enzymes catalyzed by EDC. This results in a distribution of polymer chains with one or more proteins that are covalently linked and further broadens the PDI of the protein–polymer conjugates. These observations from agarose gel electrophoresis were confirmed using TEM studies, which showed discrete particles for GOx/HRP (Fig. 3C) and nanogels of GOx–HRP–PAA (Fig. 3B) adsorbed onto GO sheets for GOx–HRP–PAA/GO (1:2) (Fig. 3A).

Zeta potential was used to gauge the electrophoretic mobilities of the samples and measure the electrostatic charges on PAA and GO surface planes. Zeta potential of single and bienzyme conjugates and conjugate

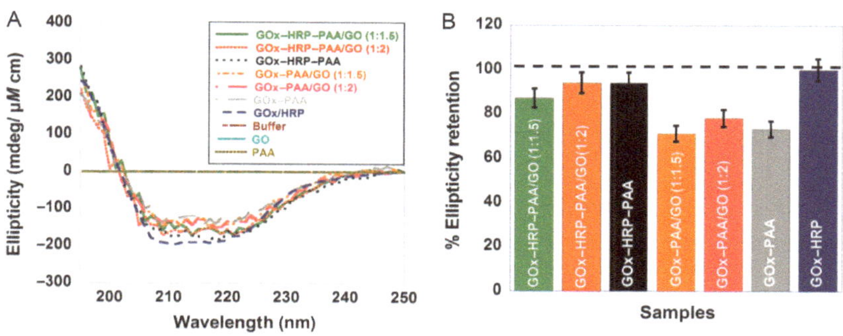

Fig. 4 Circular dichroism (CD) spectra (A) and % ellipticity retention (B) of GOx–HRP–PAA/GO (1:1.5) (*green*), GOx–HRP–PAA/GO (1:2) (*red*), GOx–HRP–PAA (*black*), GOx–PAA/GO (1:1.5) (*orange*), GOx–PAA/GO (1:2) (*purple*), GOx–PAA (*gray*), GOx–HRP (*blue*), buffer (*maroon*), GO (*light blue*), and PAA (*dark brown*). CD spectra were monitored from 190 to 250 nm in PB pH 7.4, 10 m*M*. For CD experiments, protein concentration of 1 μ*M*, for each sample, and 0.05 cm cuvette were used. Ellipticity at 222 nm in mdeg/μ*M* cm is used to calculate % ellipticity retention and ellipticity at 222 nm of GOx/HRP is referenced at 100. *Reprinted with permission from Zore, O. V., Pattammattel, A., Gnanaguru, S., Kumar, C. V., & Kasi, R. M. (2015). Bienzyme–polymer–graphene oxide quaternary hybrid biocatalysts: Efficient substrate channeling under chemically and thermally denaturing conditions.* ACS Catalysis, 5, *4979–4988. Copyright (2015) American Chemical Society.*

hybrids was dictated by PAA and GO because of their larger size when compared to single enzyme. Therefore, modification of enzyme by PAA and consecutively by GO resulted in increased negative zeta potential. Although the interaction of enzyme conjugates and GO did not result in a big shift of surface potential, it indicated that electrostatic interactions are less dominant at this interface. In this case, a possible driving force is the hydrogen bonding between carboxyl/amide groups of enzymes–PAA conjugates and oxygen-based functional groups on GO.

CD studies were carried out to test the effect of polymer encapsulation and GO adsorption on structure on enzyme(s) since the 3D structure of enzyme(s) is important for the biocatalytic properties. PAA covalent conjugation and subsequent noncovalent interaction with GO resulted in as high as 30% decrease in molar ellipticity. This could be attributed to interactions of PAA and GO with enzymes. Despite the reduction in molar ellipticity, PAA matrix and GO may be able to protect the enzyme from external stresses and help in retention of the activity. This hypothesis is tested by colorimetric assays.

4.5 Activity Studies

Catalytic activities are computed using colorimetric assays pertinent to GOx and HRP, when present in the bienzyme conjugate or within conjugate hybrids. Similar catalytic assays for GOx or HRP are also determined using single enzyme conjugates. To carry out activity studies, glucose is used as a substrate. GOx catalyzes glucose to gluconolactone and H_2O_2. In a cascade catalysis, HRP catalyzes oxidation of *o*-methoxyphenol in the presence of H_2O_2 to make dimer of *o*-methoxyphenol that is tracked using UV spectrophotometry at 470 nm. Activity studies were referenced using the initial activity of unmodified GOx/HRP at 100% at 25°C in pH 7.4 PB. The biological activities are also tested at nonambient conditions of pH (2.5–7.4), temperature (25–65°C), and denaturant (4 m*M* SDS). Activity procedures are described later (Scheme 3).

4.5.1 Equipment

1. UV spectrometer

4.5.2 Materials

1. UV cuvettes (1 cm path length)
2. Magnetic stir bar

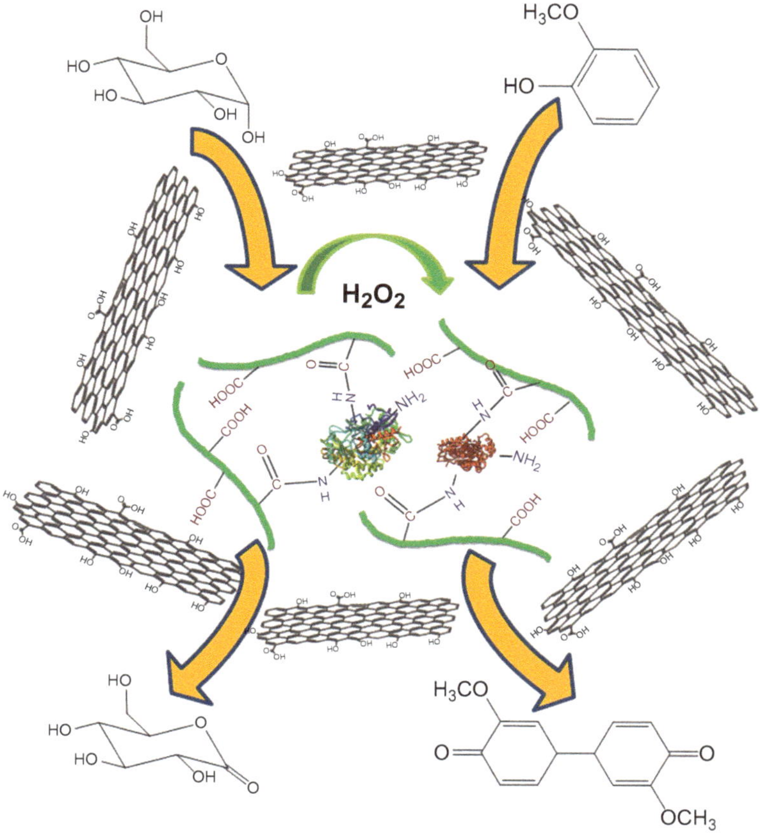

Scheme 3 Cascade catalysis conversion of glucose to gluconolactone and H_2O_2 using GOx as a catalyst. HRP catalyzes the conversion of *o*-methoxyphenol to dimer (using H_2O_2 as an oxidant), which is colorimetrically monitored. *Reprinted with permission from Zore, O. V., Pattammattel, A., Gnanaguru, S., Kumar, C. V., & Kasi, R. M. (2015). Bienzyme–polymer–graphene oxide quaternary hybrid biocatalysts: Efficient substrate channeling under chemically and thermally denaturing conditions.* ACS Catalysis, 5, *4979–4988. Copyright (2015) American Chemical Society.*

4.5.3 Activity Studies at 25°C at Multiple pH Values

1. Use UV spectrophotometer (Agilent Technologies, Santa Clara, CA) for measuring activity of the samples using known assays.
2. Samples are synthesized such that unmodified bienzyme, bienzyme conjugates, and bienzyme conjugate hybrids have enzyme concentration(s) of ~6 μ*M*.
3. Stock solutions of glucose (25 m*M*) and *o*-methoxyphenol (75 m*M*) in PB are prepared.

4. For bienzyme containing conjugates and hybrids, samples (0.5 μ*M*), 1.125 m*M* *o*-methoxyphenol, and appropriate buffer (10 m*M* each of pH 2.5 glycine/HCl, pH 4 citrate, pH 5.5 acetate, and pH 7.4 PB) were mixed.
5. HRP–PAA (0.5 μ*M*) was added to the assay mixture at the time of performing activity studies.
6. 0.3 m*M* glucose is added to the above solution and kinetic traces of the product formation are monitored at 470 nm using UV spectrophotometer.
7. Initial 10–20 s is used to calculate the initial rate of the product formation.

4.5.4 Activity Studies at 65°C and in the Presence of a Denaturant (4 mM SDS)

1. Protocol similar to Section 4.5.1 is used to carry out activity studies at 65°C or in the presence of 4 m*M* SDS.
2. Follow steps from 1 to 5 from the previous section.
3. Incubate the samples at 65°C or in the presence of 4 m*M* SDS for 10–20 min before performing the activity studies in appropriate buffer (10 m*M* each of pH 2.5 glycine/HCl, pH 4 citrate, pH 5.5 acetate, and pH 7.4 PB).
4. 0.3 m*M* glucose is added and kinetic traces of the product formation are monitored at 470 nm using UV spectrophotometer.
5. Use initial 10–20 s to calculate the rate of the product formation.

   Tip: Aqueous solution of *o*-methoxyphenol is prepared by sonication.

Fig. 5 presents the average for triplicates of kinetic traces of all the samples at 25°C in pH 7.4, 10 m*M* PB, and Fig. 6 shows the biocatalytic activity comparison for all the samples at various conditions. Polymer encapsulation resulted in activity retention from pH 2.5 to 7.4. Similar results are obtained when bienzyme conjugate is adsorbed onto GO, GOx–HRP–PAA/GO samples showed superior activity retention under harsh conditions of pH (2.5–4) and in the presence of 4 m*M* SDS. Under these conditions, unmodified bienzyme showed little or no biocatalytic activity. At high temperature of 65°C, bienzyme conjugate hybrids performed better compared to single enzyme conjugate hybrids and unmodified bienzyme. This activity retention can be attributed to hydrophilic polymer encapsulation, which allowed retention of water inside the enzyme microenvironment. Also, enhancement of activity above 100% is attributed to substrate channeling of H_2O_2 inside GO matrix (Zore et al., 2015).

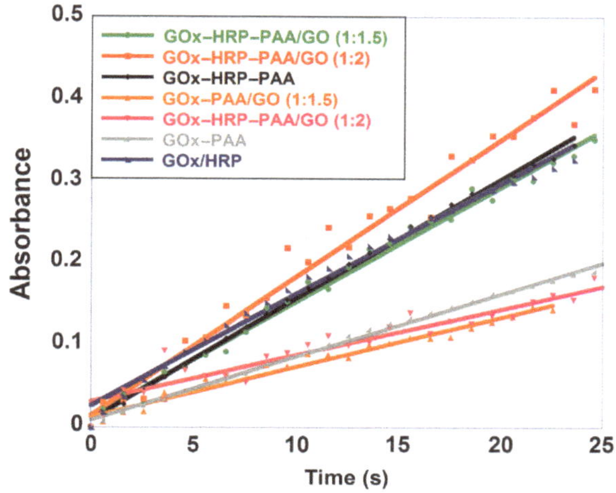

Fig. 5 Kinetic traces of GOx–HRP–PAA/GO (1:1.5) (*green*), GOx–HRP–PAA/GO (1:2) (*red*), GOx–HRP–PAA (*black*), GOx–PAA/GO (1:1.5) (*orange*), GOx–PAA/GO (1:2) (*pink*), GOx–PAA (*gray*), and GOx/HRP (*blue*) in 10 m*M*, pH 7.4 PB. Activity was established using 1.125 m*M* *o*-methoxyphenol, 0.3 m*M* glucose, and 1 μ*M* equivalent of each of GOx and HRP equivalents. *Reprinted with permission from Zore, O. V., Pattammattel, A., Gnanaguru, S., Kumar, C. V., & Kasi, R. M. (2015). Bienzyme–polymer–graphene oxide quaternary hybrid biocatalysts: Efficient substrate channeling under chemically and thermally denaturing conditions.* ACS Catalysis, 5, *4979–4988. Copyright (2015) American Chemical Society.*

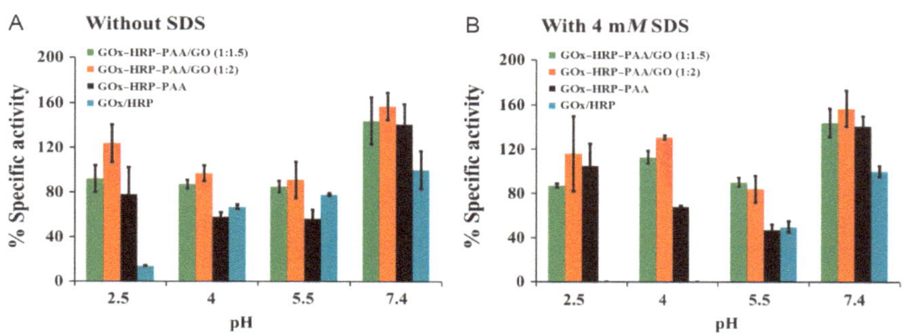

Fig. 6 (A) Activity study as a function of increasing pH. (B) Activity study at increasing pHs in the presence of 4.0 m*M* SDS (denaturant)—GOx–HRP–PAA/GO (1:1.5) (*green*), GOx–HRP–PAA/GO (1:2) (*red*), GOx–HRP–PAA (*black*), and GOx/HRP (*blue*). Initial activity of each sample (20–40 s) was measured and compared to GOx/HRP, which was referenced to 100%. *Reprinted with permission from Zore, O. V., Pattammattel, A., Gnanaguru, S., Kumar, C. V., & Kasi, R. M. (2015). Bienzyme–polymer–graphene oxide quaternary hybrid biocatalysts: Efficient substrate channeling under chemically and thermally denaturing conditions.* ACS Catalysis, 5, *4979–4988. Copyright (2015) American Chemical Society.*

5. CONCLUSIONS

A method to synthesize bienzyme–PAA conjugates that are self-assembled onto GO is presented here. These novel bienzyme–PAA–GO conjugate hybrids present impressive enzyme stabilization and catalytic activity retention under thermally and chemically challenging conditions unlike native single enzyme or bienzyme mixtures (Table 2).

In this chapter, we explained, systematically, a method to synthesize bienzyme–polymer conjugate and bienzyme–polymer conjugate hybrids using GOx, HRP, PAA, and GO. Also, the characterization methods such as agarose gel electrophoresis, CD, zeta potential, and TEM are described to confirm the structure and morphology of the conjugates. UV spectrophotometry was used to test the biological activities of the conjugates. Covalent conjugation of GOx and HRP to PAA and subsequent physical interaction with GO resulted in hybrid conjugates that showed superior catalytic properties at lower pH (2.5), high temperatures (65°C), and in the presence of denaturant (4 m*M* SDS).

For understanding the effectiveness of our system and for comparison, we computed the degree of substrate channeling (DSC). DSC is calculated by taking the ratio of activity of bienzyme conjugate hybrid to that of bienzyme at specific conditions. The DSC values are compared with relevant and important literature values and the data are presented in Table 3. To the best of our

Table 2 Different Conjugates and the Mass Percent of Enzymes, PAA, and GO Used for Synthesis

	Mass Percent				
Samples	**GOx**	**HRP**	**PAA**	**GO**	**Description**
GOx–HRP–PAA/GO (1:1.5)	13	3.3	58	25	Bienzyme conjugate hybrids
GOx–HRP–PAA/GO (1:2)	12	3.0	54	31	
GOx–HRP–PAA	18	4.5	78	0	Bienzyme conjugate
GOx–PAA/GO (1:1.5)	17	0	58	25	Single enzyme conjugate hybrids
GOx–PAA/GO (1:2)	15	0	54	31	
GOx–PAA	22	0	78	0	Single enzyme conjugate
GOx–HRP	80	20	0	0	Unbound Bienzyme

Reprinted with permission from Zore, O. V., Pattammattel, A., Gnanaguru, S., Kumar, C. V., & Kasi, R. M. (2015). Bienzyme–polymer–graphene oxide quaternary hybrid biocatalysts: Efficient substrate channeling under chemically and thermally denaturing conditions. *ACS Catalysis, 5*, 4979–4988. Copyright (2015) American Chemical Society.

Table 3 Comparison of Substrate Channeling With Current System With Important Literature Studies

No.	Enzymes Used in Multienzyme Systems	Degree of Substrate Channeling	Temperature of the Study	pH of the Study	Conc. of Surfactant (SDS) (m*M*)
1	Krebs cycle enzymes (Moehlenbrock, Toby, Waheed, & Minteer, 2010)	1–2	25	7–8	Not applicable (NA)
2	Ferredoxin and hydrogenase (Agapakis et al., 2010)	1	25	7–8	NA
3	GOx and HRP (Wilner et al., 2009)	20–30	25	7–8	NA
4	Cellulosomes (Fierobe et al., 2005)	2–4	37	6–7	NA
5	Tryptophan synthase (Hyde, Ahmed, Padlan, Miles, & Davies, 1988)	Infinite	25	6–7	NA
6	This work (GOx and HRP)	1.3	25	7.4	4.0
7	This work (GOx and HRP)	Infinite	25	2.5	4.0
8	This work (GOx and HRP)	7.5	65	7.4	NA

Reprinted with permission from Zore, O. V., Pattammattel, A., Gnanaguru, S., Kumar, C. V., & Kasi, R. M. (2015). Bienzyme–polymer–graphene oxide quaternary hybrid biocatalysts: Efficient substrate channeling under chemically and thermally denaturing conditions. *ACS Catalysis*, *5*, 4979–4988. Copyright (2015) American Chemical Society.

information, current system showed superior substrate channeling under harsh conditions of pH, temperature, and 4 m*M* SDS. The cascade biocatalyst platform is environmental friendly, biocompatible or biodegradable and is useful for biofuel cell or biobattery applications.

ACKNOWLEDGMENTS

C.V.K. and R.M.K. acknowledge financial support from the University of Connecticut Research Foundation Research Excellence Program Award 2015. C.V.K. acknowledges NSF EAGER grant (DMR-1441879).

REFERENCES

Agapakis, C. M., Ducat, D. C., Boyle, P. M., Wintermute, E. H., Way, J. C., & Silver, P. A. (2010). Insulation of a synthetic hydrogen metabolism circuit in bacteria. *Journal of Biological Engineering*, *4*, 3–18.

Dalal, S., & Gupta, M. N. (2007). Treatment of phenolic wastewater by horseradish peroxidase immobilized by bioaffinity layering. *Chemosphere*, *67*, 741–747.

Eremin, A. N., Budnikova, L. P., Sviridov, O. V., & Metelitsa, D. I. (2002) Stabilization of diluted aqueous solutions of horseradish peroxidase. *Applied Biochemistry and Microbiology*, *38*, 151–158.

Fersht, A. (1999). *Structure and mechanism in protein science: A guide to enzyme catalysis and protein folding*. New York: W. H. Freeman and Company.

Fierobe, H. P., Mingardon, F., Mechaly, A., Belaich, A., Rincon, M. T., Pages, S., et al. (2005). Action of designer cellulosomes on homogeneous versus complex substrates: Controlled incorporation of three distinct enzymes into a defined trifunctional scaffoldin. *Journal of Biological Chemistry*, *280*, 16325–16334.

Gao, Y., Yip, H.-L., Chen, K.-S., O'Malley, K. M., Acton, O., Sun, Y., et al. (2011). Surface doping of conjugated polymers by graphene oxide and its application for organic electronic devices. *Advanced Materials*, *23*, 1903–1908.

Ghimire, A., Zore, O. V., Thilakarathne, V. K., Briand, V. A., Lenehan, P. J., Lei, Y., et al. (2015). "Stable-on-the-table" biosensors: Hemoglobin-poly (acrylic acid) nanogel Bio-Electrodes with high thermal stability and enhanced electroactivity. *Sensors*, *15*, 23868–23885.

Gilabert, M. A., Fenoll, L. G., Garcia-Molina, F., Tudela, J., Garcia-Canovas, F., & Rodriguez-Lopez, J. N. (2004). Kinetic characterization of phenol and aniline derivates as substrates of peroxidase. *Biological Chemistry*, *385*, 795–800.

Gouda, M. D., Singh, S. A., Rao, A. G. A., Thakur, M. S., & Karanth, N. G. (2003). Thermal inactivation of glucose oxidase. Mechanism and stabilization using additives. *Journal of Biological Chemistry*, *278*, 24324–24333.

He, C., Liu, J., Xie, L., Zhang, Q., Li, C., Gui, D., et al. (2009). Activity and thermal stability improvements of glucose oxidase upon adsorption on core-shell PMMA-BSA nanoparticles. *Langmuir*, *25*, 13456–13460.

Hyde, C. C., Ahmed, S. A., Padlan, E. A., Miles, E. W., & Davies, D. R. (1988). Three-dimensional structure of the tryptophan synthase alpha 2 beta 2 multienzyme complex from Salmonella typhimurium. *Journal of Biological Chemistry*, *263*, 17857–17871.

Jeykumari, D. R. S., & Narayanan, S. S. (2008). Fabrication of bienzyme nanobiocomposite electrode using functionalized carbon nanotubes for biosensing applications. *Biosensors and Bioelectronics*, *23*, 1686–1693.

Jo, S. M., Lee, H. Y., & Kim, J. C. (2009). Glucose-sensitivity of liposomes incorporating conjugates of glucose oxidase and poly(N-isopropylacrylamide-co-methacrylic acid-co-octadecylacrylate). *International Journal of Biological Macromolecules*, *45*, 421–426.

Kavitha, T., Kang, I.-K., & Park, S.-Y. (2013). Poly(acrylic acid)-grafted graphene oxide as an intracellular protein carrier. *Langmuir*, *30*, 402–409.

Kirk, O., Borchert, T. V., & Fuglsang, C. C. (2002). Industrial enzyme applications. *Current Opinion in Biotechnology*, *13*, 345–351.

Kumar, C. V., & Chaudhari, A. (2000). Proteins immobilized at the galleries of layered α-zirconium phosphate: Structure and activity studies. *Journal of the American Chemical Society*, *124*, 830–837.

Lavery, C. B., MacInnis, M. C., MacDonald, M. J., Williams, J. B., Spencer, C. A., Burke, A. A., et al. (2010). Purification of peroxidase from Horseradish (Armoracia rusticana) roots. *Journal of Agricultural and Food Chemistry*, *58*, 8471–8476.

Lawrence, P. B., Billings, W. M., Miller, M. B., Pandey, B. K., Stephens, A. R., Langlois, M. I., et al. (2016). Conjugation strategy strongly impacts the conformational stability of a PEG-protein conjugate. *ACS Chemical Biology*, *11*, 1805–1809.

Lawrence, P. B., Gavrilov, Y., Matthews, S. S., Langlois, M. I., Shental-Bechor, D., Greenblatt, H. M., et al. (2014). Criteria for selecting PEGylation sites on proteins for higher thermodynamic and proteolytic stability. *Journal of the American Chemical Society*, *136*, 17547–17560.

Lele, B. S., Murata, H., Matyjaszewski, K., & Russel, A. J. (2005). Synthesis of uniform protein-polymer conjugates. *Biomacromolecules*, *6*, 3380–3387.

Moehlenbrock, M. J., Toby, T. K., Waheed, A., & Minteer, S. D. (2010). Metabolon catalyzed pyruvate/air biofuel cell. *Journal of the American Chemical Society*, *132*, 6288–6289.

Munoz-Munoz, J. L., Garcia-Molina, F., Varon, R., Rodriguez-Lopez, J. N., Garcia-Canovas, F., & Tudela, J. (2007). Kinetic characterization of the oxidation of esculetin by polyphenol oxidase and peroxidase. *Bioscience, Biotechnology, and Biochemistry*, *71*, 390–396.

Naar, J., Bourdelais, A., Tomas, C., Kubanek, J., Whitney, P. L., Flewelling, L., et al. (2002). A competitive ELISA to detect brevetoxins from Karenia Brevis (formerly Gymnodinium breve) in seawater, shellfish, and mammalian body fluid. *Environmental Health Perspectives*, *110*, 179–185.

Pattammattel, A., Puglia, M., Chakraborty, S., Deshapriya, I. K., Dutta, P. K., & Kumar, C. V. (2013). Tuning the activities and structures of enzymes bound to graphene oxide with a protein glue. *Langmuir*, *29*, 15643–15654.

Paz-Alfaro, K. J., Ruiz-Granados, Y. G., Uribe-Carvajal, S., & Sampedro, J. G. (2009). Trehalose-mediated thermal stabilization of glucose oxidase from Aspergillus niger. *Journal of Biotechnology*, *141*, 130–136.

Potts, J. R., Dreyer, D. R., Bielawski, C. W., & Ruoff, R. S. (2011). Graphene-based polymer nanocomposites. *Polymer*, *52*, 5–25.

Riccardi, C. M., Cole, K. S., Benson, K. R., Ward, J. R., Bassett, K. M., Zhang, Y., et al. (2014). Toward "stable-on-the-table" enzymes: Improving key properties of catalase by covalent conjugation with poly(acrylic acid). *Bioconjugate Chemistry*, *25*, 1501–1510.

Rocha-Martín, J., Rivas, B., Muñoz, R., Guisán, J. M., & López-Gallego, F. (2012). Rational co-immobilization of bi-enzyme cascades on porous supports and their applications in bio-redox reactions with in situ recycling of soluble cofactors. *ChemCatChem*, *4*, 1279–2188.

Shen, H., Liu, M., He, H., Zhang, L., Huang, J., Chong, Y., et al. (2012). PEGylated graphene oxide-mediated protein delivery for cell function regulation. *ACS Applied Materials & Interfaces*, *4*, 6317–6323.

Shim, J., Kim, D.-G., Kim, H. J., Lee, J. H., Baik, J.-H., & Lee, J.-C. (2014). Novel composite polymer electrolytes containing poly(ethylene glycol)-grafted graphene oxide for all-solid-state lithium-ion battery applications. *Journal of Materials Chemistry A*, *2*, 13873–13883.

Shin, K. S., Youn, H. D., Han, Y. H., Kang, S. O., & Hah, Y. C. (1993). Purification and characterisation of D-glucose oxidase from white-rot fungus Pleurotus ostreatus. *European Journal of Biochemistry*, *215*, 747–752.

Song, Y., Qu, K., Zhao, C., Ren, J., & Qu, X. (2010). Graphene oxide: Intrinsic peroxidase catalytic activity and its application to glucose detection. *Advanced Materials*, *22*, 2206–2210.

Spadiut, O., Rossetti, L., Dietzsch, C., & Herwig, C. (2012). Purification of a recombinant plant peroxidase produced in *Pichia pastoris* by a simple 2-step strategy. *Protein Expression and Purification*, *86*, 89–91.

Torchilin, V. P., Maksimenko, A. V., SMirknov, V. N., Berezin, I. V., Klibanov, A. M., & Martinek, K. (1978). The principles of enzyme stabilization. III. The effect of the length of intra-molecular cross-linkages on thermostability of enzymes. *Biochimica et Biophysica Acta*, *552*, 277–283.

Vickery, J. L., Patil, A. J., & Mann, S. (2009). Fabrication of graphene–polymer nanocomposites with higher-order three-dimensional architectures. *Advanced Materials*, *21*, 2180–2184.

Wilner, O. I., Weizmann, Y., Gill, R., Lioubashevski, O., Freeman, R., & Willner, I. (2009). Enzyme cascades activated on topologically programmed DNA scaffolds. *Nature Nanotechnology*, *4*, 249–254.

Witt, S., Wohlfahrt, G., Schomburg, G., Hecht, H. J., & Kalisz, H. M. (2000). Conserved arginine-516 of Penicillium amagasakiense glucose oxidase is essential for the efficient binding of beta-D-glucose. *Biochemistry Journal*, *347*, 553–559.

Wong, C. M., Wong, K. H., & Chen, X. D. (2008). Glucose oxidase: Natural occurrence, function, properties and industrial applications. *Applied Microbiology and Biotechnology*, *78*, 927–938.

Zore, O. V., Lenehan, P. J., Kumar, C. V., & Kasi, R. M. (2014). Efficient biocatalysis in organic media with hemoglobin and poly(acrylic acid) nanogels. *Langmuir*, *30*, 5176–5184.

Zore, O. V., Pattammattel, A., Gnanaguru, S., Kumar, C. V., & Kasi, R. M. (2015). Bienzyme–polymer–graphene oxide quaternary hybrid biocatalysts: Efficient substrate channeling under chemically and thermally denaturing conditions. *ACS Catalysis*, *5*, 4979–4988.

CHAPTER NINE

Approaches for Conjugating Tailor-Made Polymers to Proteins

Matthew Paeth[1], Jacob Stapleton[1], Melissa L. Dougherty[1], Henry Fischesser, Jerry Shepherd, Matthew McCauley, Rebecca Falatach, Richard C. Page[2], Jason A. Berberich[2], Dominik Konkolewicz[2]

Miami University, Oxford, OH, United States

[2]Corresponding authors: e-mail address: pagerc@miamioh.edu; berberj@miamioh.edu; d.konkolewicz@miamioh.edu

Contents

Abstract

A series of methods are outlined for attaching functional polymers to proteins. Polymers with good control over structure, functionality, and composition can be created using reversible addition–fragmentation chain transfer (RAFT) polymerization. These polymers

[1] These authors contributed equally.

Methods in Enzymology, Volume 590
ISSN 0076-6879
http://dx.doi.org/10.1016/bs.mie.2016.12.004

© 2017 Elsevier Inc.
All rights reserved.

can be covalently linked to enzymes and proteins using either the "grafting-to" approach, where a preformed polymer is attached to the protein surface, or the "grafting-from" approach, where the polymer is grown from the protein surface. Methods for grafting-to, or attaching the RAFT chain transfer agent to the protein surface outlined include the commonly used carbodiimide/activated ester (EDC/NHS) coupling. Methods are also outlined to graft-from the surface of the protein using RAFT polymerization. Additionally, it is possible to site specifically introduce a reactive azide group to the protein surface using enzymatic ligation as a posttranslational modification. This reactive azide group can be conjugated to an alkyne-containing polymer using highly efficient click chemistry. These robust protocols can produce protein–polymer conjugates with various architectures and functionalities. Methods are also outlined for characterization of the resulting bioconjugates.

1. INTRODUCTION

1.1 Background Information

Proteins have evolved to perform a very wide range of functions ranging from efficient catalysis of complex reactions to signaling physiological responses (Jencks, 1969; Lu, Xun, & Xie, 1998; Pickart & Fushman, 2004; Schramm, 2011; Zeyda & Stulnig, 2009). Using just 20 amino acids as monomers, proteins form higher order structures that are optimized to a vast variety of applications (Das, Dawson, & Orengo, 2015; Goncearenco & Berezovsky, 2015; Sawada & Honda, 2006). However, the introduction of other functional groups is a target of protein and enzyme science. One approach is to introduce nonnatural amino acids into the protein backbone (de Graaf, Kooijman, Hennink, & Mastrobattista, 2009; Jackson, Duffy, Hess, & Mehl, 2006; Link, Mock, & Tirrell, 2003; Zhang, Otting, & Jackson, 2013). Nonnatural amino acid incorporation has seen significant development in the past decades; however, it can be difficult to incorporate many distinct units of nonnatural amino acids (Zhang, Otting, et al., 2013). Bioconjugates between synthetic polymers and natural proteins offer a solution to this problem. Synthetic polymers can be synthesized with a large variety of functional groups and macromolecular architectures (Hawker & Wooley, 2005). By linking synthetic polymers and proteins, the defined structure and catalytic activity of the protein can be integrated with the flexibility and functionality of macromolecular architecture of the synthetic polymer (Averick, Mehl, Das, & Matyjaszewski, 2015). Additionally, introduction of synthetic polymers to the surface of proteins can stabilize the protein against various challenges including antibody recognition (Veronese, 2001), protease degradation (Cummings, Murata, Koepsel, & Russell, 2014; Falatach, Li, et al., 2015), thermal deactivation (Cummings et al.,

2014; Mancini, Lee, & Maynard, 2012), and chemical denaturation (Lucius et al., 2016). This creates a need to have efficient methods for conjugating polymers and proteins.

In general, a polymer can be attached to a protein using one of two approaches. The first is the "grafting-to" (GT) approach that has been utilized for decades (Averick et al., 2015; Roberts, Bentley, & Harris, 2012). This involves the synthesis of a polymer followed by the attachment of the already synthesized polymer to a reactive group on the protein surface as indicated in the top of Scheme 1 (Averick et al., 2015). Alternatively, "grafting-from" (GF) the protein surface can be achieved by attaching a polymerization initiating group or chain transferring agent to the surface of a protein (Averick et al., 2015; Sumerlin, 2012). Subsequently, the polymer can be grown from the protein surface typically in an aqueous medium (Averick, Simakova, et al., 2012; Boyer et al., 2007; De, Li, Gondi, & Sumerlin, 2008; Pokorski, Breitenkamp, Liepold, Qazi, & Finn, 2011; Sumerlin, 2012). This is indicated at the bottom of Scheme 1. Additionally, hybrid approaches exist where short but extendable polymers are first attached to the protein, followed by chain extension to create longer structures (Danielson et al., 2016; Falatach, McGlone, et al., 2015; Lucius et al., 2016).

Both the GT and GF approaches have distinct advantages and disadvantages. The GT approach has the advantage of allowing the polymer to be synthesized under conditions optimal for its synthesis and with full characterization of the polymer possible before attachment to the protein (Falatach, McGlone, et al., 2015). Additionally, many proteins are either difficult to produce or only stable at very low concentrations. The GT approach can

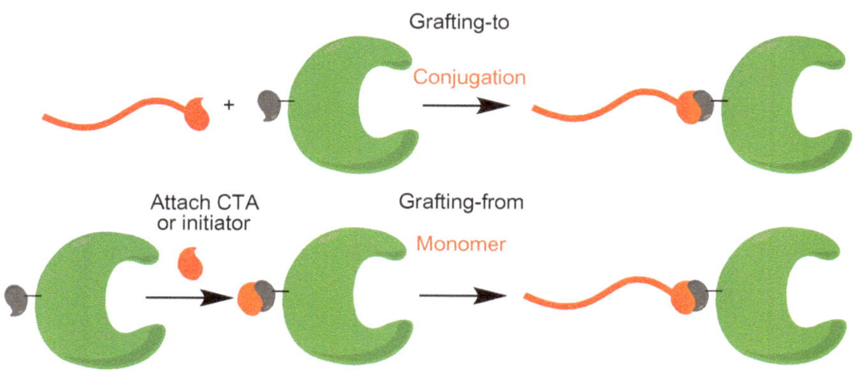

Scheme 1 The grafting-to and grafting-from approaches for protein–polymer conjugation.

promote efficient conjugation by using a large excess of the polymer to be attached to ensure an acceptable rate of conjugation. However, there are two major disadvantages to the GT approach. The first is that the unreacted polymer must be separated from the bioconjugates after reaction (Averick et al., 2015). This separation can be a significant challenge if the polymer and bioconjugate have similar molecular weights (Falatach, McGlone, et al., 2015). The second major challenge with the GT approach is that it is kinetically and thermodynamically challenging to attain high polymer grafting density especially with high-molecular-weight polymers (Averick et al., 2015). In contrast, since only small molecules are typically added to the bioconjugate, the GF approach is able to attain high grafting densities with comparative ease (Averick et al., 2015; Averick, Simakova, et al., 2012). Additionally, after conjugation only the bioconjugates and small molecules should be present, facilitating isolation of a pure bioconjugate (Averick et al., 2015). However, under certain conditions, the initiator-modified protein loses solubility or stability (Falatach, McGlone, et al., 2015). Additionally, it can be challenging to produce polymers with low dispersity by GF, since the reaction conditions are not always optimal for efficient polymerization. The ultimate choice between these approaches must be made on a case-by-case basis. Herein we provide protocols for both GT and GF methods.

General guidelines to consider are that any coupling chemistry utilized should be high yielding to maximize the likelihood of success, especially at the high dilution of functional groups characteristic of many protein solutions (Averick, Simakova, et al., 2012; Cobo, Li, Sumerlin, & Perrier, 2015). For instance, a 20-kDa protein at 1 mg/mL represents a protein concentration of 0.05 m*M*, several orders of magnitude lower than the concentration of reagents used in many organic reactions. Amide bond formation can be performed either through a carbodiimide/activated ester such as the EDC/NHS protocol (Falatach, Li, et al., 2015; Falatach, McGlone, et al., 2015; Fischer, 2010; Lucius et al., 2016) or through organic synthesis of an *N*-hydroxysuccinimide (NHS)-activated ester (Li, Bapat, Li, & Sumerlin, 2011). These amide bond forming reactions are commonly utilized since all proteins harbor an N-terminal amine, in addition to lysine residues present in the amino acid sequence, making this a universally applicable process. The pK_a of the N-terminal tends to be lower by about 1–2 pK_a units compared to the lysine residues, although local affects can alter these values, and steric effects become important determinant of attachment site if the N-terminus or lysine is not surface exposed (Berberich, Yang,

Madura, Bahar, & Russell, 2005; Brinkley, 1992). Other commonly utilized reactions include the highly efficient thiol-based chemistries. These include thiol-disulfide exchange with pyridyl disulfides, thiol-ene, and thiol-Michael additions (Cobo et al., 2015). These reactions are highly efficient, although less universal than the amine-based reactions since not all proteins contain cysteine residues, and those cysteine residues present are often used to form disulfide linkages in the native protein and thus are unavailable for thiol-based polymer attachment. Therefore, within this manuscript, we will highlight methods for polymer conjugation using amine chemistry.

Recently, "click chemistry," especially copper-catalyzed azide–alkyne coupling has received significant interest (Kolb, Finn, & Sharpless, 2001). These reactions proceed to high yield with minimal side products across various solvents and functional groups (Kolb et al., 2001). These properties make "click" reactions excellent candidates for bioconjugation (Lahann, 2009). There has been significant work on incorporating nonnatural amino acids into proteins with precise definition of the location of attachment through genetic code expansion (Averick et al., 2015; Averick, Paredes, et al., 2012; Jackson et al., 2006; Lang & Chin, 2014; Peeler et al., 2010). However, nonnatural amino acid incorporation can lead to poor yields of the nonnatural amino acid-containing protein (Young & Schultz, 2010). Recently, ligation of an azide-containing organic molecule to a lysine in a 13-residue natural amino acid sequence, known as the LAP sequence, has been reported (Liu et al., 2014; Uttamapinant et al., 2010). This ligation is achieved using a lipoid acid ligase mutant that recognizes 10-azidodecanoic acid and attaches this to the lysine residue in the LAP sequence as a posttranslational modification. The LAP sequence may be placed anywhere in the protein backbone, allowing for site-specific modification without the potential drawback of low expression yield, as has been demonstrated for green fluorescent protein (GFP) which was conjugated by poly(ethylene oxide) (Plaks, Falatach, Kastantin, Berberich, & Kaar, 2015). We highlight site-specific methods for conjugating polymers to these proteins using the lipoic acid ligase technology.

1.2 General Considerations in Polymer Synthesis, Conjugation, and Characterization

1.2.1 Safety Considerations

General safe lab practices should be implemented at all times. Monomers and organic compounds (reversible addition–fragmentation chain transfer (RAFT) chain transfer agents (CTAs), atom transfer radical polymerization

(ATRP) initiators, and ligands) are typically toxic and irritants, and all should be handled with care. Radical initiators are unstable when heated and should be handled very carefully. Monomers and radical initiators should be stored in an approved refrigerator or freezer. Where possible, manipulations should be performed in a fume hood to minimize likelihood of inhalation. All waste should be disposed of in a manner that is consistent with local requirements. This typically involves the collection of organic solvents as a separate waste.

1.2.2 Personal Protective Equipment

All methods should be implemented with appropriate personal protective equipment. These include at a minimum, approved eye protection, nitrile gloves, and closed toe shoes. Manipulations in the fume hood are recommended.

1.2.3 Materials Needed

Materials needed for polymerization: All reagents utilized in these polymerizations are available from commercial suppliers. Unless otherwise specified, all reagents are used as received. The following materials are needed for the RAFT or ATRP reactions described here: azobis(isobutyronitrile) (AIBN); 2,2′-Azobis[2-(2-imidazolin-2-yl)propane]dihydrochloride (VA-044); acrylamide (Am); *N*,*N*-dimethylacrylamide (DMAm); oligo(ethylene oxide) methylether acrylate average molecular weight 454 or 480 (OEOA); oligo(ethylene oxide)methylether methacrylate average molecular weight 475 or 500 (OEOMA); acrylic acid (AA); 2-(*N*,*N*-dimethylamino)ethyl methacrylate (DMAEMA); phosphocholine methacrylate (PCMA); carbon disulfide (CS_2); ethanethiol (EtSH); potassium hydroxide (KOH); sodium hydroxide (NaOH); 2-bromopropionic acid (BPA); propargyl alcohol; 1-(3-dimethylaminopropyl)-3-ethylcarbodiimide hydrochloride (EDC); 4-dimethylaminopyridine (DMAP); NHS; copper(II) bromide ($CuBr_2$); tris(2-pyridylmethyl)amine (TPMA); 2-hydroxyethyl bromoisobutyrate (HEBiB); bromoisobutyrylbromide (BiB); hydrochloric acid (12 *M*); sodium carbonate (Na_2CO_3); lithium chloride (LiCl); lithium bromide (LiBr); ethanol (EtOH); methanol (MeOH); acetone; water (H_2O); ice; *N*,*N*-dimethylformamide (DMF); diethylether; dichloromethane (DCM); hexanes.

The following reagents are needed for polymer conjugation and purification: 1-(3-Dimethylaminopropyl)-3-ethylcarbodiimide hydrochloride (EDC); NHS; polymer- or initiator-containing carboxylic acid; protein of interest; water; glycine; ammonium sulfate. Additionally for GF by RAFT, monomers of interest, including Am, DMAm, etc., VA-044, and a protein modified with a RAFT group are needed. For GF by ATRP: copper(II) bromide ($CuBr_2$); copper(I) bromide (CuBr); tris(2-pyridylmethyl)amine (TPMA); monomer such as OEOMA are needed. Buffers such as glycine and phosphate tris(hydroxymethyl)aminomethane are needed.

Reagents needed for incorporation of LAP sequence and click chemistry: Ligation of the azido group requires 10-azidodecanoic acid, lipoic acid ligase mutant ^{W37V}LplA, which can be created from the plasmid: pYFJ16-LplA(W37V) (Plaks et al., 2015), adenosine triphosphate (ATP), and magnesium chloride ($MgCl_2$) can optionally be added. For the click reaction copper (II) sulfate ($CuSO_4$), tris(3-hydroxypropyltriazolylmethyl)amine (THPTA), and sodium ascorbate are needed.

1.2.4 Equipment Needed

Equipment for polymer synthesis and characterization: Polymer synthesis can be performed in round bottom flasks capped with rubber septa. Improved anaerobic conditions can be achieved using Schlenk flasks. Ideal volumes for these reaction vessels are in the range of 5–25 mL. Deoxygenation can be achieved by bubbling solutions with nitrogen for 10–20 min; therefore, a source of nitrogen releasing pure nitrogen gas at approximately 2–5 psi is needed. A nitrogen manifold or Schlenk line is appropriate. Polymerizations should be performed in a temperature controlled oil bath with stirring capability. CTAs can be synthesized in 100–500 mL round bottom flasks. Addition funnels assist in slowly adding reagents. Purification is achieved by extractions between organic and aqueous solvents in 250–1000 mL separatory funnels. Solvents should be removed under reduced pressure on a rotary evaporator. A nuclear magnetic resonance (NMR) spectrometer capable of performing proton characterization is necessary to characterize starting reagents and polymers. Polymers can be characterized using size exclusion chromatography (SEC). This system should contain an in-line degasser, an isocratic pump, a minimum 1 guard and 2 analytical columns, column oven, and either a ultraviolet absorbance (UV) or refractive index detector. An autosampler attached to the SEC system will assist with high-throughput applications and analysis of multiple polymers. Polymers and protein–polymer conjugates can also be characterized using electrospray ionization mass spectrometry (ESI-MS) or matrix-assisted laser desorption ionization time of flight mass spectrometry (MALDI-ToF-MS). Monomers and radical initiators should be stored in a refrigerator or freezer.

Equipment needed for polymer conjugation and purification and characterization: Conjugation can be performed in a round bottom flask with a magnetic stirrer. Protein–polymer conjugates can be purified using ammonium sulfate precipitation, dialysis membranes, centrifugal purification tubes, or ultrafiltration continuous flow systems. A centrifuge with temperature control is necessary. A electrophoresis system capable of performing standard poly(acrylamide) gel electrophoresis (PAGE) is needed to characterize conjugates. Polymers and protein–polymer conjugates can also be characterized using ESI-MS or MALDI-ToF-MS.

Equipment needed for click conjugation: The GFP can be expressed in *Escherichia coli* using standard shaker set up, and the LAP sequence of GFEIDKVWYDLDA can be introduced into a protein using the QuickChange Lightning site-directed mutagenesis kit. Similarly, the lipoic acid ligase can be expressed in *E. coli* using a standard shaker set up from the plasmid pYFJ16-LplA(W37V). Ligation of azido decanoic acid can be achieved in a round bottom flask, followed by purification using dialysis membranes. Click chemistry can be performed in round bottom flasks, followed by purification by dialysis, centrifugal purification, or ultrafiltration. Conjugates can be analyzed using standard PAGE systems.

2. POLYMER SYNTHETIC PROCEDURES

Whether GT or GF the enzyme, polymer synthetic procedures need to follow several guidelines to ensure polymerization success (Averick, Simakova, et al., 2012; McCormick & Lowe, 2004; Sumerlin, 2012). In recent decades, reversible-deactivation radical polymerization (RDRP) methods have seen significant interest as techniques for the synthesis of well-defined polymers with predictable molecular weight, composition, and interesting architectures (Braunecker & Matyjaszewski, 2007). These RDRP reactions provide control over polymerization by introducing a dynamic equilibrium between the propagating polymeric radical and a dormant, nonradical, form of this same molecule. Nitroxide-mediated polymerization (Georges, Veregin, Kazmaier, & Hamer, 1993; Nicolas et al., 2013), RAFT polymerization (Boyer et al., 2009; Chiefari et al., 1998; Moad, Rizzardo, & Thang, 2013; Perrier & Takolpuckdee, 2005), and ATRP (Konkolewicz et al., 2014; Matyjaszewski, 2012; Wang & Matyjaszewski, 1995), have received significant attention due to their compatibility with various functional groups and relative ease of implementation. RAFT and

ATRP are particularly well suited to the creation of bioconjugates since the techniques are compatible with many water-soluble functional groups and the polymerization can be performed under mild conditions, often at or near room temperature in aqueous media (Averick, Simakova, et al., 2012; Cobo et al., 2015; Konkolewicz et al., 2014, 2012; McCormick & Lowe, 2004; Simakova, Averick, Konkolewicz, & Matyjaszewski, 2012; Sumerlin, 2012; Xu, Jung, Corrigan, & Boyer, 2014). Protein–polymer conjugates have been synthesized by these techniques utilizing both the GT and GF processes (Bontempo & Maynard, 2005; Boyer et al., 2009, 2007; Cummings et al., 2014; Grover & Maynard, 2010; Lele, Murata, Matyjaszewski, & Russell, 2005; Mancini et al., 2012; Murata, Cummings, Koepsel, & Russell, 2013; Pelegri-O'Day, Lin, & Maynard, 2014; Xu et al., 2014). A summary of the mechanisms of RAFT and ATRP processes is given in Scheme 2. Note that RAFT depends on degenerative transfer of thiocarbonylthio groups and ATRP involves transition metal, most commonly Cu, catalyzed carbon–halogen bond cleavage.

2.1 RAFT CTAs and ATRP Initiators

ATRP initiators are often commercially available, including the commonly used water-soluble initiator 2-hydroxyethyl 2-bromoisobutyrate (HEBiB). HEBiB and related water-soluble ATRP initiators have been extensively used for aqueous Cu-mediated polymerizations (Averick, Simakova, et al., 2012; Jones et al., 2016; Konkolewicz et al., 2012; Magnusson, Bersani, Salmaso, Alexander, & Caliceti, 2010; Simakova et al., 2012; Zhang, Wilson, et al., 2013). The hydroxyl group of groups like HEBiB can be modified to a carboxylic acid group that can be attached to a protein by reacting the hydroxyl group with succinic anhydride, as outlined by Averick, Simakova, et al. (2012). RAFT CTAs are commercially available; however, many of these commercially available CTAs contain a long hydrophobic chain making them less suitable for aqueous systems. All RAFT CTAs have a thiocarbonylthio group and are characterized by the

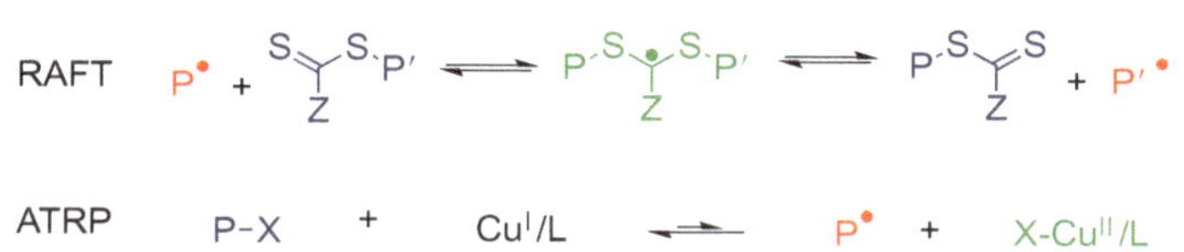

Scheme 2 Key features of the RAFT and ATRP mechanisms. Here, P and P′ represent polymer chains, Z is a group that stabilizes radicals, L is a ligand, and X is a halogen.

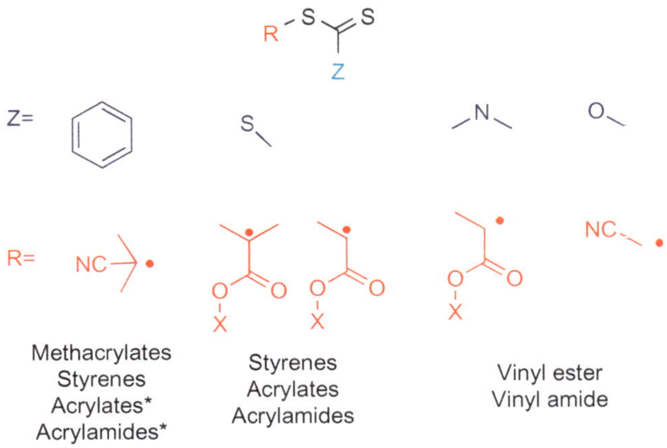

Scheme 3 General structure and specific examples of RAFT agents and the compatibility of typical R and Z groups with general monomer classes. Starred monomer is compatible with the polymerization, although better polymerized with a different RAFT agent.

Z group and the R group, as seen in Scheme 3. General compatibilities of RAFT agents R and Z groups with classes of monomers are given in Scheme 3 (Moad, Rizzardo, & Thang, 2005, 2009).

We have utilized 2-(((ethylthio)carbonothioyl)thio)propanoic acid, also referred to as (propionic acid)yl ethyl trithiocarbonate (PAETC), as a water-compatible RAFT CTA that can create synthetic polymers containing various functional groups from amides, amines, carboxylic acids, sugars, and phosphocholine (Lucius et al., 2016). In addition, alkyne reactive handles are advantageous for "click chemistry" and other postpolymerization modifications. We will outline the synthesis of a RAFT agent (propynyl propionate)yl ethyl trithiocarbonate (PYPETC), a PAETC derivative, with an alkyne group that is capable of performing "click" reactions (Konkolewicz, Gray-Weale, & Perrier, 2009).

2.1.1 Synthesis of PAETC

Potassium hydroxide (14.6 g, 0.26 mol) was dissolved in distilled water (15.0 mL) and was added dropwise to a solution of ethanethiol (18.6 mL, 0.26 mol) in acetone (150 mL) by stirring on an ice bath (Scheme 4). Carbon disulfide (16.1 mL, 0.27 mol) was added to the reaction mixture and stirred for 30 min while still on ice. The mixture was removed from the ice bath and 2-bromopropionic acid (22.0 mL, 0.25 mol) was added dropwise. The reaction mixture was left to stir overnight at room temperature. Solvent

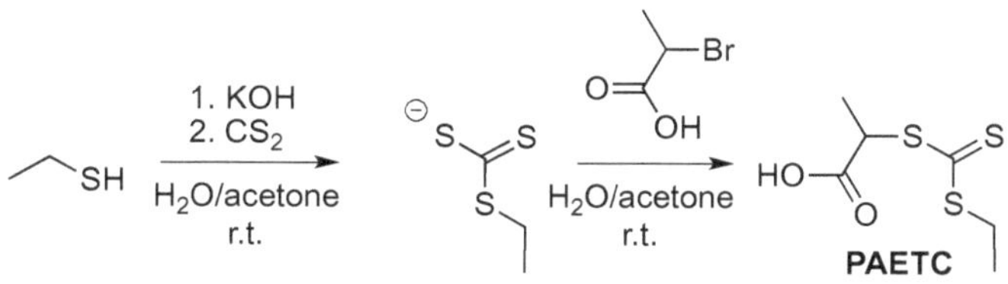

Scheme 4 Synthesis of the RAFT agent PAETC.

was removed via rotary evaporation and the residue was dissolved into 200 mL of ether first, followed by 200 mL of water. The ether layer was added onto the aqueous layer and shaken. The yellow ether layer was collected and subsequently washed with water (6 × 200 mL), and once with 200 mL brine. Solvent was removed by rotary evaporation to give 32.7 g of PAETC (0.155 mol, 59%) as a viscous yellow liquid, which solidifies upon freezing. The product was characterized as typically being greater than 95% pure by proton NMR. The most common impurity is a small amount of unreacted 2-bromopropionic acid, which is not active in RAFT polymerization and is removed during the precipitation of the polymer. Additional purification could be achieved using flash chromatography. ^{1}H NMR (300 MHz, $CDCl_3$) δ ppm 4.87 (1H, q, J=7.4 Hz, $CH_3C\underline{H}(S)$ COOH), 3.38 (2H, q, J=7.4 Hz, $CH_3C\underline{H_2}S$), 1.63 (3H, d, J=7.4 Hz, $C\underline{H_3}CH(S)COOH$), 1.36 (3H, t, J=7.4 Hz, $C\underline{H_3}CH_2S$).

Note: Ethanethiol and intermediates have strong odors. All manipulations should be completed in the fume hood and all equipment used should be rinsed thoroughly in a bleach water bath before removing from the hood.

2.1.2 Synthesis of PYPETC

PAETC (0.909 g, 4.34 mmol) was added to 50 mL of dichloromethane (Scheme 5). To that, propargyl alcohol (1.21 g, 21.58 mmol), EDC (1.65 g, 1063 mmol), and DMAP (0.5521 g, 4.52 mmol) were added and allowed to stir for 2 h on ice, then overnight at room temperature. This crude product was first washed four times with 70 mL of 0.2 *M* HCl. It was then washed once with 70 mL of deionized (DI) water, three times with 50 mL of 0.5 *M* sodium carbonate, and once more with 70 mL of DI water. The resulting product was dried under vacuum and the purity was determined using proton NMR. ^{1}H NMR (300 MHz, $CDCl_3$) δ ppm, 4.84 (1H, q, J=7.4 Hz, C(O)–C(S)$\underline{H}$–CH_3), 4.75 (2H, d, J=2.4 Hz, C≡C–$C\underline{H_2}$–O), 3.35 (2H, q, J=7.4 Hz, S–$C\underline{H_2}$–CH_3), 2.51 (1H, t, J=2.4 Hz, $\underline{H}$–C≡C), 1.63 (3H, d, J=7.4 Hz, C(S)H–$C\underline{H_3}$), 1.35 (3H, t, J=7.4 Hz, S–CH_2–$C\underline{H_3}$).

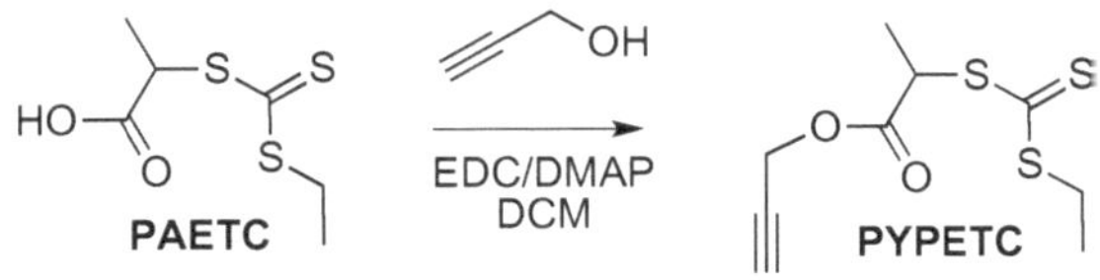

Scheme 5 Synthesis of the RAFT agent PYPETC.

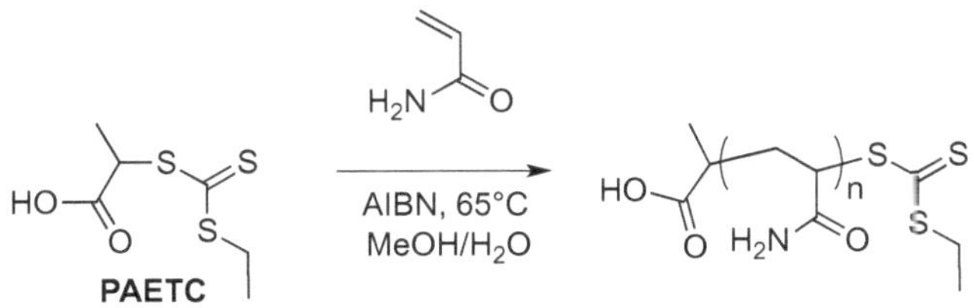

Scheme 6 RAFT polymerization of acrylamide (Am) using PAETC and the CTA.

2.2 Typical Polymerization Methods

Polymers are synthesized using RAFT polymerization here. RAFT polymerization is tolerant to functional groups and aqueous solvents. However, the polymerization can only occur under anaerobic conditions. Simple bubbling with nitrogen gas for 10–20 min is sufficient to displace oxygen.

2.2.1 Typical RAFT Polymerization of Acrylamide Using PAETC

Carboxylic acid-functionalized polymers of acrylamide (Am) may be prepared using the following protocol (Scheme 6). Am (1.7770 g, 25 mmol, 5 eq.), PAETC (1.0512 g, 5 mmol, 1 eq.), and AIBN (0.1642 g, 1 mmol, 0.2 eq.) were added to a glass vial with 1.4927 g (1.89 mL) of methanol and 1.4927 g (1.49 mL) of water. The mixture was then stirred until the contents of the vial were fully dissolved. Once dissolved, the contents were then transferred from the vial to a 25-mL round bottom flask. The reaction mixture was then purged with nitrogen for 10 min followed by the head space for 5 min to remove any oxygen. After the oxygen was removed, the reaction was placed in a 65°C oil bath to stir for 22–24 h. After polymerization, ^{1}H NMR spectra were acquired to determine monomer conversion. Monomer conversion is determined by the relative disappearance of vinyl bonds in the range 6.5–5 ppm after polymerization. This produced poly(Am) with a degree of polymerization of 5 units. To synthesize an Am polymer of other chain lengths, the same procedure can be adapted to use different amounts of monomer and solvent. Typically, the ratio of monomer to solvent is left unchanged in this procedure. The amount of water as well as methanol used was equivalent to the total mass of reagents used, making the polymerization run at a ratio of 1:1 by weight monomer plus RAFT agent:solvent. A similar procedure can be used for the synthesis of Am copolymers with functional monomers such as AA, DMAEMA, or PCMA. A similar approach may be used to create alkyne-functionalized polymers of Am by using PYPETC rather than PAETC as the CTA, keeping all molar ratios unchanged.

2.2.2 *Precipitation of Am Polymers*

All the Am polymers were precipitated by adding the reaction mixture dropwise into a vigorously stirred solution of four parts tetrahydrofuran (THF) to one part diethyl ether solution at 0°C. Additional methanol/water may be necessary to ensure that the precipitation is efficient. The ratio of nonsolvent (THF/ether) to solvent (MeOH/H_2O) for the polymer should be at least 10 parts nonsolvent to 1 part solvent.

2.2.3 *Typical RAFT Polymerization of* N,N*-Dimethylacrylamide Using PAETC*

Carboxylic acid-functionalized polymers of DMAm may be prepared using the following protocol (Scheme 7). DMAm (2.58 mL, 25 mmol, 5 eq.), PAETC (1.0512 g, 5 mmol, 1 eq.), and AIBN (0.1642 g, 1 mmol, 0.2 eq.) were added to a glass vial with 3.6937 g or 4.70 mL of ethanol (EtOH). The mixture was then stirred until the contents of the vial were fully dissolved. Once dissolved, the solution was then transferred from the vial to a 25-mL round bottom flask. The reaction mixture was then bubbled with nitrogen for 10 min followed by the head space for 5 min to remove any oxygen. After the oxygen was removed, the reaction was placed in a 65°C oil bath to stir for 24 h. ^{1}H NMR spectra were acquired to determine polymer conversion, which is typically greater than 95%. Monomer

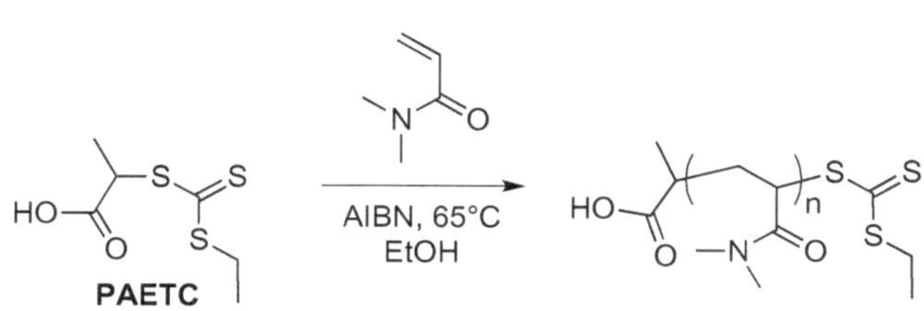

Scheme 7 RAFT polymerization of *N,N*-dimethylacrylamide (DMAm) using PAETC and the CTA.

conversion is determined by the relative disappearance of vinyl bonds in the range of 6.5–5 ppm. This produced poly(DMAm) with a degree of polymerization of 5 units. To synthesize a DMAm polymer of other chain lengths, the same procedure can be used, with different amounts of monomer and solvent. Typically, the ratio of monomer to solvent is left unchanged in this procedure. The amount of solvent used was equivalent to the total mass of reagents used. A similar procedure can be used for the synthesis of DMAm copolymers with functional monomers such as AA, DMAEMA, and PCMA. A similar approach may be used to create alkyne-functionalized polymers of DMAm by using PYPETC rather than PAETC as the CTA, keeping all molar ratios unchanged.

2.2.4 Precipitation of DMAm Polymers

All DMAm polymers were precipitated by adding the reaction mixture dropwise into a vigorously stirred solution of hexanes at 0°C. Additional ethanol may be necessary to ensure that the precipitation is efficient. The ratio of nonsolvent (hexanes) to solvent (EtOH) for the polymer should be at least 10 parts nonsolvent to 1 part solvent.

2.2.5 RAFT Polymerization With PYPETC

Polymers can be prepared by using PYPETC rather than PAETC. The polymerization can be set up by the same procedure with the same molar ratios of monomer, AIBN and PYPETC and solvent. For the polymerization of DMAm, ethanol is an appropriate solvent at a 1:1 ratio of monomer: solvent. Polymerization is performed in an oil bath 65°C for at least 12 h. Monomer conversion can be determined by ^{1}H NMR by measuring the relative disappearance of vinyl groups in the range of 6.5–5 ppm.

2.2.6 ATRP of Oligo(Ethylene Oxide)Methyl Ether Acrylate

A water-soluble polymer of oligo(ethylene oxide)methylether acrylate (OEOA, monomer average molecular weight 454) can be prepared using the following method, adapted from a method developed by Matyjaszewski et al. (Konkolewicz et al., 2012). OEOA (1.00 g, 2.22 mmol) and 20 μL of a 11.3 m*M* $CuBr_2$ plus 23 m*M* TPMA solution (0.23 μmol Cu) are placed in a vial (Scheme 8). To this mixture 60 μL of a 27.6 m*M* VA-044 solution (1.7 μmol), and 0.2 mL of a 27.8 m*M* HEBiB solution (5.72 μmol) are added. To this vial, tetraethylammonium bromide (0.1075 g, 0.5115 mmol) and water (4.0128 g) are added. This solution is homogenized by shaking for 1 min and transferred to a Schlenk flask and deoxygenated by bubbling with

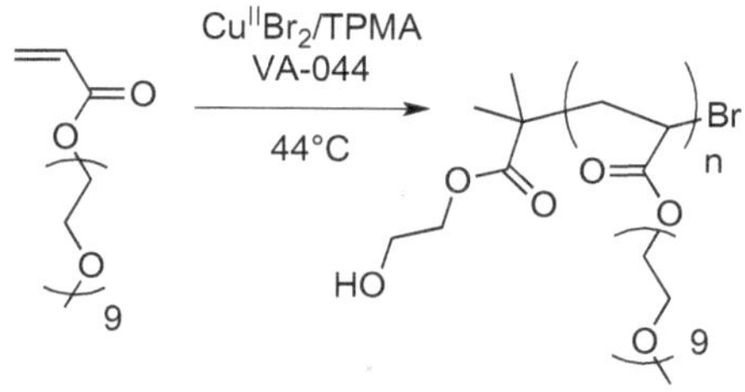

Scheme 8 ATRP of oligo(ethylene oxide)methylether acrylate using $Cu^{II}Br_2$ and tris(pyridylmethyl)amine (TPMA) as the catalyst initiated by VA-044.

nitrogen for 30 min. The reaction is commenced by placing into an oil bath at 44°C and allowed to react for 21 h, giving poly(OEOA).

2.2.6.1 Purification of OEOA Polymers

After aqueous phase polymerization, poly(OEOA) can be extracted by 4 × 50 mL washed with dichloromethane, followed by collection of the organic phases and removal of the solvent under reduced pressure. The polymer can be dissolved in acetone, and optionally passed over basic alumina to remove traces of Cu catalyst, and precipitated out of hexanes to give the pure product.

2.2.7 Tips on Polymerization

Polymerization is generally facile provided that oxygen has been rigorously excluded. However, it is often possible that conversion is below the targeted value. In that case two options exist. One is to remove excess monomer by precipitation or under reduced pressure on a rotary evaporator. The other option is to add a small amount of additional radical initiator followed by repeating the deoxygenation and polymerization.

2.2.8 Characterization of Polymers

A library of different techniques may be used to characterize the molecular weight and composition of polymers. ^1H NMR using solvents D_2O or DMSO may be used on the precipitated polymer to determine mean relative composition of two monomers in a polymer, and for polymers with relatively low-molecular-weight (number averaged molecular weight, $M_n < 20{,}000$), the M_n can be estimated by integrating the signal of monomer repeat units relative to a resonance unique to the ATRP initiator of the RAFT end groups. Gel permeation chromatography (GPC) using *N*,*N*-dimethylformamide with 0.1 wt.% LiCl or LiBr as an eluent at 50°C can be used to characterize most polymers. However, due to the poor solubility of acrylamide polymers in nonaqueous solvents, GPC can be performed using aqueous conditions, such as an aqueous eluent of 100 m*M* sodium phosphate with 0.2 vol.% trifluoroacetic acid (pH 2.5) (Falatach, Li, et al., 2015). Additionally, MALDI-ToF-MS and ESI-MS are useful techniques for characterizing polymers with moderate molecular weights. Commonly used matrices for analyzing polymers include 2,5-dihydroxybenzoic acid and α-cyano-4-hydroxycinnamic acid.

3. PROTEIN CONJUGATION METHODS—AMINE BASED

Due to the ubiquitous nature of amines in proteins, modification of amine groups is a popular method for protein modification with polymers. Here, we describe protocols for modification of two enzymes, lysozyme and chymotrypsin by GT, as well as GF procedures for lysozyme conjugates. In all cases, the attachment of the polymer to the protein occurs through an amide bond with either lysine groups on the protein surface or the N-terminal amine. In general, GT procedures are efficient for grafting lower molecular weight polymers to proteins even at relatively low protein concentrations, and attachment of very large polymers to proteins can be a challenge. In contrast, GF is well suited for creating bioconjugates with high-molecular-weight polymers or high grafting densities, although polymerization can be challenging at low protein/polymer initiator concentrations, and the initiating group can alter the solubility and stability of the protein.

3.1 GT Approaches Using In Situ EDC/NHS Coupling

The GT approaches outlined commence from a lyophilized enzyme. If the enzyme is already in a buffered aqueous solution, the procedure can be modified to add the enzyme buffer solution directly to the solids.

3.1.1 Typical Conjugation of Acrylamide Polymers to Lysozyme

Poly(Am) of average degree of polymerization (DP) = 5 units (3.7401 g, 420 eq.), 1-ethyl-3-(3-dimethylaminopropyl)carbodiimide-hydrochloride salt (EDC, 1.1261 g, 420 eq.), NHS (0.1151 g, 71.5 eq.), and lysozyme (200 mg, 1 eq., 7 amine groups) were weighed out and added to a 500-mL round bottom flask. Then 200 mL of 0.1 *M* phosphate buffer pH 8 was added. The solution was then stirred at room temperature for 2 h. Once the 2 h had passed enough 0.2 *M* glycine buffer pH 10.6 (1 mL) was added so that the final concentration of glycine buffer was 1 m*M*. Other

Am polymers of different chain lengths followed this same equivalence and concentration ratios for conjugation. If low modification is observed, the ratio of polymer, EDC, and NHS can be increased (Falatach, McGlone, et al., 2015). This protocol can be adapted to other functional polymers such as copolymers of DMAm, with DMAEMA or PCMA. Polymers with AA can also be conjugated with this procedure, although the attachment site to the protein is no longer exclusively at the α-terminus of the polymer. When attaching poly(Am) of DP = 5, this procedure leads to 2.3 polymers attached per lysozyme and essentially no unmodified protein (Falatach, McGlone, et al., 2015). Grafting efficiency and density decreases with larger polymer chain length, as is typical for GT approaches.

3.1.2 Typical Conjugation of Poly(DMAm) to Lysozyme

Poly(DMAm) of average DP = 5 (0.2364 g, 420 eq.), EDC (0.0563 g, 420 eq.), and NHS (0.0058 g, 71.5 eq.) were weighed out into a 50-mL round bottom flask. Then 10 mL of a 1 mg lysozyme to 1 mL of 0.1 *M* phosphate buffer solution pH 8 was added to the round bottom flask. The solution was allowed to stir for 2 h. Once 2 h passed enough 0.2 *M* glycine buffer pH 10.6 (0.05 mL) was added so that the final concentration of glycine buffer was 1 m*M*, to halt any side reactions. This mixture was allowed to stir for 10 more minutes and then underwent an ammonium sulfate precipitation. Other dimethylacrylamide polymers of different chain lengths followed this same equivalence and concentration ratios for conjugation. If low degree of modification is observed, the ratio of polymer, EDC and NHS can be increased. This protocol can be adapted to other functional polymers such as copolymers of DMAm, with DMAEMA or PCMA. Polymers with AA can also be conjugated with this procedure, although the attachment site to the protein is no longer exclusively at the α-terminus of the polymer.

3.1.3 Conjugation of Poly(DMAm) to Chymotrypsin

In addition to lysozyme, the in situ EDC/NHS coupling may be used to attach to other proteins including the protease chymotrypsin. Poly(DMAm) of average DP = 50 units (1.058 g, 0.21 mmol), EDC-HCl (201 mg, 1.05 mmol), and NHS (242 mg, 2.10 mmol) were dissolved in 3.25 mL of 1 × PBS and added dropwise to a stirring solution of α-chymotrypsin (3 mL of 2 mg/mL 0.0035 mmol Lys residues). The reaction was carried out at room temperature for 24 h. Once the reaction was completed, 0.2 *M* glycine buffer pH 10.6 (1 mL) was added so that the final concentration of glycine buffer was 1 m*M* to halt any side reactions.

3.1.4 Purification of Protein–Polymer Conjugates Using Ammonium Sulfate Precipitation

In order to purify both the poly(Am)-lysozyme, the poly(DMAm)-lysozyme or any other polymer conjugates, an ammonium sulfate precipitation can be used. The ammonium sulfate precipitation requires that the conjugation solution be stirring and on ice. Once the conjugation solution temperature is between 0°C and 4°C, ammonium sulfate is slowly added, about a spatula tip every 30 s, until the solution is 63% saturated. The solution is then left to stir for 20 min to ensure that the ammonium sulfate is fully dissolved. After 20 min, the solution and solids are transferred to a centrifuge tube and is centrifuged at 4°C and at 11,500 RPM for 15 min. Subsequently, the supernatant is poured off and the solid is resuspended in a 25 m*M* phosphate buffer at pH 8. This process can be repeated to improve purity.

3.1.5 Conjugate Purification Using Centrifugal Filtration

The RAFT end group has a distinctive yellow color. This allows determination of purity by examining the washed solution by UV–vis or even by eye. Protein–polymer conjugates in buffer were transferred to a 5 kDa cutoff centrifugal filter. Note that the size of the filter cutoff should be chosen to be at least two times lower than the bioconjugates molecular weight. The samples were centrifuged at 6000 RPM for 30 min. After 30 min, the solution passed through the centrifugal filter membrane is inspected. If the solution that passed through the membrane has a strong yellow color, it was centrifuged again after adding more buffer. If the solution that passed through the membrane is colorless, then the centrifugal purification process is finished. This process can be also used to concentrate the bioconjugate sample to a desired concentration.

3.2 GF Approaches Using RAFT and ATRP

We have studied GF lysozyme as a model enzyme using RAFT polymerization. Typically, this polymerization can be completed in 16 h. The methods developed start from lysozyme modified with an oligomer of Am, where the DP of Am is approximately 5 units. It has been reported in the literature that lysozyme directly modified with the small molecule CTA can also be used for chain extension (Li, Li, Yu, Bapat, & Sumerlin, 2011; Lucius et al., 2016);

however, the solubility of the lysozyme-CTA construct can be low (Falatach, McGlone, et al., 2015). Attaching a highly water-soluble oligomer decreases this concern (Falatach, McGlone, et al., 2015). Depending on the nature of the protein and the number of RAFT CTAs or ATRP initiators to be attached to the surface, direct attachment of the small molecule RAFT or ATRP group may be possible, or a short water-soluble oligomer containing the RAFT or ATRP group may be used instead.

3.2.1 Attachment of a RAFT CTA or ATRP Initiator to a Protein Using Amine Chemistry

To graft-from a protein, amine chemistry may be utilized either through EDC/NHS coupling or through direct attachment of the NHS ester (Averick, Simakova, et al., 2012; Falatach, McGlone, et al., 2015; Fischer, 2010; Li, Li, et al., 2011). The key requirements for this attachment are that the RAFT CTA or ATRP initiator has a single free carboxylic acid group (Scheme 9). This group can be transformed into a NHS ester, either in an organic transformation, followed by conjugation or in situ during EDC/NHS coupling. This NHS-activated ester then reacts with amine (lysine and N-terminal) groups on the surface of the protein.

Although the RAFT CTA can be directly attached to the protein surface (Li, Li, et al., 2011), it has been found that the higher water solubility of a short oligomer (~5 units) of acrylamide or other water-soluble polymers can be created first. This oligomer can be synthesized as outlined in Section 2 of this methods contribution and attached to the protein as outlined in Section 3.1

3.2.2 Attachment of an ATRP Initiator

A bovine serum albumin (BSA) solution was prepared at 1 mg of protein for 1 mL of 50 m*M* Borate at pH 9. A BiB solution DCM is added to a BSA solution to give a molar ratio of BSA:BiB is 10:1, 25:1, 50:1, etc., depended on desired degree of modification. Since this is a two phase system and the volume of DCM is small (~1–50 or 100 mL of buffer) vigorous mixing of

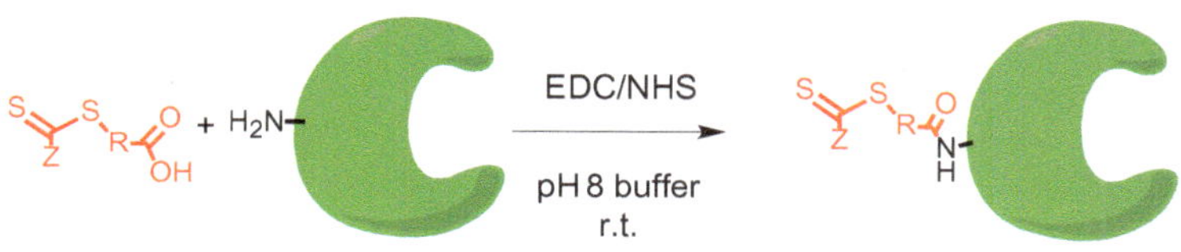

Scheme 9 Attachment of the RAFT agent or ATRP initiator to amine groups on the protein using EDC/NHS coupling. This method can be adapted to attach small molecules or polymers.

the system is required. The reaction is allowed to proceed for 2 h, followed by the removal of the DCM phase using a separatory funnel or decanting. The resulting ATRP initiator-modified protein can be dialyzed against a buffer of choice for at least two changes. An alternative approach involves the modification of HEBiB with succinic anhydride and formation of the NHS-activated ester, as outlined by Averick, Simakova, et al (2012).

3.2.3 Typical GF Using RAFT Polymerization

VA-044 (2.15 mg) and Am monomer (23.61 mg, 250 eq.) were weighed out in a vial (Scheme 10). Then 0.95 mL of a RAFT-modified lysozyme (9.4 mg protein per 1 mL) of phosphate buffer pH 6.2 were added, 0.47 eq. of RAFT agent. The solution was mixed until all materials were fully dissolved. Once dissolved, the solution was transferred from the vial to a 10-mL Schlenk flask where it was *gently* bubbled with nitrogen in the solution for 10 min and the headspace for 5 min. After the headspace and solution is purged of oxygen, it is placed in a 30°C oil bath where it is stirred for an allotted time, typically 6 h of reaction time leads to 40%–45% conversion of Am, and 16 h of polymerization time leads to 90%–95% conversion of Am. This procedure can be used to graft various functional polymers directly from the enzyme surface. At lower protein and RAFT agent concentrations, to ensure consistent kinetics VA-044 concentration can be held constant at 2.2 mg/mL. This however will lead to a greater fraction of polymers not conjugated to the protein. In contrast, the VA-044 concentration can be reduced to ensure that the vast majority of chains are connected to the protein. However, in this case, the polymerization rate will decrease, and consequently longer reaction times will be needed.

3.2.4 Typical GF Using ATRP Polymerization

The following reaction conditions were used for polymerization of oligo(ethylene oxide)methylether methacrylate (OEOMA, average molecular

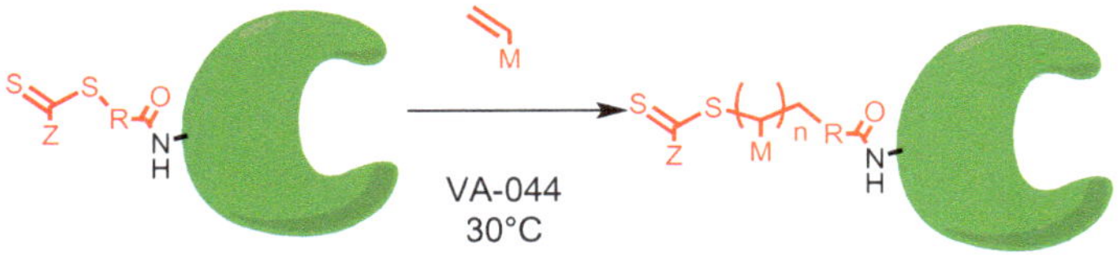

Scheme 10 Grafting from the protein surface using RAFT polymerization to grow the polymer off the surface of the protein.

weight 475) from BSA to give a desired degree of polymerization of 100 (assuming 10 initiator sites per BSA-Br): $[OEOMA_{475}]_0$:$[BSA\text{-}Br]_0$:$[CuBr]_0$:$[CuBr_2]_0$:$[TPMA]_0$ = 1000:1:7:3:10; $[OEOMA_{475}]_0$ = 0.050 *M*. The polymerization is carried out in an ice bath (0°C). Prepare a stock solution of TPMA/$CuBr_2$ in a glass vial by adding 2.5 mg $CuBr_2$ and 10.9 mg TPMA and swirling with 10.0 mL of 50 Tris–HCl buffer (pH 7.5) until dissolved. In an empty 20 mL glass vial with a magnetic stirring bar add 356 mg OEOMA, 51 mg of BSA-Br (modified with 10 initiators per protein), 12.7 mL of 50 Tris–HCl buffer (pH 7.5), and 2.0 mL of the TPMA/$CuBr_2$ stock solution. Seal the vial with a rubber septum, place in an ice bath, and bubble with nitrogen or argon gas for 15 min while stirring to remove oxygen. To another glass vial, add 10.0 mg CuBr and a stir bar. Seal the vial and purge with nitrogen or argon gas for 15 min to remove oxygen. After 15 min, add 10.0 mL of acetonitrile (which has also been purged with nitrogen gas for 15 min) to the vial-containing CuBr using a nitrogen-purged syringe. This mixture was stirred until all CuBr dissolved to form a CuBr/acetonitrile stock solution.

While the stock solutions are being purged to remove oxygen, add a stir bar to a 25-mL Schlenk flask and seal and purge with nitrogen for 15 min. After purging oxygen from the Schlenk flask, add 753 μL of the CuBr/acetonitrile stock solution to the Schlenk flask using a purged syringe. Flow nitrogen gas through the Schlenk flask to evaporate the acetonitrile (leaving solid CuBr behind). Put the Schlenk flask in an ice bath over a stir plate and add the cold solution-containing BSA-Br, OEOMA, $CuBr_2$/TPMA, and buffer using a nitrogen-purged syringe to start polymerization. Stop the reaction by opening the flask to air.

3.2.5 Tips on Conjugation

In general GT using EDC/NHS coupling is facile and easily implemented. However, if low grafting densities are achieved, this can be rectified by using higher ratios of polymer EDC and NHS to the amine groups on the polymer. Additionally, raising the pH should promote this reaction due to a larger fraction of the amines being deprotonated; however, care must be taken to ensure the protein is stable at the pH used for conjugation. In GF by RAFT, low concentrations of protein can be challenging. The protocol outlined will graft polymer form the protein; however, the process may lead to more poorly controlled and higher dispersity polymers at very low protein (and CTA) concentrations.

4. POLYMER CONJUGATION BY GT USING CLICK APPROACHES

These methods take advantage of the intrinsically high yields and efficiencies of "click" reactions (Kolb et al., 2001). Lysine chemistry is still utilized, although the lysine residue is posttranslationally modified to an azide group using the lipoic acid ligase method. Incorporation of nonnatural azide-containing amino acids is a possibility. However, in some cases, expression yields of the desired protein are relatively low. In contrast, we highlight a technique that allows azide groups to be placed with precision along the protein backbone utilizing ligase reactions as a form of posttranslational modification (Plaks et al., 2015). This allows the nonnatural azide group to be introduced with precision to a protein containing only natural amino acids.

4.1 Incorporation of a LAP Sequence Into a Protein Backbone

As outlined by Plaks et al. (2015), the 13 amino acid sequence known as the LAP sequence (GFEIDKVWYDLDA) can be inserted into the protein backbone using standard protein engineering techniques. To ensure effective folding, it is important to ensure that the LAP sequence is not placed in a region that interferes with protein folding. Typically, regions near the C- or N-terminus or internal loop regions are targeted since these are less critical for protein folding. In the protocols outlined below GFP had a LAP sequence inserted at the N-terminus or an internal loop region.

4.2 Ligation of Azide Functional Group to Protein Backbone

To ligate the 10-azidodecanoic acid to the LAP sequence, stock solutions of 10-azidodecanoic acid and ATP were prepared. 10-Azidodecanoic acid stock solution was made with the ratio of 2 μL of acid to 4.3 mL of 25 m*M* sodium phosphate buffer with 2 m*M* $MgCl_2$. The ATP was made to a concentration of 100 m*M* with the volume dependent on the final volume of the ligation. For a given reaction, the final concentrations of the components in the reaction flask were 0.1 μ*M* lipoic acid ligase, 600 μ*M* 10-azidodecanoic acid, 10 μ*M* GFP, 2 m*M* ATP, 2 m*M* $MgCl_2$, and 25 m*M* sodium phosphate buffer. This solution was covered with aluminum foil, stirred, and allowed to react at 30°C for at least 4 h. Once the reaction is complete, the mixture is dialyzed at 4°C in pH 7.0 overnight. The ligation was confirmed using MALDI-TOF analysis by a 195-Da shift in the proteins molecular weight. This enzyme-mediated ligation approach results in a single reactive azide group on the surface of the protein for each LAP sequence.

4.3 Polymer Conjugation to Protein Using Click Chemistry

The alkyne-functionalized polymer, created by RAFT polymerization starting from PYPETC, was attached to the azide groups on the protein surface using the copper-catalyzed cycloaddition (Kolb et al., 2001). Stock solutions of $CuSO_4$ (20 m*M*), tris(3-hydroxypropyltriazolylmethyl)amine (THPTA, 50 m*M*), aminoguanidine (100 m*M*), sodium ascorbate (100 m*M*), and polymer conjugate (20 m*M*) were made in sodium phosphate buffer at pH 7. The $CuSO_4$ and THPTA solutions were combined in a 1:2 ratio to create a stock solution of $CuSO_4$/THPTA. For a 1 mL reaction, 865 μL of 10 μ*M* azide-modified GFP, 15 μL of $CuSO_4$/THPTA mixture, 50 μL of aminoguanidine, 20 μL of polymer conjugate, and 50 μL sodium ascorbate were added to a round bottom flask and allowed to react at 37°C for at least 4 h. This reaction was then dialyzed in 20 m*M* sodium phosphate buffer at pH 7.

4.3.1 Tips on Click Chemistry

It is important to avoid excessively heating the click reaction mixture. Excessive heating will degrade the reagents and lead to poor conjugation efficiency. Additionally, heating proteins may lead to thermal denaturation.

5. CONJUGATE CHARACTERIZATION METHODS

One of the greatest challenges in the field of protein–polymer conjugates is their characterization. We outline several methods to assist with interpretation of protein conjugation; however, there is no currently existing method that is universal across all conjugates, simple to use, and providing fully quantitative information. Instead, PAGE, MALDI-ToF-MS, and GPC are commonly used methods. Although PAGE tends to be applicable to the majority of conjugates, the results tend to be qualitative rather than quantitative. In contrast, MALDI-MS provides quantitative data; however, obtaining reliable results in a facile manner is a challenge. Often matrix optimization needs to be performed.

5.1 Polyacrylamide Gel Electrophoresis

Combine the protein conjugate (~10 μg) with a 4× SDS loading buffer-containing 50 m*M* DTT and incubate at 70°C for 10 min. Load the samples into the wells of a polyacrylamide gel. The type of gel to use will vary depending on the molecular weight of the protein–polymer conjugate and the properties of the polymer. We have had good results using Bio-Rad Mini-PROTEAN TGX 4%–20% gradient polyacrylamide gels and Novex 12% Tris–Glycine Gels (Life Sciences). Protein bands can be stained using a coomassie dye (such as GelCode Blue protein stain). Protein–polymer conjugates prepared using PEG can be stained using barium iodide (Kurfürst, 1992). The iodine stain is composed of 1.3% (w/v) iodine, 1.0% (w/v) potassium iodide, and 2.5% (w/v) barium chloride in a 0.6 *M* HCl solution. Depending on the heterogeneity of the protein–polymer conjugates and the staining procedure used, the amount of the protein conjugate loaded onto the gel may need to be increased for good visualization of the protein–polymer bands. An example of PAGE data for protein–polymer conjugates is given in Fig. 1. Typically, a shift of mass to higher mass is anticipated after polymer conjugation. PAGE is not an ideal technique to quantitatively determine the number of polymer chains attached to the protein. However, observation of the disappearance of the native polymer band and a significant shift to higher mass is an excellent tool for qualitative confirmation of polymer conjugation.

5.2 Matrix-Assisted Laser Desorption Ionization Time of Flight Mass Spectrometry (MALDI-ToF-MS)

Before analysis, the protein samples (0.1–1 mg/mL) should be dialyzed with DI water or a volatile buffer to desalt the sample and help to remove any residual surfactants and monomers which may interfere with ionization (Amini, Dormady, Riggs, & Regnier, 2000). Matrix selection and sample preparation may need to be optimized for each sample (Gusev, Wilkinson, Proctor, & Hercules, 1995); however, sinapinic acid is effective for many protein conjugates (Berberich et al., 2005; Lucius et al., 2016). Prepare the matrix solution by combining 0.5 mL water, 0.5 mL of acetonitrile, 1 μL of trifluoroacetic acid, and 10 mg of sinapinic acid and vortex mix until dissolved. Mix 1 μL of the protein conjugate solution with 1 μL of matrix solution and spot onto the target plate. Allow sample to thoroughly dry before analysis. Other spotting techniques and matrices can be used if good spectra cannot be obtained using this method (Speicher, 2011). An example

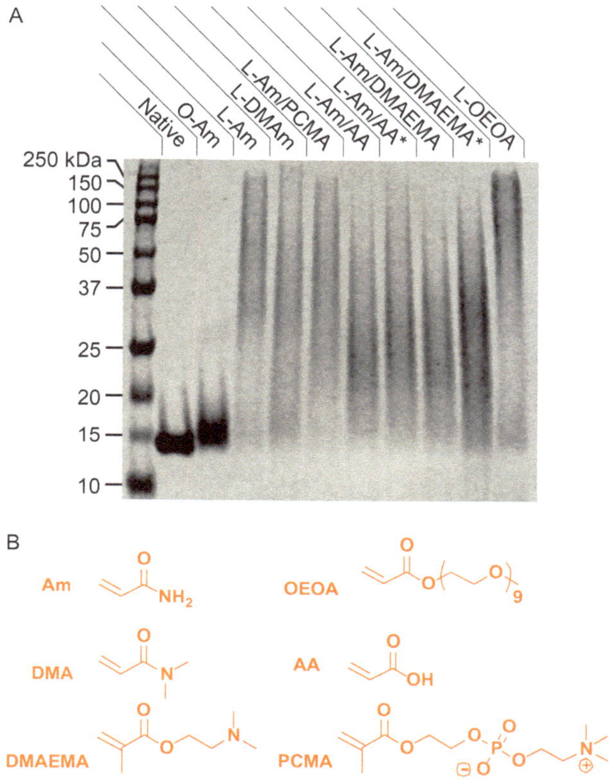

Fig. 1 PAGE data for lysozyme conjugates with various polymers. Panel (A) shows the PAGE data for the conjugates note that O-Am was prepared by grafting a RAFT synthesized DP = 5 polymer of Am using EDC/NHS coupling. All other conjugates were prepared by grafting-from the O-Am conjugate with ca. 50 units of the monomers shown in (B). *Adapted with permission from Lucius, M., Falatach, R., McGlone, C., Makaroff, K., Danielson, A., Williams, C., …, Berberich, J. A. (2016). Investigating the impact of polymer functional groups on the stability and activity of lysozyme–polymer conjugates.* Biomacromolecules, 17*(3), 1123–1134. http://10.1021/acs.biomac.5b01743. Copyright 2016 American Chemical Society.*

of MALDI-MS data for protein–polymer conjugates is given in Fig. 2. Typically, a positive shift of the *m*/*z* value and broadening of the peaks is anticipated after polymer conjugation, and this is confirmed in Fig. 2. The magnitude of the shift in *m*/*z* and correlation of this shift with the average molecular weight of the polymer will determine the number of polymer attachments.

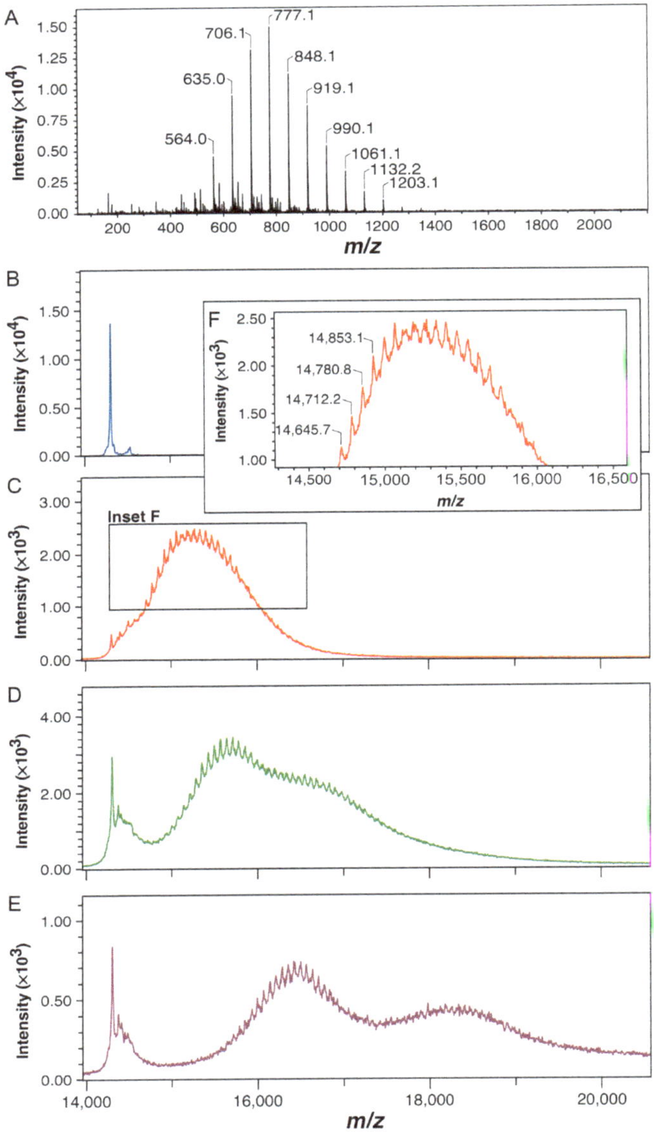

Fig. 2 (A) ESI-MS of the polymer of Am centered at approximately 7–8 units. (B) MALDI-MS of native lysozyme. (C) MALDI mass spectrum of the lysozyme conjugated with the poly(Am) with average DP ≈ 5. (D) MALDI-MS of the lysozyme conjugated with the poly (Am) with average DP ≈ 10. (E) MALDI-MS of the lysozyme conjugated with the poly(Am) with average DP ≈ 20. (F) Inset of the peak MALDI-MS of the lysozyme conjugated with the poly(Am) with average DP ≈ 5, showing that the spacing of the peak is consistent with the molar mass of Am. *Reproduced from Falatach, R., McGlone, C., Al-Abdul-Wahid, M. S., Averick, S., Page, R. C., Berberich, J. A., & Konkolewicz, D. (2015). The best of both worlds: Active enzymes by grafting-to followed by grafting-from a protein.* Chemical Communications, 51*(25), 5343–5346. http://10.1039/C4CC09287B with permission from The Royal Society of Chemistry.*

6. CONCLUSIONS

This work outlines several distinct protocols for protein modification with synthetic polymers. RDRP, including RAFT polymerization and ATRP, methods allow the synthesis of well-defined polymers containing various functional groups. ATRP and RAFT are also compatible with aqueous media, allowing polymerization to proceed under protein friendly conditions. Arguably the simplest method for protein conjugation is to create a polymer with a carboxylic acid and to link the acid to amine groups on the protein surface. This GT approach can be implemented using in situ EDC/NHS coupling and is applicable to a variety of proteins. This approach has advantages of relatively simple preparation, although high concentrations of polymer are often needed for effective conjugation, which can make purification a challenge. Alternatively, an ATRP initiator or RAFT CTA can be attached to a protein, allowing GF the protein surface to be performed, leading to facile purification. This process typically works best at relatively high protein concentrations, which can be a limitation if the protein itself is only slightly soluble in buffer. Additionally, the EDC/NHS approaches are nonspecific in the attachment site in the protein backbone. Finally, posttranslational modification of the protein using ligase approaches allows nonnatural azide groups to be introduced with precise control over location. The azide group can be combined with "click" chemistry to efficiently conjugate alkyne-containing synthetic polymers to a specific site on the protein surface. Conjugate characterization typically involves PAGE or MALDI-MS. Generally, PAGE can be used to confirm conjugation for almost any conjugate and allows qualitative interpretation. MALDI-MS can give quantitative data; however, the technique tends to require more optimization to achieve sufficient signal to noise.

REFERENCES

Amini, A., Dormady, S. J., Riggs, L., & Regnier, F. E. (2000). The impact of buffers and surfactants from micellar electrokinetic chromatography on matrix-assisted laser desorption ionization (MALDI) mass spectrometry of peptides: Effect of buffer type and concentration on mass determination by MALDI-time-of-flight mass spectrometry. *Journal of Chromatography A*, *894*(1–2), 345–355. http://dx.doi.org/10.1016/S0021-9673(00)00798-6.

Averick, S., Mehl, R. A., Das, S. R., & Matyjaszewski, K. (2015). Well-defined biohybrids using reversible-deactivation radical polymerization procedures. *Journal of Controlled Release*, *205*, 45–57. http://dx.doi.org/10.1016/j.jconrel.2014.11.030.

Averick, S. E., Paredes, E., Grahacharya, D., Woodman, B. F., Miyake-Stoner, S. J., Mehl, R. A., ... Das, S. R. (2012). A protein–polymer hybrid mediated by DNA. *Langmuir, 28*(4), 1954–1958. http://dx.doi.org/10.1021/la204077v.

Averick, S., Simakova, A., Park, S., Konkolewicz, D., Magenau, A. J. D., Mehl, R. A., & Matyjaszewski, K. (2012). ATRP under biologically relevant conditions: Grafting from a protein. *ACS Macro Letters, 1*(1), 6–10. http://dx.doi.org/10.1021/mz200020c.

Berberich, J. A., Yang, L. W., Madura, J., Bahar, I., & Russell, A. J. (2005). A stable three-enzyme creatinine biosensor. 1. Impact of structure, function and environment on PEGylated and immobilized sarcosine oxidase. *Acta Biomaterialia, 1*(2), 173–181. http://dx.doi.org/10.1016/j.actbio.2004.11.006.

Bontempo, D., & Maynard, H. D. (2005). Streptavidin as a macroinitiator for polymerization: In situ protein–polymer conjugate formation. *Journal of the American Chemical Society, 127*(18), 6508–6509. http://dx.doi.org/10.1021/ja042230+.

Boyer, C., Bulmus, V., Davis, T. P., Ladmiral, V., Liu, J., & Perrier, S. (2009). Bioapplications of RAFT polymerization. *Chemical Reviews, 109*(11), 5402–5436. http://dx.doi.org/10.1021/cr9001403.

Boyer, C., Bulmus, V., Liu, J., Davis, T. P., Stenzel, M. H., & Barner-Kowollik, C. (2007). Well-defined protein–polymer conjugates via in situ RAFT polymerization. *Journal of the American Chemical Society, 129*(22), 7145–7154. http://dx.doi.org/10.1021/ja070956a.

Braunecker, W. A., & Matyjaszewski, K. (2007). Controlled/living radical polymerization: Features, developments, and perspectives. *Progress in Polymer Science, 32*(1), 93–146. http://dx.doi.org/10.1016/j.progpolymsci.2006.11.002.

Brinkley, M. (1992). A brief survey of methods for preparing protein conjugates with dyes, haptens and crosslinking reagents. *Bioconjugate Chemistry, 3*(1), 2–13. http://dx.doi.org/10.1021/bc00013a001.

Chiefari, J., Chong, Y. K., Ercole, F., Krstina, J., Jeffery, J., Le, T. P. T., ... Thang, S. H. (1998). Living free-radical polymerization by reversible addition–fragmentation chain transfer: The RAFT process. *Macromolecules, 31*(16), 5559–5562. http://dx.doi.org/10.1021/ma9804951.

Cobo, I., Li, M., Sumerlin, B. S., & Perrier, S. (2015). Smart hybrid materials by conjugation of responsive polymers to biomacromolecules. *Nature Materials, 14*(2), 143–159. http://dx.doi.org/10.1038/nmat4106.

Cummings, C., Murata, H., Koepsel, R., & Russell, A. J. (2014). Dramatically increased pH and temperature stability of chymotrypsin using dual block polymer-based protein engineering. *Biomacromolecules, 15*(3), 763–771. http://dx.doi.org/10.1021/bm401575k.

Danielson, A. P., Van Kuren, D. B., Lucius, M. E., Makaroff, K., Williams, C., Page, R. C., ... Konkolewicz, D. (2016). Well-defined macromolecules using horseradish peroxidase as a RAFT initiase. *Macromolecular Rapid Communications, 37*(4), 362–367. http://dx.doi.org/10.1002/marc.201500633.

Das, S., Dawson, N. L., & Orengo, C. A. (2015). Diversity in protein domain superfamilies. *Current Opinion in Genetics & Development, 35*, 40–49. http://dx.doi.org/10.1016/j.gde.2015.09.005.

de Graaf, A. J., Kooijman, M., Hennink, W. E., & Mastrobattista, E. (2009). Nonnatural amino acids for site-specific protein conjugation. *Bioconjugate Chemistry, 20*(7), 1281–1295. http://dx.doi.org/10.1021/bc800294a.

De, P., Li, M., Gondi, S. R., & Sumerlin, B. S. (2008). Temperature-regulated activity of responsive polymer–protein conjugates prepared by grafting-from via RAFT polymerization. *Journal of the American Chemical Society, 130*(34), 11288–11289. http://dx.doi.org/10.1021/ja804495v.

Falatach, R., Li, S., Sloane, S., McGlone, C., Berberich, J. A., Page, R. C., ... Konkolewicz, D. (2015). Why synthesize protein–polymer conjugates? The stability and activity of chymotrypsin-polymer bioconjugates synthesized by RAFT. *Polymer, 72*, 382–386. http://dx.doi.org/10.1016/j.polymer.2015.04.010.

Falatach, R., McGlone, C., Al-Abdul-Wahid, M. S., Averick, S., Page, R. C., Berberich, J. A., & Konkolewicz, D. (2015). The best of both worlds: Active enzymes by grafting-to followed by grafting-from a protein. *Chemical Communications, 51*(25), 5343–5346. http://dx.doi.org/10.1039/C4CC09287B.

Fischer, M. J. E. (2010). Amine coupling through EDC/NHS: A practical approach. In J. N. Mol & E. M. J. Fischer (Eds.), *Surface plasmon resonance: Methods and protocols* (pp. 55–73). Totowa, NJ: Humana Press.

Georges, M. K., Veregin, R. P. N., Kazmaier, P. M., & Hamer, G. K. (1993). Narrow molecular weight resins by a free-radical polymerization process. *Macromolecules, 26*(11), 2987–2988. http://dx.doi.org/10.1021/ma00063a054.

Goncearenco, A., & Berezovsky, I. N. (2015). Protein function from its emergence to diversity in contemporary proteins. *Physical Biology, 12*(4), 045002. http://dx.doi.org/10.1088/1478-3975/12/4/045002.

Grover, G. N., & Maynard, H. D. (2010). Protein-polymer conjugates: Synthetic approaches by controlled radical polymerizations and interesting applications. *Current Opinion in Chemical Biology, 14*(6), 818–827. http://dx.doi.org/10.1016/j.cbpa.2010.10.008.

Gusev, A. I., Wilkinson, W. R., Proctor, A., & Hercules, D. M. (1995). Improvement of signal reproducibility and matrix/comatrix effects in MALDI analysis. *Analytical Chemistry, 67*(6), 1034–1041. http://dx.doi.org/10.1021/ac00102a003.

Hawker, C. J., & Wooley, K. L. (2005). The convergence of synthetic organic and polymer chemistries. *Science, 309*(5738), 1200–1205.

Jackson, J. C., Duffy, S. P., Hess, K. R., & Mehl, R. A. (2006). Improving nature's enzyme active site with genetically encoded unnatural amino acids. *Journal of the American Chemical Society, 128*(34), 11124–11127. http://dx.doi.org/10.1021/ja061099y.

Jencks, W. P. (1969). *Catalysis in chemistry and enzymology*. New York: McGraw-Hill.

Jones, G. R., Li, Z., Anastasaki, A., Lloyd, D. J., Wilson, P., Zhang, Q., & Haddleton, D. M. (2016). Rapid synthesis of well-defined polyacrylamide by aqueous Cu(0)-mediated reversible-deactivation radical polymerization. *Macromolecules, 49*(2), 483–489. http://dx.doi.org/10.1021/acs.macromol.5b01994.

Kolb, H. C., Finn, M. G., & Sharpless, K. B. (2001). Click chemistry: Diverse chemical function from a Few good reactions. *Angewandte Chemie International Edition, 40*(11), 2004–2021. http://dx.doi.org/10.1002/1521-3773(20010601)40:11<2004::AID-ANIE2004>3.0.CO;2-5.

Konkolewicz, D., Gray-Weale, A., & Perrier, S. (2009). Hyperbranched polymers by thiol-yne chemistry: From small molecules to functional polymers. *Journal of the American Chemical Society, 131*(50), 18075–18077. http://dx.doi.org/10.1021/ja908206a.

Konkolewicz, D., Krys, P., Góis, J. R., Mendonça, P. V., Zhong, M., Wang, Y., … Matyjaszewski, K. (2014). Aqueous RDRP in the presence of Cu0: The exceptional activity of CuI confirms the SARA ATRP mechanism. *Macromolecules, 47*(2), 560–570. http://dx.doi.org/10.1021/ma4022983.

Konkolewicz, D., Magenau, A. J. D., Averick, S. E., Simakova, A., He, H., & Matyjaszewski, K. (2012). ICAR ATRP with ppm Cu catalyst in water. *Macromolecules, 45*(11), 4461–4468. http://dx.doi.org/10.1021/ma300887r.

Kurfürst, M. M. (1992). Detection and molecular weight determination of polyethylene glycol-modified hirudin by staining after sodium dodecyl sulfate-polyacrylamide gel electrophoresis. *Analytical Biochemistry, 200*(2), 244–248. http://dx.doi.org/10.1016/0003-2697(92)90460-O.

Lahann, J. (Ed.), (2009). *Click chemistry for biotechnology and materials science*. West Sussex: John Wiley & Sons.

Lang, K., & Chin, J. W. (2014). Cellular incorporation of unnatural amino acids and bioorthogonal labeling of proteins. *Chemical Reviews*, *114*(9), 4764–4806. http://dx.doi.org/10.1021/cr400355w.

Lele, B. S., Murata, H., Matyjaszewski, K., & Russell, A. J. (2005). Synthesis of uniform protein–polymer conjugates. *Biomacromolecules*, *6*(6), 3380–3387. http://dx.doi.org/10.1021/bm050428w.

Li, H., Bapat, A. P., Li, M., & Sumerlin, B. S. (2011). Protein conjugation of thermoresponsive amine-reactive polymers prepared by RAFT. *Polymer Chemistry*, *2*(2), 323–327. http://dx.doi.org/10.1039/C0PY00178C.

Li, H., Li, M., Yu, X., Bapat, A. P., & Sumerlin, B. S. (2011). Block copolymer conjugates prepared by sequentially grafting from proteins via RAFT. *Polymer Chemistry*, *2*(7), 1531–1535. http://dx.doi.org/10.1039/C1PY00031D.

Link, A. J., Mock, M. L., & Tirrell, D. A. (2003). Non-canonical amino acids in protein engineering. *Current Opinion in Biotechnology*, *14*(6), 603–609. http://dx.doi.org/10.1016/j.copbio.2003.10.011.

Liu, D. S., Nivón, L. G., Richter, F., Goldman, P. J., Deerinck, T. J., Yao, J. Z., … Ting, A. Y. (2014). Computational design of a red fluorophore ligase for site-specific protein labeling in living cells. *Proceedings of the National Academy of Sciences*, *111*(43), E4551–E4559. http://dx.doi.org/10.1073/pnas.1404736111.

Lu, H. P., Xun, L., & Xie, X. S. (1998). Single-molecule enzymatic dynamics. *Science*, *282*(5395), 1877–1882. http://dx.doi.org/10.1126/science.282.5395.1877.

Lucius, M., Falatach, R., McGlone, C., Makaroff, K., Danielson, A., Williams, C., … Berberich, J. A. (2016). Investigating the impact of polymer functional groups on the stability and activity of lysozyme–polymer conjugates. *Biomacromolecules*, *17*(3), 1123–1134. http://dx.doi.org/10.1021/acs.biomac.5b01743.

Magnusson, J. P., Bersani, S., Salmaso, S., Alexander, C., & Caliceti, P. (2010). In situ growth of side-chain PEG polymers from functionalized human growth hormone—A new technique for preparation of enhanced protein–polymer conjugates. *Bioconjugate Chemistry*, *21*(4), 671–678. http://dx.doi.org/10.1021/bc900468v.

Mancini, R. J., Lee, J., & Maynard, H. D. (2012). Trehalose glycopolymers for stabilization of protein conjugates to environmental stressors. *Journal of the American Chemical Society*, *134*(20), 8474–8479. http://dx.doi.org/10.1021/ja2120234.

Matyjaszewski, K. (2012). Atom transfer radical polymerization (ATRP): Current status and future perspectives. *Macromolecules*, *45*(10), 4015–4039. http://dx.doi.org/10.1021/ma3001719.

McCormick, C. L., & Lowe, A. B. (2004). Aqueous RAFT polymerization: Recent developments in synthesis of functional water-soluble (co)polymers with controlled structures. *Accounts of Chemical Research*, *37*(5), 312–325. http://dx.doi.org/10.1021/ar0302484.

Moad, G., Rizzardo, E., & Thang, S. H. (2005). Living radical polymerization by the RAFT process. *Australian Journal of Chemistry*, *58*(6), 379–410. http://dx.doi.org/10.1071/CH05072.

Moad, G., Rizzardo, E., & Thang, S. H. (2009). Living radical polymerization by the RAFT process—A second update. *Australian Journal of Chemistry*, *62*(11), 1402–1472. http://dx.doi.org/10.1071/CH09311.

Moad, G., Rizzardo, E., & Thang, S. H. (2013). RAFT polymerization and some of its applications. *Chemistry—An Asian Journal*, *8*(8), 1634–1644. http://dx.doi.org/10.1002/asia.201300262.

Murata, H., Cummings, C. S., Koepsel, R. R., & Russell, A. J. (2013). Polymer-based protein engineering can rationally tune enzyme activity, pH-dependence, and stability. *Biomacromolecules*, *14*(6), 1919–1926. http://dx.doi.org/10.1021/bm4002816.

Nicolas, J., Guillaneuf, Y., Lefay, C., Bertin, D., Gigmes, D., & Charleux, B. (2013). Nitroxide-mediated polymerization. *Progress in Polymer Science*, *38*(1), 63–235. http://dx.doi.org/10.1016/j.progpolymsci.2012.06.002.

Peeler, J. C., Woodman, B. F., Averick, S., Miyake-Stoner, S. J., Stokes, A. L., Hess, K. R., ... Mehl, R. A. (2010). Genetically encoded initiator for polymer growth from proteins. *Journal of the American Chemical Society*, *132*(39), 13575–13577. http://dx.doi.org/10.1021/ja104493d.

Pelegri-O'Day, E. M., Lin, E.-W., & Maynard, H. D. (2014). Therapeutic protein–polymer conjugates: Advancing beyond PEGylation. *Journal of the American Chemical Society*, *136*(41), 14323–14332. http://dx.doi.org/10.1021/ja504390x.

Perrier, S., & Takolpuckdee, P. (2005). Macromolecular design via reversible addition–fragmentation chain transfer (RAFT)/xanthates (MADIX) polymerization. *Journal of Polymer Science A: Polymer Chemistry*, *43*(22), 5347–5393. http://dx.doi.org/10.1002/pola.20986.

Pickart, C. M., & Fushman, D. (2004). Polyubiquitin chains: Polymeric protein signals. *Current Opinion in Chemical Biology*, *8*(6), 610–616. http://dx.doi.org/10.1016/j.cbpa.2004.09.009.

Plaks, J. G., Falatach, R., Kastantin, M., Berberich, J. A., & Kaar, J. L. (2015). Multisite clickable modification of proteins using lipoic acid ligase. *Bioconjugate Chemistry*, *26*(6), 1104–1112. http://dx.doi.org/10.1021/acs.bioconjchem.5b00161.

Pokorski, J. K., Breitenkamp, K., Liepold, L. O., Qazi, S., & Finn, M. G. (2011). Functional virus-based polymer–protein nanoparticles by atom transfer radical polymerization. *Journal of the American Chemical Society*, *133*(24), 9242–9245. http://dx.doi.org/10.1021/ja203286n.

Roberts, M. J., Bentley, M. D., & Harris, J. M. (2012). Chemistry for peptide and protein PEGylation. *Advanced Drug Delivery Reviews*, *64*(Suppl.), 116–127. http://dx.doi.org/10.1016/j.addr.2012.09.025.

Sawada, Y., & Honda, S. (2006). Structural diversity of protein segments follows a power-law distribution. *Biophysical Journal*, *91*(4), 1213–1223. http://dx.doi.org/10.1529/biophysj.105.076661.

Schramm, V. L. (2011). Enzymatic transition states, transition-state analogs, dynamics, thermodynamics, and lifetimes. *Annual Review of Biochemistry*, *80*(1), 703–732. http://dx.doi.org/10.1146/annurev-biochem-061809-100742.

Simakova, A., Averick, S. E., Konkolewicz, D., & Matyjaszewski, K. (2012). Aqueous ARGET ATRP. *Macromolecules*, *45*(16), 6371–6379. http://dx.doi.org/10.1021/ma301303b.

Speicher, D. (2011). Mass spectrometry. In *Current protocols in protein science*. John Wiley & Sons, Inc.

Sumerlin, B. S. (2012). Proteins as initiators of controlled radical polymerization: Grafting-from via ATRP and RAFT. *ACS Macro Letters*, *1*(1), 141–145. http://dx.doi.org/10.1021/mz200176g.

Uttamapinant, C., White, K. A., Baruah, H., Thompson, S., Fernández-Suárez, M., Puthenveetil, S., & Ting, A. Y. (2010). A fluorophore ligase for site-specific protein labeling inside living cells. *Proceedings of the National Academy of Sciences*, *107*(24), 10914–10919. http://dx.doi.org/10.1073/pnas.0914067107.

Veronese, F. M. (2001). Peptide and protein PEGylation: A review of problems and solutions. *Biomaterials*, *22*(5), 405–417. http://dx.doi.org/10.1016/S0142-9612(00)00193-9.

Wang, J.-S., & Matyjaszewski, K. (1995). Controlled/"living" radical polymerization. Atom transfer radical polymerization in the presence of transition-metal complexes. *Journal of the American Chemical Society*, *117*(20), 5614–5615. http://dx.doi.org/10.1021/ja00125a035.

Xu, J., Jung, K., Corrigan, N. A., & Boyer, C. (2014). Aqueous photoinduced living/controlled polymerization: Tailoring for bioconjugation. *Chemical Science*, *5*(9), 3568–3575. http://dx.doi.org/10.1039/C4SC01309C.

Young, T. S., & Schultz, P. G. (2010). Beyond the canonical 20 amino acids Expanding the genetic lexicon. *Journal of Biological Chemistry*, *285*(15), 11039–11044. http://dx.doi.org/10.1074/jbc.R109.091306.

Zeyda, M., & Stulnig, T. M. (2009). Obesity, inflammation, and insulin resistance—A mini-review. *Gerontology*, *55*(4), 379–386.

Zhang, W. H., Otting, G., & Jackson, C. J. (2013). Protein engineering with unnatural amino acids. *Current Opinion in Structural Biology*, *23*(4), 581–587. http://dx.doi.org/10.1016/j.sbi.2013.06.009.

Zhang, Q., Wilson, P., Li, Z., McHale, R., Godfrey, J., Anastasaki, A., … Haddleton, D. M. (2013). Aqueous copper-mediated living polymerization: Exploiting rapid disproportionation of CuBr with Me6TREN. *Journal of the American Chemical Society*, *135*(19), 7355–7363. http://dx.doi.org/10.1021/ja4026402.

CHAPTER TEN

NanoArmoring of Enzymes by Polymer-Functionalized Iron Oxide Nanoparticles

Gayan Premaratne, Leslie Coats, Sadagopan Krishnan[1]
Oklahoma State University, Stillwater, OK, United States
[1]Corresponding author: e-mail address: gopan.krishnan@okstate.edu

Contents

Methods in Enzymology, Volume 590
ISSN 0076-6879
http://dx.doi.org/10.1016/bs.mie.2017.01.006
© 2017 Elsevier Inc.
All rights reserved.

Abstract

Polymer-armored enzymes loaded onto magnetic nanoparticles, as efficient nanobioreactors with enhanced properties, are described in this chapter. Polymers are useful macromolecules carrying a large number of surface charges and repeating units of desired chemical functional groups for linking enzymes onto them. Magnetic micro/nanoparticles have been widely used as enzyme carriers with the incorporation of suitable polymer layers. Synthesized iron oxide magnetic nanoparticles have been used to immobilize a peroxide-catalyzing enzyme-like heme protein: myoglobin using covalent and noncovalent strategies. The stability, scalability, and kinetics of the conjugate were studied in detail using spectroscopic and electrochemical analysis. Compared to the free myoglobin in solution, myoglobin conjugated to magnetic nanoparticles demonstrated high catalytic stability and easy recovery from the reaction medium for further use. Due to the large surface area offered by the magnetic nanoparticles, a large amount of myoglobin could be loaded with a small amount of magnetic nanoparticles. Selected examples of polymer–enzyme and polymer–magnetic nanoparticle–enzyme conjugates developed by us and others are presented in this chapter, and representative methods for making cost-effective scalable and reusable enzymatic reactors have been described.

1. INTRODUCTION

Enzymes covalently conjugated to magnetic nanoparticles (MNPs) via polymer linkers offer enhanced product yields, cost-effective utilization of enzyme, reusability feature, easy isolation of products by a rapid magnetic separation (within 20–30 s), and improved enzyme stability compared to the use of solution forms of enzymes. Covalent conjugation of myoglobin (MB) onto MNPs was effective for peroxide-mediated oxidation of the substrate, 2,2′-azino-bis(3-ethylbenzothiazoline-6-sulfonic acid) (ABTS) with 1.5 times greater product yield than MB solution prepared in buffer. The conjugates followed Michaelis–Menten kinetics displaying K_M values close to MB in the buffer, suggesting no loss of enzyme–substrate affinity as a result of conjugation with nanoparticles. Nearly 2- and 250-fold greater catalytic turnover rates were observed for the conjugate than the MB in solution and only MNPs, respectively. The results for electrochemically driven reduction of organic peroxides demonstrated a similar trend. In addition, higher electron transfer rates between electrode and immobilized MB–MNP covalent conjugate were observed than the corresponding films of electrostatically bound MB with MNPs, only MB, or only MNP-coated electrodes.

Enzymes are biocatalysts that offer high turnover rates, substrate specificity, and operate efficiently under ambient conditions (Callender & Dyer, 2014; Que & Tolman, 2008; Shoda, Uyama, Kadokawa, Kimura, & Kobayashi, 2016). Moreover, enzymes can catalyze a wide range of unusual kinetically challenging functions for various applications in synthesis, energy, biosensing, and environmental science. Despite these advantages, enzyme activity is influenced by factors such as pH, temperature, pressure, solvent nature, and ionic strength of the solution in which they are dissolved. Enzymes can become denatured at high temperatures, outside a narrow pH range, and upon exposure to nonaqueous solvents, which together limit their applications. A bioconjugation methodology to counteract some of these limitations of enzymes while retaining the intrinsic catalytic properties is to immobilize enzymes onto a specific substrate and entrapping them in a region of space such as membranes and polymeric gels (Brena, González-Pombo, & Batista-Viera, 2013). To favor maximum utilization in attractive biotechnological applications, immobilization of enzymes within a confined surface can be advantageous for repeated and continuous usage while retaining the catalytic properties (Tosa, Mori, Fuse, & Chibata, 1967).

Enzyme immobilization by definition is to attach enzymes onto a support material (Jia, Zhu, & Wang, 2003). Immobilization increases the local concentration of enzyme and yet allowing substrates to effectively diffuse toward the enzyme to undergo catalytic conversion to products over a duration of time with stability, as of enzymes in solution (Datta, Christena, & Rajaram, 2013; Liese & Hilterhaus, 2013). Immobilization depends on the functional groups present in the side chains of amino acids which can support physical adsorption, ionic interactions, or affinity-based binding, and irreversible covalent bonds via ether, thioether, amide, and carbamate bonds and so on (Brena & Batista-Viera, 2006). Adsorption is a noncovalent process, and it is easy to perform via secondary interactions including hydrogen bonding, van der Waals forces, hydrophobic interactions, and ionic bonding. Enzyme-like heme proteins such as hemoglobin, MB, and horseradish peroxidase (HRP) have been adsorbed onto modified Fe_3O_4 MNPs for catalytic applications (Peng, Liang, & Qiu, 2011; Peng, Liang, Zhang, & Qiu, 2011; Qiu, Peng, Liang, & Xia, 2010). Adsorption of enzymes can be reversed resulting in enzyme removal from the support material and this can be effected by changes in pH, ionic strength, temperature, or polarity of the solvent. Entrapment is another method of immobilizing enzymes that confines them in an optimal physicochemical microenvironment (Mohamad,

Marzuki, Buang, Huyop, & Wahab, 2015). Lipase encapsulated in silica-hardened organogels (Schuleit & Luisi, 2001), HRP in a polymer nanogel (Yan, Ge, Liu, & Ouyang, 2006), and urease encapsulated in poly(vinyl alcohol)-modified silica sol–gel materials (Tsai & Doong, 2007) are representative examples. Although entrapment can protect the enzyme from denaturation, it is an irreversible method and the slow substrate mass transfer through the membranes or gels can limit the enzyme's catalytic turnover rates and product formation. In covalent approaches, a chemical bond is formed between an enzyme and the support material via various surface functional groups that minimizes enzyme leaching from the surface while retaining catalytic properties of the enzyme. Moreover, separation of immobilized enzyme–support material from the products in solution allows reuse of the enzyme. However, scalability of a covalent strategy for large-scale applications is challenging.

Among various materials used for enzyme binding, polymers represent an excellent class of support materials for high loading density immobilization of enzymes with bioactivity in comparison to that of enzymes in dissolved buffer solutions. Polymers possess attractive physical (high molecular weight, viscosity, and film-forming ability; solvent, temperature, and pH stability) and chemical (hydrophilic, hydrophobic, ionic, and other desired functional groups) properties. These make them suitable for bioconjugation with enzymes offering dramatically improved stability, catalytic activity, and sensing property compared to the enzymes used as dissolved molecules in buffer solutions. Polymers can adsorb a high surface density of enzyme biocatalysts via covalent and noncovalent interactions controlled by the surface chemistry of the polymer and the bioconjugation conditions. These features of polymers are further exploited by functionalizing them onto high surface area MNPs in a miniature volume. Such polymer-functionalized MNPs represent a new class of materials that has recently received considerable attention in catalysis, biosensing, biomedical imaging, drug delivery, and material science. The polymer chains can be regarded as multiple arms of nanoparticles to capture and stabilize desired enzyme molecules to realize a novel biomaterial for development of superior biocatalytic systems. Additionally, a large number of nanoparticles functionalized with polymer chains can provide a vast number of sites for anchoring enzymes for catalysis and biosensing applications. This is because nanoparticles have greater surface area per unit mass when compared to micro- or macroparticles. Specifically, MNPs are unique in view of easy isolation, recovery, and purification of the attached enzyme conjugates by simple use of a magnet, and these features are not available with other inorganic nanoparticles. This is subsequently extended to easy separation of reaction products, stability of the enzymes in the conjugates under high temperatures, and a wide pH range (Kouassi, Irudayaraj, & McCarty, 2005; Xu et al., 2014).

2. NanoArmoring BY MAKING POLYMER–ENZYME CONJUGATES

Many catalytically important enzymes are difficult to handle due to their unstable nature even under ambient conditions for a prolonged duration. Fundamental protein-structure function studies of enzymes have been routinely carried out in homogeneous media at room temperature or under ice-cold conditions, in order to efficiently exploit the fast kinetics and high stereoselective catalytic properties. Along with these properties, conjugation of enzymes with various types of polymers as useful bioreactors has gained considerable attention (De, Li, Gondi, & Sumerlin, 2008). The characteristics of enzyme activity in polymer–enzyme conjugates have been recently reviewed (Gauthier & Klok, 2010), and the field is constantly growing (Herzberger et al., 2015; Hou, Yuan, Zhou, Yu, & Lu, 2016; Pelegri-O'Day, Lin, & Maynard, 2014; Pelegri-O'Day & Maynard, 2016).

Covalently functionalized chymotrypsin with a polymerization initiator to facilitate surface grafting of a temperature-responsive polymer increased the enzyme stability at high temperatures, low pH, and against proteases (Cummings, Murata, Koepsel, & Russell, 2014). Such methodologies eliminate the need for tedious and relatively more expensive molecular biology methods to alter enzyme functions. Cascade biocatalysis by a bienzyme conjugate, glucose oxidase (GOx) and HRP, with poly(acrylic acid) (PAA) polymer that was adsorbed onto a graphene oxide (GO) support has been shown (Zore, Pattammattel, Gnanaguru, Kumar, & Kasi, 2015). This bienzyme–polymer conjugate with the GO support provided higher turnover rates, sustained, and significantly improved catalytic activity at high temperatures (65°C). At low pH (2.0) even in the presence of a surfactant denaturant, the rates were significantly higher compared to the mixture of the two enzymes dissolved in buffer solution. Such cascade enzymatic reactions would facilitate better reactant conversion efficiency by in situ production and efficient use of unstable reactive intermediates or by channeling reaction pathways that are not possible when using the enzymes separately.

Site-directed incorporation of polymers onto *Geobacillus thermocatenulatus* lipase has allowed improved chemoselectivity with a several fold activity enhancement (Romero, Rivero, Guisan, & Palomo, 2013). Design of filtration membranes for reversible immobilization of bioactive enzyme–polymer conjugates has been recently demonstrated (Moridi, Corvini, & Shahgaldian, 2015). Electrocatalytic studies showing stabilization of enzyme-like heme proteins covalently cross-linked with poly(lysine) on electrode surfaces allowed for higher stability under high temperature (90°C) and a wide range of pH (pH 2–11), when used as enzyme bio-electrodes, which was not the case with the free enzymes in solutions (Guto, Kumar, & Rusling, 2007, 2008; Panchagnula, Kumar, & Rusling, 2002). Overall, the significance of polymeric materials in enzyme bio-conjugation for superior catalytic applications can be understood based on restricted mobility or decreased conformational entropy of the denatured state. Further advancement of polymer–enzyme conjugation is reported here, where the enzyme–polymer conjugates are loaded onto nanomaterials to realize miniature bioreactors for synthesis and biosensing applications.

3. MNP–POLYMER–ENZYME CONJUGATES

Different types of nanomaterials such as nanoparticles, nanotubes, nanorods, and nanocomposites have been explored to carry out the function of enzyme immobilization. The fundamental objective of nanotechnology strategies is to create a large surface area for high-density loading of enzyme molecules in a small volume and additionally allowing a favorable microenvironment for efficient conversion of reactant to products. Polymer linkers are one of the advances of immobilizing enzymes to nanomaterials where the nanomaterials are functionalized with polymers containing desired surface groups to attach enzymes (Fujigaya & Nakashima, 2015; Rozenberg & Tenne, 2008).

Among various nanomaterials, MNPs have gained unique significance for enzyme immobilization (Ansari & Husain, 2012; Kaushik et al., 2008). Bare MNPs are unstable in strongly acidic solutions, and aggregation from strong magnetic attractions between particles is another drawback. MNPs that have been polymer functionalized are favorable when trying to produce high density of enzyme bioconjugated systems in a volume-efficient manner. Polymer chains show a wide range of elasticity and offer support to the chemical makeup of surface nanoparticles. They increase the dispersible nature of MNPs in aqueous solution, and they act as chemical linkers for enzyme attachment (Xu et al., 2014). Polyethylene glycol, polyvinylpyrrolidone, poly(ethylene-*co*-vinyl acetate), poly(lactic-*co*-glycolic acid), PAA, and polyvinyl alcohol were widely used as coating materials (Lee et al., 2006).

By appropriate design of functionalized MNPs to immobilize enzymes, one can achieve cost-effective, stable, scalable, and reusable biocatalytic systems for green synthesis of fine chemicals, specialty materials, and biosensing applications. The advantages of enzyme immobilization with MNPs include convenient preparation, efficient recovery, effective reusability, easy product isolation, and long-term storage stability of enzyme-bound MNPs in comparison to the free enzyme solution bioassays. For favorable enzyme interactions, there is a need to create surface-functionalized MNPs that contain properties that allow them to dissolve in water and that contain chemical groups to undergo covalent and noncovalent interactions with desired enzyme biocatalysts (Yong et al., 2008). However, it is important that the active-site geometry is not altered or blocked during covalent modification, which would impair the biological activities of the bound enzymes.

Cross-linking β-glucosidase using glutaraldehyde to MNPs allowed the immobilized enzyme to function at higher temperatures as well as to withstand a low pH with improved storage stability (Zhou, Pan, Wei, Wang, & Liu, 2013). Electropolymerized 4,7-di(furan-2-yl)benzo[*c*][1,2,5]thiadiazole on graphite rods has been used to coimmobilize silica-coated core–shell MNPs and acetylcholinesterase for sensitive electrochemical detection of organophosphorus pesticides (Dzudzevic Cancar et al., 2016). The combination of MNPs and polymeric matrix improved the biosensor's detection sensitivity and the stability. Functionalization of MNPs with a biocompatible reactive polymer, poly(2-vinyl-4,4-dimethylazlactone), for immobilization of L-asparaginase, has been shown to provide high enzyme loading, excellent retention of enzyme activity ($\sim$96%) even after 10 reuses of the conjugate, and retention of $\sim$73% of the initial activity after 10 weeks of storage (Mu, Qiao, Qi, Dong, & Ma, 2014).

Synthesis of monodispersed MNPs is the first step to functionalize with polymers and subsequent enzyme immobilization. Several synthetic pathways have been followed to prepare Fe_3O_4 MNPs. Physical methods such as gas-phase deposition and electron-beam lithography suffer from controlling the particle size in nanometer scale. Among wet chemical methods, sol–gel synthesis, oxidation method, hydrothermal reactions, flow-injection method, electrochemical methods, and chemical coprecipitation have been used for producing superparamagnetic MNPs (Reddy, Arias, Nicolas, & Couvreur, 2012). Due to superparamagnetism, all the spins are aligned along the same direction and single particles are magnetically saturated in the presence or absence of an external magnetic field (ci Acar, Garaas, Syud, Bonitatebus, & Kulkarni, 2005). This property is advantageous in magnetic separation (e.g., enzyme conjugation to MNPs and separation), drug delivery, and magnetic resonance imaging (Mahmoudi, Sant, Wang, Laurent, & Sen, 2011). In the subsequent sections, the synthesis, characterization, and biochemical/electrocatalytic activities of polymer-functionalized MNPs with attached enzymes are discussed.

3.1 Synthesis of PAA-Functionalized MNPs (PolyMNPs) (Mak & Chen, 2004)

General safety: Safety glasses, Latex, or nitrile gloves

Equipment:

1. Beaker (50–500 mL)
2. Stir/hot plate
3. Analytical balance
4. pH meter
5. Eppendorf tubes (1.5 mL)
6. Magnet (6000 G)
7. Magnetic stirrer bar

Chemicals and reagents:

1. PAA (25%, degree of polymerization = 2000–3000) from Showa Chemical Co., Tokyo, Japan
2. $FeCl_3{\cdot}6H_2O$ from J.T. Baker Chemical Co., NJ, USA
3. $FeCl_2{\cdot}4H_2O$ from Fluka Chemie GmbH, Buchs, Switzerland
4. NH_4OH [29.6%] from Tedia Co., OH, USA
5. *N*-(3-dimethylaminopropyl)-*N*′-ethylcarbodiimide hydrochloride (EDC) from Sigma Aldrich Co., MO, USA
6. Buffer A (0.003 *M* phosphate, pH 6, 0.1 *M* NaCl)
7. Milli-Q SP ultrapure water

Methodology:

The synthetic procedure (Scheme 1) of polyMNPs by chemical coprecipitation of Fe^{2+} and Fe^{3+} ions in ammonia solution as reported in prior literature is as follows (Mak & Chen, 2004).

1. Prepare $FeCl_3$ and $FeCl_2$ (molar ratio 2:1) solutions in water with an iron ion concentration of 0.3 *M*.
2. Stir the solution at 25°C by adding NH_4OH solution (29.6%) under vigorous stirring conditions for chemical precipitation. During this process, maintain the pH of the reaction medium at about 10.

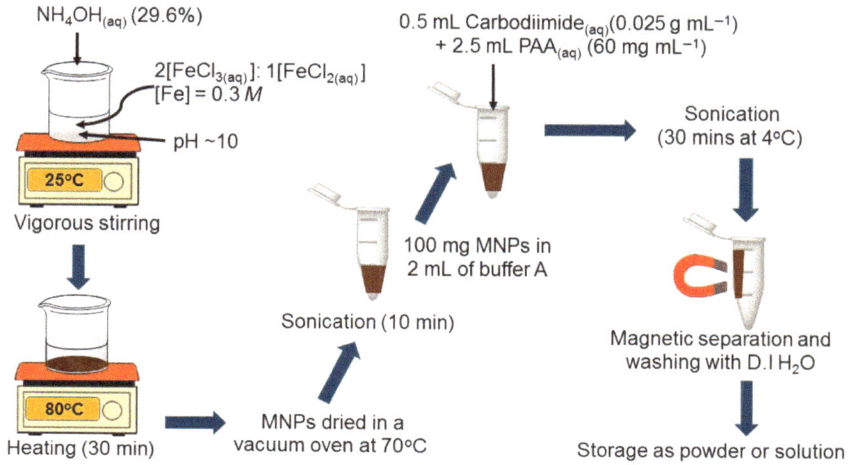

Scheme 1 Synthetic protocol for making polyMNPs.

3. Heat the precipitate at 80°C for 30 min followed by washing with deionized water and ethanol for several times.
4. Dry the precipitate well in a vacuum oven at 70°C.
5. Suspend 100 mg of the iron oxide nanoparticles in 2 mL of buffer A.
6. Sonicate the solution mixture for 10 min followed by the addition of 0.5 mL of the EDC solution (0.025 g mL^{-1} in buffer A).
7. To complete the PAA wrapping of MNPs, add 2.5 mL of PAA solution (60 mg mL^{-1} in buffer A) to the reaction mixture.
8. Sonicate the solution for 30 min at 4°C to complete the PAA wrapping of MNPs.
9. Separate the PAA-bound MNPs from the reaction using a permanent magnet.
10. Once the MNPs are settled, wash them several times with deionized water to remove any unbound PAA and extra reagents.
11. Store the nanoparticles as a dried powder or as a suspension in deionized water or buffer until further use.

3.2 Characterization of Polymer-Modified MNPs

MNPs can be tuned to interact with different chemical functionalities of polymers to enable enzyme immobilization. These would result in changes in MNP surface charge, size, and morphology that can be characterized by microscopic techniques. Fig. 1 and Table 1 present a literature reported

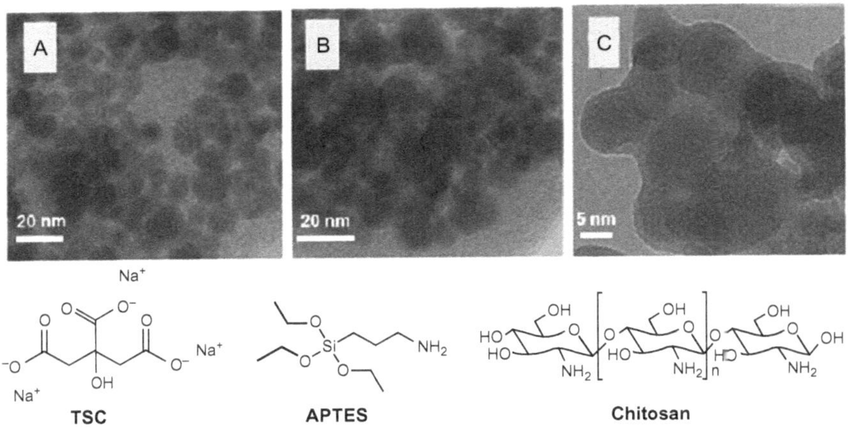

Fig. 1 TEM images of (A) TSC, (B) APTES, and (C) chitosan-modified MNPs. The chemical structures of TSC, APTES, and chitosan are shown below the respective images. *Reproduced from Bayat, N., Lopes, V. R., Sanchez-Dominguez, M., Lakshmanan, R., Rajarao, G. K., & Cristobal, S. (2015). Assessment of functionalized iron oxide nanoparticles in vitro: Introduction to integrated nanoimpact index.* Environmental Science: Nano, *2(4), 380–394, with permission from The Royal Society of Chemistry.*

Table 1 Physicochemical Characterization of MNPs Dissolved in Sterile Deionized Water With 5 m*M* NaCl (at 25°C)

MNP Modification Type	Hydrodynamic Size of the Aggregate (nm)	Zeta Potential (mV)	Polydispersity Index (PDI)
Unmodified MNPs	5845 ± 874	−27.1 ± 1.2	0.98 ± 0.03
TSC	810 ± 42	−32.1 ± 1.7	0.51 ± 0.04
APTES	5338 ± 189	4.4 ± 0.5	1.00 ± 0.00
Chitosan	1223 ± 71	40.5 ± 0.5	0.58 ± 0.00

Some contents in this table were reproduced from Bayat, N., Lopes, V. R., Sanchez-Dominguez, M., Lakshmanan, R., Rajarao, G. K., & Cristobal, S. (2015). Assessment of functionalized iron oxide nanoparticles in vitro: Introduction to integrated nanoimpact index. *Environmental Science: Nano*, *2*(4), 380–394, with permission from The Royal Society of Chemistry.

characterization data of MNPs wrapped with trisodium citrate (TSC), aminopropyl-triethoxysilane (APTES), and chitosan (Bayat et al., 2015). Transmission electron microscopy (TEM) images (TECNAI G2 Spirit Bio TWIN (FEI Company) 120 kV with a Tungsten filament), zeta potential, and hydrodynamic size characterizations (Malvern Zetasizer Nano series V5.03 (PSS0012-16 Malvern Instruments, UK) have been employed.

The polymer-modified MNPs maintained spherical or near-spherical structures as in the unmodified MNPs.

The surface amine groups of APTES and chitosan produced a net positive charge, whereas the carboxyl groups of citrate imparted a net negative charge to MNPs, as evident from zeta potential analysis. The significantly larger hydrodynamic sizes of MNPs suggest that there is aggregation and/or agglomeration due to the intrinsic magnetic properties as well as lack of suitable surfactants to stabilize the particles. Therefore, it is clear that polymer wrapping helps to minimize the degree of aggregation of single MNPs.

The design of bioconjugates of enzyme-like proteins by attaching them to PAA-functionalized MNPs for scalable and reusable bio- and electrocatalytic applications is described next.

3.3 Preparation of Covalent and Electrostatically Adsorbed MB With PolyMNPs

General safety: Safety glasses, Latex, or nitrile gloves

Equipment:

1. Eppendorf tubes (1.5 mL)
2. Micropipettes (0.1–1.0 mL)
3. Magnet (6000 G)
4. Tube rotator

Chemicals and materials:

1. PAA-functionalized MNPs (polyMNPs, 100 nm hydrodynamic diameter) from Chemicell Inc., Berlin, Germany
2. Equine heart myoglobin (MB) from Sigma Aldrich Co., MO, USA
3. Poly(ethyleneimine) (PEI) from Sigma Aldrich Co., MO, USA
4. 1-Ethyl-3-[3-dimethylaminopropyl] carbodiimide hydrochloride (EDC) from Sigma Aldrich Co., MO, USA
5. *N*-hydroxysuccinimide (NHS) from Sigma Aldrich Co., MO, USA
6. High-purity graphite disk electrodes (HPGs, geometric area 0.2 cm^2) were purchased from McMaster-Carr (Atlanta, GA)
7. Phosphate-buffered saline (PBS, 10 m*M*, pH 7.4)
8. Acetate buffer (pH 5.0)
9. Milli-Q ultrapure water

Methodology:

The procedure shown in Scheme 2 was followed to prepare the MB–polyMNP conjugates (Premaratne et al., 2016; Walgama & Krishnan, 2013).

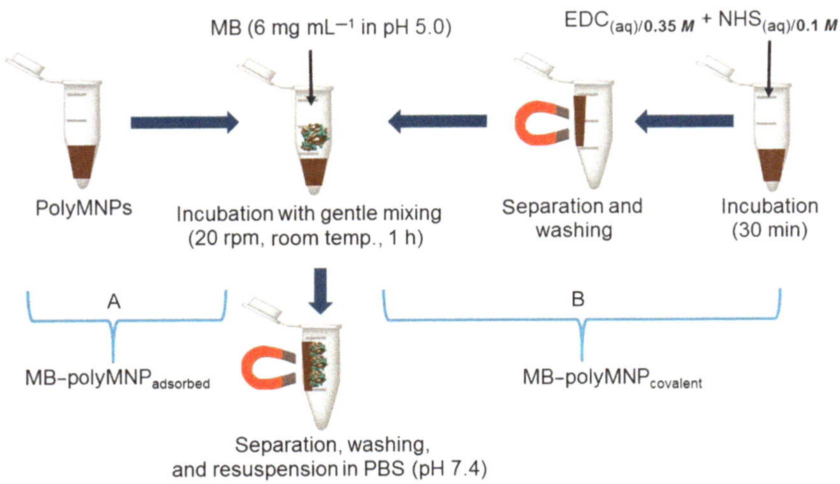

Scheme 2 Synthetic protocols for (A) MB/polyMNP$_{adsorbed}$ and (B) MB–polyMNP$_{covalent}$ preparations.

3.3.1 MB–PolyMNP$_{covalent}$ Conjugate

Covalent attachment of MB by conjugating the free amine groups of surface lysine residues to carboxyl groups of polyMNPs was conducted by the following method.

1. Add polyMNPs to freshly prepared solutions of EDC (0.35 *M*) and NHS (0.1 *M*) and gently vortex.
2. Incubate the reaction mixture for 15 min to activate the carboxyl groups via the formation of easily leaving *N*-succinimidyl ester groups.
3. Separate the activated polyMNPs from the reaction mixture by the magnet to remove unreacted EDC/NHS.
4. Wash the polyMNPs with deionized water to remove any residual EDC/NHS.
5. Suspend the particles immediately in a freshly prepared MB solution (1 mL of 6 mg mL^{-1} in pH 5.0 acetate buffer).
6. Allow the reaction mixture to gently mix in a tube rotor (rotated at 20 rpm) at room temperature for 1 h for covalent attachment to proceed.
7. Separate the MB–MNP conjugate from the reaction mixture by use of the magnet followed by two washes with PBS.
8. Finally resuspend the conjugates in PBS (pH 7.4, 1 mL) for further studies.

3.3.2 Preparation of MB/PolyMNP$_{adsorbed}$ Particles

Physisorption of MB to polyMNPs is facilitated by the electrostatic attachment of positively charged MB to negatively charged polyMNPs.

1. Maintain a net positive charge of MB by maintaining an acidic pH of the medium (MB isoelectric points are 6.8 and 7.2) (Goto & Fink, 1990; Work & Work, 1976).
2. Suspend 2 mg of polyMNPs in PBS to maintain negatively charged PAA groups (pK_a 4.3) (Leaist, 1989).
3. Add the polyMNPs to 1 mL of 6 mg mL^{-1} MB and gently vortex.
4. Incubate the reaction mixture for 2 h at room temperature.
5. Separate the physisorbed MB–polyMNPs conjugates using a magnet followed by washing with PBS, twice.
6. Resuspend the MB/polyMNP$_{adsorbed}$ in 1 mL of PBS for further use.

3.4 Fabrication of Pyrolytic Graphite Electrode for Electrochemical Studies

General safety: Safety glasses, Latex, or nitrile gloves

Chemicals:

1. Ethanol
2. Milli-Q ultrapure water
3. PEI

Materials and equipment:

1. HPG, geometric area 0.2 cm^2 disk electrodes were purchased from McMaster-Carr (Atlanta, GA).
2. P320 grit SiC polishing paper from Extec Corp., CT, USA.
3. Micropipettes (5.0–50.0 μL, 0.1–1.0 mL).

Methodology:

1. Freshly polish HPG electrodes on P320 grit SiC paper, ultrasonicate for 30 s in ethanol, followed by 30 s in DI water.
2. Dry the electrodes well with nitrogen gas.
3. Coat the electrodes immediately with 20 μL of the PEI solution (2 mg mL^{-1} in DI water) and allow to incubate for 15 min to adsorb to the electrode surface.
4. Gently rinse the excess PEI with DI water.
5. Immediately coat the electrodes with 15 μL of the prepared MB–polyMNPs conjugates (net negative charge in PBS, pH 7.4) onto the cationic polymer layer of HPG electrodes.
6. Incubate the electrodes at 4°C for 30 min for electrostatic adsorption of conjugates with the PEI layer.

7. Finally rinse the unbound conjugates using deionized water.
8. Similarly prepare two control PEI-coated HPG electrodes by coating 15 μL of MB (6 mg mL^{-1}) or polyMNPs (0.5 mg mL^{-1})

3.5 Quantitative Assessment of Biocatalytic Properties of MB–PolyMNP Conjugates

3.5.1 Oxidative ABTS Catalysis

General safety: Safety glasses, Latex, or nitrile gloves

Chemicals:

1. H_2O_2 (≥30%) from Sigma Aldrich Co., MO, USA
2. Diammonium salt of ABTS from Sigma Aldrich Co., MO, USA
3. Milli-Q SP ultrapure water

Materials and equipment:

1. Magnet (6000 G)
2. Magnetic stirrer bar
3. Glass vials (6 mL)
4. Micropipettes (20.0–200.0 μL, 0.1–1.0 mL)
5. Varian Cary 100 Bio UV–vis spectrophotometer

Methodology:

1. Take 1 mL of MB–$polyMNP_{covalent}$ (made up of 1, 5, or 10 mg of polyMNPs) conjugates.
2. Prepare fresh ABTS (20 m*M*) and H_2O_2 (10 m*M*) in deionized water.
3. Add 1500 μL of 20 m*M* ABTS solution and 300 μL of 10 m*M* H_2O_2 solution, followed by addition of 200 μL of PBS (final concentrations: 10 m*M* ABTS, 1 m*M* H_2O_2).
4. Allow the reaction to occur for 10 min at 25°C under constant stirring of 100 rpm.
5. Separate the conjugates from the reaction mixture using the magnet immediately and wash twice using PBS.
6. The resulting colored product (ABTS radical cation: $\lambda_{max} = 415$ nm) was measured by the UV–vis spectrophotometer.

3.5.2 Electrocatalytic Peroxide Reduction

General safety: Safety glasses, Latex, or nitrile gloves

Chemicals:

1. *Tert*-butylhydroperoxide (*t*-BuOOH) (≥70% in H_2O (W/V)) from Sigma Aldrich Co., MO, USA
2. PBS (pH 7.4)
3. Milli-Q ultrapure water

Materials and equipment:

1. HPG electrodes modified with conjugates
2. Ag/AgCl reference electrode
3. Pt counter electrode
4. Graduated glass syringe (100 μL, 500 μL)
5. Nitrogen/argon gas
6. Electrochemical cell
7. Electrode rotor attached to a rotation controller unit
8. CH instrument 6017 electrochemical analyzer

Methodology:

1. Fill the electrochemical cell with 15.0 mL of PBS and purge with nitrogen/argon gas for 15 min.
2. Insert the reference electrode, counter electrode, and working electrode in the electrochemical cell.
3. Use the HPG electrodes coated with MB/polyMNP$_{adsorbed}$, MB–polyMNP$_{covalent}$ (both made of 2 mg of polyMNPs), polyMNPs, or MB as working electrodes.
4. While continuously purging the cell with nitrogen/argon gas, inject various concentrations of *t*-butylhydroperoxide (*t*-BuOOH) by standard addition method (0.0, 0.01, 0.02, 0.04, 0.08, 0.10, 0.20, 0.40, 0.60, 1.0, 1.4, 2.0, 2.8, 4.4, 7.5, and 10.7 m*M*) while constantly rotating the working electrode at 1000 rpm using an electrode rotor unit.
5. For these rotation disk voltammetric experiments, use a scan rate of 0.1–0.3 V s^{-1}.

3.6 Characterization of MB–PolyMNP Conjugates

3.6.1 Fourier Transform Infrared Spectroscopy

Fourier transform infrared (FTIR) characterization is conducted using Thermo Scientific Nicolet iS50 in the attenuated total reflectance (ATR) mode. The samples to be analyzed were placed directly on the ATR diamond crystal, and 32 scans were run and averaged to obtain a good signal-to-noise ratio. The changes occurring in the surface chemical functionalities of free polyMNPs, free MB, and after covalent or noncovalent immobilization of MB onto polyMNP are shown in Fig. 2. In our study, a film of each material was drop cast on a diamond substrate, and the FTIR spectra acquired. PolyMNPs displayed a stretching vibration at 581 cm^{-1} arising from the Fe–O bonds and a broad O–H stretching band at 3032 cm^{-1} from –COOH end groups of PAA (Fig. 2a). MB film showed a –C=O stretching amide I band at 1630 cm^{-1}, an N–H bending amide II

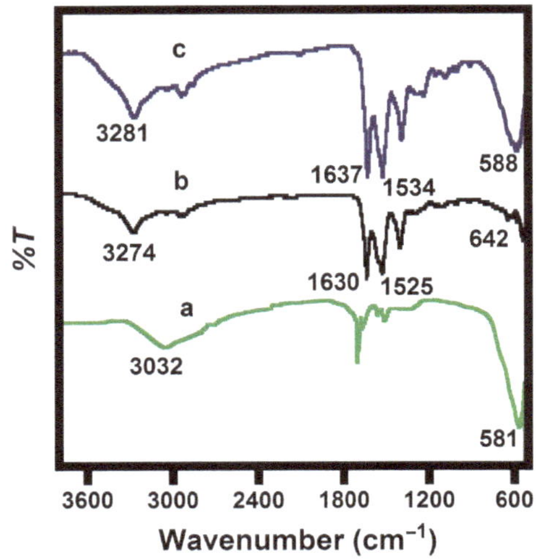

Fig. 2 FTIR spectra of (a) polyMNPs, (b) MB, and (c) MB–polyMNP conjugates. *Reproduced from Premaratne, G., Nerimetla, R., Matlock, R., Sunday, L., Koralege, R. S. H., Ramsey, J. D., et al. (2016). Stability, scalability, and reusability of a volume efficient biocatalytic system constructed on magnetic nanoparticles.* Catalysis Science & Technology, *6(7), 2361–2369, with permission from The Royal Society of Chemistry.*

band at 1525 cm^{-1}, and an N–H stretching band at 3274 cm^{-1} that are characteristics of protein amide bond vibrations (Fig. 2b). The amine I and amide II bands blue-shifted to 1637 and 1534 cm^{-1}, respectively, after covalent conjugation of MB with the polyMNP. The blue-shifted N–H stretching band to 3281 cm^{-1} suggests stronger amide bond strengths in the MB–polyMNP conjugates than the MB alone (Fig. 2c). Thus, the covalent conjugation of MB with polyMNPs was confirmed.

3.6.2 Transmission Electron Microscopy

The polyMNPs and MB–polyMNP$_{covalent}$ (1, 5, or 10 mg MNPs used) conjugates were additionally characterized by (TEM, JEOL JEM-2100). Before imaging, the conjugates were drop cast on carbon surfaces of copper grids. TEM revealed important information with regard to variations in particle size and surface morphology that are brought out by enzyme immobilization on polyMNPs (Noureddini & Gao, 2007). TEM images of only polyMNPs (Fig. 3A) and the covalent conjugates of MB–polyMNP prepared with increasing amount of polyMNP (1, 5, or 10 mg) at a constant MB amount in the reaction solution (800 μL of 6 mg mL^{-1} in pH 5.0 acetate buffer) are shown in Fig. 3B–D. During the conjugation step, MB was prepared in

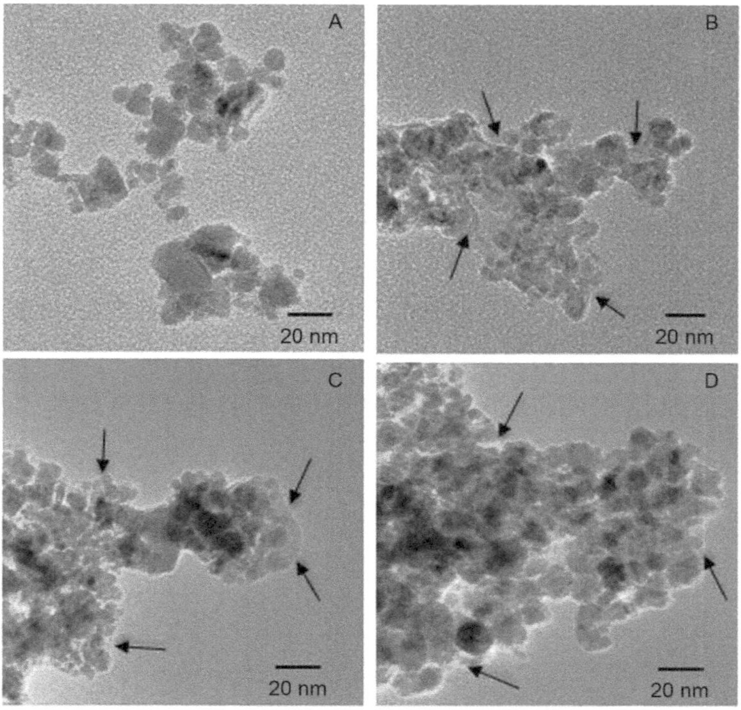

Fig. 3 TEM images of (A) free polyMNPs, (B) MB–polyMNP$_{1mg}$, (C) MB–polyMNP$_{5mg}$, and (D) MB–polyMNP$_{10mg}$ conjugates. The MB film around MNPs in each conjugate is indicated by *arrow heads. Reproduced from Premaratne, G., Nerimetla, R., Matlock, R., Sunday, L., Koralege, R. S. H., Ramsey, J. D., et al. (2016). Stability, scalability, and reusability of a volume efficient biocatalytic system constructed on magnetic nanoparticles.* Catalysis Science & Technology, *6(7), 2361–2369, with permission from The Royal Society of Chemistry.*

acidic pH 5.0 buffer to create a net positive surface charge suitable to react with negatively charged polyMNP; and after the conjugation reaction, the conjugates were resuspended in physiological pH 7.4 buffer. The conjugates are denoted as MB–polyMNP$_{1mg}$, MB–polyMNP$_{5mg}$, and MB–polyMNP$_{10mg}$, respectively (Fig. 3B–D). Here, we used varying amounts of polyMNPs to examine the conjugation efficiency of MB with the goal to maximize MB attachment with polyMNPs.

Due to drying and high vacuum conditions required for TEM measurements of the conjugates, the spherically shaped polyMNPs were much smaller in size than the actual hydrodynamic size of 100 nm diameter (Fig. 3A; Singh & Krishnan, 2015). A thin-coated layer of MB around polyMNP was observed in the conjugates but not in the polyMNPs alone. With an increase in amount of the polyMNP the coated layer of MB became more prominent. This indicated greater attachment of MB molecules with the increasing numbers of surface –COOH groups of polyMNPs. Another important observation from the TEM was the extended aggregation of MB–polyMNP conjugate with increasing polyMNP amount, suggesting more accumulation of the MB around the polyMNPs.

4. COVALENT VS NONCOVALENT IMMOBILIZATION OF PEROXIDASE ACTIVE PROTEINS ON PolyMNPs. WHAT DIFFERENCE DOES IT MAKE IN THE ELECTROCATALYTIC ACTIVITY AND KINETICS?

Understanding the mechanism of electron transfer and redox catalytic properties of redox enzyme films on electrodes is useful in designing bioreactors, biosensors, and synthesis of enzyme biocatalysts. Successfully connecting enzyme redox sites with an electrode surface to receive electrons (reducing equivalents) as well as retaining the bioactive confirmation of enzymes with activity and stability are challenges of designing electrochemical enzyme bioelectrodes. We compared the properties of the MB–polyMNP$_{covalent}$ conjugate with the corresponding noncovalent electrostatically bound MB/polyMNP$_{adsorbed}$, only polyMNPs, and only MB films coated on the polycation-adsorbed HPG electrodes. For this, PEI was used as the polycation to coat the electrode to anchor MB electrostatically as explained later: MB is negatively charged at pH 7.4 (pI = 6.8 and 7.2) (Work & Work, 1976), polyMNP is also negatively charged (pK_a of PAA group is ~4.3) (Leaist, 1989), and hence the conjugate has net negative charge. The conjugate then would electrostatically bind to the positively charged, PEI-modified electrode surface. The organization of the designed films on electrodes is shown in Fig. 4.

4.1 Electrocatalytic Properties

Electrochemically or chemically reduced heme protein films from ferric heme to ferrous heme can catalytically reduce *t*-BuOOH to *tert*-butanol (Guto & Rusling, 2005; Li, Xu, Yao, Zhu, & Chen, 2006). The increments in currents are due to the reduction of peroxides by formation of reactive MB–Fe^{IV}=O species, followed by its reduction to regeneration of MB–Fe^{III} (Li et al., 2006). Fig. 5 shows the steady-state voltage-driven *t*-BuOOH reduction currents catalyzed by the designed MB–polyMNP$_{covalent}$,

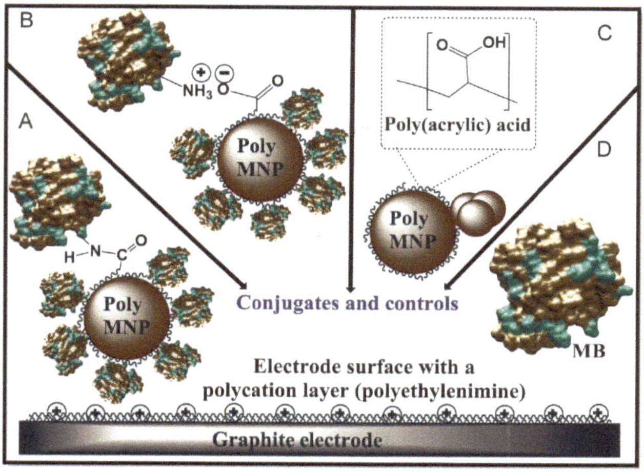

Fig. 4 Schematic illustrations of the design of (A) MB–polyMNP$_{covalent}$, (B) MB/polyMNP$_{adsorbed}$, (C) only polyMNP, and (D) only MB coated as films on PEI-adsorbed electrodes for biosensing and electrocatalytic studies.

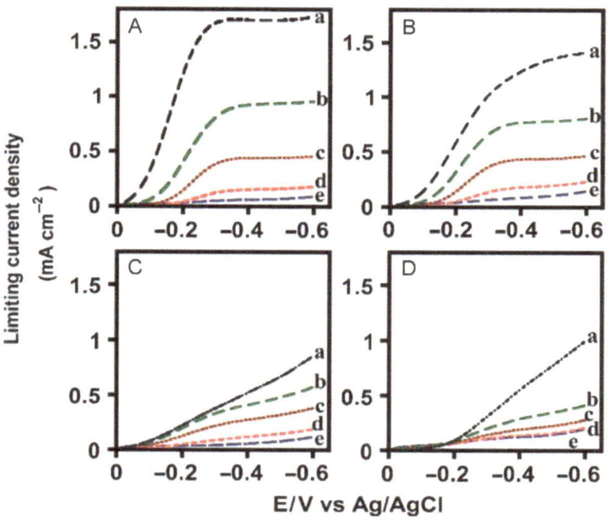

Fig. 5 Voltammograms representing steady-state catalytic reduction current densities with *t*-BuOOH concentration (a) 2.8, (b) 1.0, (c) 0.4, (d) 0.1, and (e) 0 m*M*) in phosphate-buffered saline (PBS, pH 7.4) continuously purged with N_2, 25°C for (A) MB–polyMNP$_{covalent}$, (B) MB/polyMNP$_{adsorbed}$, (C) only MB, and (D) only polyMNP films immobilized on the PEI-coated HPG electrodes at a scan rate of 0.1 V s^{-1} and 1000 rpm electrode rotation rate. The electrode rotation facilitates steady-state peroxide mass transport (by induced convection) from the bulk solution to electrode surface. *Adapted with permission from Walgama, C., & Krishnan, S. (2013). Electrocatalytic features of a heme protein attached to polymer-functionalized magnetic nanoparticles.* Analytical Chemistry, *85(23), 11420–11426, Copyright 2013 American Chemical Society.*

$MB/polyMNP_{adsorbed}$, only MB, and only polyMNP films coated on electrodes (methodology in Sections 3.3–3.5). The covalent films provided higher peroxide reduction currents than the noncovalent MB film, and more significantly than the MB film alone or the film of polyMNP alone bound on electrodes. Similar to the trend of catalytic currents, we discovered better electron transport in the film of covalent conjugate than others as measured by cyclic voltammetry. We proposed that the H-bond network formed by the amide bond linkages between the polyMNPs and MB in the case of $MB–polyMNP_{covalent}$ film facilitated better electron hopping property and electrocatalytic currents than the films of $MB/polyMNP_{adsorbed}$, only MB, and only polyMNP coated on electrodes (Walgama & Krishnan, 2013).

4.2 Electrochemical Michaelis–Menten Kinetics

We plotted the increase in reduction currents with increasing concentration of *t*-BuOOH and determined the agreement of the experimental data with Michaelis–Menten kinetics (Fig. 6). The original equation as applied to any homogeneous enzymatic bioassay can be written as

$$v = \frac{v_{\max} C_S}{K_M + C_S}$$

where K_M is the Michaelis–Menten constant (measure of enzyme–substrate affinity, lower K_M means better affinity), C_S is the molar substrate concentration in solution, v is the initial velocity of the reaction, and $v_{\max}$ is the maximum reaction velocity.

Similar to the homogeneous Michaelis–Menten kinetics for enzyme solution assays, the electrochemically driven heterogeneous enzyme catalysis can be represented by relating the catalytic current (I_{cat}) with reaction rate parameters as given by Guto and Rusling (2005), Heering, Hirst, and Armstrong (1998), and Krishnan, Schenkman, and Rusling (2011)

$$I_{cat} = \frac{nFA\Gamma(k_{cat}/K_M)C_S}{(1/K_M)C_S + 1}$$

where n is the number of electrons transferred in the catalytic reaction ($n = 2$ for peroxide reduction by heme MB), F is Faraday's constant, A is electrode area (0.2 cm^2), Γ is the electroactive enzyme amount coated on the electrode surface (usually in nmol to pmol cm^{-2}), C_S is the reactant concentration added in solution (usually in micro- to millimolar range), k_{cat} is the

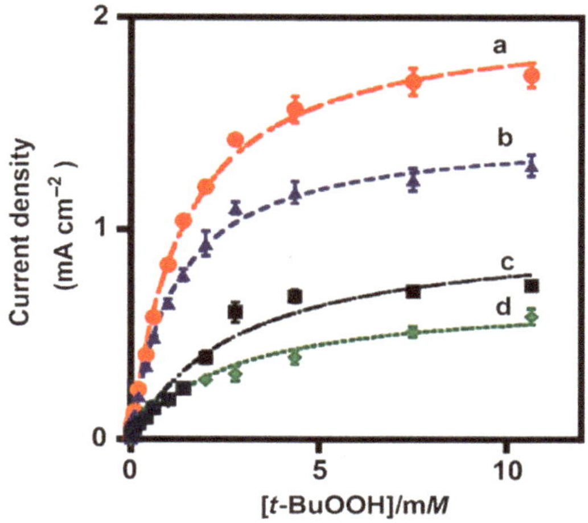

Fig. 6 Michaelis–Menten electrochemical kinetics for *t*-BuOOH reduction by (a) MB–polyMNP$_{covalent}$, (b) MB/polyMNP$_{adsorbed}$, (c) only polyMNP, and (d) only MB films (av ± SD for $N=3$ repeats). *Adapted with permission from Walgama, C., & Krishnan, S. (2013). Electrocatalytic features of a heme protein attached to polymer-functionalized magnetic nanoparticles.* Analytical Chemistry, *85(23), 11420–11426, Copyright 2013 American Chemical Society.*

catalytic rate constant (s^{-1}), and K_M is the Michaelis–Menten constant. Γ is obtained from the peak area of the MB-heme reduction current in an anaerobic (N_2 or Ar purged buffer) cyclic voltammogram that corresponds to charge in coulombs (Q) and by using the equation, $Q=nFA\Gamma$. The decreasing order of catalytic peroxide reduction current density observed among the designed MB films was (Fig. 6): MB–polyMNP$_{covalent}$ > MB/polyMNP$_{adsorbed}$ > only polyMNP > only MB.

The apparent K_M, calculated from Fig. 6, was found to be 1.36 (±0.2) m*M* for MB–polyMNP$_{covalent}$, 1.15 (±0.05) m*M* for MB/polyMNP$_{adsorbed}$, 1.6 (±0.2) m*M* for only MB, and 2.9 (±0.1) m*M* for only polyMNP films. This kinetic analysis indicates that enzyme–substrate affinity can be improved by both covalent and noncovalent conjugation of enzymes with polyMNPs compared to either MB or polyMNP film alone. The increasing order of relative catalytic reduction efficiency ($10^4\ k_{cat}/K_M$ in $M^{-1}\ s^{-1}$) was: only polyMNP (5.2 ± 0.5) < only MB (5.5 ± 0.4) < MB–polyMNP$_{covalent}$ (14.2 ± 0.7) ≈ MB/polyMNP$_{adsorbed}$ (14.8 ± 1.0).

After 1 week of storage in PBS (4°C), the MB–polyMNP$_{covalent}$ film on electrodes retained ~85% of the initial electrocatalytic current, while MB/polyMNP$_{adsorbed}$ film retained only ~60% of the initial electrocatalytic activity. Although comparable k_{cat}/K_M was observed between the covalent and electrostatic MB films on polyMNPs, the covalent film is preferred for its better stability and larger peroxide reduction currents than the corresponding adsorbed MB film. In summary, the covalent enzyme–polymer conjugation on large surface area nanoparticles support offered better electrocatalytic and stability properties than the corresponding electrostatic films. This study allowed us to further focus on the covalent conjugates to examine reusability and scalability features by both biochemical and electrochemical assays useful for catalytic applications as discussed later.

4.3 Assessment of the MB–PolyMNP$_{covalent}$ Conjugate for Stability, Scalability, and Reusability Features

In this study, we examined the catalytic stability, scalability, and reusability features of the MB–polyMNP$_{covalent}$ bioconjugate prepared with increasing amounts of polyMNPs (from 1 to 25 mg range) at a constant MB mass. The conjugates were taken in a buffer suspension to which hydrogen peroxide and substrate ABTS were added. The oxidized ABTS product was quantitated by UV–vis absorbance. The MB was retained on the surface of the polyMNP conjugate, and this eliminates interferences during product analysis after magnetic separation of the conjugates to the sides of the glass vials (Fig. 7). In contrast, reaction of free MB in solution for ABTS oxidation would require additional analytical techniques (such as chromatography) to isolate the dissolved MB component from the reaction mixture, plus the majority of the original MB activity should be retained after isolation

Fig. 7 Easy reaction procedure and product isolation from a reaction mixture containing MB–polyMNP$_{covalent}$ conjugate by use of a simple magnet.

for reuse. This adds laborious technical work, longer time, tedious analysis steps, and additional analytical instrumentation. MB–polyMNP$_{covalent}$ displayed 50% of the initial MB activity even after 4 months of storage and about 40% of the initial product was formed even after four reuses for ABTS oxidation in the presence of hydrogen peroxide. This is economical and very suitable for synthesizing sufficient products from a minimal use of precious enzymes. Together, the simplicity and enhancement of product yields by the design of enzyme–polyMNP conjugates are preferred for scalable enzyme-catalyzed bioreactors.

Fig. 8A shows the scalable amount of ABTS oxidation product catalyzed by the MB–polyMNP$_{covalent}$ conjugates prepared with increasing amounts of polyMNPs (1, 5, 10, and 25 mg) in PBS containing 10 m*M* ABTS and 1 m*M* H_2O_2. Also shown are the product levels catalyzed by only MB or only polyMNP under similar conditions. The increase in product yields with increasing polyMNP amounts used for MB conjugation is demonstrated here. Moreover, the polyMNPs conjugated MB produced higher product amounts than the individual component of polyMNPs or only MB.

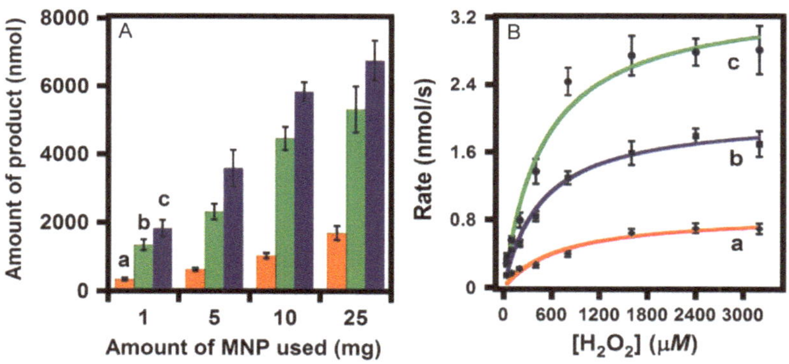

Fig. 8 (A) ABTS oxidation product yields catalyzed by (a) free polyMNPs, (b) free MB, and (c) MB–polyMNP$_{1mg}$ conjugate. The corresponding comparative data for samples of MB–polyMNP$_{5mg}$, MB–polyMNP$_{10mg}$, and MB–polyMNP$_{25mg}$ along with the respective free polyMNPs and free MB controls for each case are also shown as labeled in the *X*-axis. ABTS oxidation was carried out in PBS solution in the presence of hydrogen peroxide at 25°C, pH 7.4. (B) The Michaelis–Menten plots of (a) only polyMNPs, (b) only MB, and (c) the MB–polyMNP$_{1mg}$ conjugate suspended in buffer solutions toward ABTS oxidation for various concentrations of H_2O_2. *Reproduced from Premaratne, G., Nerimetla, R., Matlock, R., Sunday, L., Koralege, R. S. H., Ramsey, J. D., & Krishnan, S. (2016). Stability, scalability, and reusability of a volume efficient biocatalytic system constructed on magnetic nanoparticles.* Catalysis Science & Technology, *6(7), 2361–2369, with permission from The Royal Society of Chemistry.*

Table 2 Kinetic Parameters Quantified From the Michaelis–Menten Plot for Peroxide-Driven ABTS Oxidation by MB, polyMNPs, and MB–polyMNP$_{1mg}$

	PolyMNPs	MB	MB–PolyMNP$_{1mg}$
K_M (μM)	711 ± 121	482 ± 85	504 ± 91
v_{max} (nmol s^{-1})	0.9 ± 0.1	2.0 ± 0.1	3.4 ± 0.2
k_{cat} (s^{-1})	0.0006 ± 0.21 ($\times 10^{-4}$)	0.085 ± 0.012	0.153 ± 0.037

Reproduced from Premaratne, G., Nerimetla, R., Matlock, R., Sunday, L., Koralege, R. S. H., Ramsey, J. D., et al. (2016). Stability, scalability, and reusability of a volume efficient biocatalytic system constructed on magnetic nanoparticles. *Catalysis Science & Technology*, *6*(7), 2361–2369, with permission from The Royal Society of Chemistry.

Fig. 8B represents the Michaelis–Menten kinetic plots for various concentrations of H_2O_2 on ABTS oxidation catalyzed by the MB–polyMNP$_{1mg}$ conjugate or only polyMNP or only MB dispersed in solution. The kinetic parameters (K_M, v_{max}, k_{cat}) estimated from the Michaelis–Menten fit are summarized in Table 2. The strongest affinity for peroxides was observed for the Mb–polyMNP$_{1mg}$ conjugate compared to the peroxide affinity of the free MB in solution or the polyMNP alone. The k_{cat} values indicate that the covalent conjugate displayed 2- and 250-fold greater turnover rates than only MB or only polyMNP, respectively (Table 2).

4.4 Influence of PolyMNP Amount Used for the MB–PolyMNP$_{covalent}$ Conjugation in Electrochemically (Voltage) Driven Peroxide Reduction

MB–polyMNP$_{covalent}$ conjugates displayed enhanced peroxide-mediated ABTS oxidation product with increase in polyMNP amount used in the MB conjugation. Fig. 9 represents the electrochemically driven catalytic reduction of *t*-BuOOH to *t*-butanol by coated films of MB–polyMNP conjugates (made with 1, 5, or 10 mg polyMNPs), only MB, or only polyMNP on PEI-adsorbed graphite electrodes. The shift in onset catalytic reduction potentials of the voltammograms toward positive direction with increasing amounts of MNPs in the MB–polyMNP conjugates can be noted and is more positive than electrocatalysis by the free MB or only polyMNP films adsorbed on PEI-coated electrodes (a–e in Fig. 9).

The onset of peroxide reduction potentials was in the order of polyMNP (-209 ± 13 mV) < MB (-189 ± 11 mV) < MB–polyMNP$_{1mg}$ (-166 ± 9 mV) < MB–polyMNP$_{5mg}$ (-148 ± 10 mV) < MB–polyMNP$_{10mg}$ (-131 ± 7 mV) with respect to Ag/AgCl reference electrode. This property

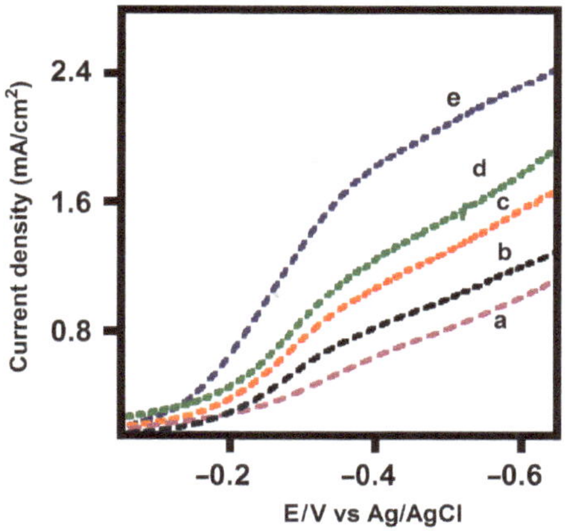

Fig. 9 Representative steady-state voltage-driven *t*-BuOOH (5 m*M*) catalytic reduction voltammograms by films of (a) free polyMNPs, (b) free MB, (c) MB–polyMNP$_{1mg}$, (d) MB–polyMNP$_{5mg}$, and (e) MB–polyMNP$_{10mg}$ conjugates adsorbed on polished PEI-coated HPG electrodes in oxygen-free PBS, 25°C, scan rate 0.3 V s^{-1}, 1000 rpm electrode rotation rate. *Reproduced from Premaratne, G., Nerimetla, R., Matlock, R., Sunday, L., Koralege, R. S. H., Ramsey, J. D., & Krishnan, S. (2016). Stability, scalability, and reusability of a volume efficient biocatalytic system constructed on magnetic nanoparticles.* Catalysis Science & Technology, *6(7), 2361–2369, with permission from The Royal Society of Chemistry.*

suggests that MB–polyMNP conjugates had an energetically favored reduction process, whereas much more negative overpotentials were required for only MB or only polyMNP films to be electrocatalytically active to reduce peroxide. This concludes that the amounts of MNP used in the preparation of MB–polyMNP conjugates have a direct influence in controlling the catalytic efficiency, by facilitating better electron transport in the films.

5. EFFECT OF MNP SIZE ON ACTIVITY AND RECOVERY OF GOx (PARK, MCCONNELL, BODDOHI, KIPPER, & JOHNSON, 2011)

In this report, three different sizes of MNPs (5, 26, and 51 nm) were synthesized and modified with 3-(aminopropyl)triethoxysilane (APTES) and glutaraldehyde to covalently immobilize GOx to the surface and study the influence of the particle size on the enzyme activity and reusability capabilities.

General safety: Safety glasses, Latex, or nitrile gloves

Chemicals and reagents:

1. GOx type II-S solid from *Aspergillus niger* (D-glucose: oxygen 1-oxidoreductase, EC 1.1.3.4) from Sigma Aldrich Co., MO, USA
2. APTES (99%) from Sigma Aldrich Co., MO, USA
3. Iron (II) chloride (99%) from Sigma Aldrich Co., MO, USA
4. Iron (III) chloride (99%) from Sigma Aldrich Co., MO, USA
5. NH_4OH (29.5%) from Fisher Scientific, PA, USA
6. Glutaraldehyde (8% aqueous) from Sigma Aldrich Co., MO, USA
7. Hydrogen peroxide (30%) from Fisher Scientific, PA, USA
8. PBS solution (10× PBS: pH 7.4, 137 m*M* NaCl, 2.7 m*M* KCl, 8 m*M* Na_2HPO_4, and 2 m*M* KH_2PO_4) from Invitrogen Co., CA, USA
9. GOx assay kit (Megazyme Inc., Wicklow, Ireland)

Materials and equipment:

1. Beaker (100–500 mL)
2. Disposable glass cuvettes
3. Analytical balance
4. Stir and hot plate
5. Magnetic stirrer bar
6. Neodymium magnets (DY0X0-N52; 14,800 G)

5.1 Synthesis of Three Different Sizes of MNPs

5.1.1 Methodology: Coprecipitation (5 nm)

1. Prepare Fe^{2+}/Fe^{3+} with a molar ratio of 0.5 (0.0125 *M*/0.025 *M*) in 51 mL of deionized water.
2. Add NH_4OH (29.5%) at a rate of 0.2 mL min^{-1} to maintain the pH at 10 (2.8–2.9 mL) with a constant stirring of the solution.
3. Allow the precipitate to be formed at room temperature.

5.1.2 Methodology: Oxidative Alkaline Hydrolysis at Room Temperature (26 nm)

1. Prepare 51 mL of a 0.05 *M* of $FeCl_2$ solution in deionized water.
2. Prepare an aqueous suspension of $Fe(OH)_2$ by adding 1 *M* KOH, adjusting to pH 7.9–8 with stirring.
3. Add the suspension to the $FeCl_2$ solution and allow for precipitate formation at room temperature for 2 h.

5.1.3 *Methodology: Oxidative Alkaline Hydrolysis at High Temperature (51 nm)*

1. Prepare 51 mL of a 0.05 *M* of $FeCl_2$ solution in deionized water and heat it for 90°C.
2. Prepare an aqueous suspension of $Fe(OH)_2$ by adding 1 *M* KOH, adjusting to pH 7.9–8 with stirring.
3. Add the suspension to the $FeCl_2$ solution and allow for precipitate formation for 2 h at 90°C.

Separate the above three precipitates using the neodymium magnet and wash 3 × with 51 mL of deionized water followed by 2 × washing with 51 mL of ethanol. Dry the precipitates and store in a cool and dry place for further use.

5.2 Preparation of GOx–MNP Conjugates

5.2.1 *Modification of the MNPs With APTES*

1. Measure 10 mg of the three different sized (5, 26, and 51 nm) MNPs, add to 10 mL of 10% APTES solution (prepared in anhydrous methanol), and disperse them by sonication.
2. Vortex the above solution overnight at room temperature.
3. Separate the APTES-bound MNPs (APTES-MNPs) by application of the magnet.
4. Wash the MNPs thrice with 10 mL of anhydrous methanol.
5. Store the APTES-MNPs at 4°C until further use.

5.2.2 *GOx Immobilization to APTES-MNPs*

1. Take 10 mg portion of APTES-MNPs and disperse in an 8% glutaraldehyde solution (pH 7.4) at room temperature for 1 h with sonication.
2. Separate the glutaraldehyde-bound MNPs by magnetic decantation and wash thrice and resuspend in 2 mL of PBS.
3. Prepare 10 mL of 1 mg mL^{-1} GOx solution in PBS, add to the glutaraldehyde-bound MNPs, and vortex the solution overnight.
4. Separate GOx-bound MNPs (GOx-APTES-MNPs) from the reaction solution by use of magnetic decantation and resuspend in 10 mL of PBS after three washes.

5.2.3 *Measurement of the Activity and Reusability of GOx-APTES-MNPs*

1. Mix β-D-glucose, *p*-hydroxybenzoic acid, 4-aminoantipyrine, and peroxidase (as specified in the assay kit) in one cuvette and add 0.5 mL of the above prepared GOx-APTES-MNPs to it.

2. Collect the conjugate at the bottom of the cuvette using a magnet.
3. Pipette out the products to a different cuvette and measure absorbance at 510 nm.
4. After each measurement, wash the conjugate thrice with PBS (pH 7.4) and separate the particles by magnetic decantation after each wash.
5. Collect the washed solutions from each wash to estimate the activity loss as a result of particle leaching.

5.3 Characterization of the Conjugates Using XPS

The conjugates can be characterized by X-ray photoelectron spectroscopy (XPS, 5800 series, Multi-Technique ESCA systems: data analyzed by XPSPeak 4.1 software). We can obtain elemental compositions from XPS spectra of MNPs, APTES-MNPs, glutaraldehyde-modified APTES-MNPs, and GOx-APTES-MNP bioconjugates to confirm whether the MNP modification was successful (Park et al., 2011). Decrease in the Fe2p signal of MNPs with successive surface modification steps (e.g., MNP-APTES, MNP-APTES-glutaraldehyde, and GOx-APTES-MNPs conjugate) was noted (Benbenishty-Shamir, Gilert, Gotman, Gutmanas, & Sukenik, 2011).

5.4 GOx-MNP Activity and Stability Assessment Upon Reuse

Using a glucose assay kit, the GOx activity was assessed. Briefly, the mechanism of the assay kit is: GOx oxidizes β-D-glucose to D-glucono-δ-lactone, which then hydrolyzes to gluconic acid and H_2O_2. The produced H_2O_2 then reacts with *p*-hydroxybenzoic acid and 4-aminoantipyrine in the presence of peroxidase to form a quinoneimine dye complex. This dye complex strongly absorbs at 510 nm. Results from the glucose assay depicted that the conjugates prepared using larger MNPs retained greater enzyme activity than the smaller particles after 10 cycles of reuse. This was attributed to the observation of better separation ability and lower aggregation of larger particles (among 50, 26, and 5 nm sizes), allowing more exposure of individual MNP and its surface functional groups for effective conjugation with the enzyme in comparison to the smaller particles. Moreover, the optimum and narrower size distribution of 26 nm MNPs was proposed to help retaining the maximum enzyme activity (Park et al., 2011).

6. SUMMARY

We discussed the incorporation of polymer linkers onto MNPs to attach enzymes for constructing volume efficient, stable, scalable, and reusable biocatalytic systems. The enzyme–polymer–nanoparticle conjugates are attractive to accomplish the stereoselective synthesis of fine chemicals and specialty materials. Enzyme conjugation with polymeric MNPs offers the advantages of straightforward preparation, efficient recovery of unreacted enzyme, and easy isolation of the conjugate by use of a small magnet within a few tens of seconds. Additionally, reusability of the enzyme–polyMNP conjugates, easy product isolation after reaction as the enzyme is bound strongly with the polyMNPs, and improved storage stability of the enzymes attached with polyMNPs when compared to free enzyme solutions in the buffer.

Moreover, ultrasensitive biosensing of analytes and antigens present in complex clinical matrices can be easily achieved by designing direct antigen–polyMNP conjugates. Also a specific capture biomolecule conjugated with polyMNPs is useful to isolate interested target analytes or antigens from a sample matrix such as serum and plasma for detection. Our laboratory has additionally made some notable contributions in this clinical biosensor research based on electrochemical and surface plasmon resonance methods that utilized polyMNPs (Niroula, Premaratne, Shojaee, Lucca, & Krishnan, 2016; Singh & Krishnan, 2015; Singh, Rodenbaugh, & Krishnan, 2016).

7. FUTURE OUTLOOK

Quantitative characterization of polymers, enzymes, polyMNPs, and those with conjugated enzymes would allow better understanding of the role of each component and how to move forward scaling up these bioconjugates catalyzed reactions in a cost-effective and -efficient manner. Representative examples include understanding the effect of surface functionalization percentage, size and shape effect of MNPs, polymer chain length, polymer type, and polydispersity on the amount of immobilization of enzyme molecules and their specific activity. Large-scale applicability for various classes of enzymes and limitations with regard to protein-structure and function would broaden our knowledge on such fascinating nanobiocatalytic materials. The reusable property of nanoparticle–polymer–enzyme conjugates and the unreacted bulk enzyme solution for further conjugation is important when considering the low-cost synthesis at an industrial scale. Furthermore, an added focus would be to develop efficient multienzyme–polymer–nanoparticle hybrids to carry out a cascade of catalytic reactions to yield products that are otherwise impossible to synthesize by using a single-enzyme system or by direct chemical routes. Similarly, combining a chemical reaction with an enzymatic reaction or vice versa to accomplish in situ generation of intermediates and final conversion to useful products would be a new impactful direction. Along with being effective biocatalysts for biomass transformation applications, enzymes coupled with MNPs can also be useful to enhance detection signals for developing high-throughput biosensors. Enzymes that participate in such a setting will be useful for biomedical applications. In conclusion, NanoArmoring of enzymes via polymer-functionalized nanoparticles has the potential to solve many long-term issues in heterogeneous catalysis for applications in energy, health, and environment, and thus represents an important state-of-the-art research area.

ACKNOWLEDGMENTS

We are grateful for funding support from the National Institute of Diabetes and Digestive and Kidney Diseases of the National Institutes of Health (Award Number R15DK103386) and Oklahoma State University. The authors thank colleagues and collaborators named in joint publications for their valuable contributions. The content is solely the responsibility of the authors and does not necessarily represent the official views of the National Institutes of Health.

REFERENCES

Ansari, S. A., & Husain, Q. (2012). Potential applications of enzymes immobilized on/in nano materials: A review. *Biotechnology Advances*, *30*(3), 512–523.

Bayat, N., Lopes, V. R., Sanchez-Dominguez, M., Lakshmanan, R., Rajarao, G. K., & Cristobal, S. (2015). Assessment of functionalized iron oxide nanoparticles in vitro: Introduction to integrated nanoimpact index. *Environmental Science: Nano*, *2*(4), 380–394.

Benbenishty-Shamir, H., Gilert, R., Gotman, I., Gutmanas, E. Y., & Sukenik C. N. (2011). Phosphonate-anchored monolayers for antibody binding to magnetic nanoparticles. *Langmuir*, *27*(19), 12082–12089.

Brena, B. M., & Batista-Viera, F. (2006). *Immobilization of enzymes: Immobilization of enzymes and cells* (2nd ed., pp. 123–124). New Jersey, NJ: Humana Press Inc.

Brena, B., González-Pombo, P., & Batista-Viera, F. (2013). Immobilization of enzymes: A literature survey. In J. M. Guisan (Ed.), *Immobilization of enzymes and cells* (3rd ed., pp. 15–31). New York: Humana Press.

Callender, R., & Dyer, R. B. (2014). The dynamical nature of enzymatic catalysis. *Accounts of Chemical Research*, *48*(2), 407–413.

ci Acar, H. Y., Garaas, R. S., Syud, F., Bonitatebus, P., & Kulkarni, A. M. (2005). Superparamagnetic nanoparticles stabilized by polymerized PEGylated coatings. *Journal of Magnetism and Magnetic Materials*, *293*(1), 1–7.
Cummings, C., Murata, H., Koepsel, R., & Russell, A. J. (2014). Dramatically increased pH and temperature stability of chymotrypsin using dual block polymer-based protein engineering. *Biomacromolecules*, *15*(3), 763–771.
Datta, S., Christena, L. R., & Rajaram, Y. R. S. (2013). Enzyme immobilization: An overview on techniques and support materials. *3 Biotech*, *3*, 1–9.
De, P., Li, M., Gondi, S. R., & Sumerlin, B. S. (2008). Temperature-regulated activity of responsive polymer—protein conjugates prepared by grafting-from via RAFT polymerization. *Journal of the American Chemical Society*, *130*(34), 11288–11289.
Dzudzevic Cancar, H., Soylemez, S., Akpinar, Y., Kesik, M., Göker, S., Gunbas, G., et al. (2016). A novel acetylcholinesterase biosensor: Core–shell magnetic nanoparticles incorporating a conjugated polymer for the detection of organophosphorus pesticides. *ACS Applied Materials & Interfaces*, *8*(12), 8058–8067.
Fujigaya, T., & Nakashima, N. (2015). Non-covalent polymer wrapping of carbon nanotubes and the role of wrapped polymers as functional dispersants. *Science and Technology of Advanced Materials*, *16*(2), 024802.
Gauthier, M. A., & Klok, H. A. (2010). Polymer–protein conjugates: An enzymatic activity perspective. *Polymer Chemistry*, *1*(9), 1352–1373.
Goto, Y., & Fink, A. L. (1990). Phase diagram for acidic conformational states of apomyoglobin. *Journal of Molecular Biology*, *214*(4), 803–805.
Guto, P. M., Kumar, C. V., & Rusling, J. F. (2007). Thermostable poly(lysine)-enzyme films for biocatalysis at 90°C. *The Journal of Physical Chemistry. B*, *111*(30), 9125–9131.
Guto, P. M., Kumar, C. V., & Rusling, J. F. (2008). Thermostable biocatalytic films of enzymes and polylysine on electrodes and nanoparticles in microemulsions. *Langmuir*, *24*(18), 10365–10370.
Guto, P. M., & Rusling, J. F. (2005). Enzyme-like kinetics of ferryloxy myoglobin formation in films on electrodes in microemulsions. *The Journal of Physical Chemistry. B*, *109*(51), 24457–24464.
Heering, H. A., Hirst, J., & Armstrong, F. A. (1998). Interpreting the catalytic voltammetry of electroactive enzymes adsorbed on electrodes. *The Journal of Physical Chemistry. B*, *102*(35), 6889–6902.
Herzberger, J., Niederer, K., Pohlit, H., Seiwert, J., Worm, M., Wurm, F. R., et al. (2015). Polymerization of ethylene oxide, propylene oxide, and other alkylene oxides: Synthesis, novel polymer architectures, and bioconjugation. *Chemical Reviews*, *116*(4), 2170–2243.
Hou, Y., Yuan, J., Zhou, Y., Yu, J., & Lu, H. (2016). A concise approach to site-specific topological protein–poly (amino acid) conjugates enabled by in situ-generated functionalities. *Journal of the American Chemical Society*, *138*(34), 10995–11000.
Jia, H., Zhu, G., & Wang, P. (2003). Catalytic behaviors of enzymes attached to nanoparticles: The effect of particle mobility. *Biotechnology and Bioengineering*, *84*(4), 406–414.
Kaushik, A., Khan, R., Solanki, P. R., Pandey, P., Alam, J., Ahmad, S., et al. (2008). Iron oxide nanoparticles–chitosan composite based glucose biosensor. *Biosensors and Bioelectronics*, *24*(4), 676–683.
Kouassi, G. K., Irudayaraj, J., & McCarty, G. (2005). Activity of glucose oxidase functionalized onto magnetic nanoparticles. *BioMagnetic Research and Technology*, *3*(1), 1.
Krishnan, S., Schenkman, J. B., & Rusling, J. F. (2011). Bioelectronic delivery of electrons to cytochrome P450 enzymes. *The Journal of Physical Chemistry. B*, *115*(26), 8371–8380.
Leaist, D. G. (1989). Coupled diffusion of weakly ionized polyelectrolytes. Polyacrylic acids in water. *Journal of Solution Chemistry*, *18*(5), 421–435.

Lee, H., Lee, E., Kim, D. K., Jang, N. K., Jeong, Y. Y., & Jon, S. (2006). Antibiofouling polymer-coated superparamagnetic iron oxide nanoparticles as potential magnetic resonance contrast agents for *in vivo* cancer imaging. *Journal of the American Chemical Society, 128*(22), 7383–7389.

Li, N., Xu, J. Z., Yao, H., Zhu, J. J., & Chen, H. Y. (2006). The direct electron transfer of myoglobin based on the electron tunneling in proteins. *The Journal of Physical Chemistry. B, 110*(23), 11561–11565.

Liese, A., & Hilterhaus, L. (2013). Evaluation of immobilized enzymes for industrial applications. *Chemical Society Reviews, 42*(15), 6236–6249.

Mahmoudi, M., Sant, S., Wang, B., Laurent, S., & Sen, T. (2011). Superparamagnetic iron oxide nanoparticles (SPIONs): Development, surface modification and applications in chemotherapy. *Advanced Drug Delivery Reviews, 63*, 24–46.

Mak, S. Y., & Chen, D. H. (2004). Fast adsorption of methylene blue on polyacrylic acid-bound iron oxide magnetic nanoparticles. *Dyes and Pigments, 61*(1), 93–98.

Mohamad, N. R., Marzuki, N. H. C., Buang, N. A., Huyop, F., & Wahab, R. A. (2015). An overview of technologies for immobilization of enzymes and surface analysis techniques for immobilized enzymes. *Biotechnology & Biotechnological Equipment, 29*(2), 205–220.

Moridi, N., Corvini, P. F. X., & Shahgaldian, P. (2015). Reversible supramolecular surface attachment of enzyme–polymer conjugates for the design of biocatalytic filtration membranes. *Angewandte Chemie, International Edition, 54*(49), 14800–14804.

Mu, X., Qiao, J., Qi, L., Dong, P., & Ma, H. (2014). Poly (2-vinyl-4, 4-dimethylazlactone)-functionalized magnetic nanoparticles as carriers for enzyme immobilization and its application. *ACS Applied Materials & Interfaces, 6*(23), 21346–21354.

Niroula, J., Premaratne, G., Shojaee, S. A., Lucca, D. A., & Krishnan, S. (2016). Combined covalent and noncovalent carboxylation of carbon nanotubes for sensitivity enhancement of clinical immunosensors. *Chemical Communications, 52*(88), 13039–13042.

Noureddini, H., & Gao, X. (2007). Characterization of sol-gel immobilized lipases. *Journal of Sol-Gel Science and Technology, 41*(1), 31–41.

Panchagnula, V., Kumar, C. V., & Rusling, J. F. (2002). Ultrathin layered myoglobin-polyion films functional and stable at acidic pH values. *Journal of the American Chemical Society, 124*(42), 12515–12521.

Park, H. J., McConnell, J. T., Boddohi, S., Kipper, M. J., & Johnson, P. A. (2011). Synthesis and characterization of enzyme–magnetic nanoparticle complexes: Effect of size on activity and recovery. *Colloids and Surfaces. B, Biointerfaces, 83*(2), 198–203.

Pelegri-O'Day, E. M., Lin, E. W., & Maynard, H. D. (2014). Therapeutic protein–polymer conjugates: Advancing beyond PEGylation. *Journal of the American Chemical Society, 136*(41), 14323–14332.

Pelegri-O'Day, E. M., & Maynard, H. D. (2016). Controlled radical polymerization as an enabling approach for the next generation of protein–polymer conjugates. *Accounts of Chemical Research, 49*(9), 1777–1785.

Peng, H. P., Liang, R. P., & Qiu, J. D. (2011a). Facile synthesis of Fe_3O_4@Al_2O_3 core–shell nanoparticles and their application to the highly specific capture of heme proteins for direct electrochemistry. *Biosensors and Bioelectronics, 26*(6), 3005–3011.

Peng, H. P., Liang, R. P., Zhang, L., & Qiu, J. D. (2011b). Sonochemical synthesis of magnetic core–shell Fe_3O_4@ZrO_2 nanoparticles and their application to the highly effective immobilization of myoglobin for direct electrochemistry. *Electrochimica Acta, 56*(11), 4231–4236.

Premaratne, G., Nerimetla, R., Matlock, R., Sunday, L., Koralege, R. S. H., Ramsey, J. D., et al. (2016). Stability, scalability, and reusability of a volume efficient biocatalytic system constructed on magnetic nanoparticles. *Catalysis Science & Technology, 6*(7), 2361–2369.

Qiu, J. D., Peng, H. P., Liang, R. P., & Xia, X. H. (2010). Facile preparation of magnetic core–shell Fe_3O_4@Au nanoparticle/myoglobin biofilm for direct electrochemistry. *Biosensors and Bioelectronics*, *25*(6), 1447–1453.

Que, L., & Tolman, W. B. (2008). Biologically inspired oxidation catalysis. *Nature*, *455*(7211), 333–340.

Reddy, L. H., Arias, J. L., Nicolas, J., & Couvreur, P. (2012). Magnetic nanoparticles: Design and characterization, toxicity and biocompatibility, pharmaceutical and biomedical applications. *Chemical Reviews*, *112*(11), 5818–5878.

Romero, O., Rivero, C. W., Guisan, J. M., & Palomo, J. M. (2013). Novel enzyme-polymer conjugates for biotechnological applications. *PeerJ*, *1*, e27.

Rozenberg, B. A., & Tenne, R. (2008). Polymer-assisted fabrication of nanoparticles and nanocomposites. *Progress in Polymer Science*, *33*(1), 40–112.

Schuleit, M., & Luisi, P. L. (2001). Enzyme immobilization in silica-hardened organogels. *Biotechnology and Bioengineering*, *72*(2), 249–253.

Shoda, S. I., Uyama, H., Kadokawa, J. I., Kimura, S., & Kobayashi, S. (2016). Enzymes as green catalysts for precision macromolecular synthesis. *Chemical Reviews*, *116*(4), 2307–2413.

Singh, V., & Krishnan, S. (2015). Voltammetric immunosensor assembled on carbon-pyrenyl nanostructures for clinical diagnosis of type of diabetes. *Analytical Chemistry*, *87*(5), 2648–2654.

Singh, V., Rodenbaugh, C., & Krishnan, S. (2016). Magnetic optical microarray imager for diagnosing type of diabetes in clinical blood serum samples. *ACS Sensors*, *1*(4), 437–443.

Tosa, T., Mori, T., Fuse, N., & Chibata, I. (1967). Studies on continuous enzyme reactions. 3. Enzymatic properties of the DEAE-cellulose-aminoacylase complex. *Enzymologia*, *32*(3), 153–168.

Tsai, H. C., & Doong, R. A. (2007). Preparation and characterization of urease-encapsulated biosensors in poly (vinyl alcohol)-modified silica sol–gel materials. *Biosensors and Bioelectronics*, *23*(1), 66–73.

Walgama, C., & Krishnan, S. (2013). Electrocatalytic features of a heme protein attached to polymer-functionalized magnetic nanoparticles. *Analytical Chemistry*, *85*(23), 11420–11426.

Work, T. S., & Work, E. (Eds.), (1976). *Techniques of sample preparation for liquid scintillation counting + isoelectric focusing laboratory techniques in biochemistry and molecular biology*. Amsterdam: North Holland Publishing Co.

Xu, J., Sun, J., Wang, Y., Sheng, J., Wang, F., & Sun, M. (2014). Application of iron magnetic nanoparticles in protein immobilization. *Molecules*, *19*(8), 11465–11486.

Yan, M., Ge, J., Liu, Z., & Ouyang, P. (2006). Encapsulation of single enzyme in nanogel with enhanced biocatalytic activity and stability. *Journal of the American Chemical Society*, *128*(34), 11008–11009.

Yong, Y., Bai, Y., Li, Y., Lin, L., Cui, Y., & Xia, C. (2008). Preparation and application of polymer-grafted magnetic nanoparticles for lipase immobilization. *Journal of Magnetism and Magnetic Materials*, *320*(19), 2350–2355.

Zhou, Y., Pan, S., Wei, X., Wang, L., & Liu, Y. (2013). Immobilization of β-glucosidase onto magnetic nanoparticles and evaluation of the enzymatic properties. *BioResources*, *8*(2), 2605–2619.

Zore, O. V., Pattammattel, A., Gnanaguru, S., Kumar, C. V., & Kasi, R. M. (2015). Bienzyme–polymer–graphene oxide quaternary hybrid biocatalysts: Efficient substrate channeling under chemically and thermally denaturing conditions. *ACS Catalysis*, *5*(9), 4979–4988.

CHAPTER ELEVEN

Expression of Cellulolytic Enzyme as a Fusion Protein That Reacts Specifically With a Polymeric Scaffold

Priya Katyal*, Yongkun Yang†, Olga Vinogradova*, Yao Lin†,1
*University of Connecticut, Storrs, CT, United States
†Polymer Program, Institute of Materials Science, University of Connecticut, Storrs, CT, United States
[1]Corresponding author: e-mail address: yao.lin@uconn.edu

Contents

Abstract

The formation of higher-order assemblies of multiple proteins or enzymes is a general mechanism to achieve more sophisticated biological function in biological systems. For example, cellulosomes are large complexes consisting of multiple cellulolytic enzymes that rely on the concerted actions of different enzymes built onto a common protein scaffold to facilitate the breakdown of the polymeric substrate, cellulose. One strategy for mimicking these highly effective nanomachines may involve the use of synthetic scaffolds that can react to and organize multiple engineered enzymes to promote synergistic action between the enzymes on the scaffold. As an example of the earlier strategy, we describe here an approach for the expression of cellulolytic enzymes with a serine esterase tag, and the rapid reaction between the tag and the end-functionalized polymers to form enzyme–polymer–enzyme multienzyme conjugates. In principle, this general and versatile supramolecular approach may be used to organize specific

Methods in Enzymology, Volume 590
ISSN 0076-6879
http://dx.doi.org/10.1016/bs.mie.2016.12.003
© 2017 Elsevier Inc.
All rights reserved.

cellulolytic enzymes onto synthetic scaffolds to form multienzyme complexes to potentially work in synergy for enhanced biological activities. Best reaction conditions, good activities of the armored cellulolytic enzymes and the design of optimal protein linker in the fusion protein are discussed in detail. If other reactive tags are included on the enzyme in future, multiple types of synergistic enzymes may be positioned at specific sites on a designed polymer scaffold that mimics the complex structure and enhanced function of natural cellulosomes. This type of nanoarmoring of multiple enzymes on a nanoscale might also enhance enzyme stability, when compared to the unprotected enzymes.

1. INTRODUCTION

The successful synthesis and characterization of fusion proteins assembled onto a polymer nanoscaffold at defined locations/geometries, as a test case for the assembly of large, catalytic supramolecular assemblies, are described here. Recent discoveries in chemical biology have revealed that higher-order assemblies of multiple intracellular proteins, also known as "signalosomes," are capable of rather sophisticated, machine-like function in signal transmission and transduction (Bienz, 2014; Wu, 2013). Formation of higher-order protein assemblies may be a general mechanism for cells to achieve and regulate specific biological functions (Norris et al., 2007; Wilson & Gitai, 2013). Interestingly, extracellular proteins and enzymes can also form higher order, structured assemblies or machines in order to accomplish more challenging tasks (Amar et al., 2008; Norris et al., 2007).

Cellulosome, for example, is a multienzyme complex comprising of a noncatalytic protein scaffold (scaffoldin) onto which a number of cellulolytic enzymes are anchored and organized into a large enzyme assembly (Bayer, Belaich, Shoham, & Lamed, 2004; Fontes & Gilbert, 2010). The presence of similar and different multiple catalytic units in the cellulosomes make them function in a concerted fashion to catalyze the conversion of highly complex, insoluble, lignocellulose polymeric substrates into soluble sugars, which is an arduous and challenging task for individual cellulolytic enzymes (Moraïs et al., 2012). The enhanced functions of multiple copies of the same enzyme and different enzymes that are present in the cellulosome make it a formidable biocatalyst to match, where the function of one enzyme could depend on the function of another. The product from one enzyme, for example, might generate the necessary reactive site for another enzyme and thus function in a concerted fashion before the biocatalyst is desorbed from the substrate surface. Mimicking such highly efficient, multicatalytic nanomachines is of considerate importance in developing new and more efficient green processes for applications in a variety of enzyme-based technologies (Mitsuzawa et al., 2009; Moraïs et al., 2012; Wilson, 2009).

However, the design and scale-up of large enzyme/protein/scaffold structures such as scaffoldins to serve as a framework to attach specific enzymes is challenging, but one promising alternative approach is the use of synthetic polymers or nanoparticles as scaffolds to organize multiple, synergistic enzymes into higher-order complexes. This can be done by specific high affinity binding or by selective chemical conjugation of the polymer scaffold to particular sites on the enzymes (Kamat et al., 2013; Sun & Chen, 2016; Zhang et al., 2013; Zheng et al., 2015). The use of polymer scaffold may have certain advantages for the assembly of multiple enzymes into functional biocatalysts over naturally occurring scaffoldins.

Numerous methods, for example, have been employed to prepare a variety of protein–polymer conjugates as described in this volume, or protein–nanoparticle conjugates in the previous volume of this series (Vol. 571, 2016) in support of this strategy, and these approaches include but not limited to: (1) the utilization of free reactive thiol, carboxyl, or amine groups that are naturally present on enzymes and proteins for coupling to specific reactive groups placed on the polymer/nanoparticle (Dieterich et al., 2007; Nilsson, Kiessling, & Raines, 2000; Rabuka, Rush, deHart, Wu, & Bertozzi, 2012); (2) incorporation of unnatural amino acids into the primary sequence of the enzyme for site-specific coupling to the scaffold (de Graaf, Kooijman, Hennink, & Mastrobattista, 2009; Dieterich et al., 2007); or (3) the use of a fusion protein carrying a protein or enzymatic tag that can covalently conjugate with multiple desired moieties on the scaffold and provides a good synthetic control to form large assemblies of protein–polymer conjugates of specific size, shape, or structure (Los et al., 2008; Rabuka et al., 2012; Sunbul & Yin, 2009; van Vught, Pieters, & Breukink, 2014). These nanoarmored enzymes in the supramolecular complex may also be more stable than individual enzymes due to the presence of polymeric scaffold surrounding the enzymes.

In this chapter, we outline the strategy of engineering a cellulase–cutinase fusion protein, and the attachment of the fusion protein to a synthetic, water-soluble polymer where the two ends of the polymer are functionalized with a chemical moiety that reacts irreversibly with cutinase. Together, these form cellulase–cutinase–polymer–cutinase–cellulase biocatalysts in an optimized enzyme to polymer molar ratio (e.g., 2:1) or they form cellulase–cutinase–polymer complexes if polymer is used in a large excess, under complete synthetic control. These supramolecular enzyme–polymer complexes are fully active, further verifying the success of the earlier strategy.

The strategy of using fusion proteins with enzymatic tag that reacts with synthetic scaffolds containing suicide inhibitors positioned at strategic locations on the polymer scaffold confers many advantages in building artificial analogs of multienzyme assemblies (Sunbul & Yin, 2009; van Vught et al., 2014). The reaction, for example, specifically takes place at the catalytic site of the enzyme, resulting in the formation of highly specific, efficient, and stable covalent adducts between the fusion protein and the scaffold. Being selective and irreversible, the resultant complexes may not require subsequent purification steps, which can be especially challenging for the preparation of large protein–scaffold complexes (Hodneland, Lee, Min, & Mrksich, 2002). Other advantages include: (1) the use of end-functionalized polymers with the suicide inhibitors attached at specific location which can have different or distinct architectures such as linear, branched, star, or comb type to systematically explore the polymer scaffold structure and enzyme arrangement on the function of the multienzyme complex (González-Toro & Thayumanavan, 2013; McRae et al., 2012; Tao et al., 2009a, 2009b) and (2) the use of mild conditions in aqueous media for the coupling of the fusion protein with the polymer scaffold, as compared to other chemical reactions (Nwe & Brechbiel, 2009). The challenge associated with the use of such fusion proteins is to express the protein of interest with specific tags, without decreasing enzymatic activity of the corresponding fusion protein.

We have used the recombinant DNA technology to form the desired fusion protein, and the technology involves manipulation of DNA at the molecular level, along with rational design of the fusion construct, and using optimal conditions for the fusion protein production. Work in this area by numerous groups have provided effective solutions in expressing engineered proteins with full retention of functionality (Lambertz et al., 2014; Miller & Baxter, 1981). We expressed target proteins with an enzymatic tag that reacts specifically, irreversibly, and rapidly with synthetic scaffolds and these were shown to facilitate a modular design, good adaptability, and in certain cases, orientation of the target proteins on the scaffolds (Gauthier & Klok, 2010; Hodneland et al., 2002; Thordarson,

Le Droumaguet, & Velonia, 2006). Controlling the enzyme position, orientation, and composition of the catalytic complex is important in optimizing function.

Herein, we demonstrate the use of cutinase, a highly stable enzyme, that can be fused to a variety of target enzymes by recombinant DNA technology, as a bridge to attach cellulase to the polymer scaffold (Fig. 1). The cutinase tag is a 22 kDa serine esterase, consisting of a Ser–His–Asp catalytic triad (Martinez, De Geus, Lauwereys, Matthyssens, & Cambillau, 1992). The catalytic serine residue at the active site of the enzyme forms an irreversible covalent conjugate with phosphonate-functionalized oligomers or polymers (Bonasio et al., 2007; Dijkstra et al., 2008; Galbiati et al., 2015; Kwon, Han, Karatan, Mrksich, & Kay, 2004; Mannesse et al., 1995; Modica, Skarpathiotis, & Mrksich, 2012).

We discuss the design of a fusion protein comprising of cutinase tag with our target enzyme, a cellulase (Cel48F, family 48), from *Clostridium cellulolyticum* cellulosome (Fontes & Gilbert, 2010; Parsiegla et al., 2000) and the method to prepare functionally active Cel48F-cutinase fusion protein that can be site specifically linked with phosphonate-functionalized polymers (Fig. 1). For simplicity, we only show the characterization and analysis performed on the reaction of the fusion proteins to bifunctionalized synthetic scaffolds. By precisely controlling the number of reactive groups on the polymers (e.g., in a brush-like polymers), one may tune the number of cellulolytic enzymes being conjugated to the scaffold. In the current design, the polymer is end-labeled with two phosphonate tags so that one or two enzyme–cutinase moieties would react with one polymer scaffold, under complete synthetic control.

Advantages associated with this approach are its modular nature to extend the strategy to numerous other enzymes and the flexibility of decorating the polymeric scaffold with different functional groups and the use of different scaffold geometries to produce a variety of geometrical structures for the assembly. Additionally, the present assembly design mimics the natural cellulosome where the dockerin domain (unit that binds to scaffoldin) is replaced by the cutinase domain and scaffoldin is replaced by the synthetic polymer. If other reactive tags (e.g., SNAP tag, Keppler et al., 2003; aldehyde tag, Rabuka et al., 2012; and dehalogenase tag, Los et al., 2008) are included for bioorthogonal conjugation, multiple types of synergistic enzymes may be positioned at specific sites along the designed polymeric scaffold to mimic the structure and function of natural cellulosomes (Fig. 2).

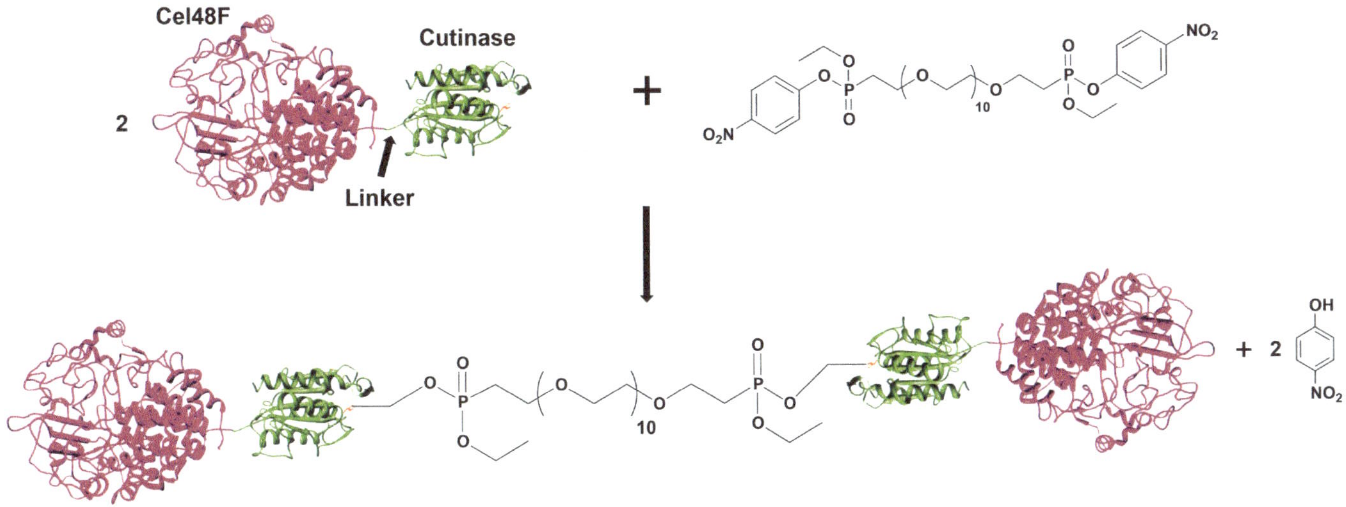

Fig. 1 Conjugation of cutinase-tagged Cel48F with bisphosphonate end-functionalized polymer. The fusion protein prepared from Cel48F (*pink*) and cutinase (*green*). A linker of (GGGGS) repeat, present within the two domains, provides the optimal flexibility to the two enzymes. The fusion protein reacts with the bifunctional ends of the polymer to form a protein–polymer–protein complex, with the release of *p* nitrophenol.

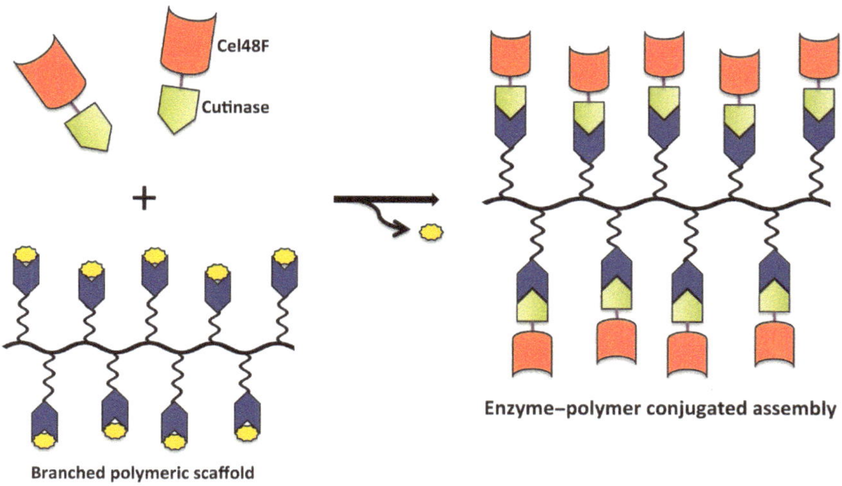

Fig. 2 Developing an artificial cellulosome by replacing a linear polymeric scaffold with a brush-like polymeric scaffold. Brush like polymers decorated with reactive moieties (shown as *blue ends*) can be tagged with cellulase–fusion proteins. The color generated by the release of the leaving groups (for example, *p*-nitrophenolate, in this specific case, shown as *yellow*) can be used to monitor the extent of conjugation in real time, as the conjugation reaction progresses.

2. MATERIALS

2.1 Personal Protective Equipments

Laboratory coat, nitrile gloves, safety glasses, fume hood, and UV blocking face shield. Properly labeled chemical waste receptacles and secondary containers, and biohazard sharps container are required as necessary.

2.2 Equipments

Heating block (Fisher Scientific Isotemp 125D), Incubator (New Brunswick Scientific), Benchtop mini centrifuge (Fisher Scientific accuSpin MicroR), 1.5 mL microcentrifuge tubes, 50 mL Falcon tubes, French Press, ÄKTA system (Amersham Biosciences), Magnetic stir bar, Stirrer, Gel electrophoresis apparatus, Gel imaging system with BioRad GelDoc system, Resource Q column (GE Healthcare #17-1177-01), Spectrophotometer, nutator mixer (Labnet Rocker 25), Micropipettors, Pipettor tips, dialysis bags, clips, precast gels, glass cuvettes, and amicon filters.

2.3 Reagents

Kanamycin, agar plates, isopropyl β-D-1-thiogalactopyranoside (IPTG), lysozyme, Terrific broth modified EZmix powder (Sigma), glycerol, Tris base, Tris maleate, sodium chloride, imidazole, calcium chloride, dithiothreitol (DTT), phenylmethylsulfonyl fluoride (PMSF), 2× Laemmli sample buffer (Bio-Rad Catalog #1610737), electrophoresis buffer, and Avicel (PH101, Fluka, Buchs, Switzerland). Origami 2 (DE3) chemically competent cells were purchased from EMD Biosciences. Ni-NTA agarose was purchased from Qiagen (Catalog #30230). Complete-Mini EDTA-free protease inhibitor tablets were purchased from Roche Applied Science.

3. METHODS

3.1 Molecular Cloning of Fusion Protein

The Cel48F-cutinase fusion protein was constructed by fusing the N-terminus of cutinase domain to the C-terminal end of catalytic domain of Cel48F, a cellulolytic enzyme (Fig. 1A). The newly constructed fusion protein has a C-terminal His tag with the two enzymes separated by a $(GGGGS)_2$ linker (Fig. 3). The N- and C-termini of Cel48F enzyme are opposite to the catalytic active site (Parsiegla et al., 2000), thus modification of either end would be suitable for conjugation. When cutinase is fused to a large cellulase domain, a flexible linker of (GGGGS) repeat at its N-terminus provides the optimal flexibility to the two enzymes, as compared to the construct without the (GGGGS) repeat. Compared to $(GGGGS)_1$, the cutinase activity is about an order of magnitude higher in the fusion protein with $(GGGGS)_2$ linker. Longer linkers of 3–4 repeats of (GGGGS) might further improve the cutinase activity within the fusion protein; however, care must be taken, as longer linkers are susceptible to proteolysis (Huang & Shusta, 2006) and may detach the enzyme Cel48F from the cutinase tag in the fusion protein or from the enzyme–polymer assembly.

The Cel48F plasmid was synthesized by GenScript Corporation, which was further ligated with DNA coding for the cutinase insert, as described previously (Modica et al., 2012). The product was cloned into pET26b vector via NdeI and XhoI, and the resulting plasmid (pCel48F-Cut) was transformed in Origami2 (DE3) strain. The transformed cells were placed on LB-agar plate supplemented with kanamycin (50 μg/mL) for 16–18 h at 37°C.

MASSPANKVY QDRFESMYSK IKDPANGYFS EQGIPYHSIE TLMVEAPDYG HVTTSEAMSY YMWLEAMHGR
FSGDFTGFDK SWSVTEQYLI PTEKDQPNTS MSRYDANKPA TYAPEFQDPS KYPSPLDTSQ PVGRDPINSQ
LTSAYGTSML YGMHWILDVD NWYGFGARAD GTSKPSYINT FQRGEQESTW ETIPQPCWDE HKFGGQYGFL
DLFTKDTGTP AKQFKYTNAP DADARAVQAT YWADQWAKEQ GKSVSTSVGK ATKMGDYLRY SFFDKYFRKI
GQPSQAGTGY DAAHYLLSWY YAWGGGIDST WSWIIGSSHN HFGYQNPFAA WVLSTDANFK PKSSNGASDW
AKSLDRQLEF YQWLQSAEGA IAGGATNSWN GRYEAVPSGT STFYGMGYVE NPVYADPGSN TWFGMQVWSM
QRVAELYYKT GDARAKKLLD KWAKWINGEI KFNADGTFQI PSTIDWEGQP DTWNPTQGYT GNANLHVKVV
NYGTDLGCAS SLANTLTYYA AKSGDETSRQ NAQKLLDAMW NNYSDSKGIS TVEQRGDYHR FLDQEVFVPA
GWTGKMPNGD VIKSGVKFID IRSKYKQDPE WQTMVAALQA GQVPTQRLHR FWAQSEFAVA NGVYAILFPD
QGPEKLLGDV DGGGGSGGGG SGLPTSNPAQ ELEARQLGRT TRDDLINGNS ASCADVIFIY ARGSTETGNL
GTLGPSIASN LESTFGKDGV WIQGVSGAYQ ATLGDNALPR GTSSAAIREM LGLFQQANTK CPDATLIAGG
YSQGAALAAA SIEDLDSAIR DKIAGTVLFG YTKNLQNRGR IPNYPADRTK VFCNTGDLIC TGSLIVAAPH
LAYGPDARGP APEFLIEKVR AVRGSAGGLE HHHHHH

Fig. 3 Sequence of Cel48F-cutinase. The Cel48F (*red*) is linked with cutinase (*green*) by the $(GGGGS)_2$ linker.

3.2 Expression and Purification

Expression

1. Pick a single colony from the agar plate and suspend it in 50 mL Terrific Broth having Kanamycin (50 μg/mL) antibiotic. Allow it to grow for 12–16 h at 37°C with shaking at 225 rpm. This is referred to as seed culture.
2. Inoculate 1 L of autoclaved Terrific Broth with 10 mL of seed culture (1:100 dilution). Grow at 37°C with shaking at 225 rpm till it reaches OD = 0.6. Transfer the flask at 18°C and allow it to cool for an hour. Induce it with 20 μ*M* IPTG. Grow at 18°C with shaking at 225 rpm for 16 h.
3. Harvest cells by centrifuging at 4000 rpm for 40 min. Discard the supernatant and store the pellet at −20°C or use it immediately for further purification.

Useful tips

- The same seed culture can be used to prepare multiple batches, when expressed the same day. The cell pellets can be stored at −80°C for long term storage.

- Lower temperature and lower IPTG concentrations help in expression of Cel48F-cutinase as a soluble protein which otherwise is expressed in inclusion bodies. Once OD = 0.6 has been obtained, the flask should be cooled by using ice or by placing in a cold room, till the shaker reaches the desired temperature.

Purification

1. Resuspend the frozen pellet in 50 m*M* Tris, pH 8.0, 300 m*M* sodium chloride (lysis buffer). Add 3 mg lysozyme, 1 tablet of EDTA-free protease inhibitor, and 1 m*M* of PMSF. Incubate the cell suspension at room temperature for 20 min.
2. Lyse the cells using French press. Repeat the cell lysis cycle for four times. Remove the insoluble cell debris by centrifugation at 8000 rpm for 1 h at 4°C.
3. Preparation of affinity column: Take 2 mL Ni-NTA agarose resin in affinity column and decant the supernatant. Continue to wash the resin 3–5 times with lysis buffer to completely remove ethanol.
4. Load the lysate onto the affinity column and incubate with Ni-NTA agarose resin at 4°C for 1 h on a nutating mixer.
5. Wash the resin sequentially with 20 mL of lysis buffer with increasing concentrations of imidazole, ranging from 5 to 500 m*M*.
6. Check the fractions with SDS-PAGE (200 V, ~35 min).
7. Pool the fractions containing the target protein and dialyze against 1 L of dialysis buffer (20 m*M* Tris, pH 8.0) at 4°C.
8. Further purify the protein using Resource Q column on FPLC. Run a linear gradient (0–100%B) in 12 CV; 1.5 mL/fraction at a flow rate of 2 mL/min, using buffer A (20 m*M* Tris, pH 7.5) and buffer B (20 m*M* Tris, pH 7.5, 1 *M* NaCl).
9. Check the fractions on SDS-PAGE (an example is shown in Fig. 4).
10. Pool the fractions that contain the desired protein.

Useful tips

- If French press is not available, the cells can be lysed using Bug Buster solution, with sonication.
- The Ni-NTA beads can be washed, equilibrated, and incubated in a Falcon tube. Alternatively, His–Trap columns can be used with liquid chromatography system.
- Handling the protein at 4°C prevents its degradation and helps its long-term stability. The protein solution can be stored in 1 mL aliquots at −20°C. The aliquots should be thawed on ice or at 4°C.

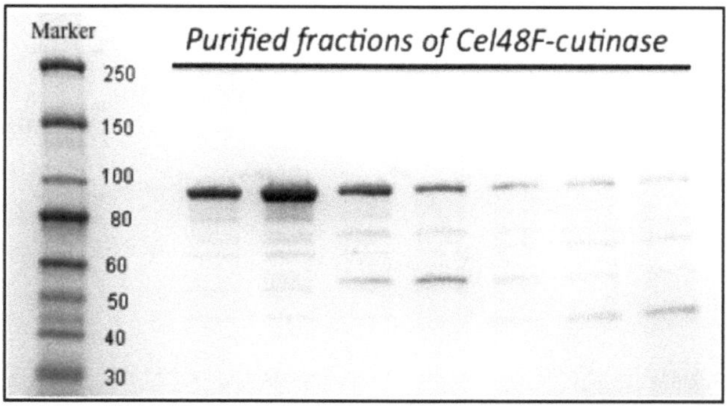

Fig. 4 SDS-PAGE gel of different fractions obtained after anion exchange of Cel48F-cutinase.

3.3 Conjugation of Cel48F-Cutinase

Conjugation of Cel48F with polymeric scaffold via the cutinase tag offers certain advantages including site specificity and formation of conjugates, which are resistant to hydrolysis. Apparently, one of the biggest challenges in conjugating polymers with cutinase-tagged fusion proteins is their access to the catalytic serine at cutinase active site. We investigated the conjugation of *p*-nitophenyl phosphonate-functionalized polymer with Cel48F-cutinase using UV–vis absorption spectroscopy. The rate constant of the reaction between the enzyme and the phosphonate polymer was obtained by following the release of *p*-nitrophenolate ($\lambda_{max} = 401$ nm) as a function of reaction time.

Kinetic study for conjugation

1. Prepare 4 μ*M* of Cel48F-cutinase in 20 m*M* Tris, pH 8, 100 m*M* NaCl.
2. Prepare 0.1 *M* stock solution of EG10 bisphosphonate polymer in DMSO. Dilute the stock solution to 200 μ*M* using the same buffer as above. At 1:50 ratio, the dimer formation is suppressed.
3. The polymer was added just before the UV analysis.
4. Monitor the release of *p*-nitrophenolate from the reaction at 401 nm using high throughput UV spectrophotometer for 15 min.

The resulting kinetic trace (Fig. 5) was fit to a pseudo first-order equation and the effective second-order rate constant was obtained by dividing these quantities by the concentration of excess ligand. The rate constant was ~37 M^{-1} s^{-1}, confirming that cutinase domain in our engineered construct is capable of reacting with the phosphonate-functionalized polymers.

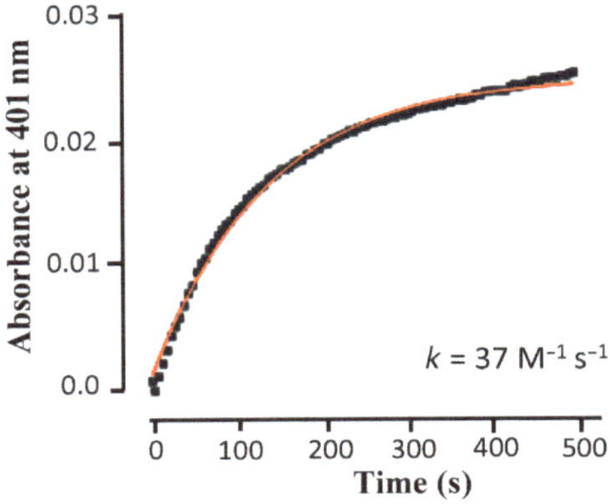

Fig. 5 Ligand functionalization of Cel48F-cutinase with EG10 bisphosphonate (polymer) and release of *p*-nitrophenolate as monitored by its absorbance at 401 nm. Data were fit to pseudo first-order kinetics (*solid line*) and divided by the concentration of excess species to yield the effective second-order rate constant.

Useful tips

- Amicon filters with 3 kDa cut-off were used for buffer exchange and caution should be taken to prevent protein aggregation during the buffer exchange.
- Hydrolysis of polymer, indicated by yellow color is observed even in the absence of protein. Thus, it is extremely important to perform a blank run (buffer with the polymer) simultaneously with the actual experiment, which is then subtracted from the data obtained from the actual run.
- The polymer should be added once everything is set. While adding the polymer, gently mix the polymer into the solution by pipetting the liquid up and down.
- A double beam UV–vis spectrophotometer or a high throughput UV–vis spectrophotometer eases data analysis.

Finding optimized conditions for the conjugation reaction

We further verified the conjugation of cutinase fusion protein with bifunctionalized polymer by varying the concentration of the polymer. We expected Cel48F-cutinase to conjugate at the two ends of the polymer linker and assemble into a dimeric assembly at low mole ratios while single site attachment is expected when the polymer is in excess.

1. Incubate 500 μL protein (5 m*M* Tris maleate, pH 6, 100 m*M* NaCl) with EG10 polymer at 25 and 37°C at different *E*:*L* ratio: 1:0, 1:1, 1:2, 2:1, 4:1, 1:10, 1:20.

2. Aliquot 50 μL of samples after 6, 24, 48, and 72 h.
3. Prepare 1:1 dilution of sample with Laemmli buffer.
4. Heat the samples at 70°C for 5 min.
5. Load 25 μL to the mini-protean gel and run at 200 V for 35 min.

SDS-PAGE analysis indicated that dimeric protein assemblies were formed at different protein: polymer ratios (Fig. 6). The lanes at 1:1, 1:2, 2:1, 4:1 ratios show the presence of an additional band of higher molecular weight, that is attributed to the dimeric protein–polymer conjugate (E_2L). When a higher concentration of polymer was used, no dimerization was observed (as in the case of 1:10 and 1:20) (data not shown). Theoretically, two moles of enzyme should react with one mole of bifunctionalized polymer so that the optimized mole ratio is 2:1. However, presumably due to some hydrolysis of the phosphonate moiety in solution, we observed an optimal ratio of 1:1, where a slightly higher concentration of polymer is required to facilitate a better yield for dimers. No dimerization was observed in the absence of polymer (lane 1:0), demonstrating that the dimer formation is a result of site-specific conjugation with the linker, resulting in the formation of enzyme–linker or enzyme–linker–enzyme conjugates.

Useful tips

- The protein solution can be diluted to the appropriate concentration and the same diluted solution can be used for all the samples by keeping the enzyme concentration constant.

3.4 Cellulase Activity Assay of the Fusion Protein

As we fused a cutinase tag onto the cellulase, it would be necessary to confirm the functionality of the protein, in particular its hydrolytic activity to breakdown cellulose. The activity of the Cel48F enzyme can be determined by high performance liquid chromatography (HPLC) analysis of the reaction products or by ferricyanide assay, using Avicel as a substrate. Avicel is hydrolyzed into simple sugars including glucose, which reacts with ferricyanide, and ferrous ammonium sulfate to give Prussian blue color. The color intensity was monitored at 690 nm, which is proportional to the amount of soluble sugars released upon Avicel hydrolysis (Park & Johnson, 1949).

Cellulase activity assay

1. Perform buffer exchange of Cel48F-cutinase with 20 m*M* Tris maleate, pH 6.0, 100 m*M* NaCl, 1 m*M* $CaCl_2$ using amicon filter (mol. wt. cut-off 30 kDa).
2. Incubate the protein with EG10 polymer at 1:1 ratio.

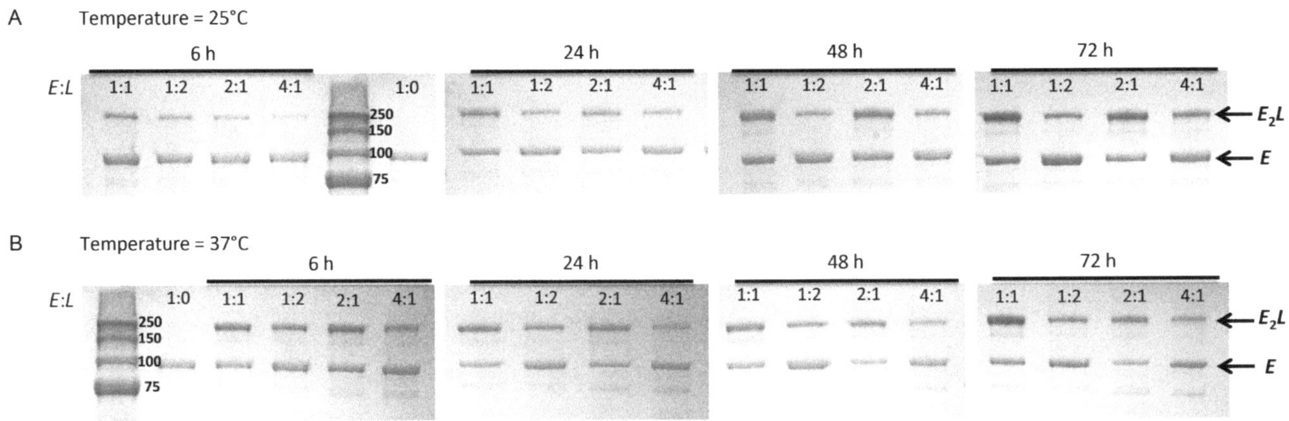

Fig. 6 SDS-PAGE analysis of Cel48F-cutinase with bifunctional polymer. The samples were incubated at different enzyme: linker (*E:L*) ratios for 6, 24, 48, and 72 h. The dimerization (E_2L) was observed at pH 6.0, at (A) 25°C and (B) 37°C. Gel images were cropped and resized for publication purpose only.

3. Incubate 1 μ*M* of protein in free and conjugated state with 4 mL of microcrystalline cellulose (Avicel PH101, Fluka, Buchs, Switzerland) at 20 mg/mL in 20 m*M* Tris maleate, pH 6.0, 1 m*M* $CaCl_2$ at 37°C.
4. Aliquot 0.6 mL at 0, 1, 5, and 24 h, centrifuge, and transfer the supernatant to a new 1.5 mL Eppendorf tube. Examine for soluble reducing sugars using ferricyanide assay.

Fig. 7 compares the absorbance at 690 nm at different time intervals by the same amount of proteins, either in the free state or in the conjugated state with bifunctionalized polymer. The hydrolytic activity of Cel48F is retained upon conjugation (red squares) and is comparable to that of the free enzyme (black diamonds).

Useful tips

- Set up the reaction in a falcon tube, which allows better mixing of cellulose.
- Cellulose tends to settle at the bottom of the tube. Before withdrawing the sample, gently mix the contents of the tube to ensure resuspension of any settled cellulose.
- The withdrawn sample is collected in a microcentrifuge tube. Centrifuge the tubes long enough so that Avicel forms a pellet at the bottom of the

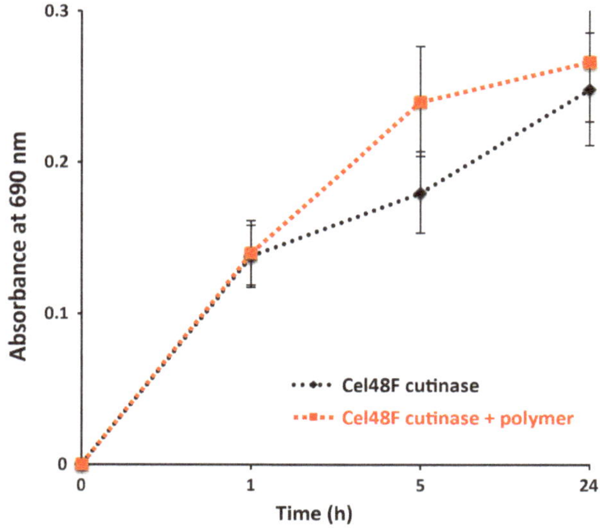

Fig. 7 Avicel hydrolysis by Cel48F-cutinase in free state (*black diamonds*) or conjugated with the polymer (1:1, *red squares*). Sugar release was followed by the ferricyanide assay after incubation at 37°C. The final enzyme concentration was 1 μ*M* in both free and conjugated states.

tube. Carefully transfer the clear supernatant to a new microcentrifuge tube. If the Avicel particles are being withdrawn with the pipette, it is advised to centrifuge the sample again.

4. FUTURE OUTLOOK

This chapter describes a strategy to create cutinase-tagged cellulase that is capable of conjugating with functionalized polymers to form higher-order enzyme–polymer nanoarmored complexes. The conjugated enzyme (Cel48F–cutinase–polymer) activity is comparable to that of its free form. The method may be easily adapted to include other enzymes, proteins, or polypeptide tags. Attaching other types of cellulases that work synergistically with Cel48F enzyme on the same scaffold may lead to enhanced enzyme activities/stabilities for cellulose to sugar conversion. Developing brush-like polymers having multiple reactive chain ends that can be used to conjugate with differently tagged cellulases, using bioorthogonal chemistry, will be interesting. The approach described here may ultimately lead to the preparation of multicatalytic nanomachines that can act in concert to carry out complex sequences of enzymatic reactions in a highly effective way.

ACKNOWLEDGMENTS

Y.L. acknowledges supports from the US Department of Energy, Office of Basic Energy Sciences (DE-SC0005039) and the US National Science Foundation (DMR-1150742).

REFERENCES

Amar, P., Legent, G., Thellier, M., Ripoll, C., Bernot, G., Nystrom, T., et al. (2008). A stochastic automaton shows how enzyme assemblies may contribute to metabolic efficiency. *BMC Systems Biology*, *2*(1), 1–13.

Bayer, E. A., Belaich, J. P., Shoham, Y., & Lamed, R. (2004). The cellulosomes: Multienzyme machines for degradation of plant cell wall polysaccharides. *Annual Review of Microbiology*, *58*, 521–554.

Bienz, M. (2014). Signalosome assembly by domains undergoing dynamic head-to-tail polymerization. *Trends in Biochemical Sciences*, *39*(10), 487–495.

Bonasio, R., Carman, C. V., Kim, E., Sage, P. T., Love, K. R., Mempel, T. R., et al. (2007). Specific and covalent labeling of a membrane protein with organic fluorochromes and quantum dots. *Proceedings of the National Academy of Sciences of the United States of America*, *104*(37), 14753–14758.

de Graaf, A. J., Kooijman, M., Hennink, W. E., & Mastrobattista, E. (2009). Nonnatural amino acids for site-specific protein conjugation. *Bioconjugate Chemistry*, *20*(7), 1281–1295.

Dieterich, D. C., Lee, J. J., Link, A. J., Graumann, J., Tirrell, D. A., & Schuman, E. M. (2007). Labeling, detection and identification of newly synthesized proteomes with bioorthogonal non-canonical amino-acid tagging. *Nature Protocols*, *2*(3), 532–540.

Dijkstra, H. P., Sprong, H., Aerts, B. N. H., Kruithof, C. A., Egmond, M. R., & Klein Gebbink, R. J. M. (2008). Selective and diagnostic labelling of serine hydrolases with reactive phosphonate inhibitors. *Organic & Biomolecular Chemistry*, *6*(3), 523–531.

Fontes, C. M., & Gilbert, H. J. (2010). Cellulosomes: Highly efficient nanomachines designed to deconstruct plant cell wall complex carbohydrates. *Annual Review of Biochemistry*, *79*, 655–681.

Galbiati, E., Cassani, M., Verderio, P., Martegani, E., Colombo, M., Tortora, P., et al. (2015). Peptide-nanoparticle ligation mediated by cutinase fusion for the development of cancer cell-targeted nanoconjugates. *Bioconjugate Chemistry*, *26*(4), 680–689.

Gauthier, M. A., & Klok, H.-A. (2010). Polymer–protein conjugates: An enzymatic activity perspective. *Polymer Chemistry*, *1*(9), 1352–1373.

González-Toro, D. C., & Thayumanavan, S. (2013). Advances in polymer and polymeric nanostructures for protein conjugation. *European Polymer Journal*, *49*(10), 2906–2918.

Hodneland, C. D., Lee, Y. S., Min, D. H., & Mrksich, M. (2002). Selective immobilization of proteins to self-assembled monolayers presenting active site-directed capture ligands. *Proceedings of the National Academy of Sciences of the United States of America*, *99*(8), 5048–5052.

Huang, D., & Shusta, E. V. (2006). A yeast platform for the production of single-chain antibody-green fluorescent protein fusions. *Applied and Environmental Microbiology*, *72*(12), 7748–7759.

Kamat, R. K., Ma, W., Yang, Y., Zhang, Y., Wang, C., Kumar, C. V., et al. (2013). Adsorption and hydrolytic activity of the polycatalytic cellulase nanocomplex on cellulose. *ACS Applied Materials & Interfaces*, *5*(17), 8486–8494.

Keppler, A., Gendreizig, S., Gronemeyer, T., Pick, H., Vogel, H., & Johnsson, K. (2003). A general method for the covalent labeling of fusion proteins with small molecules in vivo. *Nature Biotechnology*, *21*(1), 86–89.

Kwon, Y., Han, Z., Karatan, E., Mrksich, M., & Kay, B. K. (2004). Antibody arrays prepared by cutinase-mediated immobilization on self-assembled monolayers. *Analytical Chemistry*, *76*(19), 5713–5720.

Lambertz, C., Garvey, M., Klinger, J., Heesel, D., Klose, H., Fischer, R., et al. (2014). Challenges and advances in the heterologous expression of cellulolytic enzymes: A review. *Biotechnology for Biofuels*, 7, 135.

Los, G. V., Encell, L. P., McDougall, M. G., Hartzell, D. D., Karassina, N., Zimprich, C., et al. (2008). HaloTag: A novel protein labeling technology for cell imaging and protein analysis. *ACS Chemical Biology*, *3*(6), 373–382.

Mannesse, M. L., Boots, J. W., Dijkman, R., Slotboom, A. J., van der Hijden, H. T., Egmond, M. R., et al. (1995). Phosphonate analogues of triacylglycerols are potent inhibitors of lipase. *Biochimica et Biophysica Acta*, *1259*(1), 56–64.

Martinez, C., De Geus, P., Lauwereys, M., Matthyssens, G., & Cambillau, C. (1992). Fusarium solani cutinase is a lipolytic enzyme with a catalytic serine accessible to solvent. *Nature*, *356*(6370), 615–618.

McRae, S., Chen, X., Kratz, K., Samanta, D., Henchey, E., Schneider, S., et al. (2012). Pentafluorophenyl ester-functionalized phosphorylcholine polymers: Preparation of linear, two-arm, and grafted polymer–protein conjugates. *Biomacromolecules*, *13*(7), 2099–2109.

Miller, W. L., & Baxter, J. D. (1981). Synthesis of biologically active proteins by recombinant DNA technology. *Drug Development Research*, *1*(4), 435–454.

Mitsuzawa, S., Kagawa, H., Li, Y., Chan, S. L., Paavola, C. D., & Trent, J. D. (2009). The rosettazyme: A synthetic cellulosome. *Journal of Biotechnology*, *143*(2), 139–144.

Modica, J. A., Skarpathiotis, S., & Mrksich, M. (2012). Modular assembly of protein building blocks to create precisely-defined megamolecules. *Chembiochem: A European Journal of Chemical Biology*, *13*(16), 2331–2334.

Moraïs, S., Morag, E., Barak, Y., Goldman, D., Hadar, Y., Lamed, R., et al. (2012). Deconstruction of lignocellulosecbe into soluble sugars by native and designer cellulosomes. *MBio*, *3*(6), e00508–e00512.

Nilsson, B. L., Kiessling, L. L., & Raines, R. T. (2000). Staudinger ligation: A peptide from a thioester and azide. *Organic Letters*, *2*(13), 1939–1941.
Norris, V., den Blaauwen, T., Cabin-Flaman, A., Doi, R. H., Harshey, R., Janniere, L., et al. (2007). Functional taxonomy of bacterial hyperstructures. *Microbiology and Molecular Biology Reviews*, *71*, 230–253.
Nwe, K., & Brechbiel, M. W. (2009). Growing applications of "Click Chemistry" for bioconjugation in contemporary biomedical research. *Cancer Biotherapy & Radiopharmaceuticals*, *24*(3), 289–302.
Park, J. T., & Johnson, M. J. (1949). A submicrodetermination of glucose. *The Journal of Biological Chemistry*, *181*(1), 149–151.
Parsiegla, G., Reverbel-Leroy, C., Tardif, C., Belaich, J. P., Driguez, H., & Haser, R. (2000). Crystal structures of the cellulase Cel48F in complex with inhibitors and substrates give insights into its processive action. *Biochemistry*, *39*(37), 11238–11246.
Rabuka, D., Rush, J. S., deHart, G. W., Wu, P., & Bertozzi, C. R. (2012). Site-specific chemical protein conjugation using genetically encoded aldehyde tags. *Nature Protocols*, 7(6), 1052–1067.
Sun, Q., & Chen, W. (2016). HaloTag mediated artificial cellulosome assembly on a rolling circle amplification DNA template for efficient cellulose hydrolysis. *Chemical Communications (Cambridge, England)*, *52*(40), 6701–6704.
Sunbul, M., & Yin, J. (2009). Site specific protein labeling by enzymatic posttranslational modification. *Organic & Biomolecular Chemistry*, 7(17), 3361–3371.
Tao, L., Kaddis, C. S., Loo, R. R. O., Grover, G. N., Loo, J. A., & Maynard, H. D. (2009a). Synthesis of maleimide-end functionalized star polymers and multimeric protein–polymer conjugates. *Macromolecules*, *42*(21), 8028–8033.
Tao, L., Kaddis, C. S., Loo, R. R. O., Grover, G. N., Loo, J. A., & Maynard, H. D. (2009b). Synthetic approach to homodimeric protein-polymer conjugates. *Chemical Communications (Cambridge, England)*, (16), 2148–2150.
Thordarson, P., Le Droumaguet, B., & Velonia, K. (2006). Well-defined protein–polymer conjugates—Synthesis and potential applications. *Applied Microbiology and Biotechnology*, *73*(2), 243–254.
van Vught, R., Pieters, R. J., & Breukink, E. (2014). Site-specific functionalization of proteins and their applications to therapeutic antibodies. *Computational and Structural Biotechnology Journal*, *9*, e201402001.
Wilson, D. B. (2009). Cellulases and biofuels. *Current Opinion in Biotechnology*, *20*(3), 295–299.
Wilson, M. Z., & Gitai, Z. (2013). Beyond the cytoskeleton: Mesoscale assemblies and their function in spatial organization. *Current Opinion in Microbiology*, *16*(2), 177–183.
Wu, H. (2013). Higher-order assemblies in a new paradigm of signal transduction. *Cell*, *153*(2), 287–292.
Zhang, Y., Yang, Y., Ma, W., Guo, J., Lin, Y., & Wang, C. (2013). Uniform magnetic core/shell microspheres functionalized with Ni2+ – iminodiacetic acid for one step purification and immobilization of His-tagged enzymes. *ACS Applied Materials & Interfaces*, *5*(7), 2626–2633.
Zheng, J., Li, Y., Sun, Y., Yang, Y., Ding, Y., Lin, Y., et al. (2015). A generic magnetic microsphere platform with "clickable" ligands for purification and immobilization of targeted proteins. *ACS Applied Materials & Interfaces*, 7(13), 7241–7250.

CHAPTER TWELVE

Nanoarmoring of Proteins by Conjugation to Block Copolymer Micelles

Nisaraporn Suthiwangcharoen, Ramanathan Nagarajan[1]
Natick Soldier Research, Development and Engineering Center, Natick, MA, United States
[1]Corresponding author: e-mail address: ramanathan.nagarajan.civ@mail.mil

Contents

Abstract

The creation of polymer nanoparticles with protein functionality is of great interest to many applications such as targeted drug or gene delivery, diagnostic imaging, cancer theranostics, delivery of protein therapeutics, sensing chemical and biomolecular analytes in complex environments, and design of protective clothing

Methods in Enzymology, Volume 590
ISSN 0076-6879
http://dx.doi.org/10.1016/bs.mie.2017.01.013
© 2017 Elsevier Inc.
All rights reserved.

resembling a second skin. Many approaches to achieving this goal are being explored in the current literature. In this chapter, we describe a relatively simple and flexible approach of conjugating the protein to an amphiphilic block copolymer and creating polymer nanoparticles with protein functionality by taking advantage of the intrinsic self-assembly behavior of the amphiphilic block copolymer. The commercially available and biocompatible polyethylene oxide–polypropylene oxide–polyethylene oxide triblock copolymer is used as the polymer building block. For demonstrative purposes, bovine serum albumin was chosen as the protein. We determine the molecular weight of the protein–polymer conjugate and thereby the degree of conjugation using sodium dodecyl sulfate-polyacrylamide gel electrophoresis and matrix-assisted laser desorption/ionization time-of-flight mass spectrometry measurements. Retention of protein secondary structure in the conjugate was determined by circular dichroism spectroscopy, and the biological activity of the protein in the conjugated state has been evaluated by kinetic assay involving hydrolysis of an organophosphate compound. Dynamic light scattering and zeta potential measurements were used to characterize the size and charge of the protein–polymer conjugate micelle. Precise control of the size of the micelle and surface number density of the proteins on the micelle surface by coassembling with a second block copolymer have been demonstrated. These studies document a rational approach to armor the protein by conjugation with a block copolymer micelle, as a general approach.

1. INTRODUCTION

Protein–polymer conjugates have been the focus of significant research interest over the last two decades based on the premise that the conjugation would allow new properties to be created, enabling the application of conjugates in important areas such as drug delivery, protein therapeutics, and biological or chemical agent sensing (Briand, Kumar, & Kasi, 2011; De et al., 2009; Elsabahy & Wooley, 2012; Kratz & Beyer, 1998; Muszanska, Busscher, Herrmann, van der Mei, & Norde, 2011; Song et al., 2010; Yi, Batrakova, Banks, Vinogradov, & Kabanov, 2008). Many approaches to achieving this goal are being explored in the current literature (Börner, 2009; Jung & Theato, 2013; Tolstyka & Maynard, 2012). Numerous chemical methods have been established to create covalent conjugation of proteins with polymers, and many approaches continue to be explored to control and limit the degree of polymer conjugation.

An early reason for creating protein–polymer conjugates in the literature was to impart new useful properties to a protein molecule, such as reduced immunogenicity, protracted retention in circulation, and enhanced stability in the physiological environment (Abuchowski, van Es, Palczuk, & Davis, 1977). The most common polymer used to create these conjugates has been polyethylene glycol, PEG (Pasut & Veronese, 2012). In typical conjugates, multiple PEG molecules are covalently attached to a single protein molecule and cover much of the protein surface. This allows the protein to remain water soluble and also protected from any other antagonistic component it may encounter that is likely to remove the protein from the physiological system. The surface coverage provided by the conjugation allows for long-circulation times for the protein for drug delivery or therapeutic applications. However, such conjugation of multiple polymer molecules on the protein surface inevitably comes at the cost of decreasing the biological activity of the protein because the polymer attachments can significantly cover the protein active site and thereby reduce its intrinsic catalytic activity (Pasut & Veronese, 2012). In this sense, there is clear competition between the increasing stability or protection to the protein and the decrease in its catalytic activity, both being consequences of a high degree of polymer conjugation.

An alternate approach to stabilizing the protein is through creation of structures where the proteins are closely packed. The expectation is that the high number density of proteins would provide steric barriers against protein unfolding or denaturation. A possible way to realize this is by covalently attaching multiple protein molecules to the surface of a polymer nanoparticle (Boyer, Huang, Whittaker, Bulmus, & Davis, 2011). However, attaching multiple protein molecules on the surface of a preexisting polymer nanoparticle has many chemical challenges, since the chemical steps required for the process of attachment itself can affect the biological activity of the protein.

An effective approach to overcome this challenge and create a protein–polymer nanoparticle is through the mechanism of molecular self-assembly. If the protein is attached to a homopolymer, then depending upon the hydrophobicity/hydrophilicity of the polymer and of the protein, the protein–polymer conjugate molecule can potentially self-assemble to create a nanoparticle (Olsen, 2013; Thomas, Xu, & Olsen, 2012). In this approach, the possibility of self-assembly critically depends on the hydrophilic/hydrophobic balance arising from the choice of the polymer and solution conditions that determine the charged state of the protein. Hence, the self-assembly process is not independently controllable.

A facile approach to ensure independent control over the self-assembly process and to control the extent of protein functionality on the particle surface is through the selection of amphiphilic block copolymers to conjugate with the protein (Yi et al., 2008). In this case, independent control over self-assembly can be exercised entirely through the amphiphilic block copolymer without having to rely on the protein properties. Amphiphilic block copolymers with one hydrophobic block and the other hydrophilic block self-assemble in aqueous media and generate multimolecular nanoparticles of differing sizes and shapes, determined by the relative sizes of the two polymer blocks and their mutual interactions with water (Nagarajan, 1999). A widely used amphiphilic block copolymer is the symmetric triblock copolymer of polyethylene oxide–polypropylene oxide–polyethylene oxide (PEO–PPO–PEO), commercially marketed as Pluronics or poloxamers (Fig. 1). The versatility of their hydroxyl ending group enables easy chemical conjugation at that end, facilitating chemical modifications of the block copolymer (Muszanska et al., 2011; Song et al., 2010; Yi et al., 2008).

A diblock copolymer (PEO–PPO) or the symmetric triblock copolymer (PEO–PPO–PEO) undergoes self-assembly to generate multimolecular aggregates, with the hydrophobic polypropylene oxide forming the core of the micelle and the hydrophilic polyethylene oxide constituting the corona or shell of the micelle (Fig. 2). Indeed, the size and shape characteristics of self-assembled aggregates derived from Pluronics have been theoretically predicted as functions of the molecular weight and composition of the block copolymer in our earlier work (Nagarajan, 1999).

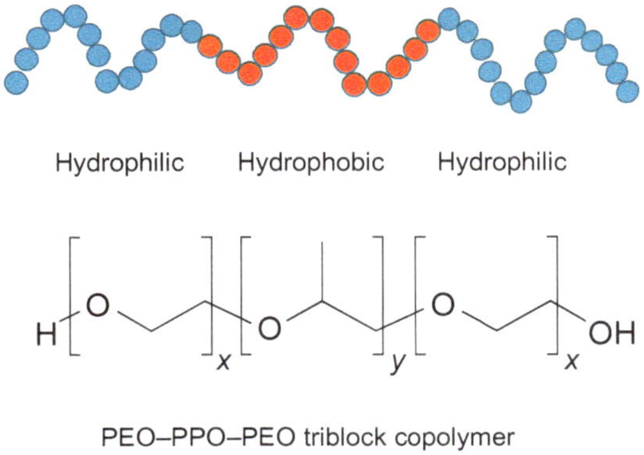

Fig. 1 Schematic structure of the symmetric triblock copolymer of polyethylene oxide–polypropylene oxide–polyethylene oxide. The polyethylene oxide end blocks usually have hydroxyl terminal groups in the commercial Pluronics.

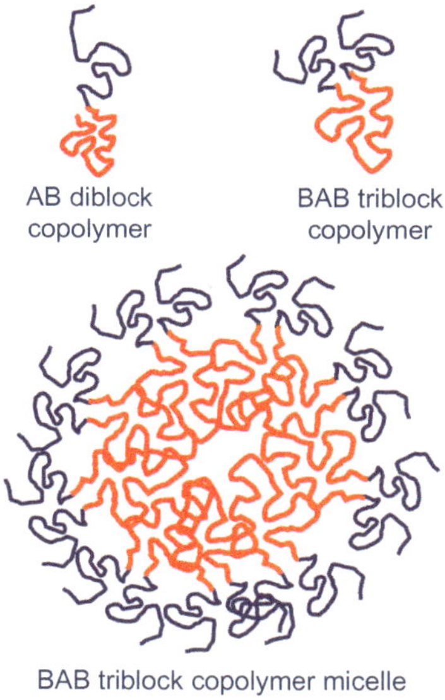

Fig. 2 Schematic of diblock and triblock copolymers and the resulting multimolecular micelle. A denotes the hydrophobic PPO block, and B denoted the hydrophilic PEO block, for Pluronic triblock copolymers. In the case of triblock copolymers, the PPO block has to fold back to ensure that both PEO end blocks are exposed to water, forming the corona region of the micelle.

The wide use of Pluronic copolymers has been facilitated by the commercial availability of a large family of triblock copolymers with molecular weights in the range 2–15 kDa and block composition in the range 20–80 wt% polyethylene oxide (Fig. 3). Further, because polypropylene oxide is not too hydrophobic and the PPO block size in the Pluronics is not too large, the Pluronics are able to spontaneously self-assemble in water. This is in contrast to the behavior of most other amphiphilic block copolymers for which the self-assembly process is not spontaneous and one has to create so-called nonequilibrium aggregates through a process involving dissolution in mixed aqueous-organic solvents followed by the removal of the organic solvent (Nagarajan, 2015). The spontaneous self-assembly behavior of Pluronics has also been a critical factor favoring their use in research and for potential applications.

In this paper, we have used the block copolymer F127 ($PEO_{100}PPO_{64}PEO_{100}$) for covalent conjugation to the protein bovine serum albumin (BSA) and the block copolymer P123 ($PEO_{20}PPO_{70}PEO_{20}$) for manipulating the size of the protein–polymer conjugate micelle and thereby

Commercially available Pluronic copolymers

Increasing weight percent of PEO block ⟶ (top to bottom)

L62 $E_6P_{35}E_6$ 2523	L72 $E_6P_{38}E_6$ 2750		L92 $E_6P_{50}E_8$ 3650		L122 $E_{11}P_{69}E_{11}$ 5000
L63 $E_9P_{32}E_9$ 2616				P103 $E_{17}P_{60}E_{17}$ 4916	P123 $E_{20}P_{70}E_{20}$ 5750
L64 $E_{13}P_{30}E_{13}$ 2854		P84 $E_{19}P_{43}E_{19}$ 4123		P104 $E_{27}P_{61}E_{27}$ 5853	
P65 $E_{19}P_{29}E_{19}$ 3325	P75 $E_{24}P_{35}E_{24}$ 4200	P85 $E_{26}P_{40}E_{26}$ 4568		P105 $E_{37}P_{56}E_{37}$ 6448	
	F77 $E_{52}P_{35}E_{52}$ 6600	F87 $E_{61}P_{40}E_{61}$ 7700			F127 $E_{100}P_{64}E_{100}$ 12,448
F68 $E_{77}P_{29}E_{77}$ 8429		F88 $E_{104}P_{39}E_{104}$ 11,375	F98 $E_{118}P_{45}E_{118}$ 12,949	F108 $E_{133}P_{50}E_{133}$ 14,554	

Increasing molecular weight of PPO block ⟶

Fig. 3 The grid showing commercially available Pluronic triblock copolymers. From *top* to *bottom* on the grid, the weight percent of PEO block increases from 20% to 80%, and from *left* to *right* on the grid, the molecular weight of the PPO block increases. Each entry includes the commercial designation for the molecule, the number of monomeric units in the PEO and PPO blocks (marked as E and P), and the overall molecular weight of the triblock copolymer. The designations L, P, and F for the molecules designate the state of their occurrence as liquid, paste, or flake.

the surface number density of the protein molecules. Indeed, we have a wide choice of Pluronic molecules available to work with, as shown on Fig. 3, giving us the option to create significant variations in the protein-functionalized nanoparticles.

Pluronics have been approved by the Food and Drug Administration (FDA) to be used as food additives, pharmaceutical ingredients, drug carriers in parenteral systems, tissue engineering, and agricultural products. Pluronic molecules can also be used as carriers in several routes of administration, including oral, cutaneous, intranasal, vaginal, rectal, ocular, and parenteral (Batrakova & Kabanov, 2008). They have also been considered for dermal and transdermal drug delivery systems.

One can now visualize conjugating a protein molecule to the Pluronic and generate a multimolecular micelle, which is a nanoparticle displaying a concentrated protein functionality at the particle surface (Fig. 4).

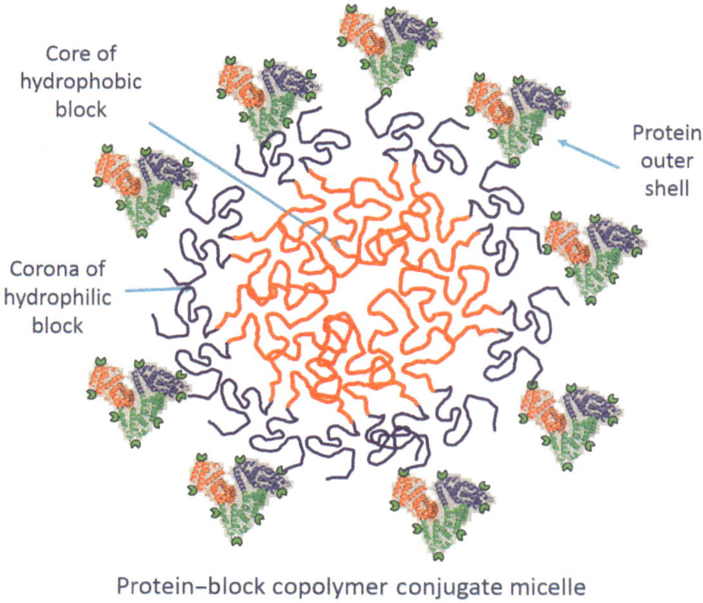

Fig. 4 Schematic of a Pluronic micelle with protein attached to the PEO end groups. The state of charge on the protein would influence the actual conformation of the micelle, with the protein being entirely exposed to water as shown in the figure, or the protein partially interacting with the PEO domain or even the PPO domain. The pac-man symbols on the protein schematically represent primary amines on the side chains of lysines which conjugate with the *N*-hydroxysuccinimide (NHS) ends, represented by the pentagon symbols, of the functionalized Pluronic molecules. The F127 micelles are about 60 nm in diameter, and the BSA molecule in the shell region adds about 17–28 nm to the diameter, based on the shapes suggested for BSA, as discussed later in Section 4.

The relative proximity of the protein molecules on the micelle surface can be expected to reduce any propensity for the unraveling of the protein structure (nanoarmoring) and such close packing can prevent denaturation of the protein and ensure longer-term stability as well as bioactivity in the aqueous phase. It is with this motivation, we have conducted this study to create and characterize protein–Pluronic conjugate molecules and the resulting self-assembled protein conjugated functional polymer nanoparticles.

We have used the conjugation approach described in this paper to construct multifunctional nanoparticles, with BSA as a model protein, folic acid as a cell-targeting moiety, and gadolinium as a contrast agent for imaging, providing multiple functionalities (Suthiwangcharoen & Nagarajan, 2014a). We have also demonstrated the enhanced stability of the enzyme organophosphate hydrolase (OPH) when it is conjugated with the Pluronic F127, providing significantly improved performance for decontaminating material surfaces exposed to organophosphate chemical agents (Suthiwangcharoen & Nagarajan, 2014b).

2. SYNTHESIS OF BLOCK COPOLYMER–PROTEIN (BSA) CONJUGATE

The synthesis of BSA conjugated block copolymer consists of three major steps shown schematically on Fig. 5. The Pluronic block copolymer is chemically inert and not ready to be conjugated to proteins. The hydroxyl end groups of the PEO blocks need to be transformed into reactive groups that could conjugate with the protein. Therefore, first, the block copolymer was activated with succinic anhydride to form carboxyl-terminated block copolymer. Second, *N*-hydroxysuccinimide (NHS) was introduced to the carboxyl-terminated block copolymer to create a good leaving group (NHS) for enzyme conjugation. Third, the amino group of lysine on the BSA reacted with the

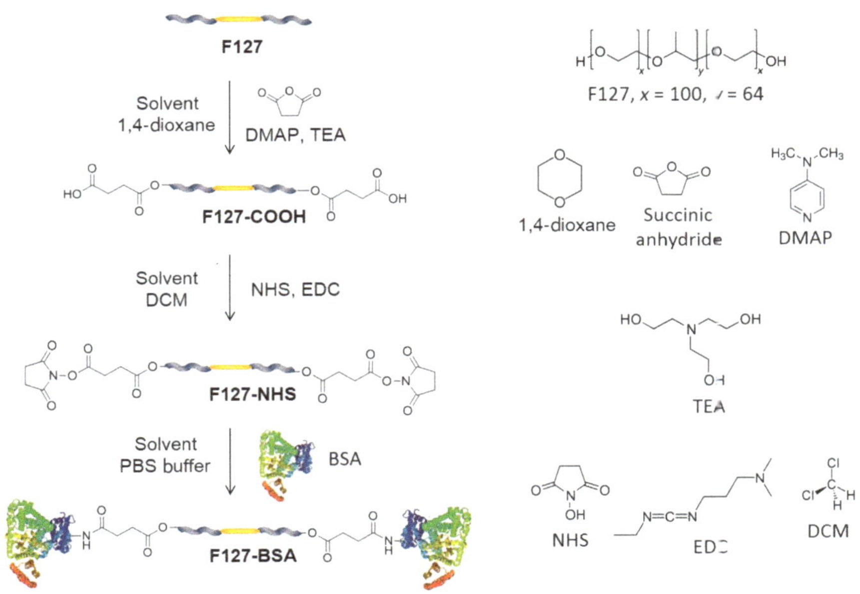

Fig. 5 Schematic of the three sequential synthetic steps involved in the preparation of bovine serum albumin conjugated with F127 block copolymer. Also shown are the chemical structures of various molecules involved in the synthesis. The synthetic steps are described in detail in the following text.

NHS group on the block copolymer via addition/elimination to create stable amide bond between the block copolymer and the BSA, creating the conjugated micelle.

2.1 Materials

A family of Pluronic PEO–PPO–PEO symmetric triblock copolymers shown on Fig. 3 were obtained as a gift from the BASF Corporation. Two molecules from this family, with commercial designations F127 ($PEO_{100}PPO_{64}PEO_{100}$, MW 12,450 Da, 70 wt% PEO) and P123 ($PEO_{20}PPO_{70}PEO_{20}$, MW 5750 Da, 30 wt% PEO), were used in our experiments. These two molecules have roughly similar lengths of the hydrophobic polypropylene oxide blocks. Other molecules from this family can also be readily used, following identical synthetic methods described here. Materials used include the following:

F127, MW 12,450 Da, 70 wt% PEO (BASF)
P123, MW 5750 Da, 30 wt% PEO (BASF)
Succinic anhydride (Sigma-Aldrich)
1,4-Dioxane (Sigma-Aldrich)
Dichloromethane, DCM (Sigma-Aldrich)
Triethylamine, TEA (Sigma-Aldrich)
Acetonitrile (Sigma-Aldrich)
Diethyl ether (Sigma-Aldrich)
BSA (Sigma-Aldrich)
1-Ethyl-3-(3-dimethylaminopropyl)carbodiimide, EDC (ACROS Chemical)
NHS (Alfa Aesar)
4-Dimethylaminopyridine, DMAP (Alfa Aesar)
Nanopure water
Phosphate saline buffer (PBS 1 ×, pH 7.4)
NuPAGE® Novex® 12% Bis-Tris Gel (Molecular Probes, Invitrogen)
MOPS (3(*N*-morpholino) propanesulfonic acid) buffer (Invitrogen)
Mark 12 standard protein (Invitrogen)
Coomassie Blue R-250
Paraoxon (Sigma-Aldrich)
Methanol (Sigma-Aldrich)

2.2 Equipment

50-mL Three-neck round-bottom flask
Addition funnel

Buchner funnel
5-mL Eppendorf tube
0.45 μm Filters (Millipore)
Rotary evaporator
Water bath
Bruker ARX 500 MHz ^{1}H-NMR spectrometer
Perkin-Elmer Spectrum One Fourier Transform Infrared (FTIR) Spectrometer
Centrifuge
Sodium dodecyl sulfate-polyacrylamide gel electrophoresis (SDS-PAGE)
Bruker Ultra-Flex II (Berman, Germany) MALDI-TOF mass spectrometer
Circular dichroism (CD) spectrophotometer (Aviv Biomedical, Inc., Lakewood, NJ)
EON ultraviolet–visible (UV–vis) microplate spectrophotometer (BioTek Instruments)
Zetasizer Nano ZS (Malvern Instruments) for light scattering and zeta potential

2.3 Safety

Chemical-resistant gloves must be worn, and nitrile gloves are recommended.
Safety glasses or chemical splash goggles are required.
Full length pants and close-toed shoes must be worn at all times.
All experiments should be performed in the hood, unless specified otherwise.

2.4 Block Copolymer Functionalization With COOH End Group

In the first step, the terminal hydroxyl groups of PEO end blocks were transformed to carboxylic acid (COOH) functionalities. The synthesis was conducted as follows:

a. In a 50-mL three-neck round-bottom flask, F127 (6.3 g; 1 mmol OH) and DMAP (122.17 mg; 1 mmol) were dissolved in 1,4-dioxane (15 mL) in the presence of TEA (139 μL) and stirred under nitrogen for 30 min.

b. Succinic anhydride solution (125 mg in 5 mL 1,4-dioxane) was then added dropwise using the addition funnel to the copolymer solution while stirring. The rate of addition was one drop for every 20–30 s.

c. The solution mixture was left stirring at room temperature for 24 h under N_2 purge.
d. The excess 1,4-dioxane was removed by rotary evaporation. Briefly, a round-bottom flask containing the sample in 1,4-dioxane was placed on the rotary evaporator. The bump trap was applied to prevent the sample from accidentally splashing into the condenser. The aspirator vacuum was turned on with the full vacuum. The flask was then lowered into the water bath with the temperature of ~70°C. Few bubbles may be evident as an indicative of the solvent being removed. Note that the temperature of the water bath and the vacuum pressure are two key parameters that control the rate of the solvent removal. The water bath temperature should not exceed the boiling point of the solvent.
e. The remaining samples were precipitated three times in cold diethyl ether while stirring.
f. The precipitate was vacuum filtered in a Buchner funnel using a coarse filter paper and dried under vacuum overnight at room temperature to give the white powder of F127-COOH.

2.5 Block Copolymer Functionalization With NHS End Group

In this step, the carboxyl-terminated F127 was converted to NHS-terminated F127. Briefly, the synthesis followed the steps below:

a. F127-COOH (500 mg; 0.08 mmol) and EDC (77 mg; 0.4 mmol) were dissolved in 5 mL DCM in a round-bottom flask.
b. After stirring the mixture for 30 min, NHS (46 mg; 0.4 mmol) was added, and the solution mixture was left stirring at room temperature for 24 h.
c. Next day, the solution mixture was precipitated three times in cold diethyl ether.
d. The precipitate was vacuum filtered in a Buchner funnel using a coarse filter paper and further dried under vacuum overnight at room temperature to give the white powder of F127-NHS.

2.6 Confirming End Functionalization of F127

The successful preparation of F127-COOH and F127-NHS was confirmed using ^{1}H-NMR (Fig. 6). ^{1}H-NMR was acquired on a Bruker ARX 500 MHz spectrometer. The NMR spectrum of F127-COOH shows a characteristic peak at ~2.6 ppm, corresponding to the protons on the CH_2–CH_2 of the succinic anhydride. For the F127-NHS, the protons on

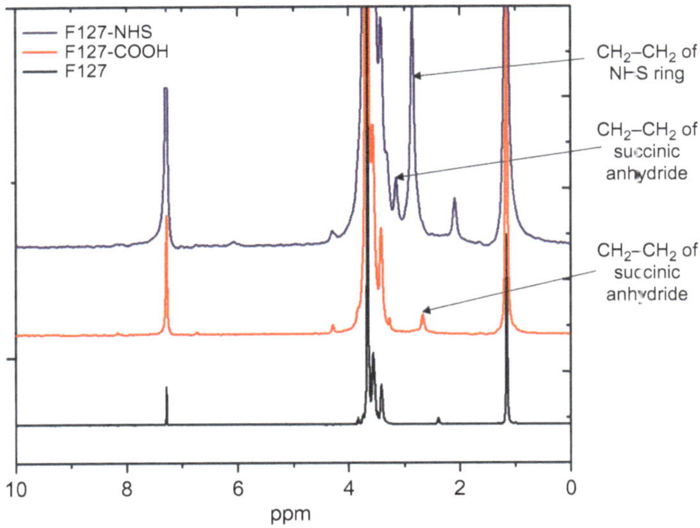

Fig. 6 ^{1}H-NMR spectra of F127, F127-NHS, and F127-COOH, respectively. The presence of COOH and NHS is confirmed by ^{1}H-NMR as explained in the text.

the CH_2–CH_2 of the succinic anhydride were deshielded to ~3.2 ppm, while a prominent peak at ~2.8 ppm corresponds to the CH_2–CH_2 of the NHS ring.

The successful preparation of F127-COOH and F127-NHS was confirmed also using Fourier transform infrared spectroscopy (FTIR) as shown on (Fig. 7). Infrared spectroscopy was recorded on a FTIR. FTIR spectrum shows a prominent peak at ~1730 cm^{-1}, corresponding to the stretching of the carbonyl group on COOH. The FTIR spectrum shows a characteristic peak of the carbonyl and amide bonds at ~1713 and 1570 cm^{-1}, respectively.

The yield of F127-COOH and F127-NHS were both greater than 80%. The yield was calculated based on the difference between weights before and after conjugation based on the well-established method described in detail (Zhang et al., 2010).

2.7 Block Copolymer Conjugation With BSA

The F127 functionalized with the NHS end group is conjugated to BSA with the following protocol. This method can be applied to other

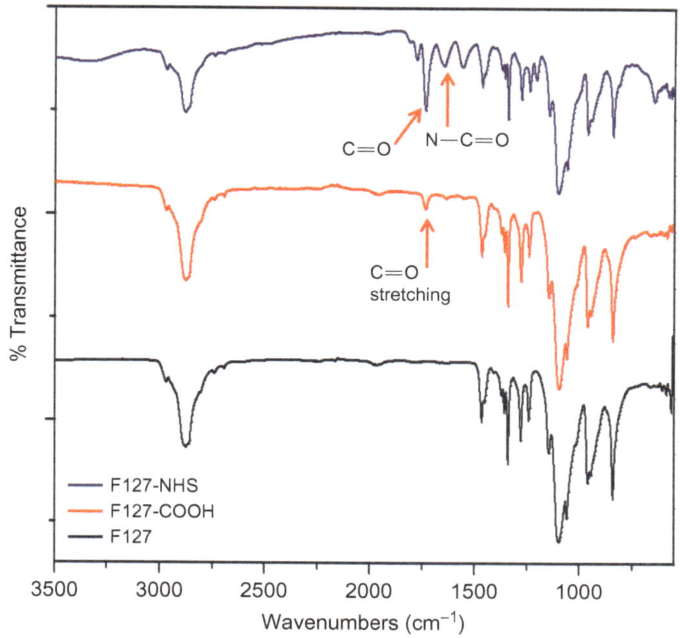

Fig. 7 FTIR spectra of F127, F127-NHS, and F127-COOH, respectively. The presence of COOH and NHS is confirmed by FTIR as explained in the text.

protein/enzyme conjugation. However, the optimal pH and ionic strength of the buffer are to be varied, depending upon the nature of the protein (i.e., isoelectric point).

a. In a 5-mL Eppendorf tube, BSA in PBS solution (30 mg mL^{-1}) was added dropwise into the F127-NHS solution (15 mg mL^{-1}) while stirring.

b. Then, the reaction mixture was allowed to stir at 4°C for 48 h.

c. The samples were centrifuged at relative centrifugal force of 8000 × *g* to 16,000 × *g* for 20 min at 4°C to remove excess BSA. The supernatant was carefully removed using a micropipette without disturbing the small pellet on the side of the tube. The pellet was then resuspended in 4°C water. To minimize the protein denaturation, the water must be pre-incubated at 4°C. Depending upon the final application, water can be replaced with an appropriate buffer. Since the sample was intended to be lyophilized immediately, water was used herein to avoid the salt content from the buffer.

d. The removal of free BSA was monitored by SDS-PAGE measurements (described in Section 3.1).
e. The purified sample was lyophilized and stored at −20°C until use.

3. CHARACTERIZATION OF BLOCK COPOLYMER–PROTEIN CONJUGATE

3.1 SDS-PAGE for Confirming Protein Conjugation

The conjugation of BSA to F127 and the purification of the conjugate from free BSA were monitored using SDS-PAGE.

a. The SDS-PAGE employed NuPAGE® Novex® 12% Bis-Tris Gel (Molecular Probes, Invitrogen) with MOPS (3(*N*-Morpholino) propanesulfonic acid) running buffer (Invitrogen). The NuPAGE® Novex® 12% Bis-Tris Gels are precast polyacrylamide gels designed to give optimal separation of small- to medium-sized proteins. The gels have a neutral pH environment that minimizes protein modifications.
b. All samples were heated at 70°C for 10 min to denature the proteins prior to electrophoresis.
c. Mark 12 protein was run in two channels, free BSA in one channel, and F127-BSA conjugate at different concentrations in three channels. Approximately 10 μg of protein is needed for each channel.
d. Constant voltage of 200 V was applied, and the gel was run for about 42 min.
e. Following electrophoresis, the gel was stained with Coomassie Blue to facilitate visualization of protein bands. The gel was stained at room temperature for 3–4 h with gentle agitation. The container must be covered to avoid contamination and to prevent evaporation of the staining solution. Appropriate nitrile gloves must be worn to prevent protein contamination from hands. After the staining was complete, the gel was destained for overnight at room temperature with the destain solution with gentle agitation. Make sure that the entire gel is covered with the destain solution. A piece of Kimwipe tied in a knot may be placed in the destain solution around the gel to facilitate the destaining process. Try to avoid laying the Kimwipe on the gel as that may cause uneven destaining. Note that the staining/destaining process can be accelerated with the use of microwave oven (optional). Detain solution: 50% (v/v) methanol in water with 10% (v/v) acetic acid.

Fig. 8 shows SDS-PAGE profiles of the following samples: Mark 12 standard protein, standard BSA, and F127-BSA at various concentrations. The major

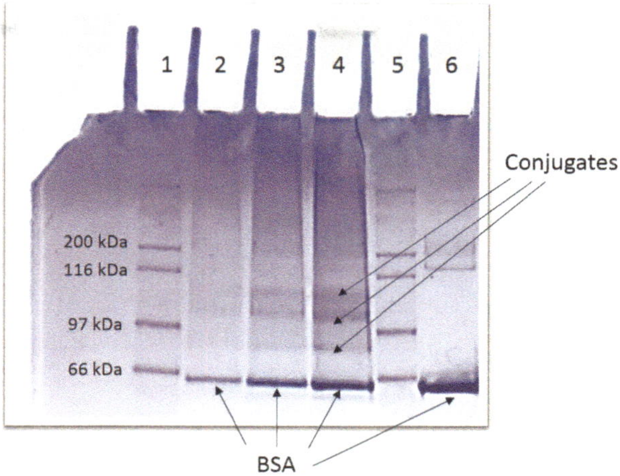

Fig. 8 SDS-PAGE analysis of Pluronic–BSA conjugates, Lanes 1 and 5—standard protein Mark 12, Lane 6—native BSA (~66 kDa), and Lanes 2, 3, and 4—F127-BSA conjugate at various concentrations during the purification process.

band at 66.5 kDa is associated with the free BSA, while the bands at a higher molecular weight are associated with F127-BSA conjugates (F127 is approximately 12.5 kDa).

For the purification process, SDS-PAGE was run at various stages of centrifugation. When no more free BSA was detected in SDS-PAGE, the product was lyophilized and stored at −20°C. The complete removal of excess BSA by centrifugation was not always consistently achieved, and it would be useful to consider alternate methods such as dialysis or liquid chromatography for this purpose.

3.2 MALDI-TOF for Molecular Weight Determination of Protein Conjugate

Molecular weight of F127-BSA after the purification step was determined using MALDI-TOF mass spectrometry (MS) analysis, conducted using a Bruker Ultra-Flex II (Berman, Germany), and equipped with solid-state smart beam.

a. The MALDI was conducted in 70% acetonitrile where the micelles will undergo disassembly to singly dispersed F127-BSA.

b. MS experiments were conducted by plate spotting of the samples, which were premixed with a sinapinic acid matrix.

c. The free BSA and F127 were used as reference samples.
d. The BSA reference and the purified F127-BSA were spotted by the dried droplet method using a 0.75 μL of a saturated sinapinic acid matrix in 70% acetonitrile/0.1% TFA and 0.75 μL of the protein samples.
e. The samples were run in linear mode with an accelerating voltage of 25 kV (Fig. 9).

In MALDI-TOF, the primary species correspond to singly ionized state of the protein, while a small amount of doubly ionized species is also generally obtained. The peaks at 66.4 and 33.3 kDa obtained for BSA represent the unmodified BSA in the singly and doubly ionized states, respectively. The peak at 79.6 kDa obtained for F127-BSA suggests that F127 was conjugated to BSA with approximately 1:1 ratio, since that will correspond to the 1:1 conjugate in its singly ionized state. The peak at 39.9 kDa obtained for F127-BSA again corresponds to the 1:1 conjugate, but in the doubly ionized state. No peak for BSA was seen in the MALDI-TOF of F127-BSA, indicating the efficiency of purification.

Because BSA has numerous lysine sites on the surface for conjugation and a number of conjugates with end-functionalized F127 are potentially

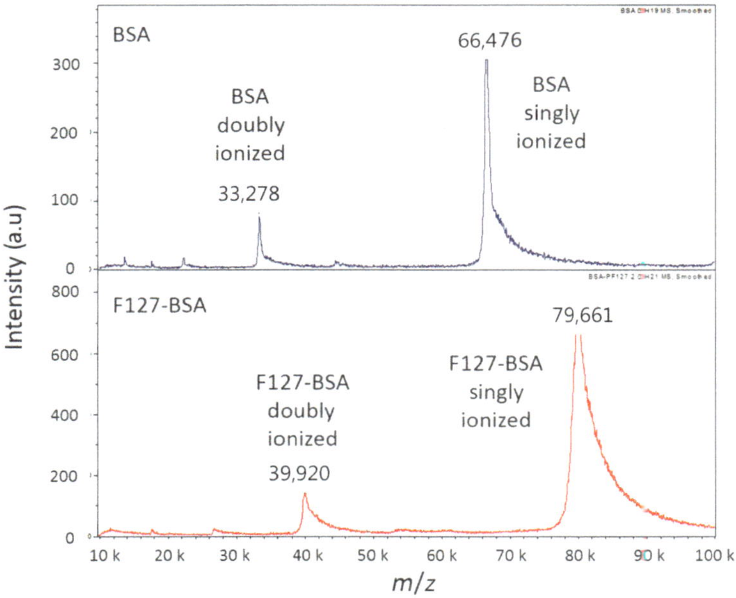

Fig. 9 MALDI-TOF mass spectrum of native unmodified BSA (*top*) and purified F127-BSA (*bottom*). The results reveal the presence of a 1:1 F127:BSA conjugate.

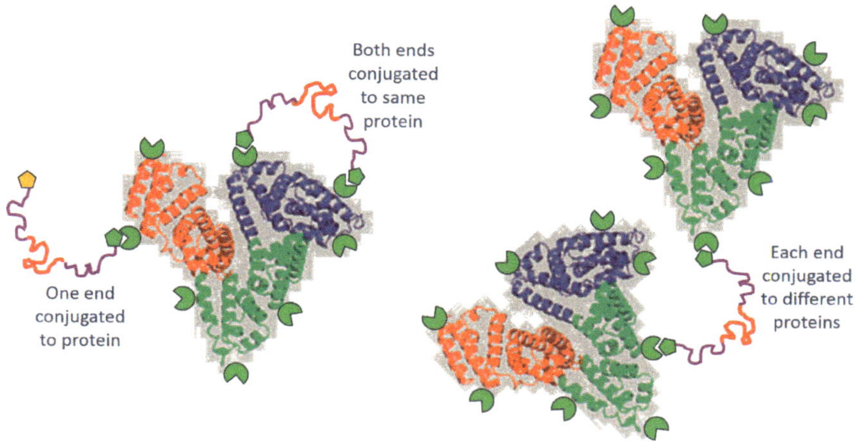

Fig. 10 Schematic of various protein–polymer conjugation states. The pac-man symbols on the protein schematically represent primary amines on the side chains of lysines which conjugate with the *N*-hydroxysuccinimide (NHS) ends, represented by the pentagon symbols, of the functionalized Pluronic molecules. *Left figure* shows the possibilities either only one end of F127 is conjugated to a molecule of BSA or both ends of F127 are conjugated to the same BSA molecule. The *right figure* shows the possibility that each end of F127 is conjugated to a different BSA molecule. Not shown are numerous other possibilities such as two or more F127 molecules conjugated to the same BSA molecule and complex conjugate networks involving multiple BSA and multiple F127 molecules.

possible (Fig. 10). The type of conjugates formed depends upon the proximity of the lysine residues on BSA, the size of the block copolymer as well as the concentrations of the protein, and the block copolymer in solution.

Some conjugates and their anticipated approximate molecular weights for singly and doubly ionized states are as follows: (i) only one end of F127 is conjugated to a molecule of BSA (1:1 conjugate, 79 and 39.5 kDa), (ii) each end of F127 is conjugated to a different BSA molecule (1:2 conjugate, 145 and 72.5 kDa), (iii) both ends of F127 are conjugated to the same BSA molecule (1:1 conjugate, 79 and 39.5 kDa), (iv) two F127 molecules are conjugated to the same BSA molecule (2:1 conjugate, 92 and 46 kDa), and (v) more complex conjugates involving multiple BSA and multiple F127 molecules. Following the purification steps, the MALDI-TOF indicates only a 1:1 conjugate in the final purified product. However, MALDI-TOF cannot resolve whether only one end of F127 or both ends of F127 are covalently bound to one BSA molecule since both options give rise to the same mass peaks.

3.3 CD and UV–Vis Spectra for Assessing Secondary Structure Changes in Conjugated Protein

The CD measurements were made using the CD spectrophotometer (Aviv Biomedical, Inc., Lakewood, NJ) with a path length of 1 mm. Measurements were carried out on BSA and F127-BSA in diluted HEPES (4-(2-hydroxyethyl)-1-piperazine ethanesulfonic acid) buffer solutions (pH 8), keeping the enzyme concentration at 100 μg mL^{-1}. Far-ultraviolet CD spectra were measured from 190 to 240 nm at 1 nm intervals. Three scans were accumulated and averaged for each spectrum after the background of diluted blank buffer was subtracted (Fig. 11).

BSA is a protein with mainly α-helix secondary structure (59% in 10 m*M* phosphate buffer, pH 7.4). Changes in its secondary structure can be easily detected by CD, which is sensitive to variations in the percentage of α-helix structure present. The CD spectrum for BSA in aqueous solution is characteristic of macromolecules with high α-helical content, monitored by the two well-defined ellipticities at 208 and 222 nm. The CD spectrum of F127-BSA essentially retains the same features as the CD spectrum of

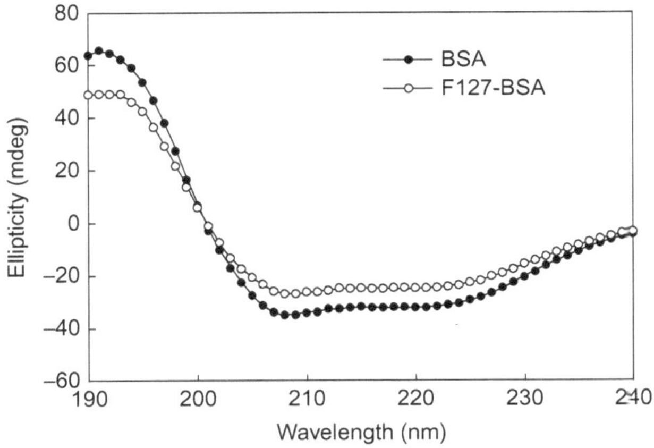

Fig. 11 Circular dichroism spectra of BSA and F127-BSA. Both spectra exhibit two minima at 208 and 222 nm, indicating that the conjugation did not significantly affect the secondary structure of the BSA. The polymer at the concentration present in the solution of the conjugate has no absorbance in the region of measurements.

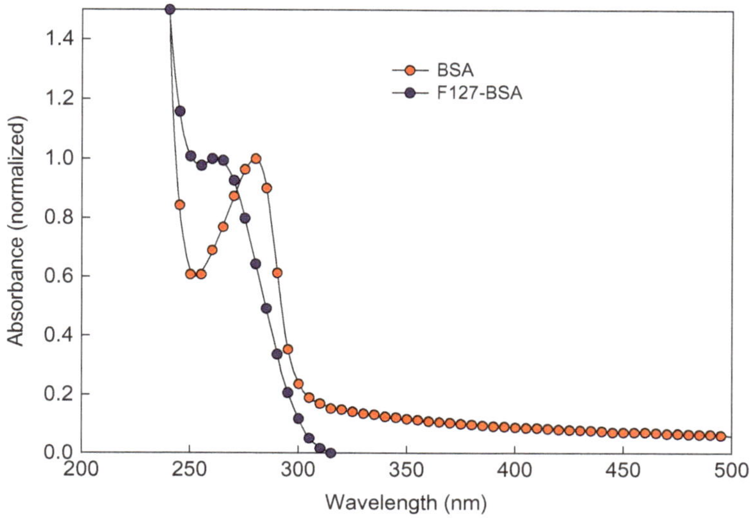

Fig. 12 UV–vis spectrum of BSA and F127-BSA. A slight blue-shift of about 3 nm was observed after the conjugation indicating a slight change of the microenvironment polarity around the tryptophan and/or tyrosine residues of BSA. Note that the spectra were normalized at the characteristic peak whose relative absorbance is set as unity, and therefore, the base lines of the two samples are different.

BSA indicating no perceptible changes in the secondary structure of the protein because of the covalent conjugation to the F127.

The UV–vis spectra of both BSA and F127-BSA are shown on Fig. 12. A blue-shift of about 3 nm was observed after the conjugation indicating a slight change of the microenvironment polarity around the tryptophan (Trp) and/or tyrosine (Tyr) residues of BSA. The interaction between the block copolymer and BSA apparently changes the polarity of microenvironment around Trp and Tyr residues of BSA but does not cause a conformational change of BSA.

3.4 Kinetic Assay to Assess Change in Biological Activity of Conjugated Protein

BSA shows hydrolytic reactivity against the organophosphate compound, paraoxon (Fig. 13) converting it to diethyl phosphate and *p*-nitrophenol. The hydrolysis product *p*-nitrophenol, under basic conditions, has a yellow color that is easily detectable by use of UV–vis spectroscopy.

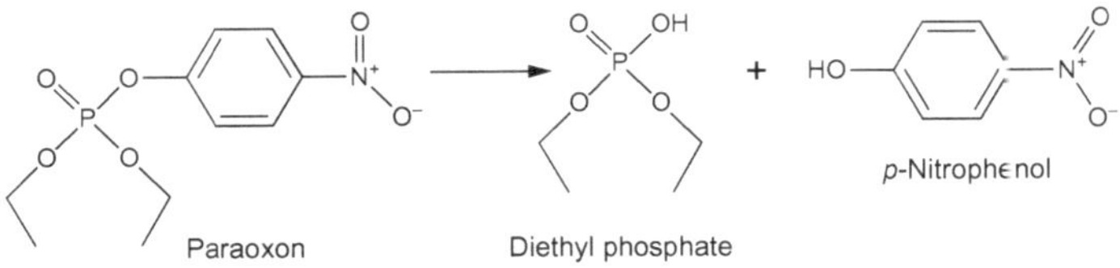

Fig. 13 Hydrolysis of paraoxon to produce diethyl phosphate and *p*-nitrophenol.

Measurements of the catalytic activity of BSA and F127-BSA were performed with an EON UV–vis microplate spectrophotometer (BioTek Instruments). The hydrolysis of paraoxon in the presence of BSA was assessed by optical density (OD) measurements at 405 nm. The appropriate blanks were subtracted from the raw measurements of OD, so that the reported OD measurements only represented the formation of the hydrolysis product, *p*-nitrophenol. The activity of BSA an F127-BSA conjugate for the hydrolysis of paraoxon was evaluated with the following protocol:

a. An aqueous solution of paraoxon containing 10% (v/v) methanol was added to BSA or F127-BSA with the same enzyme concentrations to give the final concentrations of 0.075 μg mL^{-1} of BSA and 0.1 mM paraoxon.

b. The activity levels were measured by monitoring the release of *p*-nitrophenol product spectrophotometrically at 405 nm for 10 min using EON BioTek Microplate Spectrophotometer.

c. The concentration of *p*-nitrophenol was determined based on the standard calibration curve of OD vs concentration.

d. The kinetic parameters for BSA and F127-BSA were obtained by measuring initial hydrolysis rate at different paraoxon concentrations, keeping a constant enzyme concentration (Fig. 14).

Initial rate kinetics of enzyme-catalyzed reactions follow a substrate-saturation mechanism, well known as the Michaelis–Menten kinetics (Scheme 1).

Here, E, S, ES, and P denote the enzyme, the substrate, the enzyme/substrate complex, and the product, respectively, while the various phenomenological rate constants are denoted by k_+, k_-, and k_{cat}. Under quasi steady-state approximation, the initial rate kinetics is represented by the relation

$$V = \frac{dS}{dt} = \frac{V_m S}{K_M + S}, \quad V_m = k_{cat} E_o, \quad K_M = \frac{k_- + k_{cat}}{k_+}, \quad \frac{1}{V} = \frac{1}{V_m} + \frac{K_M}{V_m S} \quad (1)$$

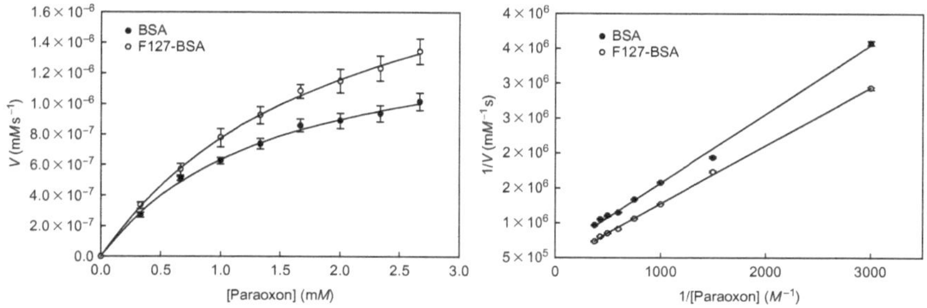

Fig. 14 Initial reaction velocity of hydrolysis of paraoxon by BSA and F127-BSA as a function of the substrate paraoxon concentration. The enzymatic hydrolysis of paraoxon was conducted in water at 37°C. The rate of the reaction was monitored at 405 nm every 30 min for 5 h. The experimental data are shown in two different forms: one plotting the velocity against the substrate concentration and the other plotting the reciprocals of those variables, known as the Lineweaver–Burk plot. The results show a small increase in enzymatic activity when BSA is conjugated to F127.

$$\mathrm{E+S}\underset{k_-}{\overset{k_+}{\rightleftharpoons}}\mathrm{ES}\overset{k_{cat}}{\rightarrow}\mathrm{E+P}$$

Scheme 1 Michaelis–Menten enzyme kinetics involving the reversible formation of enzyme-substrate complex followed by the irreversible dissociation of new product synthesized.

Table 1 Kinetic Parameters for Biocatalytic Activity of BSA and F127-BSA[a]

Parameter	BSA	F127-BSA
K_M (mM)	1.653 ± 0.305	1.884 ± 0.383
$V_m \times 10^6$ (mM s^{-1})	1.677 ± 0.225	2.249 ± 0.365
$k_{cat} \times 10^6$ (s^{-1})	8.871 ± 0.595	11.9 ± 0.386
$k_{cat}/K_M \times 10^6$ (mM^{-1} s^{-1})	5.371 ± 0.166	6.317 ± 0.083

[a]Values represent means of quintuplicate experimental data (mean ± st. dev.).

where V is the reaction velocity, $V_m = k_{cat}\ E_o$ is the maximum velocity attained at saturating substrate concentration, E_o is the total concentration of the enzyme (sum of the free-enzyme, E and that bound to the substrate, ES), and K_M is known as the Michaelis constant. The last part of Eq. (1) is the same Michaelis–Menten equation written in the double-reciprocal form commonly used to construct the Lineweaver–Burk plot to extract the constants appearing in the rate equation. The kinetic constants were determined by fitting the initial reaction velocity V obtained at various substrate concentrations [S] to the classical Michaelis–Menten equation (Eq. 1), and the results are summarized in Table 1.

4. CHARACTERIZATION OF PROTEIN–POLYMER CONJUGATE MICELLE

Protein–polymer conjugate micelles were prepared using conventional stirring. The lyophilized or solid samples of the F127, P123, and F127-BSA were dissolved in nanopure water at approximately 1% (w/v) concentration, and the micelles were spontaneously generated. The solution mixture was allowed to stir at room temperature for about 1–2 h under dark. The sample was filtered through 0.45 μm filters (Millipore) prior to the dynamic light scattering (DLS) and zeta potential measurements.

4.1 Micelle Size Determination by DLS

The hydrodynamic diameter and size distributions were measured with DLS using the Zetasizer Nano ZS (Malvern Instruments). The scattering intensity was measured at a scattering angle of 173 degree relative to the source. The deconvolution was accomplished using a nonnegatively constrained least-squares fitting algorithm.

The DLS results (Fig. 15) for F127 show one peak at about 4 nm for the nonmicellar, singly dispersed molecule of F127, and a second peak at about 60 nm for F127 micelles. The DLS data show that P123 micelles have a

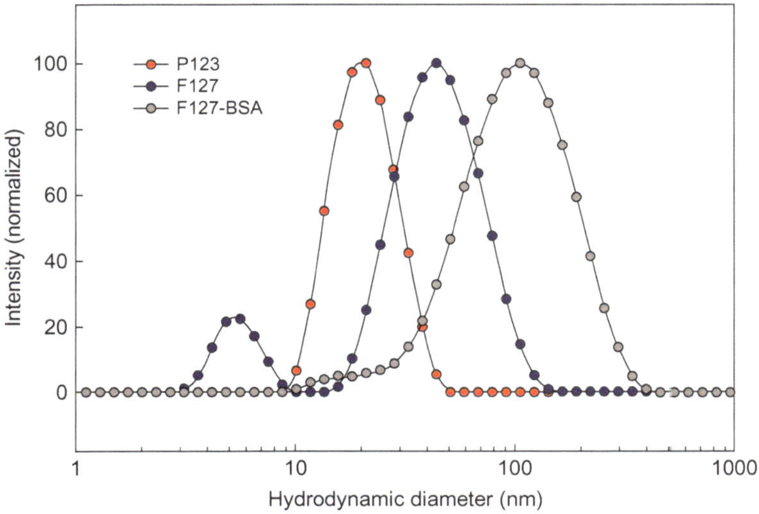

Fig. 15 Size distribution from DLS for P123, F127, and F127-BSA micelles.

diameter of about 20 nm. The F127-BSA micelles have a size of about 90 nm, and the micelle size is larger than that of the unmodified F127 micelles by about 30 nm. The shape of BSA in solution has been discussed in the literature in the context of interpreting scattering data for BSA, taking into consideration the X-ray crystallographic data available for human serum albumin (Leggio, Galantini, & Pavel, 2008; Zhang et al., 2012, 2007). The shape has been suggested to be a prolate ellipsoid with dimensions of 14 × 4 nm for the major axes, with three domains aligned along the long axis. The X-ray crystallographic data have indicated a heart-shaped structure. Recently, an equilateral, triangular prismatic shell shape with a side length of 8.4 nm and a thickness of 3.2 nm has been proposed. If either the prismatic shell or the prolate ellipsoidal shapes are considered, then the presence of the BSA at the micelle surface would account for a size increase of about 17–28 nm over the unmodified F127 micelles, accounting for most of the 30 nm size difference seen on conjugation. The remaining small size expansion could be reasonably attributed to a change in the micelle aggregation number and/or some stretching of the coronal PEO block when BSA is conjugated to the block copolymer.

The size distribution of F127 micelle shows some tailing in the large size regime (>1000 nm) and also shows increasing intensity in the μm size range (not shown on the plot). To understand this, we note that the direct result from a DLS experiment is an intensity distribution of particle sizes. The intensity distribution is weighted according to the scattering intensity of each particle fraction. For example, if we have a mixture of equal numbers of 10 and 100 nm size particles, then both will have the same peak areas in the number distribution; the 100 nm particle will have 10^3 times larger peak area in the volume distribution; and the 100 nm particle will have a 10^6 time larger peak area in the intensity distribution. Thus, the presence of even one dust particle or agglomerate can appear as a significant peak of large area in the intensity distribution, typically if the particle size appears to be 1 μm or above. No quantitative information is usually extracted from light scattering data for sizes of about 1 μm or larger, other than recognizing the presence of some dust or agglomerates. The mass of polymer in such large agglomerate would still be negligible.

4.2 Micelle Zeta Potential Measurement

The surface charge densities (zeta potential values) were measured using a Zetasizer Nano ZS (Malvern) at 25°C. The zeta potential was measured using a dip cell. All samples were diluted 100 times in water prior to the zeta potential measurements to avoid any interference from counter ions.

At pH 7.4, the block copolymers P123 and F127 were slightly negatively charged with zeta potentials of −6.8 and −8.7 mV, respectively, which is common for neutral polymeric NPs (Zhang et al., 2012). Free BSA showed a zeta potential of −11.2 mV. The carboxyl-functionalized block copolymer, F127-COOH, had a negative zeta potential of −11.4 mV, slightly larger in magnitude compared to that of nonfunctionalized F127. The protein–polymer conjugate F127-BSA had a negative zeta potential of −9.0 mV indicating that the zeta potential of the conjugate was somewhat smaller in magnitude compared to that of free BSA, and this reduction can be attributed to the presence of the block copolymer at the protein surface.

4.3 Size and Composition Control of Micelles

Block copolymers within the Pluronic family can all be mixed well with one another to create mixed aggregates. Since the polymer blocks are identical chemically, the mixing is possible in all proportions. By mixing F127-BSA with any other block copolymer, we can change the size of the mixed micelle and also the number of BSA molecules per unit surface area of the mixed micelle. In this work, we have chosen P123 (which has a PPO block comparable in length to that of F127), as the block copolymer to mix with F127-BSA.

a. Mixed micelles were prepared using conventional stirring just as the single-component micelles, but using two approaches that were found to be equivalent.

b. In one approach, the lyophilized or solid samples of the P123 and F127-BSA were mixed and then dissolved in nanopure water at approximately 1% (w/v) concentration, and the micelles were spontaneously generated.

c. The solution mixture was allowed to stir at room temperature for about 1–2 h under dark.

d. In the second approach, the lyophilized samples of the F127-BSA and the solid sample of P123 were each separately dissolved in the nanopure water and allowed to self-assemble into micelles.

e. The mixed micelle was then prepared by mixing these two solutions containing pure component micelles.

f. Again, the formation of mixed micelle was a result of spontaneous self-assembly. Since the mixed micelles are equilibrium structures, they undergo size changes as a function of the composition of the mixture.

g. In all cases, the samples were filtered through 0.45 μm filters (Millipore) prior to the DLS and zeta potential measurements.

Fig. 16 shows the size distributions for the mixed micelles consisting of different mass ratios of P123 and F127-BSA. The P123 micelles have the smallest particle sizes (about 20 nm), whereas the F127-BSA micelles have larger and more broadly dispersed sizes at about 90 nm. The DLS data show that the average size of the mixed micelles decreases as the content of P123 increases, and also there is a corresponding decrease in the polydispersity of the micelles.

While this mixed micelle size control through addition of copolymer P123 is shown for illustrative purposes, we note that other copolymers from the same family are available with ability to form aggregates of different sizes, thereby providing a convenient handle to tune the mixed micelle size.

We have also conjugated F127 also with another enzyme, OPH, since that is more catalytically active against organophosphates, not discussed here. F127-OPH conjugates showed significantly enhanced stability and higher

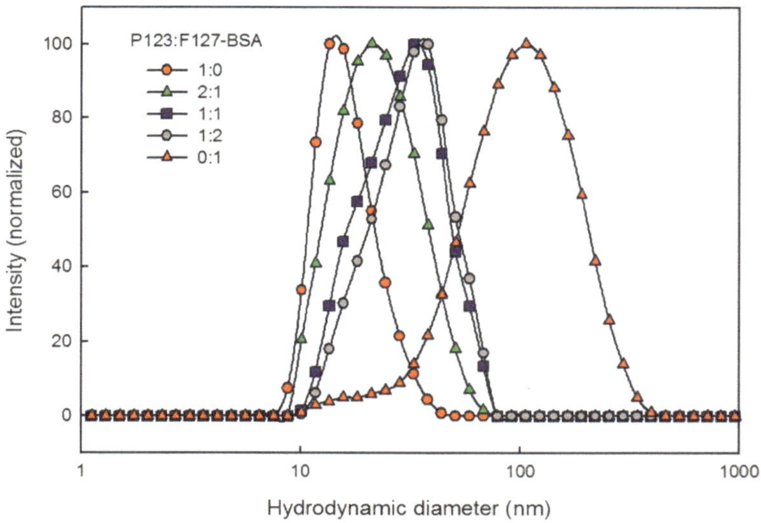

Fig. 16 Size distribution of P123+F127-BSA mixed micelles at different mass ratios starting from pure P123 micelles (1:0) to pure F127-BSA micelles (0:1). Both the average size and the polydispersity index (PDI representing the width of the size distribution) decreased as the P123 content increased.

activity as compared to the unconjugated OPH, when tested in aqueous solutions at room temperature, in aqueous solutions at higher temperatures, after multiple freeze/thaw treatments, after lyophilization, and in the presence of organic solvents.

5. CONCLUSIONS

In this study, we describe a relatively simple and flexible approach for conjugating the protein to an amphiphilic block copolymer and creating polymer nanoparticles with protein functionality by taking advantage of the intrinsic self-assembly behavior of the amphiphilic block copolymer. For demonstrative purposes, we have chosen the Pluronic block copolymers (PEO–PPO–PEO) to conjugate with BSA selected as the model protein. The molecular weight of the protein–polymer conjugate, degree of conjugation, degree retention of protein secondary structure, and biological activity of the protein in the conjugated state were delineated, as well as the size and charge of the conjugate micelles, using multiple techniques. One controls the size of the micelle and number density of the proteins on the surface by coassembling with a second block copolymer as a mixed micelle.

The synthesis approach described here offers the unique possibilities of controlling the size, composition, and shape (though not demonstrated in this study) of the protein–polymer conjugate micelles, all independent of one another through rational molecular choices. Protein–block copolymer conjugates may offer an improved system compared to free proteins for many practical applications.

ACKNOWLEDGMENTS

This work was supported by the Defense Threat Reduction Agency (DTRA) Project #BA12PHM159 and the NSRDEC In-House Laboratory Independent Research (ILIR) program. We thank Prof. Qian Wang's lab at University of South Carolina for allowing the use of their MALDI-TOF facility. N.S. conducted this research as the recipient of the National Research Council Chemical and Biological Defense (NRC/CBD) Research Associateship. Currently she works at the Dow Corning Corporation at Midland, MI.

REFERENCES

Abuchowski, A., van Es, T., Palczuk, N. C., & Davis, F. F. (1977). Alteration of immunological properties of bovine serum albumin by covalent attachment of polyethylene glycol. *The Journal of Biological Chemistry*, *252*(11), 3578–3581.

Batrakova, E. V., & Kabanov, A. V. (2008). Pluronic block copolymers: Evolution of drug delivery concept from inert nanocarriers to biological response modifiers. *Journal of Controlled Release*, *130*(2), 98–106. http://dx.doi.org/10.1016/j.jconrel.2008.04.013.

Börner, H. G. (2009). Strategies exploiting functions and self-assembly properties of bioconjugates for polymer and materials sciences. *Progress in Polymer Science, 34*(9), 811–851. http://dx.doi.org/10.1016/j.progpolymsci.2009.05.001.

Boyer, C., Huang, X., Whittaker, M. R., Bulmus, V., & Davis, T. P. (2011). An overview of protein-polymer particles. *Soft Matter,* 7(5), 1599–1614. http://dx.doi.org/10.1039/c0sm00412j.

Briand, V. A., Kumar, C. V., & Kasi, R. M. (2011). *Protein-polymer conjugates. In Encyclopedia of polymer science and technology*: Wiley. http://onlinelibrary.wiley.com/doi/10.1002/0471440264.pst325/full.

De, M., Rana, S., Akpinar, H., Miranda, O. R., Arvizo, R. R., Bunz, U. H. F., & Rotello, V. M. (2009). Sensing of proteins in human serum using conjugates of nanoparticles and green fluorescent protein. *Nature Chemistry, 1*(6), 461–465. http://dx.doi.org/10.1038/nchem.334.

Elsabahy, M., & Wooley, K. L. (2012). Design of polymeric nanoparticles for biomedical delivery applications. *Chemical Society Reviews, 41*(7), 2545–2561. http://dx.doi.org/10.1039/c2cs15327k.

Jung, B., & Theato, P. (2013). Chemical strategies for the synthesis of protein–polymer conjugates. *Advances in Polymer Science, 253,* 37–70. http://dx.doi.org/10.1007/12_2012_169.

Kratz, F., & Beyer, U. (1998). Serum proteins as drug carriers of anticancer agents: A review. *Drug Delivery, 5*(4), 281–299.

Leggio, C., Galantini, L., & Pavel, N. V. (2008). About the albumin structure in solution: Cigar expanded form versus heart normal shape. *Physical Chemistry Chemical Physics, 10*(45), 6741–6750. http://dx.doi.org/10.1039/b808938h.

Muszanska, A. K., Busscher, H. J., Herrmann, A., van der Mei, H. C., & Norde, W. (2011). Pluronic-lysozyme conjugates as anti-adhesive and antibacterial bifunctional polymers for surface coating. *Biomaterials, 32*(26), 6333–6341. http://dx.doi.org/10.1016/j.biomaterials.2011.05.016.

Nagarajan, R. (1999). Solubilization of hydrocarbons and resulting aggregate shape transitions in aqueous solutions of Pluronic (R) (PEO-PPO-PEO) block copolymers. *Colloids and Surfaces B: Biointerfaces, 16*(1–4), 55–72. http://dx.doi.org/10.1016/S0927-7765(99)00061-2.

Nagarajan, R. (2015). "Non-equilibrium" block copolymer micelles with glassy cores: A predictive approach based on theory of equilibrium micelles. *Journal of Colloid and Interface Science, 449,* 416–427. http://dx.doi.org/10.1016/j.jcis.2014.12.077.

Olsen, B. D. (2013). Self-assembly of globular-protein-containing block copolymers. *Macromolecular Chemistry and Physics, 214*(15), 1659–1668. http://dx.doi.org/10.1002/macp.201300235.

Pasut, G., & Veronese, F. M. (2012). State of the art in PEGylation: The great versatility achieved after forty years of research. *Journal of Controlled Release, 161*(2), 461–472. http://dx.doi.org/10.1016/j.jconrel.2011.10.037.

Song, H., He, R., Wang, K., Ruan, J., Bao, C., Li, N., … Cui, D. (2010). Anti-HIF-1alpha antibody-conjugated pluronic triblock copolymers encapsulated with Paclitaxel for tumor targeting therapy. *Biomaterials, 31*(8), 2302–2312. http://dx.doi.org/10.1016/j.biomaterials.2009.11.067.

Suthiwangcharoen, N., & Nagarajan, R. (2014a). Controlled design and construction of multifunctional nanoparticles by molecular self-assembly. *RSC Advances, 4*(20), 10076–10089. http://dx.doi.org/10.1039/c3ra46209a.

Suthiwangcharoen, N., & Nagarajan, R. (2014b). Enhancing enzyme stability by construction of polymer enzyme conjugate micelles for decontamination of organophosphate agents. *Biomacromolecules, 15*(4), 1142–1152. http://dx.doi.org/10.1021/bm401531d.

Thomas, C. S., Xu, L., & Olsen, B. D. (2012). Kinetically controlled nanostructure formation in self-assembled globular protein-polymer diblock copolymers. *Biomacromolecules, 13*(9), 2781–2792. http://dx.doi.org/10.1021/bm300763x.

Tolstyka, Z. P., & Maynard, H. D. (2012). Protein–polymer conjugates. In *Polymer science: A comprehensive reference* (Vol. 9, pp. 317–337). Elsevier. http://dx.doi.org/10.1016/b978-0-444-53349-4.00250-8.

Yi, X., Batrakova, E., Banks, W. A., Vinogradov, S., & Kabanov, A. V. (2008). Protein conjugation with amphiphilic block copolymers for enhanced cellular delivery. *Bioconjugate Chemistry*, *19*(5), 1071–1077. http://dx.doi.org/10.1021/bc700443k.

Zhang, W., Gilstrap, K., Wu, L., Remant Bahadur, K. C., Moss, M. A., Wang, Q., … He, X. (2010). Synthesis and characterization of thermally responsive Pluronic F127-chitosan nanocapsules for controlled release and intracellular delivery of small molecules. *ACS Nano*, *4*(11), 6747–6759. http://dx.doi.org/10.1021/nn101617n.

Zhang, F., Roosen-Runge, F., Skoda, M. W., Jacobs, R. M., Wolf, M., Callow, P., … Schreiber, F. (2012). Hydration and interactions in protein solutions containing concentrated electrolytes studied by small-angle scattering. *Physical Chemistry Chemical Physics*, *14*(7), 2483–2493. http://dx.doi.org/10.1039/c2cp23460b.

Zhang, F., Skoda, M. W. A., Jacobs, R. M. J., Martin, R. A., Martin, C. M., & Schreiber, F. (2007). Protein interactions studied by SAXS: Effect of ionic strength and protein concentration for BSA in aqueous solutions. *Journal of Physical Chemistry B*, *111*(1), 251–259. http://dx.doi.org/10.1021/jp0649955.

CHAPTER THIRTEEN

Semisynthetic Enzymes by Protein–Peptide Site-Directed Covalent Conjugation: Methods and Applications

Jose M. Palomo[1]
Institute of Catalysis (CSIC), Madrid, Spain
[1]Corresponding author: e-mail address: josempalomo@icp.csic.es

Contents

Abstract

This chapter describes the rational design and synthesis of semisynthetic lipases by site-directed incorporation of tailor-made peptides on the lipase-lid site to improve its activity, specificity, and enantioselectivity in specific biotransformations. Cysteine was genetically introduced at a particular point of the oligopeptide lid of the enzyme, and cysteine-containing peptides, complementary to the amino acid sequence on the lid site of *Geobacillus thermocatenulatus* lipase (BTL), were covalently attached on the lid of two different cysteine-BTL variants based on a fast thiol–disulfide exchange ligation followed by desulfurization. The BTL variants were initially immobilized on solid support to introduce the advantages of solid-state chemistry, such as quantitative transformations, easy purification, and recyclability. In the two different immobilized variants BTL-A193C and BTL-L230C, the cysteine was then activated with 2-dipyridyldisulfide to help the disulfide exchange with the peptide, generating the semisynthetic enzyme in high yield. Excellent results of improvement of activity and selectivity were obtained. For example, the peptide–BTL conjugate (at position 193) was 40-fold more active than

Methods in Enzymology, Volume 590
ISSN 0076-6879
http://dx.doi.org/10.1016/bs.mie.2017.01.001
© 2017 Elsevier Inc.
All rights reserved.

the corresponding unmodified enzyme for the hydrolysis of per-acetylated thymidine at pH 5, or fourfold in the desymmetrization of dimethyl-3-phenylglutarate at pH 7. The new enzyme also exhibited excellent enantioselectivity in the desymmetrization reaction with enantiomeric excess (ee) of >99% when compared to that of the unmodified enzyme (ee = 78%).

1. INTRODUCTION

Rational modification of enzymes by the introduction of nonnatural peptide fragments into its structure to improve enzyme function has been recognized as a new approach (Chooi et al., 2014; Da Silva Freitas, Mero, & Pasut, 2013; Palla, Witus, Mackenzie, Netirojjanakul, & Francis, 2015; Palomo & Filice, 2015; Zhang et al., 2016). However, most strategies for chemical modification rely directly on the natural nucleophilic side chains of amino acids, such as aspartic acid, glutamic acid, and lysines (Hackenberger & Schwarzer, 2008). These modifications are nonspecific, but in many cases, they are reasonably effective in improving enzyme stability (Cabrera, Gutarra, Guisan, & Palomo, 2010; Moreno-Pérez et al., 2016; Palomo, Fernández-Lorente, Fernández-Lafuente, & Guisán, 2007). Improvements in enzymatic activity, selectivity, enantioselectivity, or regioselectivity, however, may require the development of a site-specific approach.

Thus, many diverse strategies for accomplishing site-specific modification of enzymes and proteins have been valuable tools for chemistry and biology (Banerjee, Hart, & Cho, 2013; Chooi et al., 2014; Palomo & Filice, 2015; Palsuledesai & Distefano, 2015). Site-specific introduction of various unnatural molecules such as fluorophores, sugars, lipid chains, peptides, and bioorthogonal functional groups has produced semisynthetic enzymes for a variety of processes and applications both in vivo and in vitro (de Araújo et al., 2006; Filice, Romero, Guisan, & Palomo, 2011; Palomo, 2012; Palomo & Filice, 2015; Palsuledesai & Distefano, 2015). In this sense, a combination of molecular biology methods (site-directed mutagenesis, directed evolution, expressed protein ligation, etc.) and efficient synthetic approaches (e.g., peptide synthesis) represents an advance in technology to produce semisynthetic enzymes and proteins in large amounts (Muralidharan & Muir, 2006; Palomo, Lumbierres, & Waldmann, 2006).

Modification of the nucleophilic thiol of a unique cysteine is a widely employed strategy for site-selective bioconjugation (Deuss, Popa, Botting, Laan, & Kamer, 2010). Cysteine can be introduced at virtually any position within the primary sequence by site-directed mutagenesis and then selectively modified using disulfide reagents, for example. Furthermore, the disulfide can be efficiently transformed into a thioether by a desulfurization method (Bernardes et al., 2008).

In particular, lipases are enzymes with a peculiar catalytic mechanism, where an oligopeptide sequence blocks or allows access to the active site in a closed or an open conformation, respectively (Brady et al., 1990). This complex catalytic mechanism results in low activity and selectivity toward nonnatural substrates.

The current approach consists of site-specific incorporation of a particular peptide sequence which is complementary to that of the oligopeptide lid of a thermophilic lipase from *Geobacillus thermocatenulatus* (BTL). The target enzyme is engineered to have a single cysteine in the lid oligopeptide, and the tailor-made complementary peptides also contain a cysteine group in the sequence (Fig. 1). The chemical modifications were performed on the lipase previously immobilized on CNBr-activated Sepharose. This step introduces the advantages of solid-state chemistry such as the possibility to use an excess of peptides, quantitative transformations, and easy purification, among others (Kurth, 2010). The catalytic capacity of the new semisynthetic enzymes was improved by this approach.

2. PROTEIN–PEPTIDE SITE-DIRECTED COVALENT CONJUGATION

2.1 Materials

2.1.1 General Equipment

1. Pipettes
2. Balance
3. UV-spectrophotometer JASCO V-630 provided with magnetic stirring of the cuvettes in order to assay immobilized preparations
4. Selecta roller mixer
5. Microcentrifuge

2.1.2 Reagents

1. Sepharose® 4BCL activated with cyanogen bromide (CNBr) (GE Healthcare)
2. 2-PDS (Sigma)

A

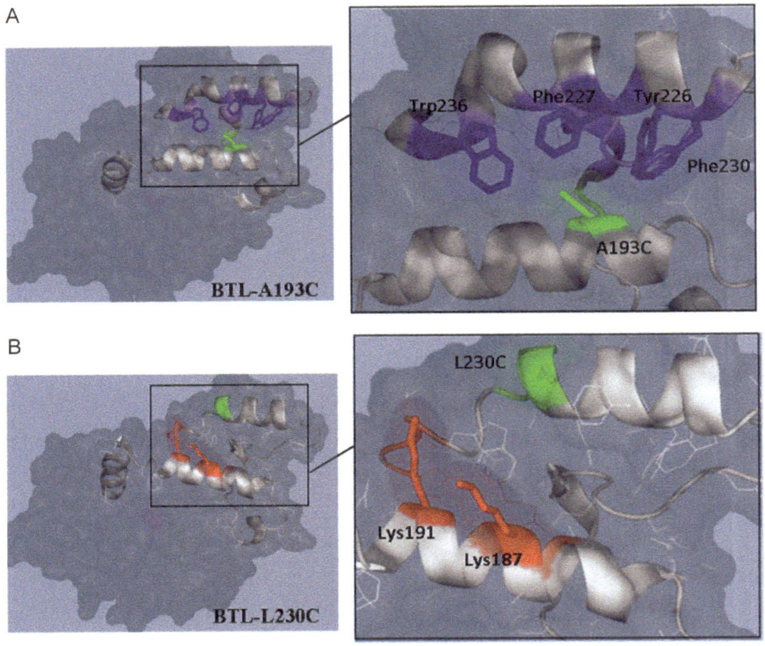

C

For A193C (**p1**) Ac-Cys-Phe-Gly-Phe-Gly-Phe-$CONH_2$
For L230C (**p2**) Ac-Cys-Asp-Asp-Asp-Asp-Asp-COOH

Fig. 1 Surface structure of lid zone of BTL mutants. (A) BTL-A193C; (B) ETL-L230C. The PDB code for BTL (open conformation) is 2W22. (C) Tailor-made cysteine containing complementary peptides.

3. Commercial peptides Ac-Cys-Phe-Gly-Phe-Gly-Phe-CONH (**p1**), Ac-Cys-Asp-Asp-Asp-Asp-Asp-COOH (**p2**) (Inbios)
4. Sodium borohydride (Sigma)
5. Tris(dimethylamino)phosphine (Sigma)
6. Phosphate-buffered saline (PBS) (100 m*M*, pH 7)
7. Sodium bicarbonate buffer (25 m*M*, pH 10)
8. Amicon Ultra 10 kDa (DyeEx columns from Qiagen)
9. Deionized water
10. Engineered lipases (BTL-A193C, BTL-L230C) were prepared by site-directed mutagenesis with a good yield, similar to that of BTL, and they have been expressed in *Escherichia coli* without a significant loss of enzyme activity (Godoy et al., 2011). All BTL variants were efficiently purified by hydrophobic chromatography on butyl-Sepharose support (Fernandez-Lorente et al., 2008)

11. Triton X-100 (Sigma)
12. Dithiothreitol (DTT)
13. Dimethylsulfoxide (DMSO)
14. CAPS buffer (Sigma)
15. *p*-Nitrophenylbutyrate (*p*-NPB) (Sigma)

2.2 Methods

1. Lipases were immobilized on CNBr-activated Sepharose prior to modification (Fig. 2). A solution of lipase (BTL-A193C or BTL-L230C) dissolved in 25 m*M* phosphate buffer pH 7 with 0.5% Triton X-100 (v/v) containing 5 mg enzyme was incubated with 1 g of support for 1 h at 25°C. Under these conditions, >95% lipase was immobilized, as followed by *p*-NPB assay.
2. The enzymatic activity of the lipase following the immobilization process was analyzed by spectrophotometrically measuring the increment in absorbance at 348 nm produced by the release of *p*-nitrophenol ($\in = 5150\ M^{-1}\ cm^{-1}$) in the hydrolysis of 0.4 m*M* *p*-NPB in 25 m*M*

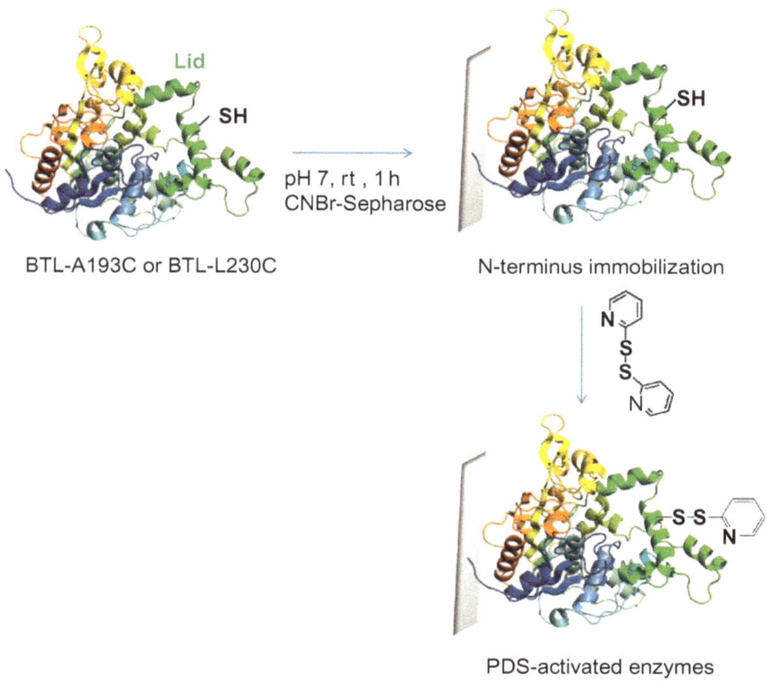

Fig. 2 Scheme of the preparation of immobilized PDS-activated BTL enzymes.

sodium phosphate at pH 7 and 25°C. To initialize the reaction, 0.05–0.2 mL of lipase solution or suspension was added to 2.5 mL of substrate solution.

3. Then, 0.2 g of the immobilized BTL variant was incubated in 2 mL of DTT solution (50 m*M* DTT in 25 m*M* sodium phosphate at pH 8) for 30 min to avoid oxidation and permit disulfide exchange. The reduced biocatalyst was washed with distilled water *until the DTT smell disappeared*. Then this biocatalyst (0.2 g) was added to 3 mL of 2-PDS solution (1.5 m*M* substrate in a mixture of DMSO (5%, v/v)–25 m*M* phosphate buffer (95%, v/v) at pH 8.0) for 1 h to obtain PDS-activated enzymes (Fig. 2).
4. A peptide solution (**p1** or **p2**) (0.5 mg mL^{-1}) was treated with sodium borohydride solution (2 mg mL^{-1}) in 25 m*M* sodium bicarbonate at pH 10.0 for 30 min in order to keep the thiol group in the reduced form. Then, pH was adjusted to 7.0 by adding diluted HCl to decompose the excess sodium borohydride.
5. Then, 0.7 mL of peptide solution was added to 2.3 mL of 500 m*M* sodium phosphate at pH 8, and 0.2 g of immobilized PDS-lipase variant has been added over 1 h. The modification was confirmed spectrophotometrically by measuring the increase of absorbance at 343 nm. The immobilized catalyst was filtered and washed with distilled water (Fig. 3).
6. A solution of Tris(dimethylamino)phosphine (1.5 m*M*, 10 mL) in 70 m*M* CAPS buffer, pH 9.5, was added to 1.0 g of immobilized lipase–peptide conjugates and maintained at room temperature for 16 h. Then, the solid was filtered and washed with distilled water. This transformed the disulfide to a thioether (Fig. 3) (Bernardes et al., 2008), irreversibly.

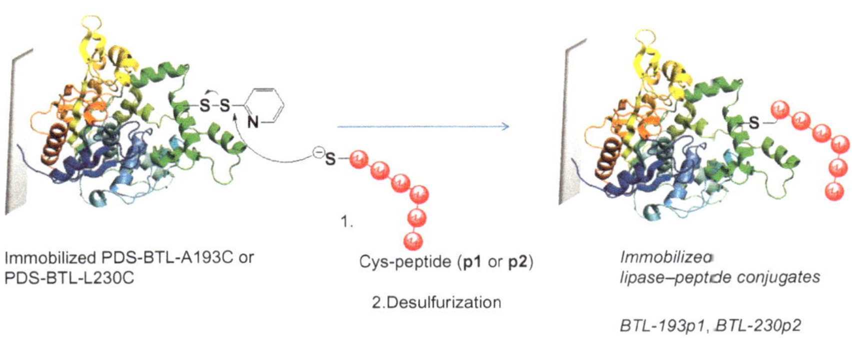

Fig. 3 Synthesis of the enzyme–peptide covalent conjugates.

3. THE NEW SEMISYNTHETIC LIPASES IN BIOTRANSFORMATIONS

The activity, specificity, and enantio- and regioselectivity of the new lipase–peptide conjugates were modulated and improved when compared to that of the unmodified BTL in most cases of the three biotransformations evaluated. Tables 1–3 show the data.

3.1 Enzymatic Hydrolysis of 3′,5′-Di-*O*-Acetylthymidine (1)

3.1.1 General Equipment

1. HPLC (Spectra Physics SP 100 coupled with a UV detector Spectra Physics SP 8450)
2. Kromasil C18 column (25 × 0.4 cm, 5 μmø)

3.1.2 Reagents

1. Thymidine (Sigma)
2. Acetic anhydride (Sigma)
3. TLC plates
4. Ethyl acetate
5. Hexane
6. Triethylamine
7. 4-(Dimethylamino)pyridine (DMAP)
8. Chloroform
9. Sodium sulfate (Na_2SO_4)

3.1.3 Methods

1. Acetylation of thymidine: Thymidine (12 mmol, 2.9 g) was dissolved in 6 mL of acetonitrile, and 4 equiv. triethylamine, 4 equiv. acetic anhydride, and a catalytic amount of DMAP have been successively added. The resulting mixture was stirred at room temperature until the reaction was complete, as demonstrated by TLC analysis, and was then diluted with chloroform and water (1:1). The organic phase was separated, washed with water (4 × 20–50 mL) and dried (Na_2SO_4), filtered, and concentrated in vacuo to afford the 3′,5′-di-*O*-acetylthymidine **1** (3.5 g, 90%). TLC (ethyl acetate/hexane, 7/3): $R_f = 0.32$.
2. Substrate **1** (5 m*M*; Table 1) was dissolved in a mixture of acetonitrile (5%, v/v) and 10 m*M* sodium acetate at pH 5.0. 0.2 g of biocatalyst was added to 2 mL of this solution at 25°C. Samples were collected at different times, and the degree of hydrolysis was analyzed by reversed-phase HPLC until 100% conversion has been achieved.

Table 1 Biocatalytic Regioselective Monodeprotection of **1** at pH 5

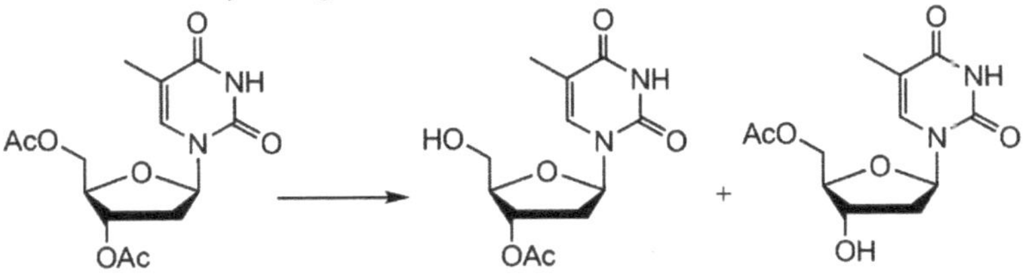

1 **1a** **1b**

Enzyme	Activity[a]	Time (h)	Yield 2 (%)[b]	Yield 3 (%)[b]	Thymidine[b]
BTL	86	80	70	4	26
BTL-193p1	354	48	89	4	7
BTL-230p2	44	162	85	8	7

[a]Specific activity was defined as: μmol min^{-1} mg$_{lip}$$^{-1}$ × 10^{-3}.
[b]Yield of the corresponding product at 100% conversion.

3. Analytical evaluation of substrate conversion was performed by HPLC using a Kromasil C18 column (25 × 0.4 cm, 5 μmø) and the following gradient program A: mixture of acetonitrile (10%, v/v) in 10 m*M* ammonium phosphate at pH 4.2; B: mixture of Milli-Q water (10%,v/v) in acetonitrile; method: 0–6 min 100% A, 6–14 min 85% A to 15% B, 14–22 min 100% A, flow: 1.0 mL min^{-1}. UV detection was performed at 260 nm. The unit of enzymatic activity was defined as micromoles of substrate hydrolyzed per minute per mg of immobilized enzyme. The monodeprotected 5-OH (**1a**) and 3-OH (**1b**) were used as pure standards. The retention time was 2.4 min for thymidine, 9.4 min for **1b** and 10.2 min for **1a** and 19 min for **1** (Table 1).
4. The semisynthetic lipase BTL-193p1 was four times more active than unmodified BTL and also improved the specificity of the enzyme. The BTL-230p2 variant improved the specificity but not the activity (Table 1).

3.2 Enzymatic Hydrolysis of 3,4,6-Tri-*O*-Acetyl-Glucal (2)

3.2.1 General Equipment

1. HPLC (Spectra Physic SP 100 coupled with an UV detector Spectra Physics SP 8450)
2. Kromasil C18 column (25 × 0.4 cm, 5 μmø)

3.2.2 Reagents

1. Per-O-acetylated glucal (Sigma)
2. Acetonitrile
3. Ammonium phosphate

3.2.3 Methods

1. Substrate **2** (2.0 m*M*) was dissolved in a mixture of acetonitrile (3%, v/v) in 10 m*M* sodium acetate at pH 5.0 (*important to avoid chemical migration*). The biocatalyst (0.4 g) was added to 2 mL of this solution at 25°C. The degree of hydrolysis was analyzed by reversed-phase HPLC until 100% conversion has been achieved.
2. Conversion was obtained by using HPLC coupled with a Kromasil C18 column (25 × 0.4 cm, 5 μmø) with acetonitrile (25%, v/v) as the mobile phase in 10 m*M* ammonium phosphate at pH 3.8 at a flow rate of 1.0 mL min^{-1}. UV detection was performed at 220 nm. The unit of enzymatic activity was defined as micromoles of substrate hydrolyzed per minute per mg of immobilized enzyme. The retention time was 6.8 min for monodeprotected 3-OH (**2a**) and 30 min for **2**.
3. Both new artificial lipases (BTL-193p1, BTL-230p2) showed better activity than the unmodified BTL in the hydrolysis of **2**, although only a slight increase in specificity was observed (Table 2).

3.3 Enzymatic Hydrolysis of Dimethyl-3-Phenylglutarate (3)

3.3.1 General Equipment

1. HPLC (Spectra Physics SP 100 coupled with an UV detector Spectra Physics SP 8450).
2. Kromasil column C8 (25 × 0.4 cm, 5 μmø)
3. Chiralcel OD-R column

Table 2 Biocatalytic Regioselective Monodeprotection of **2** at pH 5

2 **2a**

Enzyme[a]	Activity[a]	Time (h)	Yield 2a (%)[b]
BTL	5	1	70
BTL-193p1	214	1	73
BTL-230p2	63	3	50

[a]Specific activity was defined as: μmol min^{-1} mg_{lip}^{-1} × 10^{-3}.
[b]Yield of the monodeprotected **2a** at 100% conversion. The rest of the yield corresponds to the bihydrolyzed product.

3.3.2 Reagents

1. Dimethyl-3-phenylglutarate (Sigma)
2. Acetonitrile
3. Ammonium phosphate

3.3.3 Methods

1. Substrate **3** (0.5 m*M*) was dissolved in 10 m*M* sodium phosphate at pH 7.0. Then 0.25 g of immobilized preparation was added to 5 mL of this solution at 25°C. The degree of hydrolysis was analyzed by reversed-phase HPLC.
2. For these assays, a Kromasil C8 column was used with acetonitrile (35%) as mobile phase in 10 m*M* ammonium phosphate, at pH 3.0 at a flow rate of 1.0 mL min^{-1}. UV detection was performed at 225 nm The unit of enzymatic activity was defined as micromoles of substrate hydrolyzed per minute per mg of immobilized protein. The retention time was 7.4 min for **3a** and 22 min for **3**.
3. The enantiomeric excess (ee) of the released monoester **3a** was analyzed by chiral reversed-phase HPLC. The column was a Chiralcel OD-R, and the mobile phase was isocratic mixture of acetonitrile (25%, v/v) in 10 m*M* ammonium phosphate (70%, v/v) at pH 3 at a flow rate of 0.5 mL min^{-1}. The detection was performed at 210 nm and the retention times were 58.7 min for (*R*)-**3a** and 63.2 min for (*S*)-**3a**.
4. The BTL-193p1 was highly enantioselective and had fourfold higher activity (ee > 99%) than the unmodified BTL (ee 78%). The other covalent conjugate (BTL-230p2) showed similar enantioselectivity with double the activity (Table 3).

Table 3 Biocatalytic Desymmetrization of **3** at pH 7

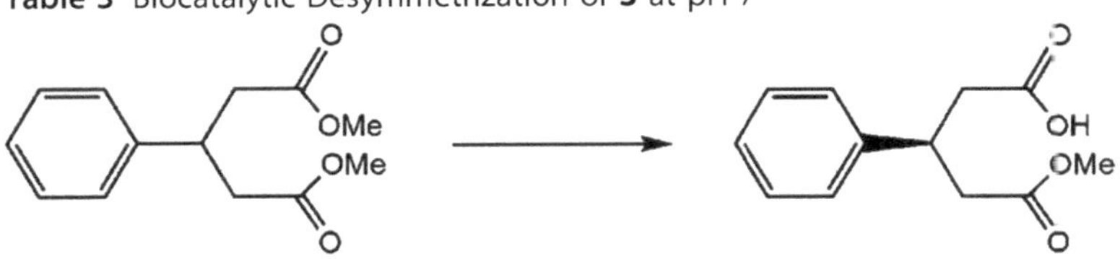

3 (*S*)-**3a**

Enzyme	Activity[a]	Time (h)	*C* (%)	Yield (%)[b]	*ee* (%)[c]
BTL	0.53	186	23	15	78
BTL-193p1	2.35	96	54	37	>99
BTL-230p2	1.13	96	26	12	73

[a]Specific activity was defined as: μmol min^{-1} $mg_{lip}{}^{-1}$ 10^{-5}.
[b]Yield of the monoester **3a**. The rest of conversion corresponds to the dicarboxylic acid.
[c]Determined by HPLC.

4. CONCLUSIONS

Site-specific incorporation of tailor-made peptides appended to the lid side of a lipase seems to promote improvement of catalytic properties. Key examples in three different biotransformations have demonstrated how the selective incorporation of a complementary peptide at a specific position of *Bacillus thermocatenulatus* lipase generated a new enzyme that is 40-fold more active or with greater regio- or specificity and extreme enantioselectivity with specific substrates, when compared with the activities/selectivities of the corresponding unmodified enzyme.

Application of this modification on a previously immobilized enzyme generates an interesting catalyst with excellent properties for industrial application and is potentially a general method for the creation of new and improved artificial enzymes.

ACKNOWLEDGMENTS

This work was supported by the Spanish National Research Council (CSIC). The author thanks the Ramon Areces Foundation for financial support.

REFERENCES

Banerjee, P. S., Hart, G. W., & Cho, J. W. (2013). Chemical approaches to study O GlcNAcylation. *Chemical Society Reviews*, *42*, 4345–4357.

Bernardes, G. J. L., Grayson, E. J., Thompson, S., Chalker, J. M., Errey, J. C., El Oualid, F., et al. (2008). From disulfide- to thioether-linked glycoproteins. *Angewandte Chemie International Edition*, *47*, 2244–2247.

Brady, L., Brzozowski, A. M., Derewenda, Z. S., Dodson, E., Dodson, G., Tolley, S., et al. (1990). A serine protease triad forms the catalytic centre of a triacylglycerol lipase. *Nature*, *343*, 767–770.

Cabrera, Z., Gutarra, M. L., Guisan, J. M., & Palomo, J. M. (2010). Highly enantioselective biocatalysts by coating immobilized lipases with polyethyleneimine. *Catalysis Communication*, *11*, 964–967.

Chooi, K. P., Galan, S. R. G., Raj, R., McCullagh, J., Mohammed, S., Jones, L. H., et al. (2014). Synthetic phosphorylation of p38α recapitulates protein kinase activity. *Journal of the American Chemical Society*, *136*, 1698–1701.

Da Silva Freitas, D., Mero, A., & Pasut, G. (2013). Chemical and enzymatic site specific pegylation of hGH. *Bioconjugate Chemistry*, *24*, 456–463.

de Araújo, A. D., Palomo, J. M., Cramer, J., Köhn, M., Schröder, H., Wacker, R., et al. (2006). Diels-Alder ligation and immobilization of peptides and proteins. *Angewandte Chemie International Edition*, *45*, 296–301.

Deuss, P. J., Popa, G., Botting, C. H., Laan, W., & Kamer, P. C. J. (2010). Highly efficient and site-selective phosphane modification of proteins through hydrazone linkage: Development of artificial metalloenzymes. *Angewandte Chemie International Edition*, *49*, 5315–5317.

Fernandez-Lorente, G., Godoy, C., Mendes, A. A., Lopez-Gallego, F., Grazu, V., de las Rivas, B., et al. (2008). Solid-phase chemical amination of bacillus thermocatenulatus lipase to improve its stabilization via covalent immobilization on highly activated glyoxyl-agarose. *Biomacromolecules, 9*, 2553–2561.

Filice, M., Romero, O., Guisan, J. M., & Palomo, J. M. (2011). trans,trans-2,4-Hexadiene incorporation on enzymes for site-specific immobilization and fluorescent labeling. *Organic & Biomolecular Chemistry, 9*, 5535–5540.

Godoy, C. A., de las Rivas, B., Grazú, V., Montes, T., Guisán, J. M., & Lopez-Gallego, F. (2011). Glyoxyl-disulfide agarose: A tailor-made support for site-directed rigidification of proteins. *Biomacromolecules, 12*, 1800–1809.

Hackenberger, C. P. R., & Schwarzer, D. (2008). Chemoselective ligation and modification strategies for peptides and proteins. *Angewandte Chemie International Edition, 47*, 10030–10074.

Kurth, M. J. (2010). Linker strategies in solid-phase organic synthesis. *Journal of the American Chemical Society, 132*, 10615.

Moreno-Pérez, S., Orrego, A. H., Romero-Fernández, M., Trobo-Maseda, L., Martins-Deoliveira, S., Munilla, R., et al. (2016). Intense PEGylation of enzyme surfaces: Relevant stabilizing effects. *Methods in Enzymology, 571*, 55–72.

Muralidharan, V., & Muir, T. W. (2006). Protein ligation: An enabling technology for the biophysical analysis of proteins. *Nature Methods, 3*, 429–438.

Palla, K. S., Witus, L. S., Mackenzie, K. J., Netirojjanakul, C., & Francis M. B. (2015). Optimization and expansion of a site-selective *N*-methylpyridinium-4-carboxaldehyde-mediated transamination for bacterially expressed proteins. *Journal of the American Chemical Society, 137*, 1123–1129.

Palomo, J. M. (2012). Click reactions in protein chemistry: From the preparation of semisynthetic enzymes to new click enzymes. *Organic & Biomolecular Chemistry, 10*, 9309–9318.

Palomo, J. M., Fernández-Lorente, G., Fernández-Lafuente, R., & Guisán, J. M. (2007). Modulation of the immobilized lipase enantioselectivity via chemical amination. *Advanced Synthesis & Catalysis, 349*, 1119–1127.

Palomo, J. M., & Filice, M. (2015). New emerging bio-catalysts design in biotransformations. *Biotechnology Advances, 33*, 605–613.

Palomo, J. M., Lumbierres, M., & Waldmann, H. (2006). Efficient solid-phase synthesis of lipopeptides employing the Ellman sulfonamide linker. *Angewandte Chemie International Edition, 45*, 477–481.

Palsuledesai, C. C., & Distefano, M. D. (2015). Protein prenylation: Enzymes, therapeutics, and biotechnology applications. *ACS Chemical Biology, 10*, 51–62.

Zhang, C., Welborn, M., Zhu, T., Yang, N. J., Santos, M. S., Van Voorhis, T., et al. (2016). π-Clamp-mediated cysteine conjugation. *Nature Chemistry, 8*, 120–128.

CHAPTER FOURTEEN

Transgultaminase-Mediated Nanoarmoring of Enzymes by PEGylation

Antonella Grigoletto, Anna Mero, Katia Maso, Gianfranco Pasut[1]
University of Padua, Padua, Italy
[1]Corresponding author: e-mail address: gianfranco.pasut@unipd.it

Contents

Abstract

PEGylation, the covalent attachment of polyethylene glycol to bioactive molecules, is one of the leading approaches used to prolong pharmacokinetics, to improve the stability, and to reduce the immunogenicity of therapeutic proteins. PEG-conjugated products are associated with better therapy outcomes and improved patient compliance. Widely applied in clinical practice, the technology is mainly used to modify proteins, peptides, and oligonucleotides but also other drug delivery systems such as the liposomal one. Undergoing continuous attempts to optimize therapeutic efficacy and to tune the formation of conjugates, a number of different PEGylation processes are now available to researchers for protein conjugation. Although the possibility of obtaining highly homogeneous conjugate mixtures, preferably formed by a single monoconjugate, from a chemical conjugation reaction continues to be limited, several

Methods in Enzymology, Volume 590
ISSN 0076-6879
http://dx.doi.org/10.1016/bs.mie.2017.01.002
© 2017 Elsevier Inc.
All rights reserved.

enzymatic conjugation approaches have recently been investigated to address this need. PEGylation mediated by microbial transglutaminase and its many advantages and modifications are outlined in detail in the current work permitting interested readers to perform site-specific protein derivatization to glutamines or lysines.

ABBREVIATIONS

ACN acetonitrile
α-LA alpha-lactalbumin
DMSO dimethyl sulfoxide
ESI-MS electrospray-ionization mass spectrometry
G-CSF granulocyte colony-stimulating factor
IEX ionic exchange chromatography
MALDI-MS matrix-assisted laser-desorption ionization mass spectrometry
MeOH methanol
MS/MS tandem mass spectrometry
MW molecular weight
NH_2OH hydroxylamine
PEG poly(ethylene glycol)
RPC reversed phase chromatography
sCT salmon calcitonin
SEC size exclusion chromatography
TFA trifluoroacetic acid
UF/DF ultra-/diafiltration

1. INTRODUCTION

The covalent conjugation of a hydrophilic polymer, and especially polyethylene glycol (PEG), is a well established, widely employed strategy for improving the therapeutic performance of bioactive substances such as proteins, peptides, small drugs, and oligonucleotides (Turecek, Bossard, Schoetens, & Ivens, 2016). In the majority of cases, PEGylation is used as a means to prolong the half-life of usually rapidly cleared drugs and to reduce immunogenicity. The former is achieved by decreasing the rates of both kidney clearance and degradation and the latter by the shielding effect of polymer chains on the protein surface. From a pharmacological point of view, a PEGylated protein can ultimately achieve a better response due to improved stability and prolonged blood concentration and circulation times that lead to higher patient compliance in view of less frequent drug administration scheduling (Jevševar, Kunstelj, & Gaberc Porekar, 2010; Mero & Veronese, 2008). In addition, lower immunogenicity means that

PEG-protein conjugates can be utilized despite the fact that the protein itself would elicit a strong immunogenic response. A number of different chemical protein PEGylation strategies have been developed, and recent advances in modern biotechnology have produced important tools to develop enzyme-mediated polymer coupling (Pasut & Veronese, 2012). It is in this context that the current paper will focus on the use of microbial transglutaminase (TGase).

1.1 Transglutaminase

TGases constitute a large family of enzymes ubiquitously present in many living organisms, including mammals, invertebrates, plants, and microorganisms. In humans, TGases are found in most tissues and body fluids, including the liver, epidermis, the prostate, hair follicles, and blood. One of the best known in humans is plasma coagulation enzyme factor XIIIA that crosslinks between fibrin molecules to form an insoluble clot to stop bleeding. TGases can catalyze an acyl transfer reaction between the γ-carboxyamide group of a protein-bound glutamine (Gln) residue (acyl donor) and a variety of unbranched primary amines (acyl acceptors) such as the ε-amino group of lysines. In fact, it has been demonstrated that only few Gln(s) have been substrates for the enzyme, whereas this specificity is much less with lysine residues (Folk, 1983). When the ε-amino group of a protein-bound lysyl residue is involved, two polypeptide chains are covalently connected through ε-(γ-glutamyl)lysine isopeptide bonds (Scheme 1). Human TGase is a calcium dependent, monomeric protein of 75 kDa expressed in the majority of cells and tissues and implicated in a wide variety of cellular functions such as extracellular matrix remodeling, cell differentiation, tumor growth, and apoptosis (Griffin, Casadio, & Bergamini, 2002).

TGases have been gaining increasing attention because they can be used to tag a glutamine group of protein with several chemical entities (including other proteins or polymers) that comprise an amino group. They are thus versatile reagents for modifying proteins in biochemical experiments.

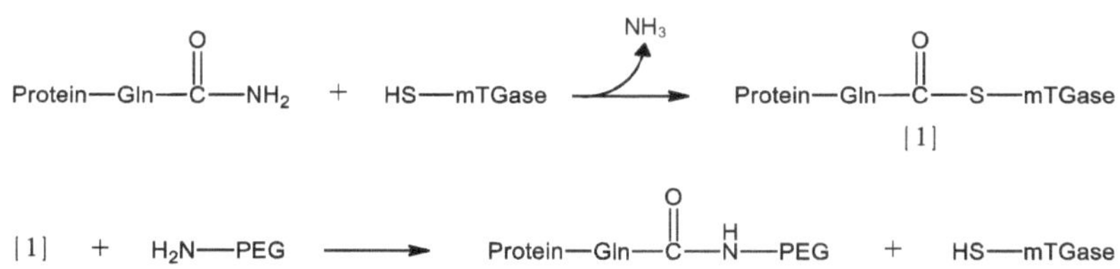

Scheme 1 Mechanism of catalytic reaction of mTGase.

In recent years, TGase activity has also been detected in bacteria, and some of the genes encoding TGase have been successfully cloned from several microorganisms (Ando et al., 1989; Washizu et al., 1994). Microbial TGase (mTGase) is derived from a variant of *Streptomyces mobaraense* whose 331-residue polypeptide chain (Kanaji et al., 1993) and 3D structure (Kashiwagi et al., 2002) have been described. mTGase exhibits no significant homology to mammalian TGases, is calcium-independent, and has a molecular weight (MW) of 37.9 kDa, nearly half that of factor XIII-like TGases (Yokohama, Nio, & Kikuchi, 2004). These characteristics as well as a higher reaction rate with respect to other TGases mean that the enzyme can be exploited as a biochemical tool for different industrial applications. mTGase is, for example, widely used to improve the physical and textural properties of protein-rich foods (Zhu, Rinzema, & Tramper, 1995).

mTGase has also been investigated for the site-specific PEGylation of pharmaceutical proteins using PEG-NH_2 as acyl acceptor. mTGase-mediated PEGylation presents two important advantages: (i) it can modify glutamine, a residue that is not modifiable by any other chemical method, and (ii) it can produce highly homogenous conjugates because only one or two glutamines usually present in a protein can be a substrate for the enzyme, and this greatly reduces the number of potential positional isomers. Sato et al. developed a novel methodology for enzymatic PEGylation by selectively coupling PEG-NH_2 to the Gln74 of interleukin-2 (Sato, Hayashi, Yamada, Yatagai, & Takahara, 2001). Examining several proteins, Fontana et al. noted that the site selectivity of mTGase is determined by strict requirements of flexibility in the peptide region comprising the Gln substrate. Gln must, in particular, be located in a highly flexible backbone to be mTGase substrate (Fontana, Spolaore, Mero, & Veronese, 2008). We have also reported examples of mTGase modification with pharmaceutical proteins such as granulocyte colony-stimulating factor (G-CSF), hGH, and interferon and also a new approach using mTGase in the presence of organic cosolvents (Mero, Clementi, & Pasut, 2011; Mero, Schiavon, Veronese, & Pasut, 2011). More specifically, although mTGase is highly selective, there are proteins or peptides (e.g., hGH and salmon calcitonin (sCT)) that can present more than one Gln as a mTGase substrate, thus preventing them from attaining a single monoconjugate with these molecules. Several different strategies have been investigated to increase mTGase's specificity in obtaining monoconjugates even in these cases. Several mTGase muteins have, for example, been screened to identify those that are highly selective for conjugating PEG only at Gln141 of hGH (Zhao et al., 2010). In another approach, a small bifunctional orthogonal spacer has been preferentially linked to Gln141 of hGH by optimizing the reaction conditions of pH, the temperature, enzyme/substrate (E/S) ratio, and reactant concentrations. The desired Gln141-spacer hGH isomer was then purified by cationic exchange and subsequently coupled to a PEG derivative specifically reactive toward the spacer (Buchardt, Selvig, Nielsen, & Langeland, 2010). These two methods still present some disadvantages linked to the necessity, in one case, of developing protein specific mTGase muteins and, in the other, of performing multiple reaction and purification steps. We recently proposed two alternative methods leading to the formation of only monoconjugates with mTGase: (i) using organic cosolvents in the reaction buffer and (ii) selectively immobilizing the enzyme on the solid beads of an inert polysaccharide resin.

Several mTGase-mediated strategies for the selective PEGylation of proteins and peptides that have been used in our laboratory or described in the literature are outlined in the current work.

2. MATERIALS AND REAGENTS

2.1 General Equipment

1. Personal safety equipment: laboratory coat, latex or nitrile gloves, and fume hood.
2. Chemical waste receptacles.
3. Ultrapure water with a resistivity of 18.2 $M\Omega$ cm, for example, from a Milli-Q purification system (Millipore, Billerica, MA, USA).
4. An UV–visible spectrophotometer (mod. Evolution 201, Thermo Fisher Scientific, Walthman, MA, USA).
5. A benchtop mini centrifuge (Scilogex D3024, Rocky Hill, CT, USA).
6. 1.5 and 2.0-mL microcentrifuge tubes (Eppendorf, Hamburg, Germany).
7. Glass vials.
8. 20, 100, 200, and 1000-μL volume pipettes (Gilson, Middleton, WI, USA).
9. A thermomixer (Eppendorf, Hamburg, Germany).
10. An analytical balance (Gibertini Elettronica srl, Milano, Italy).
11. A high performance liquid chromatography (e.g., HPLC 1260, Agilent Tech., Santa Clara, CA, USA).
12. A lyophilizer (Thermo Fisher Scientific, Waltham, MA, USA).
13. A speed-vac system (Thermo Fisher Scientific, Waltham, MA, USA).
14. Salts, reagents, solvents, protein colorimetric assays, centrifugal filter units, and dialysis membranes and all materials, unless otherwise specified, were purchased by Sigma Aldrich (Saint Louis, MO, USA).

2.2 Immobilization of mTGase on Aminolink Coupling Resin

2.2.1 Equipment

1. Bottom-capped column with filter containing the required volume of resin slurry. (An empty PD10 column could be very useful for this protocol.)
2. End-over-end rocking (mod. 708, Asaal srl, Milano, Italy).
3. Centrifuge tube filters with cellulose acetate membrane, pore size 0.22 μm (Corning® Costar® Spin-X®, Corning, NY, USA).

2.2.2 Reagents

1. mTGase from *Streptomyces mobaraensis* (ACTIVA MP, Ajinomoto Co., Tokyo, Japan, or from other suppliers).
2. AminoLink Coupling Resin (Thermo Scientific, Waltham, MA, USA).
3. Coupling buffer: 0.1 *M* sodium phosphate, 0.15 *M* NaCl, pH 7.2.
4. Sodium cyanoborohydride ($NaCNBH_3$).
5. Quenching buffer: 1 *M* Tris–HCl, pH 7.4.
6. Wash solution: 1 *M* NaCl.
7. Protein colorimetric assay, such as Bradford, Lowry, or BCA assay reagents (on the basis what you have in your lab).

2.3 Activity Assay

2.3.1 Reagents

1. mTGase from *S. mobaraensis* (ACTIVA MP, Ajinomoto Co., or from other suppliers).
2. Immobilized mTGase.
3. Coupling buffer: 0.1 *M* sodium phosphate, 0.15 *M* NaCl, pH 7.2.
4. Substrate solution: 0.2 *M* Tris–acetate, pH 6, 0.1 *M* hydroxylamine (NH_2OH), 30 m*M* CBZ-*n*-glutaminylglycine, Z-Gln-Gly.
5. Reaction buffer: 0.2 *M* Tris–acetate, pH 6, 0.1 *M* NH_2OH.
6. Stop solution: ferric chloride (5% w/v) in HCl 0.1 N, trichloroacetic acid (12% w/v), HCl 37%.

2.4 Determination of Kinetic Parameters

2.4.1 Reagents

1. Coupling buffer: 0.1 *M* sodium phosphate, 0.15 *M* NaCl, pH 7.2.
2. Enzyme solution: free (1–2.5 mg/mL) and immobilized enzyme (1–2.5 mg/mL) in coupling buffer.

3. Substrate (Z-Gln-Gly) Solutions: 1, 5, 7.5, 10, 12.5, 15, 30 mmol/L of Z-Gln-Gly in 0.2 *M* Tris–acetate, pH 6, 0.1 *M* NH_2OH.
4. Stop solution: ferric chloride (5% w/v) in HCl 0.1 N, trichloroacetic acid (12% w/v), HCl 37%.

2.5 Protein PEGylation

2.5.1 Equipment

1. An HPLC system such as the Agilent 1260 series system.
2. A C18 column such as Phenomenex Jupiter (250 × 4.6 mm, 5 μm) (Torrance, CA, USA).
3. A size exclusion chromatography (SEC) Column such as Agilent Zorbax GF-250 (250 × 4.6 mm, 5 μm) or Phenomenex Biosep-SEC 3000 (300 × 4.6 mm, 5 μm).
4. An IEX Column such as TSK-gel, SP-5PW (7.5 mm × 7.5 cm, 5 μm) (King of Prussia, PA, USA) or Dionex ProPac SAX-10 (2 × 250 mm, 5 μm) (Sunnyvale, CA, USA).
5. A Float-A-Lyzer.
6. Amicon Ultra centrifugal filter units.
7. A lyophilization system.

2.5.2 Reagents

1. A given protein X.
2. Bradford, Lowry, or BCA assay reagents (on the basis what you have in your lab).
3. mTGase from *S. mobaraensis* (ACTIVA MP, Ajinomoto Co., or from other suppliers).
4. Reaction buffer: 0.1 *M* sodium phosphate pH 7.5 or PBS pH 7.4.
5. Polymer: PEG-NH_2 (NOF Europe, Grobbendonk, Belgium) or CBZ-Gln-Gly-PEG (PEG-Gln).
6. Iodoacetamide (IA) or *N*-ethylmaleimide (NEM).

2.6 Synthesis of PEG-Gln

2.6.1 Equipment

1. A 100-mL flask.
2. A dialysis membrane at the desired length.
3. A lyophilization system.
4. NMR (400 MHz, Bruker BioSpin GmbH, Rheinstetten, Germany).

2.6.2 Reagents

1. A reaction buffer: 0.1 *M* borate buffer/ACN (acetonitrile) (3:2) mixture pH 8.0.
2. CBZ-*N*-glutaminylglycine, Z-Gln-Gly (Sigma Aldrich).
3. 1-Ethyl-3-(3-dimethylaminopropyl)carbodiimide (EDC).
4. 1-Hydroxybenzotriazole (HOBT).
5. Triethylamine (TEA).
6. PEG-NH_2.
7. Succinic anhydride.
8. TNBS assay reagents (2,4,6-trinitrobenzenesulfonic acid (1% in DMSO), glycylglycine, 0.2 *M* borate buffer pH 9.3, 0.2 *M* borate buffer pH 8).

2.7 Conjugation Site Determination

2.7.1 Equipment

1. An HPLC system such as the Agilent 1260 series system.
2. C18 Column such as Phenomenex Jupiter (250 × 4.6 mm, 5 μm).
3. PepClean C-18 Spin columns (Pierce, Rockford, IL).
4. UPLC-Q-TOF mass spectrometry (e.g., UPLC 1290 Agilent and Xevo G2-S Q-TOF Waters).

2.7.2 Reagents

1. Tris(2-carboxyethyl)phosphine hydrochloride (TCEP).
2. Iodoacetamide.
3. Denaturating buffer: 6 *M* guanidinium·HCl, 50 m*M* Tris–HCl (pH 9.0).
4. Urea.
5. Trypsin sequencing grade.

3. METHODS

3.1 Immobilizing mTGase on Aminolink Coupling Resin

mTGase was immobilized to a beaded agarose support. The immobilization reaction was achieved using a reductive alkylation involving the N-terminus of the enzyme and the aldehyde groups on the resin bead surface. At the slightly acidic pH of the coupling buffer, the enzyme's N-terminal amino group is preferentially immobilized in order to avoid interference with its active site. Fig. 1 shows a diagram that illustrates the coupling strategy.

1. Solubilize mTGase in 2–3 mL of coupling buffer by gentle shaking.
2. Centrifuge the sample at 6000 rpm for 5 min and determine the exact enzyme concentration by measuring the UV–vis absorption at 280 nm in the supernatant (A_{280} 1%: 1.9 mL cm^{-1} mg^{-1}).

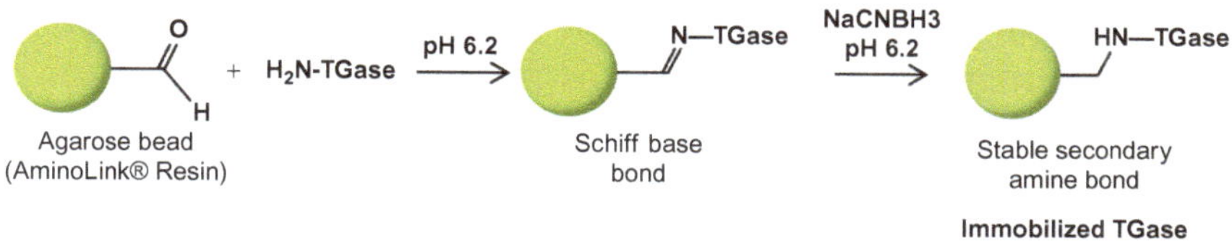

Fig. 1 A diagram illustrating mTGase immobilization.

3. Equilibrate the upright column to room temperature and allow the resin to settle. The volume of slurry to use depends on the amount of enzyme; 4 mL of slurry should generally be used for 10 mg of enzyme (to dispense the resin, invert the bottle of slurry several times before pipetting the desired volume with a cut plastic pipette tip). Drain the storage solution and equilibrate the column with three resin-bed volume of coupling buffer, allowing also the contents to drain.
4. Add the enzyme solution and, in a fume hood, 40 μL of $NaCNBH_3$ dissolved in 1 *M* NaOH in order to obtain a final concentration of 50 m*M*. Gently mix the reaction by end-over-end rocking for 6 h at room temperature.
5. Open the column in a fume hood, drain the contents into a collection tube, determine the coupling efficiency, and wash the resin with two resin-bed volumes of coupling buffer.
6. To block the remaining active sites, wash the resin with two resin-bed volumes of quenching buffer, replace the bottom cap, and, in a fume hood, add 40 μL of $NaCNBH_3$ dissolved in 1 *M* NaOH in order to obtain a final concentration of 50 m*M*. Mix the reaction by end-over-end rocking for 30 min at room temperature.
7. After the contents are carefully drained in a fume hood, wash the column with 10 resin-bed volumes of wash solution in order to remove all non-bound enzyme and transfer the resin with the immobilized mTGase in a glass vial using a cut pipette plastic tip.
8. Calculate the coupling efficiency by comparing the protein concentrations of the unbound fraction to the starting sample. Calculate the concentration of the immobilized mTGase using the BCA colorimetric assay and filter the solution containing the resin with 0.22-μm centrifuge tube filters before determining the absorbance.
9. For subsequent experiments, the immobilized mTGase is dispensed by gently shaking the vial and withdrawing the desired volume of enzyme with a cut pipette plastic tip.

3.2 Activity Assay

3.2.1 Determination of mTGase Activity

TGase activity was determined using the hydroxamate formation assay based on the method described by Grossowicz et al. with CBZ-*N*-glutaminylglycine (Z-Gln-Gly) substrate (Grossowicz, Wainfan, Borek, & Waelsch, 1950).

3.2.2 Determination of mTGase Activity

1. Prepare a solution or dispersion of free or immobilized mTGase, respectively, in the reaction buffer at the enzyme concentration of 1 mg/mL.
2. Prepare the testing solutions in accordance with directions in Table 1 (repeat the experiment in duplicate).
3. Incubate the tubes for 10 min at 37°C in a thermomixer.
4. Add 0.25 mL of stop solution and measure the absorbance of the samples at 525 nm. Carefully filter the solution containing the resin with 0.22-μm centrifuge tube filter for the sample with immobilized mTGase before determining absorbance.
5. The specific enzymatic activity of mTGase is obtained through the following formula:

$$\text{Units/mg} = (\text{Abs}_S) \times (\text{Vf})/(\varepsilon_M \times (\text{mg enzyme}/\text{RM}_{\text{mL}}) \times 10)$$

Abs_s = absorbance of mTGase sample
Vf = final volume of the reaction (510 μL)
ε_M = molar extinction coefficient of hydroxamate (M^{-1} cm^{-1})
mg enzyme = mg of enzyme added to the reaction mixture
RM_{mL} = volume of reaction mixture (260 μL).

Table 1 Testing Solutions for the Determination mTGase Activity

Blank	Sample	Sample
10 μL of coupling buffer	10 μL of free mTGase solution	10 μL of immobilized mTGase solution
250 μL of substrate solution	250 μL of substrate solution	250 μL of substrate solution

3.2.3 Determination of Thermostability of Free and Immobilized mTGase

1. Prepare several tubes containing about 100 μL of solution or dispersion of free or immobilized mTGase, respectively, in the reaction buffer at the enzyme concentration of 1 mg/mL.
2. Incubate the tubes for 3 h at 20, 30, 40, 50, 60, 70, 80°C in a thermomixer.
3. Prepare the testing solutions in accordance with directions in Table 2 (repeat the experiment in duplicate).

   Add 0.25 mL of stop solution and measure the absorbance of the samples at 525 nm. Carefully filter the solution containing the resin with 0.22-μm centrifuge tube filter for the sample with immobilized mTGase before determining absorbance.

   The enzymatic activity of free mTGase (or immobilized enzyme) at each temperature is calculated considering as 100% that presents the higher absorbance value than all the observed temperature. The following formula is applied:

$$\text{Residual activity}\,(\%) = (\text{Abs}_{X^\circ C}/\,\text{Abs}_T) \times 100$$

   $\text{Abs}_{X^\circ C}$ = absorbance of activity test of mTGase at each temperature.
   Abs_T = higher absorbance value of activity test of mTGase at a determined temperature.

3.2.4 Determining Relative Activity in the Presence of Organic Cosolvents

1. Prepare a solution of mTGase in coupling buffer at a concentration of 30 mg/mL.

Table 2 Testing Solutions for the Determination mTGase Activity

Blank	Sample	Sample
10 μL of coupling buffer	10 μL of free mTGase solution incubated at each temperature	10 μL of immobilized mTGase solution incubated at each temperature
250 μL of coupling buffer	250 μL of coupling buffer	250 μL of coupling buffer

Table 3 Testing Solutions for mTGase Activity Determination

Blank	Sample	Reference Sample
10 μL of coupling buffer	10 μL of mTGase solution	10 μL of mTGase solution
250 μL of substrate solution	250 μL of substrate solution	250 μL of substrate solution
500 μL of reaction buffer	500 μL of a solution of the organic cosolvent and the reaction buffer	500 μL of reaction buffer

2. Prepare a solution of an organic solvent (methanol, ethanol, DMSO (dimethyl sulfoxide)) in the reaction buffer to reach the final percentage of organic solvent desired in the testing solution reported in Table 3.
3. Prepare the testing solutions using the measurements indicated in Table 3 (repeat the experiment in duplicate).
4. Incubate the tubes for 10 min at 37°C in the thermomixer.
5. Add 0.25 mL of stop solution and measure the absorbance at 525 nm.
6. The residual enzymatic activity of mTGase in the presence of an organic cosolvent is calculated considering the enzymatic activity of mTGase in the reaction buffer as 100% without organic cosolvents, and it is obtained using the following formula:

$$\text{Residual activity}\,(\%) = (\text{Abs}_C/\,\text{Abs}_R) \times 100$$

Abs_C = absorbance of activity test of mTGase in the presence of organic cosolvent

Abs_R = absorbance of activity test of reference mTGase in reaction buffer

3.3 Determination of Kinetic Parameters for Free and Immobilized Enzymes

1. Prepare 14 centrifuge tubes of 1.5 mL containing 0.08 mL of stop solution.
2. Prepare 1.2 mL of substrate solutions and add 0.045 mL of 1–2.5 mg/mL of free or immobilized mTGase (to be repeated with the different substrate concentrations).
3. Incubate the solution attained at point 2 at 37°C and under 300 rpm shaking in a thermomixer. At predetermined incubation time-points (10, 20, 40, 60, 90, 120, 180, 300, 600, 900, 1200, 1500 s) withdraw

0.08 mL of the sample and add the substrate solution to the stop solutions prepared at point 1. (Carefully filter the solution containing the tubes with 0.22-μm centrifuge tube filters before determining the absorbance at 525 nm.)

4. Read the absorbance at 525 nm for each time-point.
5. Plot the absorbance values vs time for each substrate concentration and determine the initial velocity. Then express the initial velocity as nmol of hydroxamate generated in a minute (extinction coefficient: 920 M^{-1} cm^{-1}).
6. For simplicity, make a double-reciprocal plot of the data, that is, a graph of $1/V_0$ vs $1/[S]$. A linear regression should be obtained: a Lineweaver-Burke plot which should make it possible to determine the K_m and V_{max} values for the enzymatic reaction. The slope of the linear regression is K_m/V_{max}, and the ordinate axis-intercept is 1/Vmax.

3.4 Protein PEGylation

3.4.1 Protein PEGylation to Gln by mTGase-Mediated Conjugation

1. Prepare a solution of protein of your interest (about 1–4 mg/mL) in the reaction buffer (Different aqueous buffers can be used such as phosphate, borate buffer, or HEPES, but do not use *N*-(tris-hydroxymethyl)-aminomethane or any primary amine containing salt because they will compete with PEG-NH_2 for coupling.) and determine protein concentration by absorption at 280 nm or other suitable method (such as the Bradford, Lowry, BCA, etc. assays).
2. Calculate a molar excess of PEG-NH_2 in the range of 1:10 and 1:20 with respect to the protein.
3. Dissolve the polymer in the reaction buffer in the minimum volume needed.
4. Add PEG-NH_2 to the protein solution and withdraw a small aliquot of reaction mixture as reference to time zero of the reaction.
5. Prepare a stock solution of mTGase by solubilizing the enzyme in the reaction buffer at the concentration of 60 mg/mL.
6. Add the mTGase to the reaction mixture in order to obtain the E/S ratio required (E/S by weight ratio usually ranges from 1/50 to 1/25).
7. Incubate the reaction solution at 25 or 37°C under stirring on the basis of protein stability (the enzyme is more active at 37°C but sometimes protein can tend to precipitate at this pH value. In that case, it is better to carry out the reaction at 25°C).

8. Withdraw small quantities of reaction mixture to monitor the progress of the reaction at predetermined times and analyze the solution using a suitable method such as SEC or RP-HPLC (as PEG does not have UV–vis absorption, the area under the peak of the conjugate corresponds to the amount of protein modified by PEG. Use the area under the peak of the reference at time zero corresponding to the unconjugated protein to calculate the yield of conjugate formation).
9. At the end of the incubation time (to reach the maximum conversion the incubation time varies depending on the starting proteins' characteristics), quench the solution by adding IA or NEM at a quenching agent/mTGase molar ratio of 30/1. (In most cases, 4 h of incubation has proven to be sufficient to reach the maximum yield of conversion.) Purify the reaction mixture using a suitable chromatographic method and buffer exchange the purified PEG-protein conjugate in a suitable buffer for protein storage (dialysis or ultrafiltration can be used).
10. Characterize the conjugate using one of several techniques such as SDS-PAGE, RP-HPLC, MALDI-TOF mass spectrometry, circular dichroism, bioactivity testing, etc. (refer to Mero, Fang, Pasut, Veronese, & Viegas 2012).

3.4.1.1 Practical Example

42.6 mg of PEG20kDa-NH_2 or 10.6 mg of PEG5kDa-NH_2 (10-fold molar excess over the protein), dissolved in 0.8 mL of 10 m*M* phosphate, were added to 1 mL solution of 4 mg/mL G-CSF (MW 18792.8 Da) dissolved in 10 m*M* acetate buffer containing 5% sorbitol, pH 4.7. Before adding mTGase, 10 μL of reaction mixture was taken as time zero reference. Then 16 mg of mTGase powder (1% w/w powder, corresponding to 0.16 mg of enzyme in total) was dissolved in 0.2 mL of reaction buffer and added to the G-CSF/PEG solution. The exact enzyme concentration was determined by UV considering the extinction coefficient of TGase: 1.89 mL mg^{-1} cm^{-1}. The E/S ratio was 1/25 (w/w). The reaction solution was incubated for 4 h at 25°C under stirring and conjugate formation was monitored in RP-HPLC with a Jupiter C18 column (250 × 4.6 mm, 300 Å, 5 μm; Phenomenex, USA) eluted with a gradient of water (A) and ACN (B), both containing 0.1% trifluorocetic acid (TFA), at the flow rate of 1.0 mL/min (gradient: 0′ 40%B, 25′ 70%B, 27′ 90%B) and monitoring the effluent at 226 nm (Fig. 2, the chromatographic conditions listed here are as example; for a different protein the elution must be optimized using a case-by-case approach). After 4 h the reaction mixture was purified using the same conditions. The peak of the conjugate was collected and dialyzed against 10 m*M*

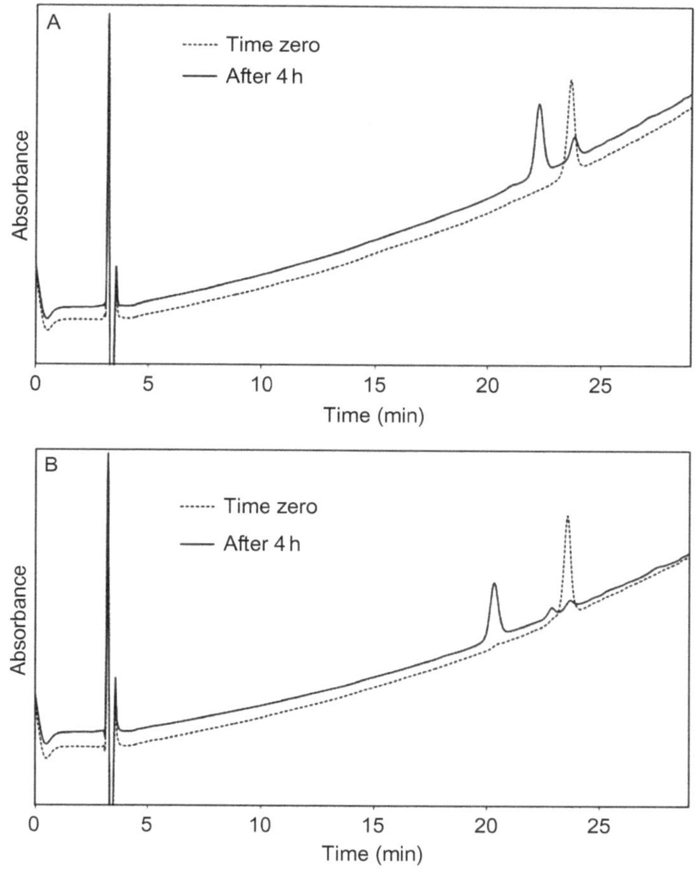

Fig. 2 RP-HPLC analysis of PEG5kDa-NH_2 (A) and PEG20kDa-NH_2 (B) conjugation reaction to G-CSF via mTGase. Phenomenex C18 column (5 μ; 300 Å; 250 × 4.6 mm); eluent A: H_2O + 0.1% TFA; eluent B: ACN + 0.1% TFA; gradient: 0′ 40%B, 25′ 70%B, 27′ 90%B; detection at 226 nm; injection 10 μL; flow rate 1 mL/min; sample concentration 0.2 mg/mL in protein. The product eluting at 20.56 min corresponds to PEG-protein conjugate, whereas the free protein elutes at 24.75 min.

acetic acid, 5% sorbitol pH 4.6 using a 3.5 kDa cut-off membrane. Finally, PEGylated protein concentration was determined by UV–vis spectroscopy at 280 nm.

3.4.2 Synthesis of PEG-Gln

1. Prepare a 9.77 m*M* solution of Z-Q-G in the reaction buffer.
2. Add EDC and HOBT to reach a final concentration of 40 m*M* for both reagents.

3. After stirring for 1 h, add PEG-NH_2 to the reaction solution at a final concentration of 2.5 m*M* and adjust the pH to 8 with TEA.
4. Let the reaction proceed at room temperature for 18 h under stirring.
5. Add succinic anhydride to the mixture at a final molarity of 5 m*M* in order to quench the reaction.
6. Dialyze the reaction solution for 24 h against Milli-Q water and then lyophilize the product (dialysis is effective if the outside buffer is frequently changed and is at least 500–1000 folds the volume of the sample).
7. Verify both the absence of free amino groups by TNBS according to Snyder and Sobocinski assay (Snyder & Sobocinski, 1975) and the product identity by ^{1}H-NMR.

3.4.3 Protein PEGylation With PEG-Gln

1. Prepare a solution of your target protein (about 1–4 mg/mL) in the reaction buffer (different aqueous buffers can be used (e.g., phosphate, borate buffer, HEPES), but do not use *N*-(tris-hydroxymethyl)-aminomethane or any primary amine containing salt because they will compete with PEG-NH_2 for coupling) and determine the protein concentration by measuring the absorption at 280 nm or by using another suitable method (such as Bradford, Lowry, BCA, etc. assay).
2. Calculate a molar excess of PEG-Gln in the range of 1:10 and 1:20 with respect to the protein.
3. Dissolve the polymer in the reaction buffer in the minimum volume needed.
4. Add PEG-Gln to the protein solution and withdraw a small aliquot of reaction mixture as reference to time zero of the reaction.
5. Prepare a stock solution of mTGase by solubilizing the enzyme in the reaction buffer at a concentration of 60 mg/mL.
6. Add the mTGase to the reaction mixture in order to obtain the E/S ratio required (E/S weight ratio usually ranges from 1/50 to 1/25).
7. Incubate the reaction solution at 25 or 37°C under stirring.
8. Withdraw a small quantity of reaction mixture to monitor the progress of the reaction at predetermined times and analyze the solution using a suitable method such as SEC or RP-HPLC (as PEG does not have UV–vis absorption, the area under the peak of the conjugate corresponds to the amount of protein modified by PEG. Use the area under the peak of the reference at time zero corresponding to the unconjugated protein to calculate the yield of conjugate formation).

9. At the end of the incubation time (the incubation time needed to reach the maximum conversion varies depending on the starting protein's characteristics), quench the solution by adding IA or NEM at a quenching agent/mTGase molar ratio of 30/1.
10. Purify the reaction mixture using an appropriate chromatographic method and buffer exchange the purified PEG-protein conjugate in a suitable buffer for protein storage (dialysis is effective if the outside buffer is frequently changed and is 500–1000 folds the volume of the sample).
11. Characterize the conjugate using any of several techniques such as SDS-PAGE, RP-HPLC, MALDI-TOF mass spectrometry, circular dichroism, bioactivity testing, etc. (refer to Mero et al., 2012; Mero et al., 2016).

3.4.3.1 Practical Example

The reaction conditions were described in Section 3.4.1. The aliquots of the reaction mixtures were analyzed by RP-HPLC on an Agilent series 1260 HPLC with online UV detection from Agilent Technologies. RP-HPLC analyses were performed using a RP-C18 Phenomenex column (Jupiter C18, 300 Å, 5 μm, 250 × 4.60 mm) applying a gradient of ACN, containing 0.1% TFA and water, 0.1% TFA from 5% to 40% of ACN in 5 min and from 40% to 70% in 25 min. The column was eluted at a flow rate of 1 mL min^{-1}, and the absorbance was recorded at 226 nm (Fig. 3A). The reaction mixtures of G-CSF presenting the highest degree of protein PEGylation were loaded on a TSK-gel SP 5-PW column (300 Å, 7 μm, 7.5 × 0.75 cm; TOSO-HAS) for purification. The chromatographic separation was performed using an Agilent series 1260 HPLC. The cation exchange column was eluted with a gradient of buffer A (10 m*M* sodium phosphate pH 4.7) and B (0.1 *M* sodium phosphate, 0.1 *M* NaCl pH 4.85). After the sample was injected, the column was eluted with 5% B for 10 min and then with a gradient of B from 5% to 65% in 50 min and from 65% to 100% in 20 min. The column was then washed for 10 min with 100% B and for 5 min with 5% B. The flow rate was 1 mL min^{-1}, and the effluent absorbance was recorded at 226 nm (Fig. 3B). The G-CSF conjugate peak was collected from the cation exchange column and dialyzed at 4°C against 10 m*M* sodium acetate buffer, pH 4.7 containing 5% sorbitol using a dialysis membrane with a MWCO of 6–8000 Da (Spectra/Por Dialysis Membrane). Then the solution was concentrated using Amicon Ultra-4 10 K (Millipore).

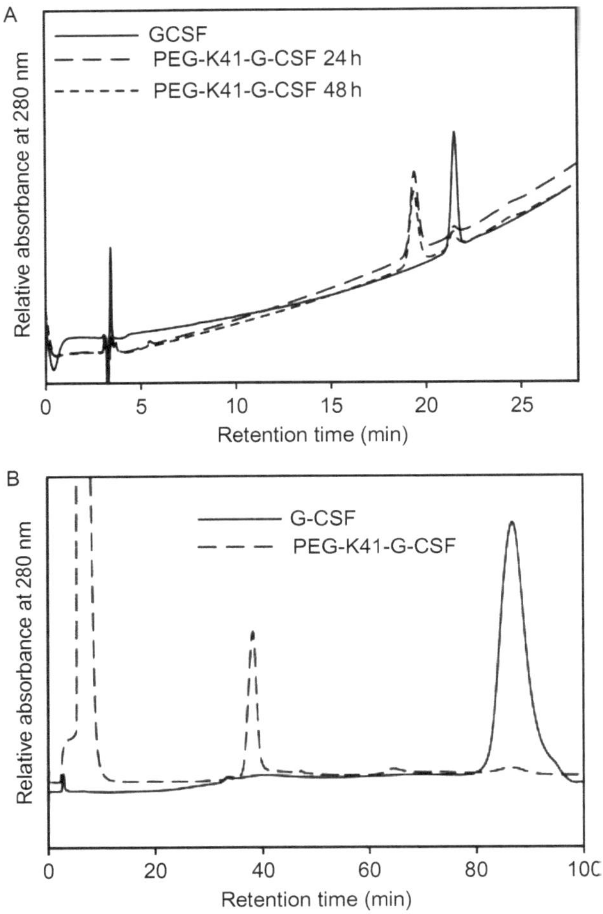

Fig. 3 (A) RP-HPLC profiles of reaction mixture after 0, 24, and 48 h. The peak eluting at 21 min is free G-CSF, whereas the peaks eluting at 19.85 min correspond to the PEGylated derivative. (B) Purification of PEG-G-CSF by CEX-HPLC. The peak eluting at 35 min corresponds to the PEGylated derivative, the G-CSF elutes at 86 min. *Adapted from Mero, A., Grigoletto, A., Maso, K., Yoshioka, H., Rosato, A., & Pasut, G. (2016). Site-selective enzymatic chemistry for polymer conjugation to protein lysine residues: PEGylation of G-CSF at lysine-41.* Polymer Chemistry, *7, 6545–6553.*

The purified product was also analyzed by MALDI-MS analysis. Mass spectrometry analyses were performed using a REFLEX time-of-flight instrument (4800 Plus MALDI-TOF/TOF, AB Sciex, Framingham, Massachusetts, USA) equipped with a SCOUT ion source operating in positive linear mode. The ions generated by a pulsed UV laser beam (nitrogen laser, λ 337 nm) were accelerated to 25 kV. A saturated solution of sinapinic acid in water/ACN (1:1, v/v) was used as a matrix and mixed with the samples dissolved in 0.1% TFA aqueous solution at a v/v ratio 1:1. The MALDI-TOF spectrum of the conjugate product presents a product with a mass of 39843.5 Da, which corresponds to the G-CSF conjugated to one chain of PEG20 kDa (Fig. 4).

3.4.4 Protein PEGylation in the Presence of Organic Cosolvents

1. Prepare several protein solutions at the same protein concentration (i.e., 1 mg/mL) in the reaction buffer (different aqueous buffers can be used (e.g., phosphate, borate buffer, HEPES), but do not use *N*-(tris-hydroxymethyl)-aminomethane or any primary amine containing salt because they will compete with PEG-NH_2 for coupling) with different percentages of an organic solvent suitable for the target protein (i.e., methanol, ethanol, trifluoroethanol, etc.).

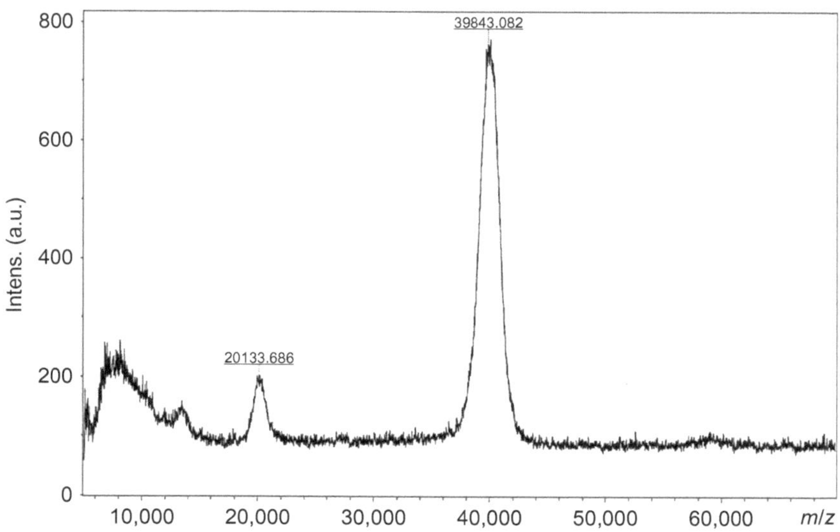

Fig. 4 MALDI-TOF mass spectrum of PEG-ZG-Q-G-CSF. The peak at 39843.082 kDa is the PEGylated derivative, and the peak at approximately 20133.686 kDa is free PEG-ZQG. *Adapted from Mero, A., Grigoletto, A., Maso, K., Yoshioka, H., Rosato, A., & Pasut, G. (2016). Site-selective enzymatic chemistry for polymer conjugation to protein lysine residues: PEGylation of G-CSF at lysine-41.* Polymer Chemistry, *7, 6545–6553.*

2. Determine the protein conformation by circular dichroism or by fluorescence analysis in relation to the different percentages of the organic cosolvent.
3. Select an organic cosolvent and a percentage that induces a change in the protein's secondary structure.
4. Verify the mTGase activity in the selected buffer with the organic cosolvent as reported in Section 3.2.2.
5. Prepare a solution of your target protein (about 1–4 mg/mL) in the reaction buffer (different aqueous buffers can be used (e.g., phosphate, borate buffer, HEPES), but do not use *N*-(tris-hydroxymethyl)-aminomethane or any primary amine containing salt because they will compete with PEG-NH_2 for coupling) containing the selected percentage of organic cosolvent and determine the protein concentration by measuring the absorption at 280 nm or by using other suitable methods such as Bradford, Lowry, assay, BCA, etc. assay
6. Calculate the molar excess of PEG-NH_2 in the range of 1:10 and 1:20 with respect to the protein.
7. Dissolve the polymer in the reaction buffer in the minimum volume needed.
8. Add PEG-NH_2 to the protein solution and withdraw a small aliquot of reaction mixture as reference to time zero of the reaction.
9. Prepare a stock solution of mTGase by solubilizing the enzyme in the reaction buffer at the concentration of 60 mg/mL.
10. Add the mTGase to the reaction mixture in order to obtain the E/S ratio required (E/S weight ratio usually ranges from 1/50 to 1/25.).
11. Incubate the reaction solution at 25 or 37°C under stirring.
12. Withdraw small quantities of reaction mixture and monitor the progress of the reaction at predetermined times by analyzing the solution using a suitable method such as SEC or RP-HPLC (as PEG does not have UV–vis absorption, the area under the peak of the conjugate corresponds to the amount of protein modified by PEG. Use the area under the peak of the reference at time zero corresponding to the unconjugated protein to calculate the yield of conjugate formation).
13. At the end of the incubation time (the incubation time to reach the maximum conversion varies depending on the starting protein characteristics), quench the solution by adding IA or NEM at a quenching agent/mTGase molar ratio of 30/1.
14. Purify the reaction mixture using a suitable chromatographic method and buffer exchange the purified PEG-protein conjugate in a suitable

buffer for protein storage (dialysis is effective if the outside buffer is frequently changed and is 500–1000 folds the volume of the sample).

15. Characterize the conjugate by several techniques such as SDS-PAGE, RP-HPLC, MALDI-TOF mass spectrometry, circular dichroism, bioactivity testing, etc. (refer to Mero et al., 2012; Mero, Clementi, et al., 2011; Mero, Schiavon, et al., 2011).

3.4.4.1 Practical Experiment

Boc–PEG556Da–NH_2 was added (10-fold molar excess with respect to the protein) to a solution of sCT (1 mg/mL) in 100 m*M* phosphate buffer pH 7 containing 30% (v/v) of DMSO or 60% (v/v) of MeOH (or EtOH 50%). The polypeptide conformation was evaluated by CD and Fluorescence. Far-UV CD spectra were measured on a Jasco J-700 spectropolarimeter equipped with a Peltier temperature control unit at 20°C. Measurements of native sCT were made in 10 m*M* phosphate buffer pH 7.0 containing MeOH (60%, v/v) or EtOH (50%) at the concentration solutions of 0.1 mg/mL as spectrophotometrically determined at 280 nm (absorption coefficients: 0.443 mL cm^{-1} mg^{-1} for sCT). The spectra were collected over the wavelength range of 190–250 nm with an average of two scans. The data at each wavelength were averaged for 8 s. The sample cell path length was 1 mm. The CD data were converted to mean residue ellipticity, expressed in deg cm^2 $dmol^{-1}$ by applying the following formula:

$$\Theta = \Theta\text{obs}(\text{MRW})/10\,L[\text{C}]$$

where Θ is the observed ellipticity in degrees, MRW is the mean residue weight of the peptide (molecular weight divided by the number of residues), [C] is the peptide concentration in mg/mL, and *L* is the optical path length in centimeters.

Fluorescence emission spectra were measured with Jasco FP-650 spectrofluorometer equipped with a thermostatted cell holder at 25°C. sCT was dissolved in 10 m*M* phosphate buffer pH 7 (final protein conc. 0.1 mg/mL) containing DMSO (30%), MeOH (60%), or EtOH (50%). sCT tyrosine was excited at 275 nm, and the emission recorded in the range of 295–400 nm. Data were collected at 1 nm increments with a slit width of 5 nm.

Far-UV-CD spectra showed that sCT had a disordered structure in water; while in 60% (v/v) of MeOH the α-helix content reached about 60% (Fig. 5A). In the case of DMSO, the FUV-CD spectrum of sCT was not evaluated due to the strong absorbance of DMSO at low

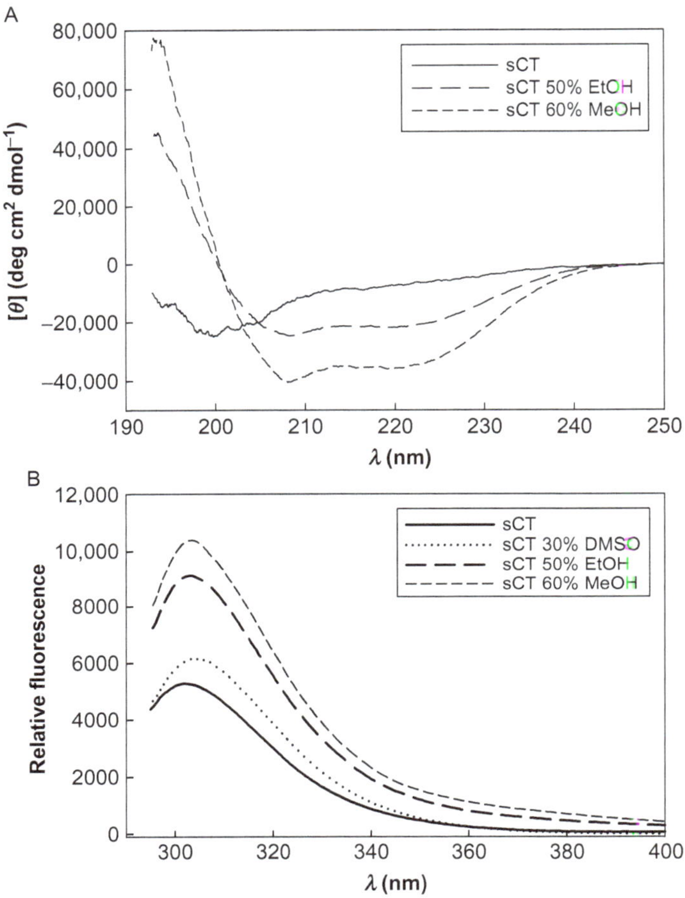

Fig. 5 (A) Effect of cosolvents on FUV-CD spectra of sCT. (B) Effect of cosolvents on intrinsic fluorescence emission spectra of sCT. *Adapted from Mero, A., Clementi, C., & Pasut, G. (2011). Covalent conjugation of poly(ethylene glycol) to proteins and peptides: Strategies and methods.* Methods Molecular Biology, 751, *95–129; Mero, A., Schiavon, M., Veronese, F. M., & Pasut, G. (2011). A new method to increase selectivity of transglutaminase mediated PEGylation of salmon calcitonin and human growth hormone.* Journal of Controlled Release, 54, *27–34.*

wavelengths. In this case fluorescence analyses were carried out as a complementary method.

For sCT a slight red shift took place in the presence of DMSO, from 302 to 305 nm (Fig. 5B), indicating a modification in the surrounding environment of the lonely tyrosine (Tyr22). Evaluations in ethanol or methanol mixtures showed an increase in intrinsic Tyr fluorescence, but it was not accompanied by a shift in the emission spectra. After evaluation of protein conformation, sCT was modified by PEG using TGase enzyme (firstly evaluating the activity of the enzyme in DMSO and MeOH).

mTGase was added at an E/S ratio of 1:30 (w/w). The reaction solution was incubated for 3 h at room temperature under stirring and then analyzed by RP-HPLC using a Jupiter C18 (250 × 4.6 mm, 300 Å, 5 μm; Phenomenex, USA) with a linear gradient of 10%–70% (v/v) ACN in water, both containing 0.05% TFA, over 25 min, followed by an isocratic wash at 90% ACN. The effluent was monitored by measuring the absorbance at 226 nm (the chromatographic conditions listed here are as example; for a different protein the elution must be optimized using a case-by-case approach). The peak of the conjugate was collected, lyophilized, and analyzed using ESI-TOF mass spectrometry.

sCT contains two glutamines, Gln 14 and Gln 20. As the polypeptide presents a random coil structure in physiological conditions, we demonstrated that both glutamines are substrates for mTGase. When the reaction was carried out in the presence of organic solvents, which promoted the formation in the peptide of α-helix or β-sheet structures, only one Gln 20 was selectively PEGylated by mTGase. Under these conditions Gln 14 is located at the inner core of a structured portion, while Gln 20 is located in a border region where it is still accessible to the enzyme.

3.4.5 Protein PEGylation With PEG-NH_2 via Immobilized mTGase

1. Prepare a solution of your target protein (about 1–5 mg/mL) in the reaction buffer (different aqueous buffers can be used (e.g., phosphate, borate buffer, HEPES), but do not use *N*-(tris-hydroxymethyl)-aminomethane or any primary amine containing salt because they will compete with PEG-NH_2 for coupling) containing the selected percentage of organic cosolvent and determine the protein concentration by measuring the absorption at 280 nm or by using another suitable method such as Bradford, Lowry, BCA, etc. assay.
2. Calculate a molar excess of PEG-NH_2 in the range of 1:5 and 1:10 with respect to the protein.
3. Dissolve the polymer in the reaction buffer in the minimum volume needed.
4. Add PEG-NH_2 to the protein solution and withdraw a small quantity of reaction mixture as reference to time zero of the reaction.
5. Add immobilized mTGase from a stock dispersion with a known activity, freshly determined as reported above, (as a general rule if a

free mTGase is used at E/S weight ratio of 1:25–50, the immobilized mTGase should usually be utilized at 1:10–20) (add the immobilized mTGase with a cut plastic pipette tip).

6. Incubate the reaction mixture at 37°C under gentle stirring for at least 18 h. Follow the reaction progress at predetermined times, at 4, 6, and 18 h using RP-HPLC or SEC analysis with respect to a zero reference collected at point 4. Carefully remove the resin from the solution after filtration or centrifugation for the HPLC injection.
7. Stop the reaction at the end of incubation time by filtering the reaction mixture with 0.22-μm centrifuge tube filter or by centrifuging to separate the resin from the solution.
8. Purify the conjugated protein using a suitable chromatographic method (in most cases, ion exchange chromatography is used) and dialyze the PEG-protein conjugate to exchange the buffer with one in which the protein is well stored (dialysis is effective if the outside buffer is frequently changed and is 500–1000 folds the volume of the sample). If necessary, concentrate the volume of purified conjugates by Amicon Ultra centrifugal filters (when free or PEGylated proteins are UF/DF, avoid reducing the volume of the protein solution below to 500 μL in order to prevent protein aggregation/absorption) before characterizing the PEGylated products by SDS-PAGE, RP-HPLC, MALDI-TOF mass spectrometry, and circular dichroism. If the protein needs to be lyophilized for better storage, dialyze the conjugate against water and lyophilize it with a cryoprotectant (use of cryoprotectant must be considered on a case-by-case basis depending on the protein being studied).

3.4.5.1 Practical Experiment

PEG5kDa-NH_2 was added (10-fold molar excess with respect to the protein) to a solution of α-LA (1 mg/mL) in 20 m*M* Na_2HPO_4, 5 m*M* EDTA, pH 7.5. Free and immobilized mTGases at an E/S ratio of 1:50 and 1:20 (w/w) were, respectively, separately added to two different protein solutions. Reaction solutions were incubated for 18 h at room temperature under stirring and then analyzed using SEC-HPLC (the chromatographic conditions listed here are as example; for a different protein the elution must be optimized using a case-by-case approach) with a Zorbax GF-250 column (4.6 × 250 mm; Agilent Technologies), eluted with 20 m*M* phosphate buffer, 130 m*M* NaCl (pH 7) at a flow rate of 0.5 mL/min and the effluent was monitored by measuring the absorbance at 280 nm. The elution profiles were similar; there was a peak at 5.3 min corresponding to the free protein

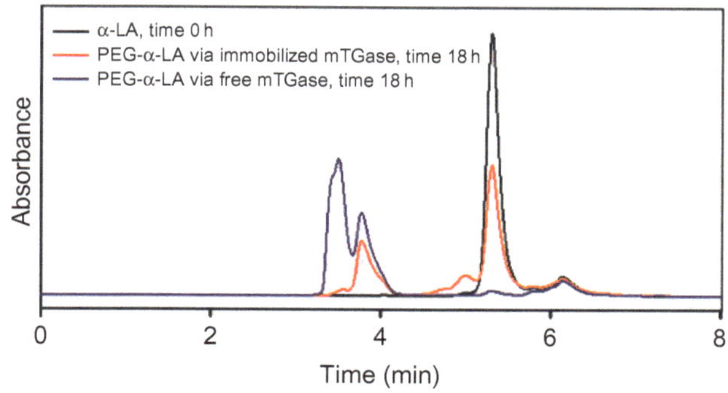

Fig. 6 SEC-HPLC profile of the reaction mixtures of PEG-α-LA after 18 h at pH 7.5 in the presence of immobilized mTGase (*red line*) and free mTGase (*blue line*) in comparison to α-LA (*black line*).

that disappeared during the course of the reaction with a concomitant appearance of a new product at lower retention times. In particular, the reaction mixture in the presence of the free enzyme led to the formation of two products with a complete disappearance of the free protein (Fig. 6). Interestingly, the reaction with immobilized enzyme yielded only one product (Fig. 6). The result can explained in two ways: (i) the steric entanglement between the PEG chains and the agarose beads carrying mTGase favors the formation of homogenous monoconjugates, avoiding biconjugation, but also (ii) a slower kinetic of reaction of immobilized enzyme can determine a slow formation of biconjugate that can be avoided stopping the reaction before its formation.

Overall, this result demonstrated that immobilized mTGase is a valid alternative to free mTGase. In fact, the immobilized enzyme can be eliminated by spinning down the agarose beads through centrifugation simplifying the purification protocol and decreasing the cost of purification.

In conclusion, it is important to keep in mind that each pharmaceutical protein with several Gln residues can be a mTGase substrate and employing the correct protocol (free-, immobilized mTGase or the presence of cosolvents) might allow the obtainment of well-defined and pure mono-PEGylated isomer product.

3.4.6 Conjugation Site Determination

1. Free and PEGylated protein (200 μg) are dissolved in denaturating buffer to reach a protein concentration of 1.0 mg/mL.

2. TCEP is added to the protein solution to reduce the disulfide bridges at the final concentration of 5 m*M*, and the reaction mixture is kept at 37°C for 1 h.
3. IA is added to the final concentration of 25 m*M* of the reduced protein. *S*-alkylation is allowed to proceed for 30 min at 37°C in the dark.
4. Protein samples are purified by RP-HPLC on a Phenomenex Jupiter C18 column (250 × 4.6 mm; 5 μm) at a flow rate of 1.0 mL/min, detection at 226 nm, eluted with a solvent gradient of water/ACN both containing 0.1% (v/v) TFA. Gradient 0′–5% ACN, 5′–40% ACN, 25′–80% ACN, 27′–90% ACN, 30′–5% ACN (the chromatographic conditions listed here are as example; for a different protein the elution must be optimized using a case-by-case approach). The collected peaks are lyophilized.
5. The reduced and *S*-carboxamidomethylated samples of free and PEGylated protein are dissolved in 8 *M* urea and diluted in 50 m*M* phosphate buffer (pH 7.9) to reach a final protein concentration of 1 mg/mL and urea concentration of 0.8 *M*.
6. An aliquot of trypsin is then added at an E/S ratio of 1:50 (w/w), and the proteolysis is allowed to proceed at 37°C overnight (the trypsin must be of sequencing grade).
7. The digestion mixtures, desalted by PepClean C-18 Spin columns, are analyzed using an LC-MS instrument (i.e., UPLC-Q-TOF) with a Grace Vydac TP C18 column (150 × 1 mm; 5 μm), maintained at 32°C, flow rate 0.05 mL/min, detection at 280 nm, eluted with a solvent gradient of water/ACN both containing 0.1% formic acid. Gradient 3′–3% ACN, 24′–80% ACN, 28′–80% ACN, 29′–3% ACN, 35′–3% (the chromatographic conditions listed here are as example; for a different protein the elution must be optimized using a case-by-case approach).
8. The peaks obtained in LC-MS, each corresponding to a peptide fragment of the native and PEGylated protein, are compared. The masses of all the peptide fragments of native and conjugated protein are determined experimentally and must be very close to the theoretical expected values (a reference digestion table can be found in on-line softwares such as ExPASy by uploading the sequence of the target protein and the enzyme used for proteolysis).
9. The peak that is not identified in the PEGylated digest but is present in the native protein digest contains the site of PEG coupling (if a missed

fragment contains more than one glutamine, MS/MS analysis must be carried out) (Mero et al., 2016).

3.4.6.1 Practical Experiment

PEG-Z-QG-G-CSF (also termed PEG-K41-G-CSF), those synthesis and purification was reported in Section 3.4.3, was proteolytic digested to determine the site of PEG conjugation in comparison with proteolytic digestion profile of G-CSF. After following the protocol above reported the peptide fragments were analyzed by Xevo G2-S Q-Tof spectrometer. The instrument was operated in the ESI-positive ion, resolution mode and with a detection window between 50 and 2000 *m/z*. Source parameters were: capillary (kV) 1.5, sampling cone voltage 30.0 V, and source offset of 80 V. MS^E acquisition was performed by alternating two MS data functions: one for acquisition of the peptide mass spectra with the collision cell at low energy (6 V) and the other for the collection of the peptide fragmentation spectra with the collision cell at elevated energy (linear ramp 20–40 V). Analyses were performed with LockSpray™ using a solution of 1 ng μL^{-1} Leu-Enk in 50:50 ACN/water containing 0.1% formic acid, sampled every 45 s. MS^E data were processed with the BiopharmaLynx 1.3.4 Software (Waters) setting trypsin as a digest reagent and five missed cleavages. The MS ion intensity threshold was set to 250 counts, and the MS^E threshold was set to 100 counts. Both MS mass match tolerance and MS^E mass match tolerance were set to 15 ppm. Conjugated peptides were confirmed by at least four MS/MS *b*/*y* fragment ions. ESI-LC-MS^E peptide mapping of the digests made it possible to achieve a yield sequence coverage of 35.6% and 39.7% for PEGylated and free protein, respectively. This low sequence coverage can be attributed to the fact that trypsin digestion of G-CSF forms the 42–147 fragment (11.5 kDa) that was undetected by the ESI-LC-MS^E method used. This fragment is nevertheless irrelevant because it does not contain free Lys residues. The fragments identified are outlined in Table 4. Three out of four fragments containing Lys residues (K17, K24, K35) were present in the digested mixture of the native and PEGylated proteins, while fragment 36–41 was not present in PEGylated G-CSF, thus confirming the selective modification of K41. Providing further evidence of selective modification, MS^E uncovered the presence of 1–17 and 24–35 fragments for both products with almost the complete *b* and *y* series, while the *b*/*y* fragment ions for the 36–41 fragments were present only in the G-CSF profile.

Table 4 Tryptic Peptides of PEGylated and Native G-CSF With Molecular Masses Determined by LC-MS^E

Tryptic Peptide	Peptide Sequence	Molecular Mass (Da)		
		Found in G-CSF[a]	Found in PEG-G-CSF[a]	Calculated[b]
1–17	MTPLGPASSLPQSFLLK	1785.9698	1785.9728	1785.9698
18–23	CLEQVR	803.3960	803.3962	803.4032
24–35	KIQGDGAALQEK	1256.6725	1256.6776	1256.7633
36–41	LCATYK	745.3684	—[c]	754.3756
149–167	AGGVLVASHLQSFLEVSYR	2032.0741	2032.0825	2032.0776
149–170	AGGVLVASHLQSFLEVSYRVLR	2400.3276	2400.3342	2400.3345
171–175	HLAQP	546.3023	546.3020	564.3020

[a]Experimental molecular mass determined by LC-MSE.
[b]Molecular mass calculated by Expasy.
[c]Fragment 36–41 was not detected in the mass spectrum of the tryptic digest of PEGylated G-CSF.
Adapted from Mero, A., Grigoletto, A., Maso, K., Yoshioka, H., Rosato, A., & Pasut, G. (2016). Site-selective enzymatic chemistry for polymer conjugation to protein lysine residues: PEGylation of G-CSF at lysine-41. *Polymer Chemistry*, 7, 6545–6553.

REFERENCES

Ando, H., Adachi, M., Umeda, K., Matsuura, A., Nonaka, M., Uchio, R., et al. (1989). Purification and characterization of a novel transglutaminase derived from microorganisms. *Agricultural and Biological Chemistry*, *53*, 2613–2617.

Buchardt, H., Selvig, P. F., Nielsen, J., & Langeland, N. (2010). Transglutaminase-mediated methods for site-selective modification of human growth hormone. *Peptide Science*, *94*, 229–235.

Folk, J. E. (1983). Mechanism and basis for specificity of transglutaminase-catalyzed ε-(−γ-glutamyl)lysine bond formation. *Advances in Enzymology and Related Areas of Molecular Biology*, *54*, 1–56.

Fontana, A., Spolaore, B., Mero, A., & Veronese, F. M. (2008). Site-specific modification and PEGylation of pharmaceutical proteins mediated by transglutaminase. *Advanced Drug Delivery Reviews*, *60*, 13–28.

Griffin, R., Casadio, R., & Bergamini, C. M. (2002). Tranglutaminases: Nature's biological glues. *Biochemistry Journal*, *368*, 377–396.

Grossowicz, N., Wainfan, E., Borek, E., & Waelsch, H. (1950). The enzymatic formation of hydroxamic acids from glutamine and asparagines. *The Journal of Biological Chemistry*, *187*, 111–125.

Jevševar, S., Kunstelj, M., & Gaberc Porekar, V. (2010). PEGylation of therapeutic proteins. *Journal of Biotechnology*, *5*, 113–128.

Kanaji, T., Ozaki, H., Takao, T., Kawajiri, H., Ide, H., Motoki, M., et al. (1993). Primary structure of microbial transglutaminase from Streptoverticillium sp. strain s-8112. *The Journal of Biological Chemistry*, *268*, 11565–11572.

Kashiwagi, T., Yokoyama, K., Ishikawa, K., Ono, K., Ejima, D., Matui, H., et al. (2002). Crystal structure of microbial transglutaminase from Streptoverticillium mobaraense. *The Journal of Biological Chemistry*, *277*, 44252–44260.

Mero, A., Clementi, C., & Pasut, G. (2011). Covalent conjugation of poly(ethylene glycol) to proteins and peptides: Strategies and methods. *Methods in Molecular Biology*, *751*, 95–129.

Mero, A., Fang, Z., Pasut, G., Veronese, F. M., & Viegas, T. X. (2012). Selective conjugation of poly(2-ethyl 2-oxazoline) to granulocyte colony stimulating factor. *Journal of Controlled Release*, *159*, 353–361.

Mero, A., Grigoletto, A., Maso, K., Yoshioka, H., Rosato, A., & Pasut, G. (2016). Site-selective enzymatic chemistry for polymer conjugation to protein lysine residues: PEGylation of G-CSF at lysine-41. *Polymer Chemistry*, *7*, 6545–6553.

Mero, A., Schiavon, M., Veronese, F. M., & Pasut, G. (2011). A new method to increase selectivity of transglutaminase mediated PEGylation of salmon calcitonin and human growth hormone. *Journal of Controlled Release*, *54*, 27–34.

Mero, A., & Veronese, F. M. (2008). The impact of PEGylation on biological therapies. *BioDrugs: Clinical Immunotherapeutics, Biopharmaceuticals and Gene Therapy*, *22*, 315–329.

Pasut, G., & Veronese, F. M. (2012). State of the art in PEGylation: The great versatility achieved after forty years of research. *Journal of Controlled Release*, *161*, 461–472.

Sato, H., Hayashi, E., Yamada, N., Yatagai, M., & Takahara, Y. (2001). Further studies on the site-specific protein modification by microbial transglutaminase. *Bioconjugate Chemistry*, *12*, 701–710.

Snyder, S. L., & Sobocinski, P. Z. (1975). An improved 2,4,6,-trinitrobenzenesulfonic acid method for the determination of amines. *Analytical Biochemistry*, *64*, 284–288.

Turecek, P. L., Bossard, M. J., Schoetens, F., & Ivens, I. A. (2016). PEGylation of biopharmaceuticals: A review of chemistry and nonclinical safety information of approved drugs. *Journal of Pharmaceutical Sciences*, *105*, 460–475.

Washizu, K., Ando, K., Koiked, S., Hiros, S., Matsuura, A., Akagi, H., et al. (1994). Molecular cloning of the gene for microbial transglutaminase from Streptoverticillium and its expression in Streptomyces lividans. *Bioscience, Biotechnology, and Biochemistry*, *58*, 82–87.

Yokohama, K., Nio, N., & Kikuchi, Y. (2004). Properties and applications of microbial transglutaminases. *Applied Microbiology and Biotechnology*, *64*, 447–454.

Zhao, X., Shaw, A. C., Wang, J., Chang, C. C., Deng, J., & Su, J. (2010). A novel high-throughput screening method for microbial transglutaminases with high specificity toward Gln141 of human growth hormone. *Journal of Biomolecular Screening*, *15*, 206–212.

Zhu, Y., Rinzema, A., & Tramper, J. (1995). Microbial transglutaminases: A review of its production and application in food processing. *Applied Microbiology and Biotechnology*, *44*, 277–282.

CHAPTER FIFTEEN

Polymer-Based Protein Engineering: Synthesis and Characterization of Armored, High Graft Density Polymer–Protein Conjugates

Sheiliza Carmali*,†, Hironobu Murata*,†, Chad Cummings*,†, Krzysztof Matyjaszewski*,†, Alan J. Russell*,†,1

*Center for Polymer-Based Protein Engineering, ICES, Carnegie Mellon University, Pittsburgh, PA, United States

†Carnegie Mellon University, Pittsburgh, PA, United States

[1]Corresponding author: e-mail address: alanrussell@cmu.edu

Contents

Methods in Enzymology, Volume 590
ISSN 0076-6879
http://dx.doi.org/10.1016/bs.mie.2016.12.005

© 2017 Elsevier Inc.
All rights reserved.

Abstract

Atom transfer radical polymerization (ATRP) from the surface of a protein can generate remarkably dense polymer shells that serve as armor and rationally tune protein function. Using straightforward chemistry, it is possible to covalently couple or display multiple small molecule initiators onto a protein surface. The chemistry is fine-tuned to be sequence specific (if one desires a single targeted site) at controlled density. Once the initiator is anchored on the protein surface, ATRP is used to grow polymers on protein surface, in situ. The technique is so powerful that a single-protein polymer conjugate molecule can contain more than 90% polymer coating by weight. If desired, stimuli-responsive polymers can be "grown" from the initiated sites to prepare enzyme conjugates that respond to external triggers such as temperature or pH, while still maintaining enzyme activity and stability. Herein, we focus mainly on the synthesis of chymotrypsin–polymer conjugates. Control of the number of covalently coupled initiator sites by changing the stoichiometric ratio between enzyme and the initiator during the synthesis of protein–initiator complexes allowed fine-tuning of the grafting density. For example, very high grafting density chymotrypsin conjugates were prepared from protein–initiator complexes to grow the temperature-responsive polymers, poly(*N*-isopropylacrylamide), and poly[*N*,*N*′-dimethyl(methacryloyloxyethyl) ammonium propane sulfonate]. Controlled growth of polymers from protein surfaces enables one to predictably manipulate enzyme kinetics and stability without the need for molecular biology-dependent mutagenesis.

1. INTRODUCTION

Protein-based products have been used commercially for centuries in industries such as food, fuel, paper, and therapeutics. Chemical activity under ambient conditions is the key to why proteins are ubiquitous enablers of health and wealth creation. However, proteins have evolved to be mostly unstable, since living systems control their environment by deciding whether or not to replace proteins they have previously made. Instability and specificity of enzymes toward only those reactions that nature intended have combined to limit where enzymes can be applied. The field of enzyme engineering has been evolving to rationally redesign nature to meet our needs. For the most part, enzyme engineering is accomplished either by molecular biology-based tools to redesign the protein or by exquisite control of formulation. Enzymes in particular have been engineered for enhanced activity, stability, and specificity but with only a limited success.

An enzyme catalyzes a reaction by removing a substrate from a solvent (typically water), converting it into a product, and then returning that product to the solvent. One can engineer this catalytic process by changing the enzyme (Peracchi, 2001; Turner, 2009), the solvent (Dordick, 1991; Wang et al., 2016), or the interface between the two (the protein surface) (Russell & Fersht, 1987; Russell, Thomas, & Fersht, 1987). Over the last 20 years, we have learned how to lock the enzyme in an active conformation by using enzymes as comonomers in polymerization reactions. We have generated remarkably stable urethane and acrylate protein–polymer networks (Drevon et al., 2002; Drevon, Hartleib, Scharff, Rüterjans, & Russell, 2001; Drevon, Urbanke, & Russell, 2003; LeJeune & Russell, 1996). The urethane network approach, in particular, has been an important tool in expanding the use of enzymes to degrade toxic compounds (LeJeune, Wild, & Russell, 1998). Proteins have previously been reacted with preexisting polymers. In particular, there is an extensive literature describing the generation of proteins that have had poly(ethylene glycol) (PEG) grafted to their surface (Jevševar, Kunstelj, & Porekar, 2010; Turecek, Bossard, Schoetens, & Ivens, 2016; Veronese & Pasut, 2005).

Protein and nucleic acid syntheses in the laboratory demonstrated that the exquisite control of functionality found in living systems is achieved through controlled sequential polymerization of small molecules. We, therefore, began to explore how to create controlled polymerization initiation sites on the surface of enzymes and use them to build armor around the enzyme to enhance specific properties of enzymes by a rational approach.

1.1 "Grafting To" vs "Grafting From"

As described earlier, protein–polymer conjugates in which polymers are covalently coupled to the surface of a protein can be prepared in two contrasting ways: "grafting to" or "grafting from" (Fig. 1). The more conventional "grafting to" approach consists of the use of presynthesized, end-functionalized polymers that are coupled to accessible complementary functionalities such as amino acid side chains or end termini on the protein surface. PEG is a nontoxic, nonimmunogenic, and nonbiodegradable polymer that has been extensively used to overcome some of the limitations associated with proteins (Abuchowski, van Es, Palczuk, & Davis, 1977; Davis, 2002; Foster, 2010; Koussoroplis et al., 2014; Mei et al., 2010). PEG is thought to impart stealth properties on the protein by reducing immunogenicity and increasing in vivo stability by slowing renal clearance and degradation. PEGylation leads to an apparent increase in weight and size and also provides steric hindrance toward degrading agents (Veronese

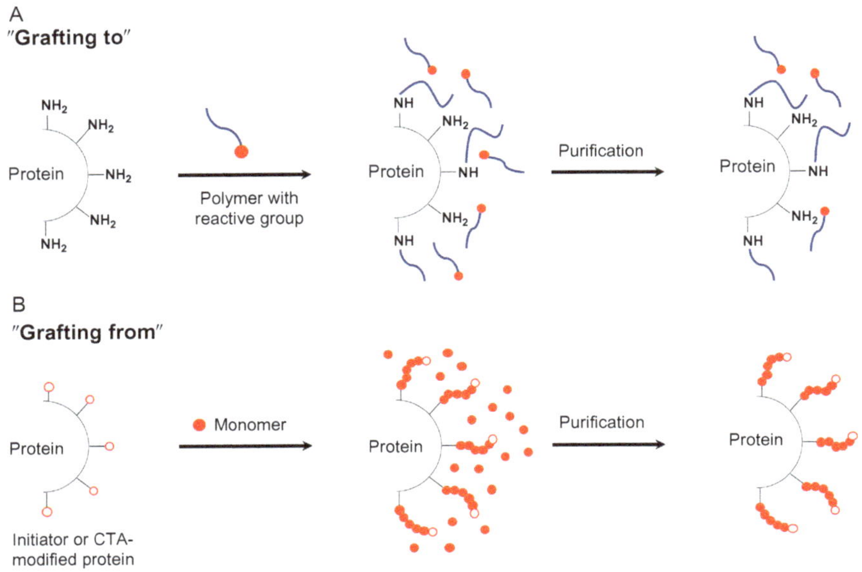

Fig. 1 Schematic representation of "grafting to" and "grafting from" methods of conjugation.

& Pasut, 2005). As a result, a decrease in kidney clearance and proteolysis is observed for PEG–protein conjugates, leading to longer circulation times. The water-soluble nature of PEG also improves protein solubility and stability by reducing propensity for protein aggregation (Harris, 2013).

Since the Food and Drug Administration (FDA) approval of the first PEG–protein conjugate in the early 1990s, more than 10 PEGylated biopharmaceuticals are now available on the market (Alconcel, Baas, & Maynard, 2011). PEG-interferon alfa-2b (PEG-Intron) is currently used as first-line treatment for chronic hepatitis C (Manns et al., 2001). PEGasys (PEG-interferon alfa-2a), a highly effective therapy for chronic hepatitis B, can provide sustained remission in a significant number of patients following a 48-week period (Santantonio & Fasano, 2014).

Despite the widespread use of PEG as the "gold standard" in protein–polymer conjugate therapeutics, PEG is not without disadvantages; PEG can elicit immune response (Ishida & Kiwada, 2013), lead to accumulation in tissues and accelerated blood clearance upon repeated exposure (Qi & Chilkoti, 2015; Yamaoka, Tabata, & Ikada, 1994; Yang & Lai, 2015). Additionally, the nonbiodegradability of PEG can form vacuoles in the liver, kidney, and spleen after administration (Yamaoka et al., 1994). To address these issues, degradable polymers have been developed, in particular, hydroxyethyl starch (HES). The conjugation of HES to erythropoietin (EPO) showed comparable in vitro and in vivo activities to PEGylated-EPO (Mircera) and a threefold increase in circulation time when compared to native EPO (Pelegri-O'Day, Lin, & Maynard, 2014). HESylation of G-CSF and interferon-α have also shown comparable results (Pelegri-O'Day et al., 2014).

While a promising and fruitful approach, the main limitation of "grafting to" is the extent of modification that can occur on a protein surface. Typically, a higher degree of modification will extend the circulation half-life and reduce the likelihood of antigenicity (Gaertner & Offord, 1996). However, in the "grafting to" method, once a polymer chain is covalently attached to a protein surface, the steric impediment of the polymers chains will prevent further modification. Previous studies with chymotrypsin (CT) have shown that the extent of PEGylation was limited to a maximum of nine polymer chains (5 kDa each) per enzyme molecule (Rodríguez-Martínez et al., 2008). Assuming that CT has an approximate surface area of 50 nm^2, this represents a polymer density of one potential polymer per 6 nm^2 (0.18 chains/nm^2). Naturally, as the polymer molecular weight increases the ability to react more than a few chains per protein molecule diminishes rapidly. For a 25-kDa PEG, one would expect to react just one chain per 50 nm^2. To further increase the polymer density without restraints on the molecular weight of the polymer, an alternative grafting method would be required.

Controlled radical polymerization has been used to initiate polymerization directly from the surface of proteins in a "grafting from approach." As seen in Fig. 2, atom transfer radical polymerization (ATRP) has been increasingly used for protein surface modification (Campbell et al., 2016; Cummings, Murata, Koepsel, & Russell, 2013, 2014; Cummings, Murata, Matyjaszewski, & Russell, 2016; Gao et al., 2009; Heredia et al., 2005; Lele, Murata, Matyjaszewski, & Russell, 2005; Nicolas, San Miguel, Mantovani, & Haddleton, 2006; Qi, Amiram, Gao, McCafferty, & Chilkoti, 2013). Along with reversible-addition fragmentation chain transfer (RAFT) (De, Li, Gondi, & Sumerlin, 2008; He and He, 2009; Liu et al., 2007), these techniques provide a large library of monomers and these can be used under biologically relevant reaction conditions (aqueous solvent and ambient temperature).

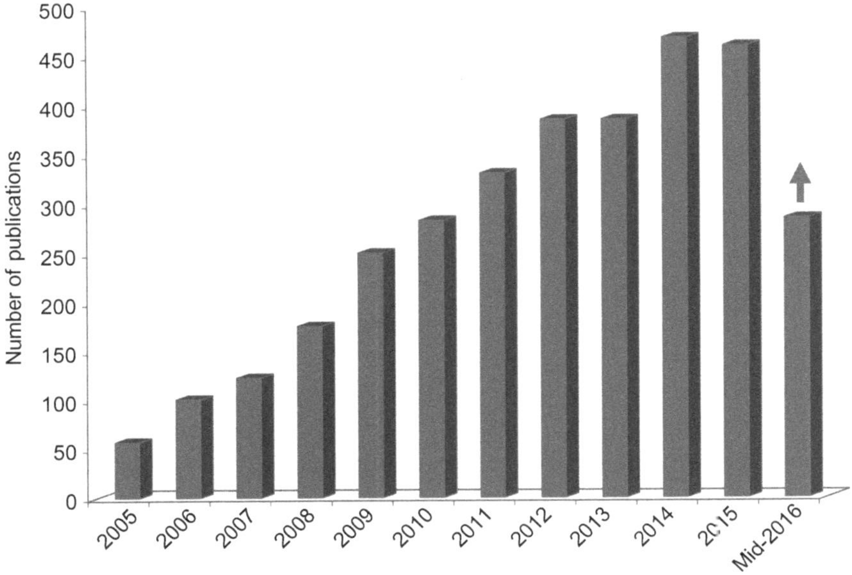

Fig. 2 Plot showing number of papers published on protein-initiated ATRP in the last decade. Over the last 10 years, protein surface-initiated ATRP has led to publication of over 3000 peer-reviewed papers. Data obtained from https://www.scopus.com using keywords "protein ATRP" and "initiated." Search was specified to have keywords within text but not in references.

1.2 The Rational Approach to Protein–Polymer Conjugate Synthesis

1.2.1 Targeting Site of Modification

The rational design of conjugates requires the control of several factors including site of modification, grafting density, and polymer selection. The most common sites on a protein for modification include α and ε amino groups of lysine residues (Kalkhof & Sinz, 2008; Nakamura, Kawai, Kitamoto, Osawa, & Kato, 2009), N-terminal amine group (Chan et al., 2012; Dawson, Muir, Clark-Lewis, & Kent, 1994), histidines (Chen et al., 2003), thiol group of cysteines (Chalker, Bernardes, Lin, & Davis, 2009; Nathani et al., 2013; Shen et al., 2012), and sulfur atoms released from disulfide bonds (Brocchini et al., 2008; Smith et al., 2010). Bioconjugation has recently focused on site specificity upon modification. Site-specific control over the location on the protein prevents heterogeneity, ensuring that protein bioactivity and structure are not compromised. One site-selective approach includes cysteine-targeted modification (Gualberto, 2012;

Senter & Sievers, 2012; Verma et al., 2012). Cysteines are less abundant in proteins, typically exist as a disulfide pair and when free, are frequently not surface accessible (Lins, Thomas, & Brasseur, 2003).

Targeted approaches also include the insertion of noncanonical amino acids at specific sites of recombinant proteins using genetic code expansion (Averick, Paredes, et al., 2012; Peeler et al., 2010). Using this strategy, translational components were evolved to genetically encode 4-(2′-bromoisobutyramido)-phenylalanine initiator into green fluorescent protein (GFP) (Peeler et al., 2010). GFP–initiator complex was then polymerized with oligo(ethylene oxide) monomethyl ether methacrylate ($OEO_{300}MA$), generating a GFP-poly(oligo(ethylene glycol) methyl ether methacrylate) conjugate with one polymer chain "grown from" the selected initiator site (Peeler et al., 2010). This method allows for the homogenous preparation of protein conjugates with polymers in a quantitative and site-specific approach.

1.2.2 Tuning Grafting Density

"Grafting from" lysine residues have been extensively explored with ATRP and RAFT techniques to grow the polymers (Table 1). The numerous surface accessible lysine residues present in proteins enable the synthesis of extremely dense polymer coats, effectively "nanoarmoring" the proteins.

As can be observed in Table 1, the majority of the initiators used to date have suffered from low solubility in aqueous solution. Poorly water-soluble initiators require immobilization reactions to be conducted in mixtures of water and organic solvents (Lele et al., 2005), which can lead to inactivation and/or the denaturation of enzymes (Averick, Simakova, et al., 2012; Magnusson, Bersani, Salmaso, Alexander, & Caliceti, 2010; Nicolas et al., 2006; Yaşayan et al., 2011). These initiators also do not allow control over the extent of modification (Lele et al., 2005). To improve upon the initiator–protein reaction, water-soluble ATRP initiators have been described (Averick, Simakova, et al., 2012; Cummings et al., 2013). Typically, initiators contain a protein-reactive moiety, such as an *N*-hydroxysuccinimide ester, and a transfer agent (typically a halogen, such as Br, Cl, or F) that initiates the controlled polymerization reaction.

Immobilization reactions with water-soluble ATRP initiator have been shown to be highly efficient and lead to the preparation of high-density protein–polymer conjugates (Cummings et al., 2013, 2014; Murata, Cummings, Koepsel, & Russell, 2013). Grafting density, as depicted in Fig. 3, refers to the density of polymer chains per unit surface area of protein.

Table 1 Examples of ATRP Initiators and RAFT Agents Employed in the Synthesis of Bioconjugates (Le Droumaguet & Nicolas, 2010)

Initiator	Solubility	Polymer	Total Sites Modified (%)	References
ATRP				
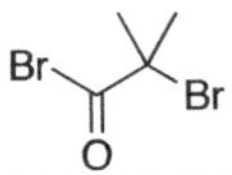	PBS/CH_2Cl_2	p(mPEGMA)	50	Lele et al. (2005)
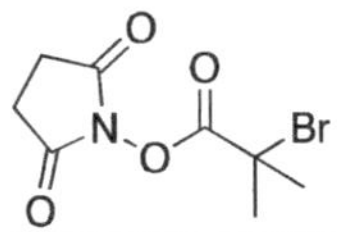	Dry DMSO	p(mPEGMA)	85	Lecolley et al. (2004)
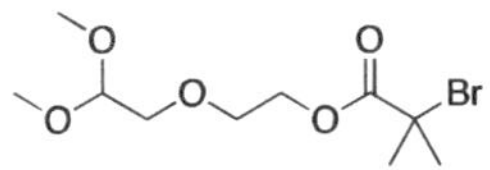	Toluene	p(mPEGMA)	15	Tao, Mantovani, Lecolley, and Haddleton (2004)
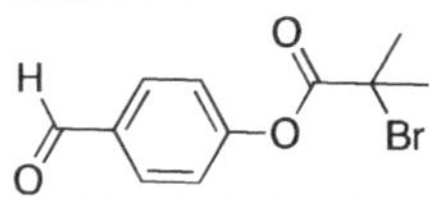	DMSO/MeOH	MPC	15	Samanta et al. (2008)
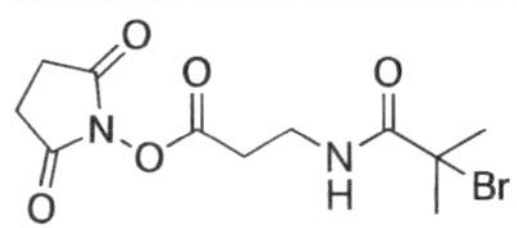	PBS buffer	pNIPAm	90	Cummings et al. (2013)
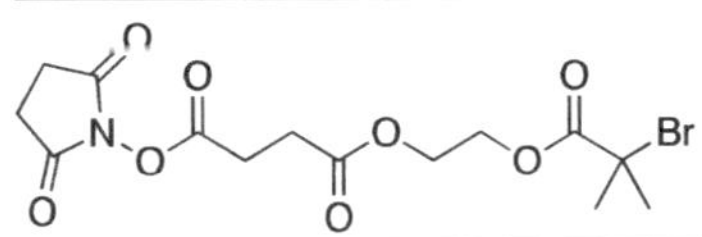	H_2O PBS buffer	p(mPEGMA)	50	Averick, Simakova, et al. (2012)

RAFT				
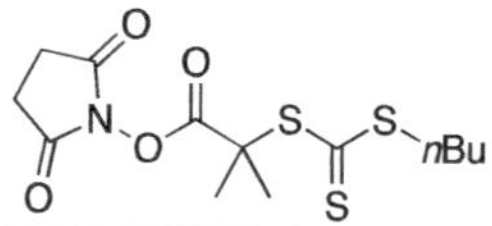	H_2O/MeOH	pNIPAm-*b*-pDMA	30	Li, Li, Yu, Bapat, and Sumerlin (2011)
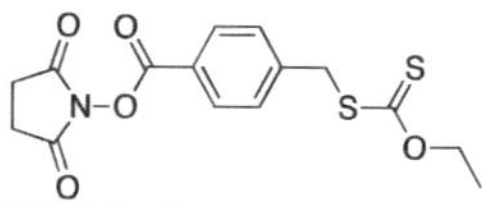	Dry DMSO	PVP	90	McDowall, Chen, and Stenzel (2008)
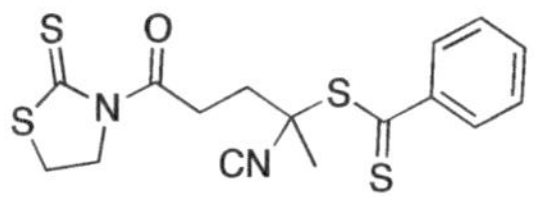	*N*,*N*-dimethylacetamide	pHPMA	~90	Xu, Boyer, Bulmus, and Davis (2009)

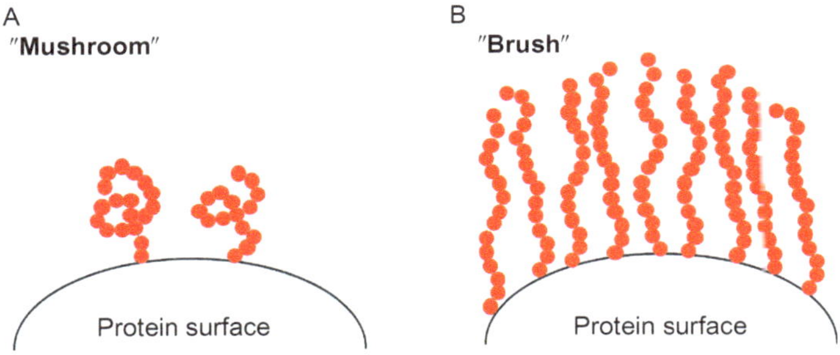

Fig. 3 Protein–polymer conformations with (A) low grafting density and (B) high grafting density.

This density will dictate the conformational structure that the polymers will adopt on the protein surface (Brittain & Minko, 2007). At low grafting densities, each polymer chain is isolated, adopting a diffuse "mushroom" conformation. This is often the case for "grafted to" conjugates. Once a first polymer chain is "grafted to" the protein surface steric hindrance will often limit further polymer binding to nearby sites on the protein surface.

At low surface coverage, the polymer will armor the protein by wrapping around the molecular surface of the protein (Li, Manjula, & Acharya, 2006). For conjugates in which the majority of the targeting residues are modified, the increase in grafting density leads the polymers to assume a "brush" conformation, when the polymer chains stretch away from the interface to avoid overlap and provide a more effective coverage of the protein surface. This results in enhanced stability, extreme protease resistance, and better ability to evade clearance by immune cells (Yang et al., 2014).

Grafting density can also be fine-tuned using surface-initiated ATRP. Changes in the stoichiometric ratio, pH, and reaction time can allow protein–initiator complexes to be prepared to controlled extents of modification. This can be particularly useful in situations where lysine residues can be involved or close to the active site of an enzyme and modification can lead to a decrease in activity (Lucius et al., 2016). Initiator immobilization can also be used to gain knowledge on the reactivity of lysine residues in a given protein. Lysine residues are potent nucleophiles under basic conditions but can show changes in reactivity as a function of their surface properties and surrounding environment (Hnízda, Šantrůček, Šanda, Strohalm, & Kodíček, 2008). Structural and chemical characterization of protein–initiator complexes can lead to a clearer understanding of differences in reactivity between lysines and how to modulate and/or site-direct modification at a given lysine residue.

1.2.3 Polymer Selection

It is of much current interest to grow polymers with their own unique properties from the surface of proteins. One important class of polymers that have been recently explored in protein modification are stimuli-responsive polymers (Hoffman & Stayton, 2007). These polymers can respond to external triggers, such as pH (Lackey et al., 1999; Strozyk, Chanana, Pastoriza-Santos, Pérez-Juste, & Liz-Marzán, 2012; Zhang, Vanparijs, Louage, De Geest, & Hoogenboom, 2014), temperature (Cummings et al., 2013, 2014), or light (Wu, Ng, Kuan, & Weil, 2015) by changing their conformation or chemical composition. This reversible response is used as an on–off switch for protein bioactivity, solubility, or stability, depending on the application. Specific polymer choices for tailored applications, such as increased substrate affinity (Keefe & Jiang, 2012), enhanced activity in the GI tract (Fuhrmann et al., 2013), higher activity at nonnative pH (Murata et al., 2013), or increased solubility in organic solvents (Cummings et al., 2016; Konieczny, Krumm, Doert, Neufeld, & Tiller, 2014; Konieczny, Leurs, & Tiller, 2015; Zore, Lenehan, Kumar, & Kasi, 2014), are becoming more common as knowledge about polymer and protein interactions progresses.

Herein, we place emphasis on how polymer-based protein engineering can be used for the nanoarmoring of proteins with a model protein, α-CT. To achieve maximum reaction efficiency between protein and initiator, water-soluble ATRP initiators are used. The number of covalently attached initiators to the protein surface can be controlled by varying the stoichiometric ratio between protein and initiator to allow fine-tuning of the grafting density in protein conjugates. Structural differences in initiators can also be useful in modulating grafting density. Stimuli-responsive polymers can then be "grown from" the surfaces of these high-density CT–initiator complexes as a way to predictably alter enzyme structure and function. We have found that ATRP reaction conditions can be manipulated to predictably tune polymer chain length and molecular weight while maintaining low dispersity and high uniformity.

2. PROTOCOLS

General lab safety procedures and personal protective equipments (e.g., lab coat and safety glasses) should be used during all experiments.

2.1 Equipment

- NewClassic Mettler Toledo balance
- Corning pH meter 430
- PerSeptive Voyager STR MS for MALDI-TOF-MS
- Waters 2695 Series with a data processor for gel permeation chromatography (GPC) analysis
- Malvern Zetasizer nano-ZS used to measure hydrodynamic radius
- Perkin Elmer Lambda 45 UV–vis spectrophotometer for protein assays

2.2 Measurements

^{1}H NMR spectra were recorded on a spectrometer (300 MHz, Bruker Avance) with deuterium oxide (D_2O), DMSO-d_6, and $CDCl_3$.

Matrix-assisted laser desorption ionization–time-of-flight mass spectroscopy (MALDI-TOF MS) measurements were recorded using a PerSeptive Voyager STR MS with nitrogen laser (337 nm) and 20 kV accelerating voltage. Sinapinic acid (20 mg/mL) in 50% acetonitrile with 0.1% trifluoroacetic acid was used as matrix. A gold sample plate was used for all samples. Apomyoglobin, cytochrome *c*, and aldolase were used as calibration samples. Degree of modification was determined by subtracting the CT conjugates m/z values from native CT m/z and dividing by the molecular weight of the initiator.

Number and weight average molecular weights (M_n and M_w) and the dispersity (M_w/M_n) were estimated by GPC on a Waters 2695 Series with a data processor, equipped with three columns (Waters Ultrahydrogel Linear, 500 and 250). Detection was conducted using a refractive index detector and detailed conditions for each analyzed polymer can be found in Table 2.

Dynamic light scattering (DLS) data were collected on a Malvern Zetasizer nano-ZS. Sample concentration was kept at 1.0 mg/mL. The hydrodynamic diameter of the native CT and the conjugate were measured three times (12 run to each measurement) at various pH.

UV–vis spectra were obtained and used for enzyme activity determination using a UV–vis spectrometer (Lambda 2, PerkinElmer) with a temperature-controlled cell holder.

2.3 Materials

α-CT from bovine pancreas (type II), copper (I) bromide, 1,1,4,7,10,10-hexamethyltriethylenetetramine (HMTETA), *N*-succinyl-L-alanine-L-alanine-L-proline-L-phenylalanine-*p*-nitoroanilide (Suc-AAPF-pNA), [2-(methacryloyloxy)ethyl]dimethyl-(3-sulfopropyl) ammonium hydroxide (DMAPS), [2-(methacryloylamino)ethyl]dimethyl-(3-sulfopropyl) ammonium

Table 2 Gel Permeation Chromatography Conditions for the Characterization of Cleaved Polymers pQA, pDMAEMA, pNIPAm, pDMAPS, and pSBAm

Polymer	Eluent	Flow Rate (mL/min)	Temperature	Standards for Calibration
GPC conditions				
pQA	0.1 *M* PBS with 0.2 vol.% TFA (pH 2.5)	0.5	r.t.	Poly (ethylene) glycol
pDMAEMA	0.1 *M* PBS with 0.2 vol.% TFA (pH 2.5)	0.5	r.t	Poly (ethylene) glycol
pDMAPS	80% 0.1 *M* PBS (pH 9.0)/ 20% acetonitrile with 0.01 vol.% NaN_3	1.0	r.t.	Polystyrene sulfonate
pNIPAm	DMF with 50 m*M* LiBr	1.0	50°C	Poly (ethylene) glycol
pSBAm	80% 0.1 *M* PBS (pH 9.0)/ 20% acetonitrile with 0.01 vol.% NaN_3	1.0	r.t.	Polystyrene sulfonate

hydroxide (sulfobetaine methacrylamide), pepsin from porcine stomach mucosa, copper (I) chloride, copper (I) bromide, and 4-hydroxy-3,5-dimethoxycinnamic acid (sinapinic acid) were purchased from Sigma Aldrich (St. Louis, MO) and used without further purification. *N*-isopropylacrylamide was purchased from Sigma Aldrich (St. Louis, MO) and purified by recrystallization using hexane. Me_6TREN was synthesized as described previously (Ciampolini & Nardi, 1966). Quaternary ammonium (QA) monomer (2-(dimethylethylammonium)ethyl methacrylate) was synthesized according to modified procedure previously reported (Tsarevsky, Pintauer, & Matyjaszewski, 2004). Dialysis tubing (molecular weight cut-off, 25, 15, and 1.0 kDa, Spectra/Por®, Spectrum Laboratories Inc., CA) for conjugate isolation was purchased from Fisher Scientific (Pittsburgh, PA).

2.4 Reaction Between ATRP Initiator and CT

2.4.1 Reagents

- *N*-2-bromo-2-methylpropanoyl-*β*-alanine *N'*-oxysuccinimide ester, synthesized as described before (Murata et al., 2013)
- 2,5-Dioxopyrrolidin-1-yl 3-(2-chloropropanamido)propanoate, synthesized as per described (Cummings et al., 2014)

- α-CT from bovine pancreas
- 0.1 *M* sodium phosphate buffer, pH 8.0
- Dialysis tubing MWCO 15,000 Da

2.4.2 Method

Two ATRP initiator molecules with distinct structural properties can be used to prepare CT–initiator conjugates. Bromine functionalized ATRP initiator was used primarily to prepare protein–homopolymer conjugates (Cummings et al., 2013; Murata et al., 2013), while chlorine-functionalized initiator is better suited for the preparation of protein–block–polymer conjugates (Cummings et al., 2014).

N-2-bromo-2-methylpropanoyl-*β*-alanine *N′*-oxysuccinimide ester (469 mg, 1.4 mmol) or 2,5-dioxopyrrolidin-1-yl 3-(2-chloropropanamido) propanoate (388 mg, 1.4 mmol) and CT (1.0 g, 0.04 mmol protein, 0.56 mmol $-NH_2$ group in lysine residues) were dissolved in 100 mL of 0.1 *M* sodium phosphate buffer, pH 8.0. The solution was stirred at 4°C for 3 h and then dialyzed against deionized water using dialysis tubing with a molecular mass cut-off of 15 kDa for 24 h at 4°C. After dialysis, CT-initiator was lyophilized.

Estimation of the extent of immobilization in CT–initiator conjugates can be achieved by several techniques. Spectrophotometric measurement of the residual amino groups using trinitrobenzene sulfonic acid (McGoff, Baziotis, & Maskiewicz, 1988) allows quantitation by comparison with a standard curve generated with an amine-containing compound prepared at various concentrations. Similarly, a fluorescamine protein dye can also be used (Udenfriend et al., 1972). Fluorescamine reacts more rapidly with primary amines than trinitrobenzene sulfone acid and provides high sensitivity for quantitation with low sample amounts. However, with this assay there is considerable protein-to-protein variation and the use of a purified protein as a standard is recommended.

Using ^1H NMR spectroscopy, quantitative comparison of the NMR peaks (integration of peaks) for the initiator molecule to that of the protein can also allow estimation of the degree of initiator immobilization. Highly modified CT–initiator complexes can allow a better visualization of the NMR peaks, and therefore more accurate quantification, for the initiator. For the characterization of CT–initiator conjugates, MALDI-TOF was used.

2.5 Surface Initiated ATRP From CT-Initiator

2.5.1 Reagents

- CT–initiator conjugate, prepared in Section 2.4.
- 0.1 *M* sodium phosphate buffer, pH 6.0
- 2-(Dimethylethylammonium) ethyl methacrylate (QA monomer)
- 3-Dimethyl(methacryloyloxyethyl) ammonium propane sulfonate (DMAPS)
- *N*-isopropylacrylamide (NIPAm)
- 2-(Dimethylamino) ethyl methacrylate (DMAEMA)
- 1,1,4,7,10,10-HMTETA
- Tris [2-(dimethylamino) ethyl] amine (Me_6TREN)
- Copper (I) bromide (Cu^IBr)
- Copper (I) chloride (Cu^ICl)
- Dialysis tubing MWCO 25,000 Da
- Deionized water

2.5.2 Method

2.5.2.1 Homopolymer Synthesis

For the synthesis of CT–pQA conjugates, solutions of QA monomer (284 mg, 1.07 mmol (for CT-pQA_{27}); 567 mg, 2.13 mmol (for CT-pQA_{54}); 1.13 g, 4.26 mmol (for CT-pQA_{108}); and 2.27 g, 8.52 mmol (for CT-pQA_{198})) and CT–initiator complex (100 mg, 0.043 mmol of initiator groups) in deionized water (30 mL) were sealed and bubbled with argon in an ice bath for 50 min. Deoxygenated catalyst solutions of HMTETA (24 μL, 0.2 mmol) and Cu^IBr (13 mg, 0.2 mmol) in deionized water (10 mL) were added to the conjugation reaction under argon bubbling. The mixture was sealed and stirred in the refrigerator for 4 h. CT–pQA conjugates were isolated by dialysis with a 25 kDa molecular mass cut-off dialysis tube in deionized water in a refrigerator for 24 h and then lyophilized.

To synthesize CT–pDMAPS conjugates, CT-initiator (50 mg, 0.024 mmol initiator) and DMAPS (335 mg, 1.2 mmol) were dissolved in sodium phosphate buffer (20 mL, pH 6.0). In a separate flask, HMTETA (33 mL, 0.12 mmol) was dissolved in deionized water (10 mL) and bubbled with argon for 10 min. Cu^IBr (17 mg, 0.12 mmol) was added to the HMTETA solution and argon was bubbled for an additional 50 min prior to addition of the copper catalyst solution. The solution was then stirred for 18 h at 4°C. Lastly, the solution was purified using dialysis tubing with a molecular mass cut-off of 25 kDa for 48 h against deionized water at 4°C and then lyophilized.

For CT-pNIPAm synthesis, CT-initiator (50 mg, 0.024 mmol initiator) and NIPAm (271 mg, 2.4 mmol) were dissolved in deionized water (20 mL). In a separate flask, Me_6TREN (32 μL, 0.12 mmol) was dissolved in deionized water (10 mL) and bubbled with argon for 10 min. Cu^IBr (17 mg, 0.12 mmol) was added to the Me_6TREN solution and argon was bubbled for an additional 10 min. The procedure for CT-pNIPAm synthesis from this point forward was as described earlier for CT-pDMAPS synthesis.

For CT-pDMAEMA synthesis, CT-initiator (100 mg, 0.046 mmol of initiator groups) and a solution of DMAEMA (1.35 mL, 8 0 mmol) in deionized water (30 mL) were sealed and bubbled with argon in an ice bath for 50 min. Deoxygenated catalyst solutions of HMTETA (55 μL, 0.2 mmol) and Cu^IBr (29 mg, 0.2 mmol) in deionized water (10 mL) were then added to the conjugation reactor under argon bubbling. The mixture was sealed and stirred for 18 h at 4°C to avoid self-polymerization of the DMAEMA. CT–pDMAEMA conjugates were isolated by dialysis with a 25 kDa molecular mass cut-off dialysis tube in deionized water in a refrigerator for 24 h and then lyophilized.

Note: Solutions of Cu^I can disproportionate in aqueous solution. The addition of Cu^I to the ligand solution should be done quickly and under argon atmosphere. The formation of insoluble precipitates indicates the presence of Cu^0 as a product of disproportionation.

2.5.2.2 Block Polymer Synthesis

To synthesize CT-pSBAm-*block*-pNIPAm conjugates, CT-initiator (50 mg, 0.029 mmol initiator) and SBAm (335 mg, 1.2 mmol (for CT-$pSBAm_{35}$); 525 mg, 1.8 mmol (for CT-$pSBAm_{50}$); and 701 mg, 2.4 mmol (for CT-$pSBAm_{90}$)) were dissolved in 0.1 *M* sodium phosphate buffer (20 mL, pH 6.0) with 35 mg NaCl (30 m*M*). In a separate flask, Me_6TREN (33 μL, 0.12 mmol) was dissolved in deionized water (5 mL) and bubbled with argon for 10 min. Cu^ICl (17 mg, 0.12 mmol) is added to the Me_6TREN solution and argon is bubbled for an additional 50 min prior to the addition of the copper catalyst solution to the monomer solution. After combining the two solutions, the reaction mixture is stirred for 5 h at 25°C until the reaction is stopped by exposing the solution to air. Finally, the solution is purified using dialysis tubing with molecular mass cut-off 25 kDa for 48 h against deionized water at 4°C and then lyophilized.

Following initial synthesis of CT–pSBAm conjugates of different lengths, pNIPAm was grown from CT-pSBAm using chain extension to yield CT-pSBAm-*block*-pNIPAm conjugates. CT-pSBAm conjugates (200 mg, 0.02 mmol of initiator for CT-pSBAm$_{35}$; 280 mg, 0.02 mmol of initiator for CT-pSBAm$_{50}$; 350 mg, 0.01 mmol of initiator for CT-pSBAm$_{90}$) and NIPAm (108 mg, 0.96 mmol; 163 mg, 1.44 mmol; 135 mg, 1.2 mmol, respectively) were dissolved in 0.1 *M* sodium phosphate buffer (20 mL, pH 6.0) with 35 mg NaCl (30 m*M*) and bubbled with argon. In a separate flask, Me_6TREN (10.7 μL, 0.05 mmol; 10.7 μL, 0.05 mmol; 6.4 μL, 0.03 mmol, respectively) was dissolved in deionized water (5 mL) and bubbled with argon for 10 min. $Cu^{I}Cl$ (4 mg, 0.04 mmol; 4 mg, 0.04 mmol; 2.4 mg, 0.03 mmol, respectively) was added to the Me_6TREN solution and argon was bubbled for an additional 50 min. The Me_6TREN/CuCl solution was quickly transferred to the CT-pSBAm/NIPAm solution and reaction was allowed to proceed for 5 h at 25°C. The reaction was stopped by quenching with air and the reaction mixture purified using dialysis tubing with molecular mass cut-off 25 kDa for 48 h against deionized water at 4°C and then lyophilized.

Note: Argon bubbling of aqueous enzyme solution can strongly enhance enzyme inactivation. Care should be taken to use a low flow rate of argon to prevent loss of catalytic activity. Under the described conditions, CT has been found to retain full activity.

2.6 Polymer Cleavage From CT Surface and Characterization

2.6.1 Reagents

- CT–polymer conjugates, prepared in Section 2.5
- 6N hydrochloric acid
- 4.5N *p*-toluene sulfonic acid
- Dialysis tubing MWCO 1000 Da

2.6.2 Method

pQA, pDMAPS, pNIPAm, and pDMAEMA were cleaved from the surface of CT using acid hydrolysis. For pQA and pDMAEMA, CT conjugates (10–20 mg) and 6N HCl (2–3 mL) were placed in separate hydrolysis tubes. After three freeze–pump–thaw cycles, hydrolysis was performed at 110°C for 24 h in vacuum. For pDMAPS, CT-conjugate was incubated (15 mg/mL) in 6N HCl at 110°C under vacuum for 24 h. For homopolymer and block polymer pNIPAm, CT-pNIPAm, and CT-pSBAm-*block*-pNIPAm (20 mg/mL) conjugate were each incubated separately in 4.5N *p*-toluene sulfonic acid at 80°C under vacuum for 72 h. Following incubation, all hydrolyzed samples were isolated from CT using dialysis tubing (MWCO 1 kDa) for 48 h and lyophilized.

Number and weight average molecular weights (M_n and M_w) and the dispersity (M_w/M_n) for homopolymers were estimated by GFC (Table 2). For pSBAm-*block*-pNIPAm cleaved from CT, M_n was calculated by quantitatively comparing NMR peaks (integration of peaks) of copolymer to cleaved first block pSBAm NMR spectra.

2.7 CT and CT-Conjugate Biocatalytic Activity

2.7.1 Reagents

- α-CT from bovine pancreas
- CT–polymer conjugates, prepared in Section 2.5
- *N*-succinyl-Ala-Ala-Pro-Phe *p*-nitroanilide (Suc-AAPF-pNA)
- 0.1 *M* phosphate buffer, pH 8.0

2.7.2 Method

In a 1 mL cuvette, 0.1 *M* sodium phosphate buffer (810–990 μL, pH 8.0, incubated at 25°C), substrate (0–180 μL, 6 mg/mL in DMSO ($0–1.2 \times 10^{-3}$ *M*)), and enzyme (10 μL, 0.1 mg enzyme/mL 0.1 *M* pH 8.0 sodium phosphate buffer (4×10^{-8} *M*)) were mixed. The rate of the hydrolysis was determined by recording the increase in absorbance at 412 nm for the first 30 s after mixing. K_M and k_{cat} values were calculated using EnzFitter software when plotting substrate concentration vs initial rate.

2.8 Inhibitor Binding

2.8.1 Reagents

- α-CT from bovine pancreas
- CT-pQA, prepared in Section 2.5
- Aprotinin (AP, bovine, recombinant, expressed in *Nicotiana*)
- Bowman–Birk CT inhibitor from *Glycine max* (GM)

2.8.2 Method

The effect of the dense cationic pQA shell surrounding the surface of CT toward inhibitor binding of positively and negatively charged inhibitors was examined. At the tested pH, AP, and GM had contrasting surface charges; these protein inhibitors had a positive and negative surface charge, respectively.

Enzyme inhibition was assayed by measuring the hydrolytic activities of conjugates, which contained a fixed concentration of protease and varying amounts of inhibitor, including a control with no inhibitor. Inhibitors and peptide substrate were simultaneously added to the native and conjugate solutions immediately before measuring initial rates. The initial rate of hydrolysis of the peptide substrate was monitored by recording the increase in absorption at 412 nm. To determine the dissociation constant of the enzyme–inhibitor complexes (K_i), $V_{max\ app}$, and K_M app were determined by nonlinear curve fitting (equation for Michaelis–Menten parameters) of plots of initial rate vs substrate concentration using the EnzFitter software. The dissociation constants were calculated from secondary plot of K_M app/$V_{max\ app}$ vs initial inhibitor concentration.

AP was not as effective as an inhibitor for CT–pQA conjugates when compared to that of native CT, but GM was able to inhibit activity of CT–pQA conjugates more quickly than that of native CT (Fig. 4).

It is hypothesized that electrostatic attraction and repulsion are responsible for the change in inhibition kinetics seen for CT–pQA conjugates. As AP was positively charged, electrostatic repulsion between the high-density cationic pQA shell surrounding the enzyme and AP likely decreased the inhibitor concentration in the vicinity of the binding site. Conversely, electrostatic attraction between pQA and the negatively charged GM likely increased the inhibitor concentration near the enzyme.

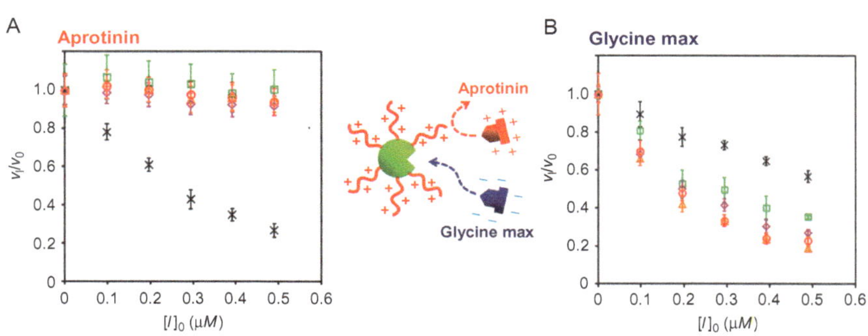

Fig. 4 Impact of polymer-based protein engineering on the inhibitor binding to chymotrypsin–pQA conjugates. Concentration dependence of aprotinin (AP) (A) and Bowman–Birk trypsin–chymotrypsin inhibitor from Glycine max (GM) (B) on the relative enzymatic activity of conjugates (CT-pQA_{27}, *open square*; CT-pQA_{54}, *open triangle*; CT-pQA_{108}, *open diamond*; CT-pQA_{198}, *open circle*), and native CT (*closed circle*). The *inset picture* shows the hypothesized effect of electrostatic attraction and repulsion on inhibitor binding.

Table 3 Apparent K_M and V_{max} Values for Native Chymotrypsin and Chymotrypsin-pQA$_{198}$ Incubated With Aprotinin or Glycine Max at 25°C in 0.1 *M* Sodium Phosphate Buffer (pH 8.0)

	Aprotinin				Glycine Max			
	Native CT		CT-pQA$_{198}$		Native CT		CT-pQA$_{198}$	
$[I]_0$ (μ*M*)	0	0.098	0	0.98	0	0.098	0	0.098
K_M (μ*M*)	85.9	174.9	46.4	69.6	78.7	120.5	46.4	120.5
V_{max} (μ*M*/s)	1.13	1.17	0.79	0.77	1.10	1.18	0.79	0.68
K_M/V_{max} (s^{-1})	76.3	149.2	58.9	90.1	71.8	102.2	58.9	177.4

AP as a competitive inhibitor of CT, functions by competing for the enzyme active site with the natural substrate, will increase K_M values while not affecting k_{cat} during inhibition. GM, a mixed noncompetitive inhibitor, can inhibit by binding to the active site and other locations on the enzyme's surface, causing both K_M and k_{cat} to change during inhibition. CT–pQA conjugates showed the same changes in K_M and k_{cat} as would be expected with each type of inhibition (Table 3).

2.9 LCST/UCST Determination

2.9.1 Reagents

- CT-pDMAPS, CT-pNIPAm, and CT-pSBAM-*block*-pNIPAm conjugates, prepared in Section 2.5
- 0.1 *M* phosphate buffer, pH 8.0

2.9.2 Method

CT-pDMAPS, CT-pNIPAm, and CT-pSBAm-*block*-pNIPAm (2–3 mg polymer/mL each) are dissolved in 0.1 *M* phosphate buffer (pH 8.0). CT-pNIPAm samples are heated from 20 to 35°C and CT-pDMAPS samples are cooled from 30 to 5°C at ±1°C/min. Block conjugates are cooled from 25 to 1°C and then heated up to 40°C at a rate of ±1°C/min. The absorbance at 490 nm is measured in 1°C increments and LCST/UCST temperature calculated from the inflection point on the temperature vs absorbance curves.

2.10 Dynamic Light Scattering

2.10.1 Reagents

- CT conjugates, prepared in Section 2.5
- 0.1 *M* phosphate buffer, pH 8.0

Table 4 Hydrodynamic Diameter Values Obtained by Dynamic Light Scattering for Native Chymotrypsin and Chymotrypsin Conjugates, Measured at Room Temperature in 0.1 *M* Phosphate Buffer, pH 8.0

Sample	D_h **(nm)**
Chymotrypsin	4.4 ± 1.3
CT-pQA_{28}	13.7 ± 3.3
CT-pQA_{54}	14.7 ± 4.5
CT-pQA_{108}	19.2 ± 2.4
CT-pQA_{198}	28.6 ± 6.1
CT-p$DMAPS_{37}$	6.5
CT-p$NIPAm_{82}$	8
CT-p$DMAEMA_{31}$	13.0 ± 2.3
CT-p$DMAEMA_{59}$	18.6 ± 3.3
CT-p$DMAEMA_{94}$	25.2 ± 3.8
CT-p$DMAEMA_{147}$	34.3 ± 3.3
CT-p$SBAm_{35}$	24.2 ± 2.5
CT-p$SBAm_{50}$	22.7 ± 1.9
CT-p$SBAm_{90}$	23.3 ± 1.0
CT-p$SBAm_{35}$-*block*-p$NIPAm_{39}$	46.1 ± 4.5
CT-p$SBAm_{50}$-*block*-p$NIPAm_{67}$	47.6 ± 8.7
CT-p$SBAm_{90}$-*block*-p$NIPAm_{100}$	64.1 ± 4.5

2.10.2 Method

CT conjugates (0.5–3 mg/mL) samples were dissolved in 0.1 *M* phosphate buffer, pH 8.0, and then filtered using a 0.45 m*M* cellulose filter. A Malvern Zetasizer nano-ZS is used to measure hydrodynamic radius (R_h). Each sample was measured in triplicate or greater at each specified temperature.

Typical DLS values for native CT and CT conjugates measured room temperature can be found in Table 4.

2.11 CT-pDMAEMA Activity in Acetonitrile

2.11.1 Reagents

- CT–pDMAEMA conjugate, prepared in Section 2.5

- *N*-acetyl-L-phenylalanine *S*-thiophenyl ester (APTE) and *N*-acetyl-L-phenylalanine propyl ester (APPE), synthesized as per described (Han, Jung, Kim, & Kim, 2004)
- 4,4′-Dithiodipyridine (DTDP)
- Anhydrous acetonitrile

2.11.2 Method

The kinetics of CT-catalyzed transesterification and hydrolysis of APTE was examined with CT–pDMAEMA conjugate. In a model reaction, the transesterification of APTE with 1-propanol in acetonitrile, catalyzed by CT-pDMAEMA produced APPE. CT-pDMAEMA also catalyzed the hydrolysis of APTE, resulting in *N*-acetyl phenylalanine (AP). As a result of both the transesterification and hydrolysis reactions, thiophenol was liberated then subsequently detected and quantified with colorimetric analysis using DTDP.

For the colorimetric analysis of transesterification and hydrolysis of APTE by CT-pDMAEMA, substrate solution was made by adding APTE (0–500 μL of 200 m*M* in dried acetonitrile, 0–100 m*M*) to dried acetonitrile (0–500 μL). The substrate solution (500 μL) was then added to 500 μL of CT-pDMAEMA (0.22 mg/mL (0.018 mg of CT/mL) $[E]_0 = 0.7$ *M*), DTDP (0.22 mg/mL, $[\mathrm{DTDP}]_0 = 0.98$ m*M*), dried 1-propanol (0–414 μL/mL, $[\mathrm{ProOH}]_0 = 0$–5000 m*M*), and water (10–30 μL/mL, [water] = 500–1500 m*M*) solution in dried acetonitrile. The initial rate of transesterification and hydrolysis of APTE at 30°C was monitored by recording the increasing in absorbance at 324 nm ($\varepsilon_{324} = 11980\ \mathrm{M}^{-1}\ \mathrm{cm}^{-1}$) using a UV–vis spectrometer. Michaelis–Menten parameters (V_{max}, k_{cat}, and K_M) were determined by using the EnzFitter software.

For HPLC analysis, APTE (60 mg, 20 m*M*) was added to a solution of CT-pDMAEMA (2.2 mg, 0.7 μ*M* of CT) with different amounts of dried 1-propanol (0–5000 m*M*) and water (500–1500 m*M*) in dried acetonitrile (10 mL) in screw-capped glass vials, and incubated at 30°C. Aliquots were removed at 1–2 h intervals and initial rates were determined from the linear progress by comparison with calibration of AP and APPE.

2.12 Enzyme Thermal Stability

2.12.1 Reagents

- α-CT from bovine pancreas
- CT-pDMAPS, CT-pNIPAm, and CT-pSBAm-*block*-pNIPAm conjugates, prepared in Section 2.5

- *N*-succinyl-Ala-Ala-Pro-Phe *p*-nitroanilide (Suc-AAPF-pNA)
- 0.1 *M* phosphate buffer, pH 8.0

2.12.2 Method

Native CT and CT conjugates (1 mg enzyme/mL) were dissolved in 0.1 *M* sodium phosphate buffer, pH 8.0, and incubated in a water bath at either 25°C or 40°C. At various time points, aliquots are removed and diluted to 0.1 mg enzyme/mL using 0.1 *M* phosphate buffer, pH 8.0. Residual activity was calculated as the percentage of activity remaining relative to the activity at time zero. Substrate (Suc-AAPF-pNA) concentration was kept constant at 288 m*M* for each sample and time point.

2.13 In Vitro Gastric Acid Stability

2.13.1 Reagents

- α-CT from bovine pancreas
- CT-pSBAm-*block*-pNIPAm conjugates prepared in Section 2.5
- *N*-succinyl-Ala-Ala-Pro-Phe *p*-nitroanilide (Suc-AAPF-pNA)
- 0.1 *M* phosphate buffer, pH 8.0
- 167 m*M* HCl aq., pH 1

2.13.2 Method

Native CT and CT conjugates were incubated at 4 μ*M* in 167 m*M* HCl at 37°C in 50 μL aliquots. Aliquots were removed at specified time points and residual activity was measured at 25°C in 0.1 *M* sodium phosphate buffer (pH 8.0) with Suc-AAPF-pNA as substrate (288 μ*M*). Each time point was measured in duplicate and residual activity was calculated as the ratio of activity remaining from time zero.

2.14 Stability to Pepsin Degradation

2.14.1 Reagents

- α-CT from bovine pancreas
- CT-pSBAm-*block*-pNIPAm conjugates, prepared in Section 2.5
- *N*-succinyl-Ala-Ala-Pro-Phe *p*-nitroanilide (Suc-AAPF-pNA)
- 0.1 *M* phosphate buffer, pH 8.0
- 16 n*M* pepsin

2.14.2 Method

Native CT and CT conjugates (4 μ*M*) were incubated in 167 m*M* HCl with 16 n*M* pepsin at 37°C in 50 μL aliquots. Samples were retrieved at specified time points and residual activity was measured in 0.1 *M* sodium phosphate buffer (pH 8.0) at 25°C with Suc-AAPF-pNA as substrate (238 μ*M*). Each time point was measured in duplicate and residual activity was calculated as the ratio of activity remaining from time zero. As a control, pepsin (16 n*M*) bioactivity toward Suc-AAPF-pNA was measured at pH 8.0 and no product formation was observed.

Fig. 5 depicts the rate of inactivation for CT-pSBAm-*block*-pNIPAm conjugates at 37°C in 0.1 *M* sodium phosphate buffer, pH 8.0, under acidic conditions and in the presence of pepsin.

3. PROTEIN–POLYMER-BASED ENGINEERING

3.1 Controlling Grafting Density With Initiator Immobilization

As described earlier, one main constraint in "grafting to" protein–polymer conjugates is the preparation of high grafting density conjugates. Once a polymer chain reacts with a protein, it is progressively more difficult to attach additional polymer chains due to steric impediments. In contrast, by reacting a small initiator molecule to a protein and then "growing from" these initiating sites, the steric restraints are decreased. To illustrate, CT was reacted with bromine-functionalized ATRP initiator and monodisperse *N*-hydroxysuccinimide PEG with 12 repeating units ($dPEG_{12}$, molar mass 685.75 Da) at pH 8.0 at 4°C. To determine the efficiency of the reaction between CT and initiator or $dPEG_{12}$, the increase in molecular weight was measured using MALDI-TOF-MS. The stoichiometric ratio between protein and initiator or $dPEG_{12}$ was varied and showed to provide modulation of the grafting density, as shown in Fig. 6. This control in grafting density was more visible with the initiator, as the slight increase in stoichiometry led to more pronounced amine modifications when compared to $dPEG_{12}$. Water-soluble bromine initiator used in a 3 *M* excess with respect to the total number of free amines resulted, on an average, in the maximum modification of 12 out of the 15 available amine residues (Cummings et al., 2013). In contrast, $dPEG_{12}$ reactions only led to the modification of six amine residues, elucidating to the steric hindrance effect occurring in "grafting to" methods. Considering grafting density on CT surface, immobilization with bromine initiator gave a higher grafting density, with the potential to have one polymer chain per 4 nm^2 of protein surface (0.24 chains/nm^2) (Murata, Cummings, Koepsel, & Russell, 2014), while the grafting density for $dPEG_{12}$ was one PEG chain per 8 nm^2 (0.12 chains/nm^2).

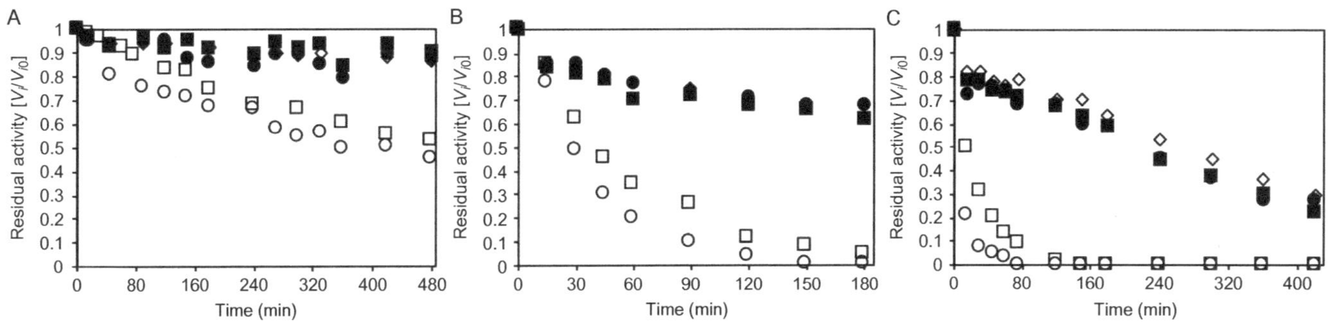

Fig. 5 Rate of irreversible inactivation for CT-pSBAm-*block*-pNIPAm conjugates (CT-50/50, *open diamonds*; CT-75/75, *closed squares*; and CT-100/100, *closed circles*) and native CT (*open circles*) at 37°C in (A) 0.1 *M* sodium phosphate buffer (pH 8.0), (B) 167 m*M* HCl (pH 1), and (C) 167 m*M* HCl with 19 n*M* pepsin. Residual activity was calculated as the activity remaining from $t=0$. All assays were conducted at 25°C.

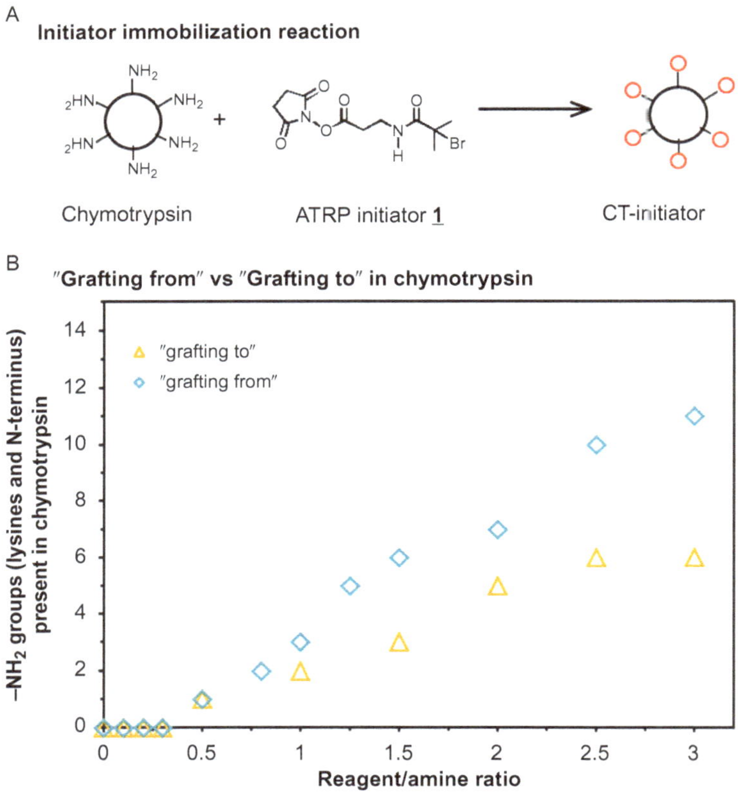

Fig. 6 (A) Initiator immobilization reaction with chymotrypsin and (B) plot showing the extent of modification of chymotrypsin using ATRP bromine initiator and monodisperse $dPEG_{12}$. Chymotrypsin (5 mg/mL) was incubated with ATRP initiator or $dPEG_{12}$ in a range of stoichiometry (0.1–3 equiv.) in 0.1 *M* phosphate buffer, pH 8.0. Reactions were carried out for 3 h at 4°C. The extent of modification was determined by MALDI-TOF-MS.

In addition to steric effects, the structure of the initiator can also be used to control grafting density. Immobilization reactions with CT using three equivalents of chlorine-functionalized initiator at pH 8.0 and 4°C showed a higher extent of modification and consequently a higher grafting density (0.28 chains/nm^2) when compared to that of the bromine initiator, under the same reaction conditions. Experimentally, chlorine initiator allowed the modification of 14 out of 15 of the available amines (lysine residues and N-terminus) (Cummings et al., 2014). This difference in degree of modification can be attributed to electron-withdrawing effects in both ATRP initiators. The subtle variation in electronegativity of the halogen

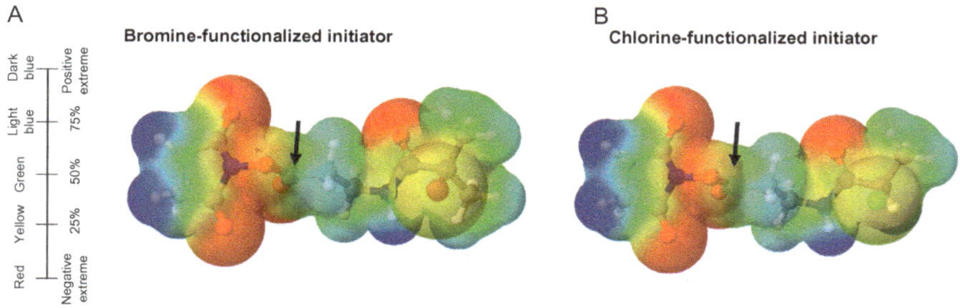

Fig. 7 Electrostatic potential maps for (A) bromine-functionalized and (B) chlorine-functionalized ATRP initiator, determined using semiempirical calculations performed with MOPAC software (Stewart, 2013). Areas in *red* denote high electron density, areas in *blue* denote low electron density, *arrow* depicts reactive center (Cox & Williams, 1981).

atoms can contribute with an inductive effect that can alter the reactive center of the ATRP initiator (Fukui, 1982; Reed, 1992).

Semiempirical calculations were conducted to provide insight into the electronic properties of both initiators. It was hypothesized that the slightly higher electronegativity of the chlorine initiator could increase the electrophilic nature of the initiator, providing a higher reactivity than observed with the bromine initiator. Fig. 7 depicts the electrostatic potential maps for bromine and chlorine initiators. Interestingly, both initiators show similar electron distribution which would indicate that the structural differences between initiators would not impact significantly the immobilization reaction.

However, a more in-depth understanding of the variations of the electron density between both initiators was achieved by analysis of the relative charges on the electrophilic carbonyl group where the nucleophilic addition would occur. The sp^2 carbon atom has a lower electron density, and consequently it is more electrophilic in chlorine bearing initiator than in bromine bearing initiator (0.529 and 0.531 C, respectively), and this can account for the increased reactivity observed. The ability to modulate density by stoichiometric changes and structural and chemical characteristics of the initiator provides further control in protein modification that can be tailored for specific applications.

3.2 High-Density CT–Polymer Conjugates

Having shown that high-density CT–initiator complexes can be synthesized efficiently, polymers can then be "grown from" these initiating sites using ATRP. In this approach, the geometry of the polymerization reaction yields high-density polymer shells around the protein surface and separation of unreacted monomer is facilitated. As a consequence of the high grafting density, the polymers are in the "brush" conformation and can respond to their environment by conformational changes, imparting unique properties to the conjugates.

Some examples are high-density CT conjugates with poly[*N*, *N*′-dimethyl(methacryloyloxyethyl) ammonium propane sulfonate] (CT-pDMAPS) and poly(*N*-isopropylacrylamide) (CT-pNIPAm) (Cummings et al., 2013). Both pDMAPS and pNIPAm respond to changes in temperature by predictable alterations in polymer folding. pNIPAm has a lower critical solution temperature (LCST), above which the polymer experiences a reversible collapse and becomes hydrophobic (Schild, 1992). The same reversible change is seen for pDMAPS, except that the polymer is immiscible below the upper critical solution temperature (UCST) (Chen et al., 2000). These characteristics were explored in CT-pDMAPS and CT-pNIPAm.

LCST and UCST of CT bioconjugates were closely related to those of the free polymers and consequently, the polymers on the protein surface had similar temperature responsiveness (Fig. 8) (Cummings et al., 2013). The conformational transitions between extended and collapsed state retained bioactivity but a higher substrate affinity was also observed after polymer collapse (Cummings et al., 2013).

For CT-pDMAPs, hydrophilic pDMAPS could have interacted with the hydrophobic substrate used in this study, increasing the local concentration of substrate near the hydrophobic substrate pocket and lowering K_M (Cummings et al., 2013; Keefe & Jiang, 2012). Below the UCST, this increased substrate affinity was still observed but to a lower extent. It is possible that the collapsed nature of pDMAPS could have restricted the access of the substrate to the active site via steric hindrance. Similarly above the LCST of CT-pNIPAm was also in its collapsed, hydrophobic state, and also

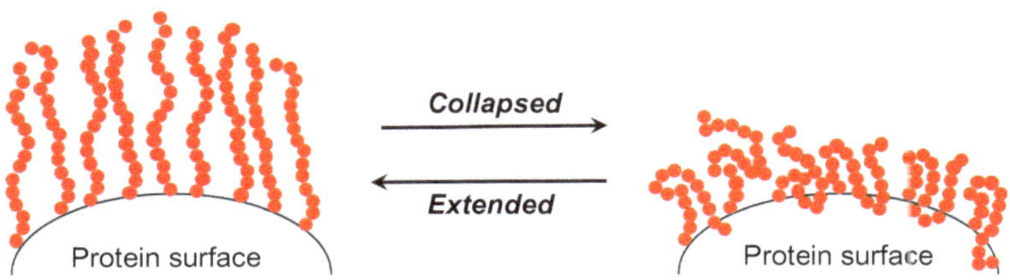

Fig. 8 Polymer conformational transitions as a result of temperature changes in high-density conjugates CT-pDMAPS and pNIPAm.

showed higher K_M due to steric hindrance. The more hydrophobic nature of pNIPAm would lead to a stronger association with the hydrophobic model substrate, reflected in a higher K_M.

4. CONCLUSIONS

Protein–polymer conjugates have led to the emergence of a multibillion dollar biotechnology industry. In most cases, proteins have been stabilized or altered by covalent coupling of the preformed polymers to the surface of the protein. This approach is inherently limited by a lack of exquisite control of polymer architecture, site of attachment, and density of chain attachment. Furthermore, polymers typically used (e.g., PEG) do not impart any additional functionality to the protein. Surface-initiated ATRP polymer-based protein engineering allows the design of conjugates in a rational and controlled manner. Grafting density, site of modification, and the "grown" polymer are variables that can be altered and tuned to improve and enhance structure–function properties inherent to proteins. Using PBPE, high-density stimuli-responsive polymers were "grown" from multiple ATRP initiating sites on the surface of CT. These conjugates had predefined molecular weights, compositions, architectures, and narrow molecular weight distributions. In summary, PBPE provides an elegant strategy to tailor bioactivity and stability through the design of protein conjugates in a controlled and rational manner.

ACKNOWLEDGMENT

The authors acknowledge financial support provided by the Carnegie Mellon University Center for Polymer-based Protein Engineering.

REFERENCES

Abuchowski, A., van Es, T., Palczuk, N. C., & Davis, F. F. (1977). Alteration of immunological properties of bovine serum albumin by covalent attachment of polyethylene glycol. *The Journal of Biological Chemistry*, *252*(11), 3578–3581.

Alconcel, S. N. S., Baas, A. S., & Maynard, H. D. (2011). FDA-approved poly(ethylene glycol)–protein conjugate drugs. *Polymer Chemistry*, *2*(7), 1442.

Averick, S. E., Paredes, E., Grahacharya, D., Woodman, B. F., Miyake-Stoner, S. J., Mehl, R. A., … Das, S. R. (2012). A protein–polymer hybrid mediated by DNA. *Langmuir*, *28*(4), 1954–1958.

Averick, S., Simakova, A., Park, S., Konkolewicz, D., Magenau, A. J. D., Mehl, R. A., & Matyjaszewski, K. (2012). Atrp under biologically relevant conditions: Grafting from a protein. *ACS Macro Letters*, *1*(1), 6–10.

Brittain, W. J., & Minko, S. (2007). A structural definition of polymer brushes. *Journal of Polymer Science Part A: Polymer Chemistry*, *45*(16), 3505–3512.

Brocchini, S., Godwin, A., Balan, S., Choi, J.-W., Zloh, M., & Shaunak, S. (2008). Disulfide bridge based pegylation of proteins. *Advanced Drug Delivery Reviews*, *60*(1), 3–12.
Campbell, A. S., Murata, H., Carmali, S., Matyjaszewski, K., Islam, M. F., & Russell, A. J. (2016). Polymer-based protein engineering grown ferrocene-containing redox polymers improve current generation in an enzymatic biofuel cell. *Biosensors and Bioelectronics*, *86*, 446–453.
Chalker, J. M., Bernardes, G. J. L., Lin, Y. A., & Davis, B. G. (2009). Chemical modification of proteins at cysteine: Opportunities in chemistry and biology. *Chemistry—An Asian Journal*, *4*(5), 630–640.
Chan, A. O.-Y., Ho, C.-M., Chong, H.-C., Leung, Y.-C., Huang, J.-S., Wong, M.-K., & Che, C.-M. (2012). Modification of n-terminal α-amino groups of peptides and proteins using ketenes. *Journal of the American Chemical Society*, *134*(5), 2589–2598.
Chen, G., Heim, A., Riether, D., Yee, D., Milgrom, Y., Gawinowicz, M. A., & Sames, D. (2003). Reactivity of functional groups on the protein surface: Development of epoxide probes for protein labeling. *Journal of the American Chemical Society*, *125*(27), 8130–8133.
Chen, L., Honma, Y., Mizutani, T., Liaw, D. J., Gong, J. P., & Osada, Y. (2000). Effects of polyelectrolyte complexation on the UCST of zwitterionic polymer. *Polymer*, *41*(1), 141–147.
Ciampolini, M., & Nardi, N. (1966). Five-coordinated high-spin complexes of bivalent cobalt, nickel, and copper with tris(2-dimethylaminoethyl)amine. *Inorganic Chemistry*, *5*(1), 41–44.
Cox, S. R., & Williams, D. E. (1981). Representation of the molecular electrostatic potential by a net atomic charge model. *Journal of Computational Chemistry*, *2*(3), 304–323.
Cummings, C., Murata, H., Koepsel, R., & Russell, A. J. (2013). Tailoring enzyme activity and stability using polymer-based protein engineering. *Biomaterials*, *34*(30), 7437–7443.
Cummings, C., Murata, H., Koepsel, R., & Russell, A. J. (2014). Dramatically increased ph and temperature stability of chymotrypsin using dual block polymer-based protein engineering. *Biomacromolecules*, *15*(3), 763–771.
Cummings, C. S., Murata, H., Matyjaszewski, K., & Russell, A. J. (2016). Polymer-based protein engineering enables molecular dissolution of chymotrypsin in acetonitrile. *ACS Macro Letters*, *5*(4), 493–497.
Davis, F. F. (2002). The origin of pegnology. *Advanced Drug Delivery Reviews*, *54*(4), 457–458.
Dawson, P. E., Muir, T. W., Clark-Lewis, I., & Kent, S. (1994). Synthesis of proteins by native chemical ligation. *Science*, *266*(5186), 776.
De, P., Li, M., Gondi, S. R., & Sumerlin, B. S. (2008). Temperature-regulated activity of responsive polymer–protein conjugates prepared by grafting-from via raft polymerization. *Journal of the American Chemical Society*, *130*, 11288–11289.
Dordick, J. S. (1991). Non-aqueous enzymology. *Current Opinion in Biotechnology*, *2*(3), 401–407.
Drevon, G. F., Danielmeier, K., Federspiel, W., Stolz, D. B., Wicks, D. A. W., Yu, P. C., & Russell, A. J. (2002). High-activity enzyme polyurethane coatings. *Biotechnology and Bioengineering*, *79*(7), 785–794.
Drevon, G. F., Hartleib, J., Scharff, E., Rüterjans, H., & Russell, A. J. (2001). Thermoinactivation of diisopropylfluorophosphatase-containing polyurethane polymers. *Biomacromolecules*, *2*(3), 664–671.
Drevon, G. F., Urbanke, C., & Russell, A. J. (2003). Enzyme-containing michael-adduct-based coatings. *Biomacromolecules*, *4*(3), 675–682.
Foster, G. R. (2010). Pegylated interferons for the treatment of chronic hepatitis c. *Drugs*, *70*(2), 147–165.

Fuhrmann, G., Grotzky, A., Lukic, R., Matoori, S., Luciani, P., Yu, H., … Leroux, J.-C. (2013). Sustained gastrointestinal activity of dendronized polymer–enzyme conjugates. *Nature Chemistry, 5*, 582–589.

Fukui, K. (1982). Role of frontier orbitals in chemical reactions. *Science, 218*(4574), 747–754.

Gaertner, H. F., & Offord, R. E. (1996). Site-specific attachment of functionalized poly(ethylene glycol) to the amino terminus of proteins. *Bioconjugate Chemistry*, 7(1), 38–44.

Gao, W., Liu, W., Mackay, J. A., Zalutsky, M. R., Toone, E. J., & Chilkoti, A. (2009). In situ growth of a stoichiometric peg-like conjugate at a protein's n-terminus with significantly improved pharmacokinetics. *Proceedings of the National Academy of Sciences of the United States of America, 106*(36), 15231–15236.

Gualberto, A. (2012). Brentuximab vedotin (sgn-35), an antibody-drug conjugate for the treatment of cd30-positive malignancies. *Expert Opinion on Investigational Drugs, 21*, 205–216.

Han, M. S., Jung, S. O., Kim, M.-J., & Kim, D. H. (2004). Fluorometric assay protocol for protease-catalyzed transesterification reactions in organic solvents. *The Journal of Organic Chemistry, 69*(8), 2853–2855.

Harris, J. M. (2013). Introduction to biotechnical and biomedical applications of poly(ethylene glycol). In J. M. Harris (Ed.), *Poly(ethylene glycol) chemistry: Biotechnical and biomedical applications* (pp. 1–14). USA: Springer.

He, P., & He, L. (2009). Synthesis of surface-anchored DNA–polymer bioconjugates using reversible addition–fragmentation chain transfer polymerization. *Biomacromolecules, 10*(7), 1804–1809.

Heredia, K. L., Bontempo, D., Ly, T., Byers, J. T., Halstenberg, S., & Maynard, H. D. (2005). In situ preparation of protein-"smart" polymer conjugates with retention of bioactivity. *Journal of the American Chemical Society, 127*, 16955–16960.

Hnízda, A., Šantrůček, J., Šanda, M., Strohalm, M., & Kodíček, M. (2008). Reactivity of histidine and lysine side-chains with diethylpyrocarbonate—A method to identify surface exposed residues in proteins. *Journal of Biochemical and Biophysical Methods, 70*(6), 1091–1097.

Hoffman, A. S., & Stayton, P. S. (2007). Conjugates of stimuli-responsive polymers and proteins. *Progress in Polymer Science, 32*(8–9), 922–932.

Ishida, T., & Kiwada, H. (2013). Anti-polyethyleneglycol antibody response to pegylated substances. *Biological and Pharmaceutical Bulletin, 36*(6), 889–891.

Jevševar, S., Kunstelj, M., & Porekar, V. G. (2010). Pegylation of therapeutic proteins. *Biotechnology Journal, 5*(1), 113–128.

Kalkhof, S., & Sinz, A. (2008). Chances and pitfalls of chemical cross-linking with amine-reactive n-hydroxysuccinimide esters. *Analytical and Bioanalytical Chemistry, 392*(1), 305–312.

Keefe, A. J., & Jiang, S. (2012). Poly(zwitterionic) protein conjugates offer increased stability without sacrificing binding affinity or bioactivity. *Nature Chemistry, 4*, 59–63.

Konieczny, S., Krumm, C., Doert, D., Neufeld, K., & Tiller, J. C. (2014). Investigations on the activity of poly(2-oxazoline) enzyme conjugates dissolved in organic solvents. *Journal of Biotechnology, 181*, 55–63.

Konieczny, S., Leurs, M., & Tiller, J. C. (2015). Polymer enzyme conjugates as chiral ligands for sharpless dihydroxylation of alkenes in organic solvents. *ChemBioChem, 16*(1), 83–90.

Koussoroplis, S. J., Paulissen, G., Tyteca, D., Goldansaz, H., Todoroff, J., Barilly, C., … Vanbever, R. (2014). Pegylation of antibody fragments greatly increases their local residence time following delivery to the respiratory tract. *Journal of Controlled Release, 187*, 91–100.

Lackey, C. A., Murthy, N., Press, O. W., Tirrell, D. A., Hoffman, A. S., & Stayton, P. S. (1999). Hemolytic activity of pH-responsive polymer-streptavidin bioconjugates. *Bioconjugate Chemistry, 10*, 401–405.

Li, H., Li, M., Yu, X., Bapat, A. P., & Sumerlin, B. S. (2011). Block copolymer conjugates prepared by sequentially grafting from proteinsvia raft. *Polymer Chemistry*, *2*(7), 1531–1535.
Lecolley, F., Tao, L., Mantovani, G., Durkin, I., Lautru, S., & Haddleton, D. M. (2004). A new approach to bioconjugates for proteins and peptides ("pegylation") utilising living radical polymerisation. *Chemical Communications (Cambridge)*, (18), 2026–2027.
Le Droumaguet, B., & Nicolas, J. (2010). Recent advances in the design of bioconjugates from controlled/living radical polymerization. *Polymer Chemistry*, *1*(5), 563–598.
LeJeune, K. E., & Russell, A. J. (1996). Covalent binding of a nerve agent hydrolyzing enzyme within polyurethane foams. *Biotechnology and Bioengineering*, *51*, 450–457.
LeJeune, K. E., Wild, J. R., & Russell, A. J. (1998). Nerve agents degraded by enzymatic foams. *Nature*, *395*(6697), 27–28.
Lele, B. S., Murata, H., Matyjaszewski, K., & Russell, A. J. (2005). Synthesis of uniform protein–polymer conjugates. *Biomacromolecules*, *6*, 3380–3387.
Li, D., Manjula, B. N., & Acharya, A. S. (2006). Extension arm facilitated pegylation of hemoglobin: Correlation of the properties with the extent of pegylation. *The Protein Journal*, *25*(4), 263–274.
Lins, L., Thomas, A., & Brasseur, R. (2003). Analysis of accessible surface of residues in proteins. *Protein Science*, *12*, 1406–1417.
Liu, J., Bulmus, V., Herlambang, D. L., Barner-Kowollik, C., Stenzel, M. H., & Davis, T. P. (2007). In situ formation of protein–polymer conjugates through reversible addition fragmentation chain transfer polymerization. *Angewandte Chemie (International Ed. in English)*, *46*(17), 3099–3103.
Lucius, M., Falatach, R., McGlone, C., Makaroff, K., Danielson, A., Williams, C., … Berberich, J. A. (2016). Investigating the impact of polymer functional groups on the stability and activity of lysozyme–polymer conjugates. *Biomacromolecules*, *17*(3), 1123–1134.
Magnusson, J. P., Bersani, S., Salmaso, S., Alexander, C., & Caliceti, P. (2010) In situ growth of side-chain peg polymers from functionalized human growth hormones a new technique for preparation of enhanced protein–polymer conjugates. *Bioconjugate Chemistry*, *21*, 671–678.
Manns, M. P., McHutchison, J. G., Gordon, S. C., Rustgi, V. K., Shiffman, M., Reindollar, R., … Albrecht, J. K. (2001). Peginterferon alfa-2b plus ribavirin compared with interferon alfa-2b plus ribavirin for initial treatment of chronic hepatitis c: A randomized trial. *The Lancet*, *358*(9286), 958–965.
McDowall, L., Chen, G., & Stenzel, M. H. (2008). Synthesis of seven-arm poly(vinyl pyrrolidone) star polymers with lysozyme core prepared by madix/raft polymerization. *Macromolecular Rapid Communications*, *29*(20), 1666–1671.
McGoff, P., Baziotis, A., & Maskiewicz, R. (1988). Analysis of polyethylene glycol modified superoxide dismutase by chromatographic, electrophoretic, light scattering, chemical and enzymatic methods. *Chemical and Pharmaceutical Bulletin*, *36*(8), 3079–3091.
Mei, B., Pan, C., Jiang, H., Tjandra, H., Strauss, J., Chen, Y., … Murphy J. E. (2010). Rational design of a fully active, long-acting pegylated factor VIII for hemophilia a treatment. *Blood*, *116*(2), 270–279.
Murata, H., Cummings, C. S., Koepsel, R. R., & Russell, A. J. (2013). Polymer-based protein engineering can rationally tune enzyme activity, pH-dependence, and stability. *Biomacromolecules*, *14*(6), 1919–1926.
Murata, H., Cummings, C. S., Koepsel, R. R., & Russell, A. J. (2014). Rational tailoring of substrate and inhibitor affinity via atrp polymer-based protein engineering. *Biomacromolecules*, *15*(7), 2817–2823.
Nakamura, T., Kawai, Y., Kitamoto, N., Osawa, T., & Kato, Y. (2009). Covalent modification of lysine residues by allyl isothiocyanate in physiological conditions: Plausible

transformation of isothiocyanate from thiol to amine. *Chemical Research in Toxicology, 22*(3), 536–542.

Nathani, R. I., Chudasama, V., Ryan, C. P., Moody, P. R., Morgan, R. E., Fitzmaurice, R. J., … Caddick, S. (2013). Reversible protein affinity-labelling using bromomaleimide-based reagents. *Organic and Biomolecular Chemistry, 11*(15), 2408–2411.

Nicolas, J., San Miguel, V., Mantovani, G., & Haddleton, D. M. (2006). Fluorescently tagged polymer bioconjugates from protein derived macroinitiators. *Chemical Communications (Cambridge)*, (45), 4697–4699.

Peeler, J. C., Woodman, B. F., Averick, S., Miyake-Stoner, S. J., Stokes, A. L., Hess, K. R., … Mehl, R. A. (2010). Genetically encoded initiator for polymer growth from proteins. *Journal of the American Chemical Society, 132*(39), 13575–13577.

Pelegri-O'Day, E. M., Lin, E.-W., & Maynard, H. D. (2014). Therapeutic protein–polymer conjugates: Advancing beyond pegylation. *Journal of the American Chemical Society, 136*(41), 14323–14332.

Peracchi, A. (2001). Enzyme catalysis: Removing chemically 'essential' residues by site-directed mutagenesis. *Trends in Biochemical Sciences, 26*(8), 497–503.

Qi, Y., Amiram, M., Gao, W., McCafferty, D. G., & Chilkoti, A. (2013). Sortase-catalyzed initiator attachment enables high yield growth of a stealth polymer from the c terminus of a protein. *Macromolecular Rapid Communications, 34*(15), 1256–1260.

Qi, Y., & Chilkoti, A. (2015). Protein–polymer conjugation—moving beyond pegylation. *Current Opinion in Chemical Biology, 28*, 181–193.

Reed, J. L. (1992). Electronegativity and atomic charge. *Journal of Chemical Education, 69*(10), 785.

Rodríguez-Martínez, J. A., Solá, R. J., Castillo, B., Cintrón-Colón, H. R., Rivera-Rivera, I., Barletta, G., & Griebenow, K. (2008). Stabilization of α-chymotrypsin upon pegylation correlates with reduced structural dynamics. *Biotechnology and Bioengineering, 101*(6), 1142–1149.

Russell, A. J., & Fersht, A. R. (1987). Rational modification of enzyme catalysis by engineering surface charge. *Nature, 328*(6130), 496–500.

Russell, A. J., Thomas, P. G., & Fersht, A. R. (1987). Electrostatic effects on modification of charged groups in the active site cleft of subtilisin by protein engineering. *Journal of Molecular Biology, 193*(4), 803–813.

Samanta, D., McRae, S., Cooper, B., Hu, Y., Emrick, T., Pratt, J., & Charles, S. A. (2008). End-functionalized phosphorylcholine methacrylates and their use in protein conjugation. *Biomacromolecules, 9*(10), 2891–2897.

Santantonio, T. A., & Fasano, M. (2014). Chronic hepatitis b: Advances in treatment. *World Journal of Hepatology, 6*(5), 284–292.

Schild, H. G. (1992). Poly(n-isopropylacrylamide): Experiment, theory and application. *Progress in Polymer Science, 17*(2), 163–249.

Senter, P. D., & Sievers, E. L. (2012). The discovery and development of brentuximab vedotin for use in relapsed Hodgkin lymphoma and systemic anaplastic large cell lymphoma. *Nature Biotechnology, 30*, 631–637.

Shen, B.-Q., Xu, K., Liu, L., Raab, H., Bhakta, S., Kenrick, M., … Junutula, J. R. (2012). Conjugation site modulates the in vivo stability and therapeutic activity of antibody-drug conjugates. *Nature Biotechnology, 30*(2), 184–189.

Smith, M. E. B., Schumacher, F. F., Ryan, C. P., Tedaldi, L. M., Papaioannou, D., Waksman, G., … Baker, J. R. (2010). Protein modification, bioconjugation, and disulfide bridging using bromomaleimides. *Journal of the American Chemical Society, 132*(6), 1960–1965.

Stewart, J. J. P. (2013). *Mopac 2012: Stewart computational chemistry*. Version 7.263W. http://OpenMOPAC.net.

Strozyk, M. S., Chanana, M., Pastoriza-Santos, I., Pérez-Juste, J., & Liz-Marzán, L. M. (2012). Protein/polymer-based dual-responsive gold nanoparticles with pH-dependent thermal sensitivity. *Advanced Functional Materials*, *22*(7), 1436–1444.

Tao, L., Mantovani, G., Lecolley, F., & Haddleton, D. M. (2004). α-Aldehyde terminally functional methacrylic polymers from living radical polymerization: Application in protein conjugation "pegylation". *Journal of American Chemical Society*, *126*(41), 13220–13221.

Tsarevsky, N. V., Pintauer, T., & Matyjaszewski, K. (2004). Deactivation efficiency and degree of control over polymerization in atrp in protic solvents. *Macromolecules*, *37*(26), 9768–9778.

Turecek, P. L., Bossard, M. J., Schoetens, F., & Ivens, I. A. (2016). Pegylation of biopharmaceuticals: A review of chemistry and nonclinical safety information of approved drugs. *Journal of Pharmaceutical Sciences*, *105*(2), 460–475.

Turner, N. J. (2009). Directed evolution drives the next generation of biocatalysts. *Nature Chemical Biology*, *5*(8), 567–573.

Udenfriend, S., Stein, S., Böhlen, P., Dairman, W., Leimgruber, W., & Weigele, M. (1972). Fluorescamine: A reagent for assay of amino acids, peptides, proteins, and primary amines in the picomole range. *Science*, *178*(4063), 871–872.

Verma, S., Miles, D., Gianni, L., Krop, I. E., Welslau, M., Baselga, J., … EMILIA Study Group. (2012). Trastuzumab emtansine for her2-positive advanced breast cancer. *The New England Journal of Medicine*, *367*, 1783–1791.

Veronese, F. M., & Pasut, G. (2005). Pegylation, successful approach to drug delivery. *Drug Discovery Today*, *10*(21), 1451–1458.

Wang, S., Meng, X., Zhou, H., Liu, Y., Secundo, F., & Liu, Y. (2016). Enzyme stability and activity in non-aqueous reaction systems: A mini review. *Catalysts*, *6*(2) 32.

Wu, Y., Ng, D. Y., Kuan, S. L., & Weil, T. (2015). Protein–polymer therapeutics: A macromolecular perspective. *Biomaterials Science*, *3*(2), 214–230.

Xu, J., Boyer, C., Bulmus, V., & Davis, T. P. (2009). Synthesis of dendritic carbohydrate end-functional polymers via raft: Versatile multi-functional precursors for bioconjugations. *Journal of Polymer Science Part A: Polymer Chemistry*, *47*(17), 4302–4313.

Yamaoka, T., Tabata, Y., & Ikada, Y. (1994). Distribution and tissue uptake of poly(ethylene glycol) with different molecular weights after intravenous administration to mice. *Journal of Pharmaceutical Sciences*, *83*(4), 601–606.

Yang, Q., Jones, S. W., Parker, C. L., Zamboni, W. C., Bear, J. E., & Lai, S. K. (2014). Evading immune cell uptake and clearance requires peg grafting at densities substantially exceeding the minimum for brush conformation. *Molecular Pharmaceutics*, *11*(4), 1250–1258.

Yang, Q., & Lai, S. K. (2015). Anti-peg immunity: Emergence, characteristics, and unaddressed questions. *Wiley Interdisciplinary Reviews. Nanomedicine and Nanobiotechnology*, 7(5), 655–677.

Yaşayan, G., Saeed, A. O., Fernández-Trillo, F., Allen, S., Davies, M. C., Jangher, A., … Alexander, C. (2011). Responsive hybrid block co-polymer conjugates of proteins–controlled architecture to modulate substrate specificity and solution behaviour. *Polymer Chemistry*, *2*(7), 1567.

Zhang, Q., Vanparijs, N., Louage, B., De Geest, B. G., & Hoogenboom, R. (2014). Dual pH- and temperature-responsive raft-based block co-polymer micelles and polymer–protein conjugates with transient solubility. *Polymer Chemistry*, *5*(4), 1140–1144.

Zore, O. V., Lenehan, P. J., Kumar, C. V., & Kasi, R. M. (2014). Efficient biocatalysis in organic media with hemoglobin and poly(acrylic acid) nanogels. *Langmuir*, *30*(18), 5176–5184.

CHAPTER SIXTEEN

Nano-Armoring of Enzymes: Rational Design of Polymer-Wrapped Enzymes

Kishore Raghupathi, Sankaran Thayumanavan[1]
University of Massachusetts, Amherst, MA, United States
[1]Corresponding author: e-mail address: thai@chem.umass.edu

Contents

Abstract

The formulation in which therapeutic proteins are administered plays a key role in retaining their biological activity. Enzyme wrapping, using synthetic polymers, is a strategy employed to provide enzymes with lower immunogenicity, longer circulation times, and better targeting capabilities. Protein–polymer complexation methods, involving covalent, noncovalent, and electrostatic interactions, that can provide means to develop formulations for retaining enzyme stability are discussed in this chapter. Amphiphilic self-cross-linkable polymer was used to encapsulate capsase-3 enzyme in the nanogel, while inverse emulsion polymerization method was used to entrap α-glucosidase enzyme in the nanogel. These nanogels were characterized by dynamic

Methods in Enzymology, Volume 590
ISSN 0076-6879
http://dx.doi.org/10.1016/bs.mie.2017.01.007
© 2017 Elsevier Inc.
All rights reserved.

light scattering, transmission electron microscopy, and gel electrophoresis. Upon release of caspase-3 enzyme from polymeric nanogel, it retained nearly 86% of its original activity. Similarly, α-glucosidase that was encased in the acid cleavable polymeric nanogel exhibited substantial activity after release under acidic conditions (pH 5, 48 h). Nano-armoring of the enzymes were nearly complete and provided high yields of the encased enzyme.

1. INTRODUCTION

Enzymes utilize their selectivity in biochemical catalysis to play a vital role in most processes in living organisms. Therefore, a deviation from their optimal activity can lead to serious physiological abnormalities and sometimes it can be life threatening (Vellodi, 2005). While their exquisitely high specificity accounts for precisely controlling biochemical processes, this specificity also makes it very difficult to correct, in case of their deficiency. Taking this into consideration, recombinant protein/enzymes technologies have been developed to "directly" supplement a deficient enzyme or to interfere with or augment an abnormal biological process (Leader, Baca, & Golan, 2008). However, systemic administration of these therapeutic enzymes as such is often ineffective, as they lead to immunogenicity and cytotoxicity (Brange, Andersen, Laursen, Meyn, & Rasmussen, 1997; Curatolo et al., 1997; Hermeling, Crommelin, Schellekens, & Jiskoot, 2004). Moreover, these proteins can also undergo clearance by renal and reticuloendothelial systems (Poznansky & Juliano, 1984), leaving only minimal fraction of the administered dose reaching the target site (Desnick & Schuchman, 2012).

To address these challenges, several formulation strategies have been laid out for delivering proteins more effectively into biological systems. These methods include PEG-ylation of proteins (Brocchini et al., 2008; Joralemon, McRae, & Emrick, 2010; Nischan & Hackenberger, 2014; Pasut & Veronese, 2012), noncovalent and electrostatic methods to complex proteins with polymers (Ayame, Morimoto, & Akiyoshi, 2008; Ghosh et al., 2010; González-Toro, Ryu, Chacko, Zhuang, & Thayumanavan, 2012; Lee et al., 2008; Mero, Ishino, Chaiken, Veronese, & Pasut, 2011; Mueller et al., 2012), encapsulation of proteins/enzymes using inorganic networks (Dyal et al., 2003; Luckarift, Spain, Naik, & Stone, 2004), encapsulation of enzymes in liposomal vesicles (Dai, Wang, Zhao, Li, & Wang, 2006), and entrapment in hydrogels (Azagarsamy, Alge, Radhakrishnan,

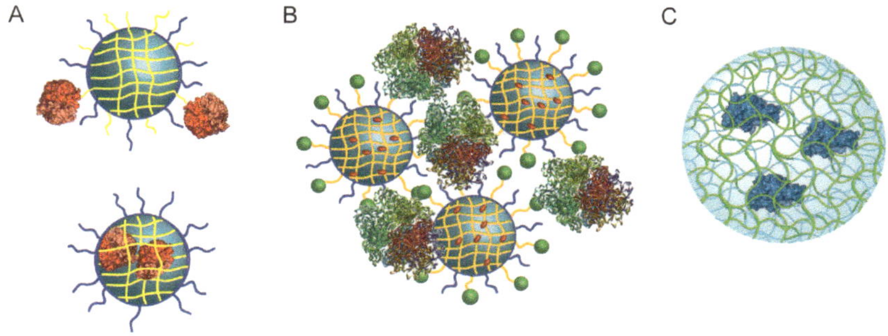

Fig. 1 Cartoon representing different types of enzyme wrapping using polymers. (A) Covalent conjugation. (B) Electrostatic complexation. (C) Noncovalent entrapment in hydrogel network.

Tibbitt, & Anseth, 2012; Cohen et al., 2008; Molla et al., 2014; Murthy et al., 2003; Vermonden, Censi, & Hennink, 2012).

In this chapter, some of these methods will be described in detail where enzymes are wrapped or conjugated with synthetic polymers (González-Toro et al., 2012; Matsumoto, González-Toro, Chacko, Maynard, & Thayumanavan, 2013; Molla et al., 2014; Ventura et al., 2015) (Fig. 1). This approach of modifying enzymes using polymers can be termed as nano-armoring of enzymes, since the resulting formulation of polymer–enzyme complex has nanoscale (10–100 nm) dimensions. The method of enzyme wrapping using polymers can be divided into three categories based on the nature of interaction between enzymes and polymers:

(i) Covalent conjugation
(ii) Electrostatic complexation
(iii) Noncovalent entrapment

2. COVALENT CONJUGATION

This method involves the formation of a covalent bond between certain functional groups of the enzyme and the polymer (Heredia & Maynard, 2007). Although in some cases covalent modification of enzymes is associated with decreased biological activity (Fishburn, 2008), it can be minimized either by site-specific modification of enzymes (Nischan & Hackenberger, 2014), where its active site is not affected or by using covalent modification which is reversible (Ventura et al., 2015). In this chapter, we will particularly focus on reversible covalent modification of enzyme using polymers, where the use of organic solvents is limited or completely avoided, because organic solvents have been implicated in enzyme denaturation during the conjugation process (Babu, Moradian, & Douglas, 2001; Knubovets, Osterhout, & Klibanov, 1999).

Covalent conjugation between enzyme and polymer through a disulfide bond is an attractive option due to its good conjugating ability in aqueous conditions and its reversible nature (ability to cleave under the reducing conditions of cell cytosol; Fig. 2). The reversible nature of this bond allows the enzyme to tracelessly dissociate from the polymer (i.e., without any remnant functional groups of the polymer in the protein), thereby retaining its original structure and function. However, this approach is more suitable for enzymes that are targeted to the cytosol such as caspases, cytosolic protein kinases, cytochromes *c*, and RNase.

The thiol reactive polymer that will be discussed here is an amphiphilic random copolymer (PEGMA–PDSMA polymer) consisting of a hydrophilic polyethylene glycol-based and pyridyldisulfide-based monomers (Ryu et al., 2010; Ventura et al., 2015). This polymer is a suitable candidate for enzyme wrapping for the following reasons: (i) the polymer is water-dispersible, (ii) it is nontoxic and can potentially be a safe pharmaceutical excipient (Ryu et al., 2010), and (iii) the pyridyl disulfide group of the polymer can readily react with free cysteine thiols of the proteins.

The covalent conjugation between an enzyme and the polymer can be executed by utilizing the surface accessible cysteines on the enzyme and thiol reactive pyridyl disulfide monomers of the polymer. One limitation is that, not all enzymes have surface accessible cysteine groups to react with the polymer. However, there exist some enzymes that have free cysteines at

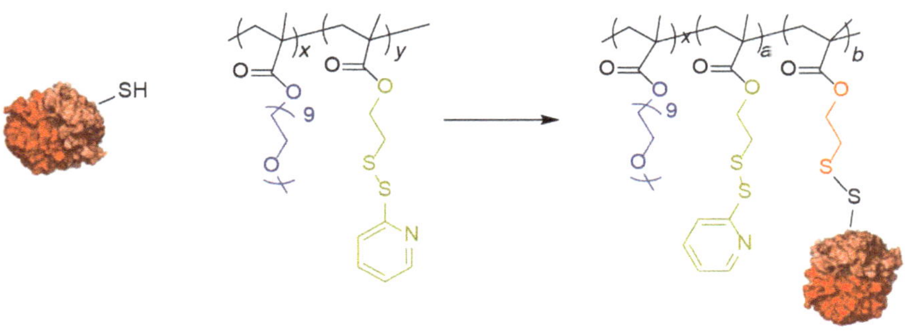

Fig. 2 Scheme involving PEGMA–PDSMA polymer conjugating with protein containing a thiol by a disulfide exchange reaction.

their active site (Chang & Yang, 2000), and conjugation of a polymer to those cysteines might adversely affect catalytic activity. However, if the target delivery site is the cytosol, the enzyme can detach itself from the polymer through cleavage of the disulfide bond under the reducing conditions of the cytosol (Balendiran, Dabur, & Fraser, 2004) and release the free enzyme. This process should then restore the active site functionalities and thus restore the enzymatic function.

2.1 Synthesis of Thiol Reactive (PEGMA:PDSMA) Polymer

The monomer (Ghosh, Basu, & Thayumanavan, 2006) and the polymer (Ventura et al., 2015) can be synthesized by following literature protocols. An overview of the polymer synthesis is shown later:

Equipment and chemicals

Poly(ethylene glycol)monomethylether methacrylate (PEGMA; MW 475), 2,2′-dithiodipyridine, 2,2′-azobis-(2-methylpropionitrile) (AIBN), 4-cyano-4-(phenylcarbonothioylthio) pentanoic acid (chain transfer agent), and D,L-dithiothreitol (DTT) from Sigma-Aldrich, and pyridyl disulfide ethyl methacrylate (PDSMA) were prepared by a reported procedure (Ghosh et al., 2006).

Equipment needed

Schlenk flask, magnetic stir bar, stirrer, hotplate, sand/oil bath, freeze–pump–thaw setup.

Procedure

(1) **i.** To synthesize a 1:1 PEGMA:PDSMA polymer, weigh 537 mg of PDSMA and 1.00 g of PEGMA into a 10-mL Schlenk flask with a magnetic stir bar.

ii. To synthesize a 3:7 PEGMA:PDSMA polymer, weigh 749 mg of PDSMA and 598 mg of PEGMA in a 10-mL Schlenk flask with a magnetic stir bar.

(2) Add 1.2 mg of AIBN and 21 mg of 4-cyano-4-(phenylcarbonothioylthio) pentanoic acid in 3 mL of THF.

(3) Close the Schlenk flask and perform freeze–pump–thaw cycles (three cycles) to degas the reaction mixture with an argon inflow in the final cycle.

(4) Seal the reaction vessel and place it in a preheated oil/sand bath at 70°C with stirring at 350 rpm for 12 h.

(5) Pour the reaction mixture in 20 mL cold diethyl ether to obtain the polymer as a precipitate.

2.2 Protein–Polymer Conjugation Utilizing Surface Accessible Cysteines of Proteins

This approach is viable if the protein of interest has enough surface accessible cysteine groups in order to react with the polymer. As an example, a detailed method to conjugate caspase-3 enzyme with PEGMA–PDSMA polymer is described here.

2.2.1 Determination of Surface Accessible Cysteine Residues on the Protein

Method overview

To quantify the amount of surface accessible cysteine residues on the protein, Ellman's reagent (5,5′-dithiobis(2-nitrobenzoic acid)/DTNB) can be used (Fig. 3). DTNB, upon reaction with free thiols of proteins, generates a colored product 2-nitro-5-thiobenzoate (TNB) that absorbs at 412 nm. For every mole of thiol that reacts with DTNB, one mole of TNB is produced, and by quantifying the amount of TNB, the number of free cysteines present per mole of protein can be estimated.

Equipment and chemicals

Ellman's reagent (5,5′-dithiobis(2-nitrobenzoic acid)/DTNB) from Sigma-Aldrich, Protein (caspase-3), 1.5-mL Eppendorf microcentrifuge tubes, 20 m*M* Tris pH 8.0 buffer, UV–visible spectrophotometer.

Procedure

(1) Prepare a protein stock solution of 10 μ*M* concentration in Tris pH 8.0 buffer.

(2) Prepare Ellman's reagent stock solution of 500 μ*M* concentration in Tris pH 8.0 buffer.

(3) Mix 50 μL of protein stock solution with 50 μL of Ellman's reagent solution in a 1.5-mL microcentrifuge tube and incubate for 30 min at room temperature.

(4) Make the volume of this reaction mixture to 1 mL using Tris pH 8.0 buffer and measure the absorbance at 412 nm.

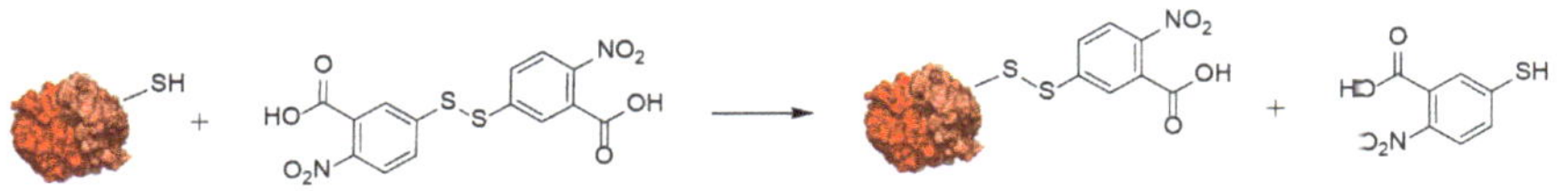

Fig. 3 Reaction scheme involved in quantifying the surface thiol groups on protein using Ellman's reagent.

(5) The number of reactive cysteines per protein can be calculated by fitting the absorption values into Beer–Lambert's equation: $A = \varepsilon.c.l$ where A = Absorbance, ε = molar extinction coefficient of TNB which is 14,150 M^{-1} cm^{-1}, c = concentration of TNB (which is equal to the concentration of free cysteines of the protein), and l is the absorption path length. The number of accessible cysteines per protein can be calculated by dividing the number of moles of cysteines (c) over number of moles of protein.

2.2.2 Reaction of Enzyme With Thiol Reactive Polymer and Synthesis of Nanogel Around the Protein

Once the number of surface accessible cysteines on the enzyme is estimated, we can calculate the amount of polymer to be used for conjugation. Typically, large excess of the polymer (molar excess of thiol reactive PDS functional groups on the polymer) should be used in order to completely engage the surface accessible cysteines on protein in conjugation with polymer.

Method overview

In this example, 0.06 mg of caspase-3 (0.002 μ*M* of enzyme and 0.01 μ*M* of total cysteines) is reacted with 3 mg of PEGMA:PDSMA polymer (containing 4 μ*M* of PDS) to form a protein–polymer conjugate. To create a polymeric network around the enzyme, self-cross-linking ability of the polymer will be utilized.

Taking the above example in perspective, 3 mg of polymer has about 4 μ*M* of PDS units. In order to achieve 20% cross-link density, i.e., to utilize 20% of PDS in polymer for cross-linking, 10 mole% DTT (0.4 μ*M*) should be added. This is because each DTT can cleave two PDS units on the polymer to result in one disulfide linkage. This cross-linking process will result in a polymeric network around the enzyme creating a nano-armor (Fig. 4). The additional unreacted PDS units on the polymer can be utilized to surface functionalize with thiol terminated PEG to provide better circulation ability or a thiol containing ligand for active targeting.

Equipment and chemicals

PBS buffer (pH 7.4), 3 mg/mL PEGMA:PDSMA polymer stock solution in PBS pH 7.4 buffer, stir plate, 7-mL glass vial and stir bar, caspase-3 stock solution 0.6 mg/mL PBS, 1 mg/mL DTT stock solution in PBS (pH 7.4), PEG thiol of MW1000 Da from Lysan Bio Inc. (stock solution of 20 mg/mL in PBS), fridge/cold room with a magnetic stir plate at 4°C.

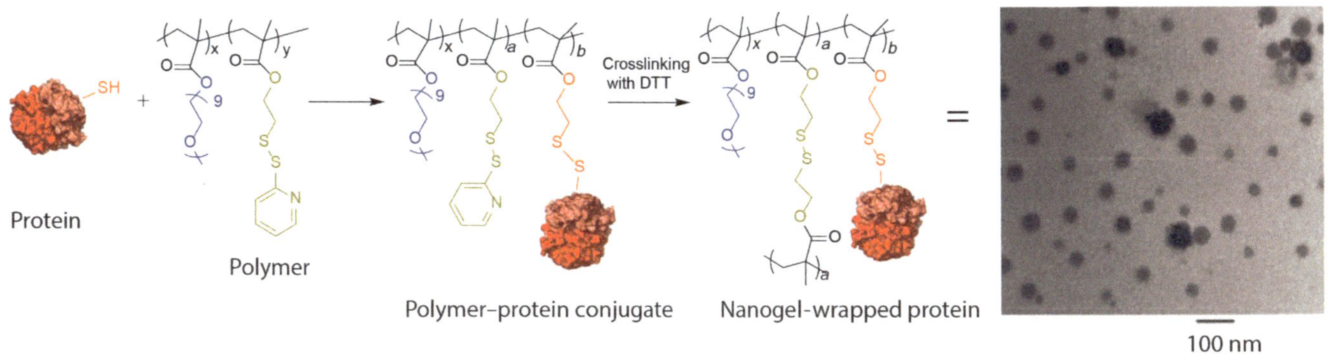

Fig. 4 Reaction scheme involving protein conjugation with polymer and self-cross-linking of polymer to result in a nanogel and its TEM image (Ventura et al., 2015).

Procedure

(1) Take 1 mL of polymer stock solution in a 7-mL glass vial with an appropriate stir bar and place it on a magnetic stir plate.
(2) Stir the polymer solution at 350 rpm at 20°C for 15 min.
(3) Add 100 μL (0.06 mg) of caspase-3 stock drop wise.
(4) Stir the solution for 1 h at 20°C.
(5) To this solution, add 62 μL of DTT stock solution (0.06 mg of DTT).
(6) Stir the solution for an additional 1 h at 20°C.
(7) Add 159 μL PEG thiol stock solution dropwise and stir the solution for an additional 24 h at 4°C.

Tips on enzyme conjugation

(1) Allow gentle stirring and avoid air bubbles inside the reaction to avoid protein denaturation.
(2) For the dropwise addition of the small volumes, use capillary pipette tips.

2.2.3 Separation of Unreacted Enzyme and By-Products From the Enzyme–Polymer Complex

The disulfide exchange reaction to form protein–polymer conjugate, polymer self-cross-linking, and nanogel surface with PEG thiol will result in pyridothione and oxidized DTT by-products which are removed by dialysis. The complete removal of pyridothione by-product was evaluated by monitoring its corresponding absorption peak at 340 nm.

Equipment and chemicals

7000 Da MWCO snake skin dialysis membrane and Amicon Ultra 100,000 Da MWCO centrifugal filters from Fisher Scientific, ultracentrifuge, PBS (pH 7.4) buffer, 1 L beaker with a magnetic stir bar, fridge/cold room with a magnetic stir plate at 4°C.

Procedure

(1) Take the nanogel sample into 7000 Da MWCO snake skin dialysis membrane and dialyze against 1 L PBS (pH 7.4) buffer in a beaker for 24 h by replacing PBS every 6 h.
(2) Collect the sample after dialysis and dialyze again using Amicon Ultra 100,000 Da MWCO centrifugal filters to remove unconjugated enzyme and to concentrate the sample to the desired volume.

2.2.4 Determination of Enzyme–Polymer Complexation and Enzyme Release

To verify the enzyme–polymer conjugation, SDS-PAGE technique can be utilized. Under nonreducing conditions, the conjugate, being high-molecular weight complex will have a lower mobility compared to just the protein by itself. However, the conjugate, upon incubation under reducing conditions, protein dissociates from the polymer and can be monitored by the appearance of a protein band.

Equipment and chemicals

SDS gel electrophoresis running buffer, DTT, 16% acrylamide gel, gel electrophoresis cell, enzyme-encapsulated nanogel solution (prepared in the previous steps).

Procedure

(1) To test the enzyme encapsulation, take 50 μg of nanogel in a microcentrifuge tube and heat it at 90°C for 5 min (DTT which is normally used for protein denaturation should not be used).

(2) To test the release of entrapped enzyme, take 50 μg of nanogel in a microcentrifuge tube and incubate it in 100 m*M* DTT solution for 1 h at 25°C. Later, heat it at 90°C for 5 min.

(3) Analyze the samples by SDS-PAGE using 16% acrylamide gel and using coomassie blue for staining (Fig. 5).

2.2.5 Enzyme Activity Study

Equipment and chemicals

Nanogel stock solution (with estimated 100 n*M* caspase-3 concentration based on protein feed during its preparation), 100 n*M* caspase-3 enzyme stock solution in caspase-3 buffer, Amicon Ultra 3000 Da MWCO centrifugal filters from Fisher Scientific, ultracentrifuge, caspase-3 buffer (20 m*M* HEPES pH 7.5, 5 m*M* $CaCl_2$, 150 m*M* NaCl, and 10% PEG 400), 1 *M* and 50 m*M* DTT stock solutions in caspase-3 buffer, fluorogenic substrate (*N*-acetyl Asp-Glu-Val-Asp-7-amino-4-methylcoumarin) from

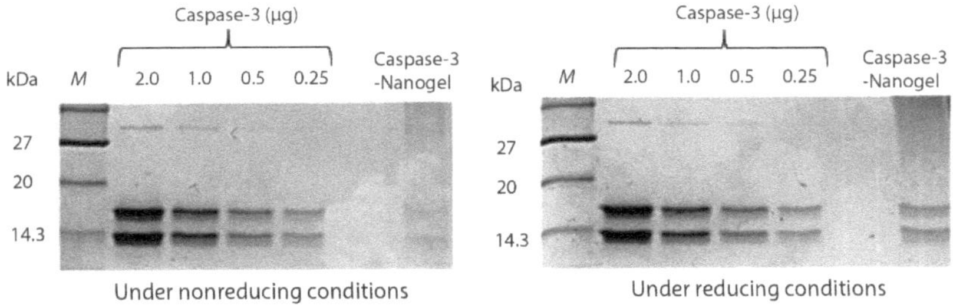

Fig. 5 SDS gel electrophoresis evaluation of protein–polymer complex under reducing and nonreducing conditions (Ventura et al., 2015).

Enzo Lifesciences (prepare 1 m*M* stock solution in caspase-3 buffer), DTT, temperature adjustable fluorescent plate reader (Fig. 6).

Procedure

(1) Take 90 μL of nanogel stock solution in a microcentrifuge tube and add 10 μL of 1 *M* DTT (to make a final DTT concentration of 100 m*M*).

(2) As a negative control, take 90 μL of nanogel stock solution and add 10 μL of 50 m*M* DTT (to make a final DTT concentration of 0.5 m*M*).

(3) Take 90 μL of 100 n*M* enzyme stock solution in two separate tubes and add 10 μL of 1 *M* DTT and 50 m*M* DTT as a reference for 100% enzyme activity.

(4) Incubate the samples at 25°C for 1 h.

(5) Take 90 μL of the each of the samples in a 96-black well plate.

(6) Add 10 μL of 1 m*M* Ac-DEVD-AMC substrate solution.

(7) Record the kinetic fluorescence by using λ_{Ex}: 365 nm/λ_{Em}: 495 nm for 7 min at 37°C and plot the slope of the kinetic curve over time to measure the enzyme activity.

2.3 Appending a Reactive Thiol to Protein

The success of the above methodology to conjugate enzyme with polymers using cysteines usually depends on the number of surface accessible cysteine units on the enzyme and their reactivity. However, this method is not very

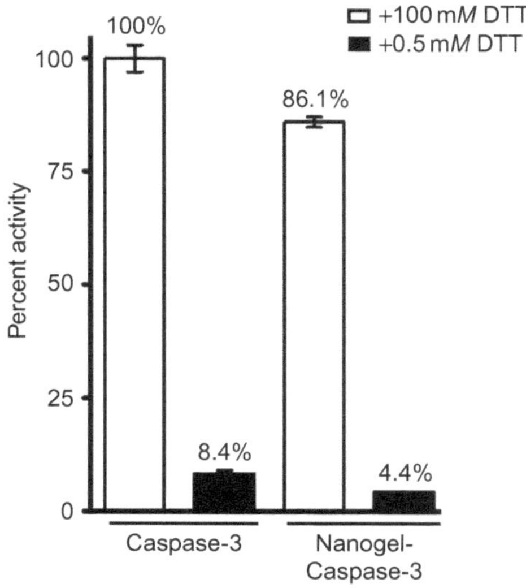

Fig. 6 Activity recovery of enzyme after release from the nanogel (Ventura et al., 2015).

efficient with proteins having less accessible cysteine units. To circumvent this, an alternative approach can be used where proteins are modified to append a thiol functional group on their surface (Matsumoto et al., 2013) (Fig. 7). This was achieved by following well-established protocols such as using SATP (*N*-succinimidyl-*S*-acetylthio propionate) or SATA (*N*-succinimidyl *S*-acetylthioacetate) to modify the lysine-based primary amines present on the protein surface.

Equipment and chemicals

SATP from Fisher Scientific (stock solution of 16 mg of SATP in 1 mL DMF, 50 m*M* sodium phosphate (pH 7.5) buffer solution with 1 m*M* EDTA), 0.5 *M* hydroxylamine hydrochloride in 50 m*M* sodium phosphate (pH 7.5) buffer solution with 25 m*M* EDTA, Amicon Ultra 3000 Da MWCO centrifugal filters from Fisher Scientific, FPLC set up with superose 6, 10/300 column with temperature set up at 4°C, elution solvent for FPLC (50 m*M* sodium phosphate (pH 7.5) buffer containing 15 m*M* NaCl), UV–visible spectrophotometer.

Procedure

(1) Weigh 5 mg of protein of interest and dissolve it in 0.5 mL of pH 7.5 50 m*M* sodium phosphate solution with 1 m*M* EDTA in a microcentrifuge tube.

(2) To that solution, add 5 μL of SATP stock solution, mix it gently, and incubate for 1 h at 20°C.

(3) Remove the unreacted SATP by ultrafiltration using Amicon Ultra centrifugal filters with suitable molecular weight cutoff (lower than the molecular weight of protein).

(4) Resuspend the concentrated protein in the ultracentrifuge tube in 0.5 mL of pH 7.5 PBS.

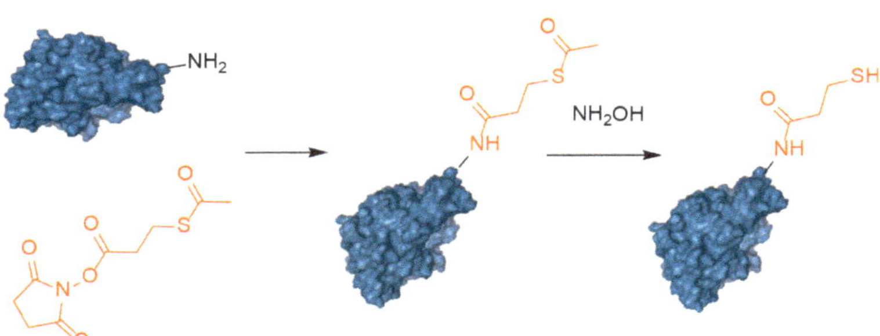

Fig. 7 Reaction scheme involving modification of amines on the protein to thiols using a SATP reagent.

(5) For the deprotection of acetylated thiols, add 50 μL of hydroxylamine hydrochloride stock solution to the protein solution and incubate for 2 h at 20°C.
(6) Use FPLC to purify the protein and concentrate it to the desired volume using ultrafiltration.
(7) Calculate the amount of free thiol using Ellman's assay described in Section 2.2.1.
(8) Once the thiol moiety is generated on the protein, the method described earlier (Sections 2.2.2–2.2.3) should be followed for polymer conjugation and nanogel synthesis.

3. ELECTROSTATIC COMPLEXATION

As discussed earlier, the disulfide mode of covalent conjugation would be an ideal choice if the target site to deliver the enzyme is cell cytosol. This is because of the cytosolic reducing environment, which can detach the polymer from the enzyme. However, if the enzyme's target site of delivery is noncytosolic (such as lysosomes) where no reducing environment is present, to cleave the polymer form the enzyme, disulfide conjugation strategy may not be a suitable choice. For example, β-galactosidase is an enzyme that is present in the lysosomes. The deficiency of this enzyme is associated with lysosomal storage diseases such as gangliosidosis, Morquio syndrome B, and galactosialidosis. Enzyme replacement therapies have been considered to supplement these deficient enzymes into the lysosomes (Desnick & Schuchman, 2012). Since lysosomes do not possess a reducing environment necessary to shed the disulfide based polymer wrapping around the enzyme, alternate methods of enzyme wrapping such as electrostatic complexation and entrapment of protein inside a polymeric matrix by noncovalent means could be of great use.

Proteins possess either a positive or a negative surface charge depending on their amino acid composition and pH of the solution in which they are present. Electrostatic complexation method involves utilizing the surface charge of a protein to complex with a polymer having a complementary charge. The example of electrostatic complexation that will be described here is based on enzyme β-galactosidase (β-gal) and polymeric nanogel derived from PEGMA:PDSMA amphiphilic random polymers (González-Toro et al., 2012). The isoelectric point (pI) of β-gal is 4.8, which suggests that at physiological pH (pH 7.4) the enzyme will have a net negative surface charge and therefore it can be complexed with a nanogel having a net positive charge. The protein complexation with nanogel involves three steps as follows:

(i) Synthesis of nanogels
(ii) Surface charge generation on the nanogels
(iii) Electrostatic complexation of the protein with nanogels

3.1 Synthesis of Polymer and Nanogels

To synthesize PEGMA:PDSMA polymer 3:7 ratio, the method described in Section 2.1 can be adopted. The polymer, when dissolved in an aqueous solution, forms nanomicellar assemblies and has an ability to encapsulate hydrophobic guest molecules. By using a calculated amount of reducing agent (DTT), these micellar assemblies are crosslinked to form nanogels.

3.1.1 Encapsulation of Hydrophobic Guests by Polymeric Micelle

Hydrophobic guest encapsulation by the amphiphilic polymer can be achieved by cosolvent method. A suitable choice of organic solvent should be made after considering solubility and stability of the hydrophobic molecules. Here, we describe the encapsulation procedure using hydrophobic DiI dye that is soluble in acetone as an example.

Equipment and chemicals

DiI dye (1,1′-dioctadecyl-3,3,3′,3′-tetramethylindocarbocyanine perchlorate ("DiI"; DiIC_{18}(3))) purchased from Fisher Scientific, acetone, PEGMA:PDSMA (3:7) polymer stock solution in PBS (pH 7.4) buffer, stir plate, 7-mL glass vial and stir bar, 1 mg/mL DTT stock solution in PBS (pH 7.4).

Procedure

(1) Prepare a 2-mg/mL DiI stock solution in acetone (depending on the solubility of the hydrophobic molecule an organic solvent must be chosen which is miscible in water and has low boiling point).
(2) Take 1 mL of polymer stock solution in a 7-mL vial with a magnetic stir bar and stir the solution at 350 rpm.
(3) To the polymeric solution, add 100 μL of hydrophobic stock solution dropwise under constant stirring. Care should be taken to ensure that the volume ratio of organic to aqueous solution is kept to a minimum (around 0.1 or less).
(4) Allow this mixture to stir in open air for 6 h to evaporate the organic solvent to form a polymeric micelle with guest molecules localized in its core.

(5) The polymeric micelle solution should be passed through a 0.45-μm nylon filter to remove the un-encapsulated hydrophobic molecules.

3.1.2 Cross-linking of the Polymeric Micelle

One mole of DTT can react with two moles of PDS functionalities in the polymer by a disulfide exchange reaction to generate one mole disulfide cross-links. Therefore, to attain 10% cross-link density (utilize 10 mole% of PDS in polymer) 5 mole% of DTT relative to the amount of PDS functionalities should be used.

In the current example which has PEGMA:PDSMA in 3:7 ratio, every 10 mg of polymer contains 0.009 mmoles of PEGMA monomer (MW 473 g/mol) and 0.021 mmoles of PDSMA monomer (MW 255 g/mol). To make a 10% cross-link nanogel, the amount of DTT that is required will be 5% relative to the PDS, which is 0.001 mmoles.

Equipment and chemicals

10 mg/mL polymeric micelle stock solution prepared in the previous step, sodium phosphate buffer (5 m*M* pH 7.4), stir plate, 7-mL glass vial and stir bar, 1 mg/mL DTT stock solution in phosphate buffer 5 m*M* (pH 7.4), 3000 Da MWCO snake skin dialysis membrane (purchased from Fisher Scientific), fridge/cold room with a magnetic stir plate at 4°C.

Procedure

(1) Take 1 mL of polymeric micelle solution which was prepared previously (10 mg/mL polymer concentration) in a vial with a magnetic stir bar.

(2) Stir the solution at 350 rpm and to it add 154 μL of DTT stock solution dropwise.

(3) Allow the reaction to stir for at least 1 h at 25°C.

(4) Monitor the progress of reaction by measuring the absorption peak at 340 nm every 30 min (resulting from the pyridothione by-product).

(5) The cross-linking reaction is complete if there is no further increase in the absorbance at 340 nm over time.

(6) Take the reaction mixture into 3000 Da MWCO snake skin dialysis membrane and dialyze it against 1 L sodium phosphate buffer (5 m*M* pH 7.4) solution for 24 h at 4°C by replacing the dialysis buffer every 6 h to remove pyridothione by-product produced in the cross-linking reaction.

3.2 Modification of Surface Charge on Nanogels

The nanogels prepared so far using the above method do not possess any surface charge. However, the PDS functional groups present on these cross-linked nanogels provide an opportunity to modify their surface using a charged ligand. Depending on the surface charge of the protein, a complementary surface charge on the nanogel can be easily generated using this method (Fig. 8).

In this example, β-galactosidase possesses a negative surface charge at neutral pH. To facilitate its electrostatic complexation with nanogels, a positive surface charge should be incorporated on the nanogels. In case the enzyme possesses a positive surface charge, a negative surface charge needs to be generated on the nanogel.

To generate a positive surface charge on the nanogel, a ligand such as triarginine peptide with a cysteine on the C-terminus (CRRR) can be used (González-Toro et al., 2012) (Fig. 8). The thiol group on cysteine will react with the reactive PDS functional groups of the nanogel thus, appending the positive triarginine group on the surface of the nanogel. The conjugation between nanogel and CRRR can be monitored by the evolution of the pyridothione group via its absorption spectrum, in addition to the surface charge measurements.

3.2.1 Surface Modification of Nanogel Using CRRR Peptide

The acceptable amount of ligand that can be conjugated to the nanogels depends on the available PDS functional groups on the nanogel. Therefore, while calculating the amount of ligand that can be conjugated to the nanogel, the cross-link density of nanogel (amount of PDS groups utilized from the polymer to make a nanogel) must be taken into account.

In this example, 10 mg of polymer (which contains 0.021 mmol of PDS monomer) was used to make a nanogel with 10% cross-link density (consumed 0.0021 mmol of PDS on polymer during cross-linking) as described in Section 3.1.2. Therefore, the number of available PDS groups on the resulting nanogel can be calculated by subtracting the moles of PDS groups consumed in cross-linking from the moles of PDS in the polymer which would be 0.021–0.0021 = 0.0189 mmol. Thus, the amount of thiol containing ligands that can be added for decorating nanogels must not exceed 0.0189 mmol.

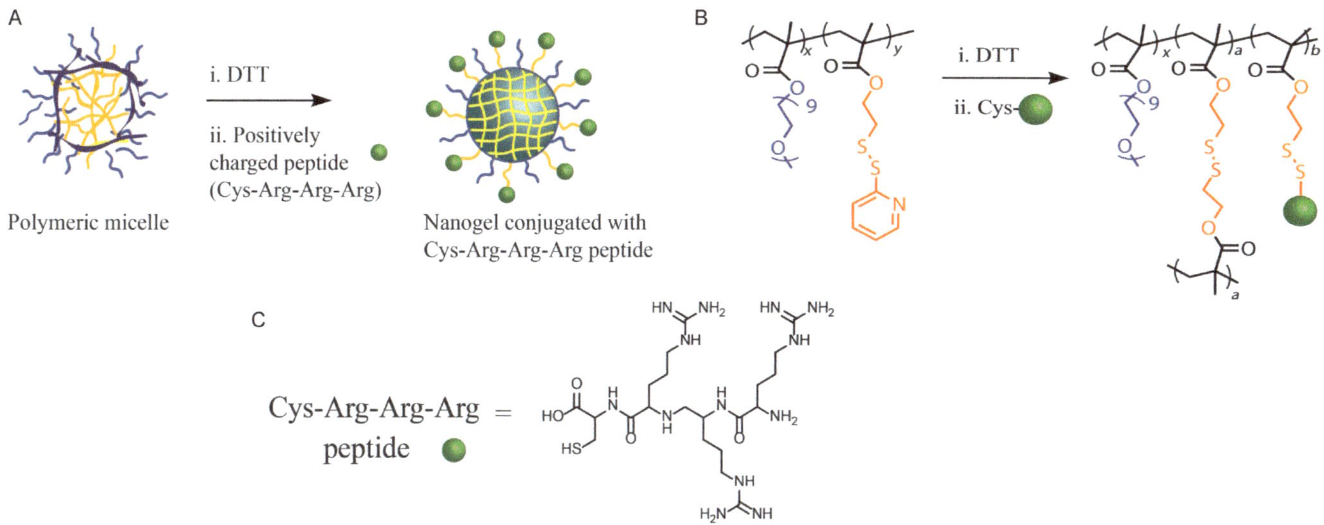

Fig. 8 (A) Cartoon and (B) structural representation of nanogel synthesis and its surface modification with (C) triarginine peptide for generating a positive charge on nanogel (González-Toro et al., 2012).

Equipment and chemicals

CRRR peptide (synthesized using a reported procedure (González-Toro et al., 2012) or available in commercial sources such as GenScript Corporation), Amicon Ultra 10,000 Da MWCO centrifugal filters, ultracentrifuge, stir plate, 7-mL glass vial and stir bar, nanogel stock solution (prepared in the previous step), PBS (pH 7.4) buffer.

Procedure

(1) Take 1 mL of 10 mg/mL nanogel solution in a 7-mL glass vial with a magnetic stir bar placed on a stirrer with stirring speed of 350 rpm.

(2) Add 5 mg of CRRR peptide (0.0084 mmoles) to the nanogel solution and allow it to stir for 12 h at 25°C.

(3) Transfer the reaction mixture into Amicon Ultra 10 kDa MWCO and dialyze the solution to remove unreacted CRRR peptide and pyridothione by-product.

(4) Resuspend the peptide conjugated nanogel solution in phosphate buffer to make it to the volume of desired concentration.

3.3 Electrostatic Complexation of Protein With Nanogels and Its Characterization

The complementary surface charge of protein and nanogel can be utilized to achieve electrostatic complexation to result in the formation of nanogel–protein aggregates. The complexation ratio between the nanogel and protein needs to be optimized by carefully monitoring the properties of the resulting nanoaggregates. A spectroscopic method based on fluorescence resonance energy transfer (FRET) can be used to optimize this complexation ratio. This can be achieved when nanogel and protein are labeled with two different dyes which are FRET partners. Fluorescein and DiI dyes make one such FRET pair, where the absorption spectrum of the DiI (acceptor) overlaps with the emission spectrum of fluorescein (donor).

In this example, DiI dye (FRET acceptor) is encapsulated inside the nanogel, and fluorescein dye (FRET donor) is conjugated to β-galactosidase by labeling with FITC. When nanogel and *β*-gal form a complex, the associated proximity of these dyes lead to FRET observation, which is characterized by a decrease in the emission intensity of donor and increase in intensity of the acceptor. By keeping the amount of one complexing partner constant which is β-gal in this case (FRET donor), the nanogel (FRET acceptor) is slowly added in increments. The emission intensity of the donor (fluorescein) will continue to decrease until a stable complex is formed and further addition of nanogel will not result in any more decrease in the emission intensity of donor (Fig. 9).

Equipment and chemicals

Fluoroscein isothiocyanate isomer I from Sigma-Aldrich, stock solution in DMSO solvent (4 mg/mL), 0.1 *M* sodium bicarbonate solution (pH 9.0), β-galactosidase protein, size exclusion chromatography system with Sephadex G-25 as stationary phase and phosphate buffer (5 m*M*, pH 7.4) as mobile phase, UV–visible spectrophotometer, stir plate, 7-mL glass vial and stir bar.

Procedure to label the enzyme with dye

(1) Take 2.5 mg of β-gal in 900 μL of 0.1 *M* sodium bicarbonate solution in a 7-mL glass vial with a magnetic stir bar.
(2) Add 250 μL of FITC stock solution.
(3) Wrap the vial with aluminum foil to protect it from light and allow the mixture for gentle stirring for 2 h.
(4) Purify the protein using SEC and concentrate using ultrafiltration using Amicon Ultra 30,000 Da MWCO tubes.
(5) Calculate the protein concentration using UV–vis absorption spectroscopy and adjust the concentration of protein to the desired amount using sodium phosphate buffer (5 m*M* pH 7.4).

Procedure to complex protein with nanogel

(1) Prepare 2 mg/mL stock solution of nanogel–CRRR using sodium phosphate buffer (5 m*M* pH 7.4).
(2) Prepare 1 mg/mL stock solution of fluorescein-labeled β-gal solution in sodium phosphate buffer (5 m*M* pH 7.4).
(3) Prepare a range of complex solutions by using volume ratios between β-gal:nanogel–CRRR of 1:1–1:6 and incubate them for 1 h at 25°C.
(4) Monitor emission spectra of the formed complex solutions by using excitation wavelength of 480 nm to identify the stable complexation ratio.
(5) Identify the stable complexation ratio such that, increase in the amount of NG–CRRR addition does not result in decrease in the emission of fluorescein.

The success of this electrostatic complexation can also be verified by studying the surface charge and the size of the aggregates (Fig. 10). Since the nanogel surface is positively charged, upon formation of nanoaggregate, a drift in the surface charge from positive to negative zeta potential can be observed. Similarly, complexes have a larger size compared to nanogel and protein by itself.

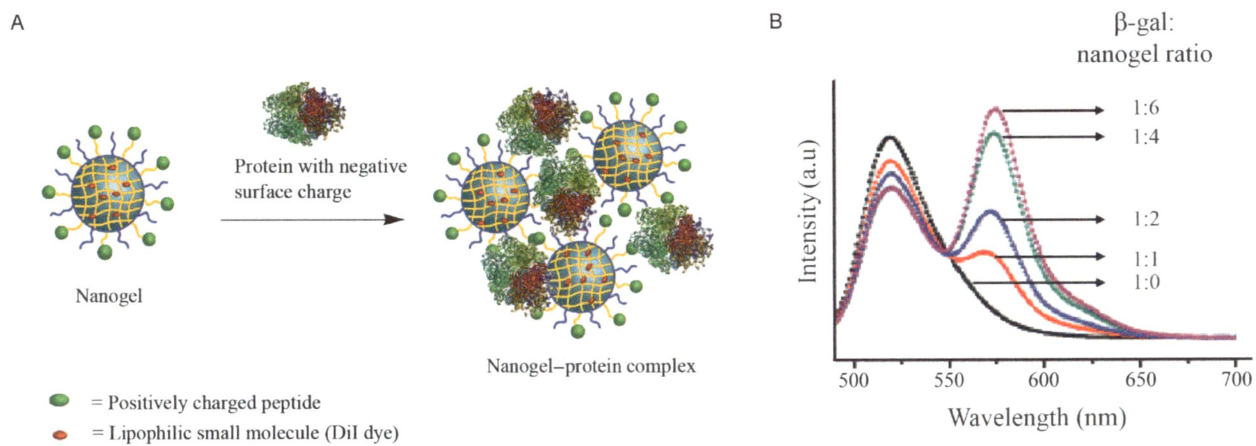

Fig. 9 (A) Pictorial representation of protein complexing with nanogel. (B) FRET study to optimize the complexation ratio between nanogel and protein (González-Toro et al., 2012).

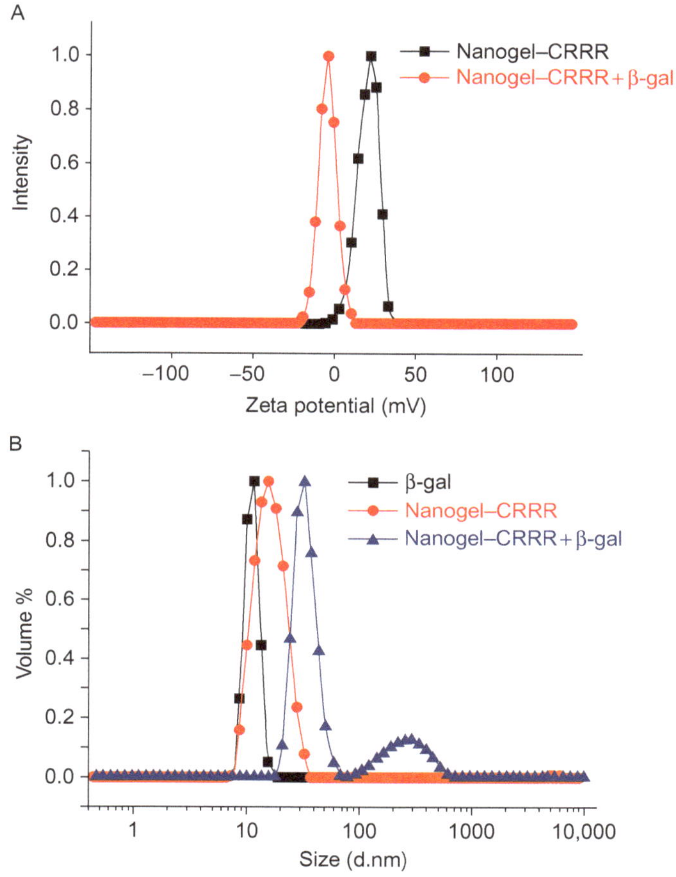

Fig. 10 Evaluation of complexation between nanogel and protein with (A) change in surface charge and (B) increase in size (González-Toro et al., 2012).

4. NONCOVALENT ENTRAPMENT

Both the covalent conjugation and electrostatic complexation methods for nano-armoring of an enzyme rely on the physical and chemical properties of the protein surface. In the case of covalent conjugation, the density of reactive functional groups on the protein surface, which can participate in covalent bond formation, determines the success of the method. However, in electrostatic complexation, the extent of surface charge of the protein will ultimately decide the stability of the complex formed with the polymer.

Depending on the structural and physical properties of a particular protein, one of these methods can be more promising over the other. However, for proteins with less accessible functional groups on their surface and with less overall surface charge, noncovalent entrapment method would be a favorable choice. This method is based on inverse emulsion polymerization through which a hydrogel matrix can be synthesized around the protein to provide a nano-armor.

This methodology of protein entrapment in nanoscale hydrogels involves the following steps:

(i) Preparation of inverse emulsion solution and nanogel synthesis
(ii) Nanogel extraction
(iii) Protein release and activity

4.1 Preparation of Inverse Emulsion Solution and Nanogel Synthesis

In this method of inverse emulsion polymerization, heptane and Brij L4 were chosen as the continuous phase and surfactant, respectively. Addition of a small amount of water relative to the amount of organic solvent to this mixture will result in an inverse miniemulsion which contains nanoscale water droplets stabilized by the surfactant in bulk organic phase.

The aqueous solution used in this process contains all the necessary starting materials to synthesize a polymeric nanogel matrix such as monomers, cross-linkers, free radical initiators along with the protein cargo that needs to be encapsulated inside the nanogel. The monomer is usually an acrylate or acrylamide derivative which results in linear chains upon polymerization, while the cross-linker based on bisacrylate or bisacrylamide derivatives can form a polymeric network resulting in a nanogel while trapping the protein in its network.

The ability of nanogel to release the entrapped protein cargo will depend on the responsive behavior of cross-linker used. Literature precedents show that the choice of cross-linkers depend on the desired stimuli-responsive features required on the nanogels such as pH (Standley et al., 2004) or redox (Ashwinkumar, Maya, & Jayakumar, 2014), light (Azagarsamy et al., 2012). In this example, nanogel synthesis using a pH-responsive cross-linker, which is a derivative of β-thiopropionate ester (Molla et al., 2014), will be described (Fig. 11).

The feed molar ratio between the monomer and the cross-linker will determine the extent of cross-linked network within the nanogel. This ratio can be varied to fine tune the controlled release of the encapsulated cargo

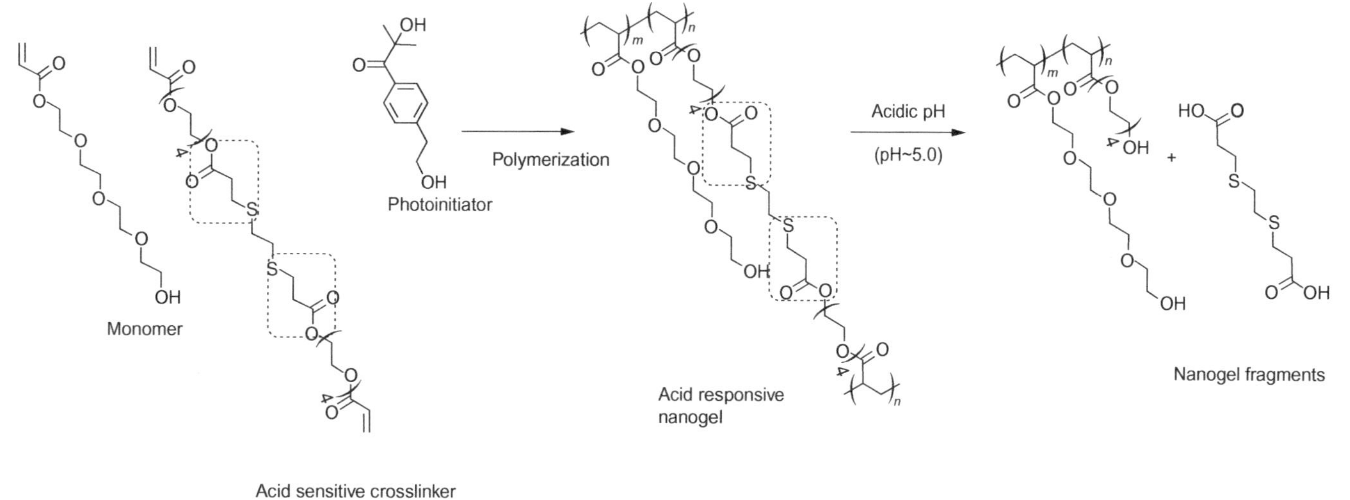

Fig. 11 Synthetic scheme involved in nanogel synthesis and its responsive behavior toward acidic pH (Molla et al., 2014).

(protein) from the nanogel. In this example, the molar ratio between monomer and cross-linker was about 90:10. This is sufficient to make stable nanogels, even though there are some other examples in the literature, where 97.5:2.5 was used to make nanogels which demonstrate that nanogel formation requires only a small amount of cross-linker relative to the amount of monomer (Azagarsamy et al., 2012). The scheme for nanogel synthesis using inverse emulsion polymerization is shown in Fig. 12.

To carry out the polymerization, there are different choices of free radical initiators that can be used, the criterion for which is that it must be soluble in aqueous solution along with monomer and cross-linker. For example, a photoinitiator such as Irgacure (2-hydroxy-4′-(2-hydroxyethoxy)-2-methylpropiophenone) can be used such that the nanogel formation can be triggered simply by exposure to UV light. Alternatively, ammonium persulfate initiator can be used, where TEMED (tetramethylethylenediamine) catalyst is used to trigger polymerization reaction.

Equipment and chemicals

Brij L4 surfactant (Sigma-Aldrich), heptane, *n*-butanol, tetraethylene glycol acrylate monomer, and β-thioester diacrylate cross-linker synthesized as per reported procedure (Molla et al., 2014), α-glucosidase enzyme (GAA), 5.5 mg/mL stock solution of Irgacure (2-hydroxy-4′-(2-hydroxyethoxy)-2-methylpropiophenone) from Sigma-Aldrich, stir plate, 20 mL vial with magnetic stir bar, vortex mixer, bath sonicator, argon gas line with 18-G needle, UV chamber equipped with stirring plate.

Procedure

(1) Take 0.6 g of Brij L4 surfactant and 5 mL of heptane in a 20-mL glass vial with a magnetic stir bar and mix thoroughly until a clear solution is obtained.

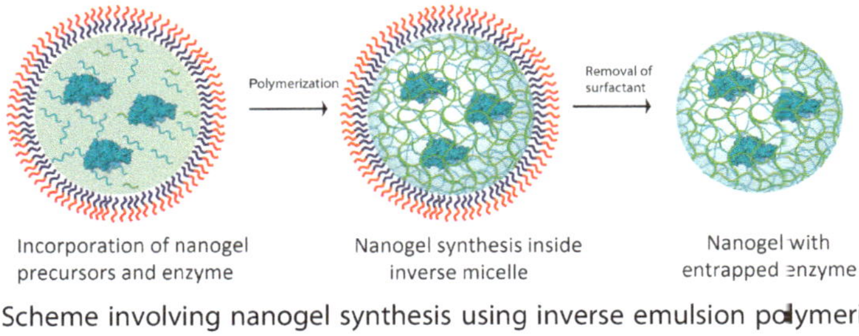

Fig. 12 Scheme involving nanogel synthesis using inverse emulsion polymerization.

(2) In a microcentrifuge tube take 150 mg of tetraethylene glycol acrylate monomer (0.604 mmol), 18 mg of pH-responsive β-thioester diacrylate cross-linker (0.06 mmol), 100 μL of initiator stock solution, 2 mg of α glucosidase and mix them thoroughly using vortex.
(3) Mix the polymer precursors and protein (step 2) into the surfactant in heptane (step 1) and vortex for 5 min followed by sonication for an additional 5 min to form a miniemulsion.
(4) Under constant stirring at 500 rpm, purge the miniemulsion solution with argon for about 10 min to remove oxygen and then close the cap of the vial.
(5) Place the vial in a UV chamber with 365 nm light source for 20 min under constant stirring at 500 rpm to form nanogel.

4.2 Nanogel Extraction

The next step in this procedure involves extraction of nanogels from the bulk organic solvent, which can be carried out using two different methods: solvent extraction method and gravimetric extraction method. The purified nanogels are characterized by dynamic light scattering and TEM (Fig. 13).

4.2.1 Solvent Extraction Method

Equipment and chemicals

PBS (pH 7.4 buffer), *n*-butanol, 50-mL centrifuge tubes made of polypropylene material, centrifuge, Pasteur pipettes, vortex mixer.

Procedure

(1) In a 50-mL centrifuge tube measure 16 mL of PBS (pH 7.4) buffer.
(2) To it, add the nanogel solution prepared in the previous step and mix it thoroughly using vortex for about a minute.
(3) To it, add 2 mL of *n*-butanol and centrifuge for 15 min at 3000 × *g*.
(4) Discard the organic layer using a Pasteur pipette
(5) Perform steps 3 and 4 twice to remove any surfactant present in the aqueous solution
(6) Take small amount of aqueous solution (about 300 μL) and lyophilize it to calculate the concentration of nanogel using the same volume of PBS buffer as blank

Tip

For enzymes that are fragile, the amount of time involved in vortexing and sonication must be minimized as it could damage their structure to result in a loss of its enzymatic activity.

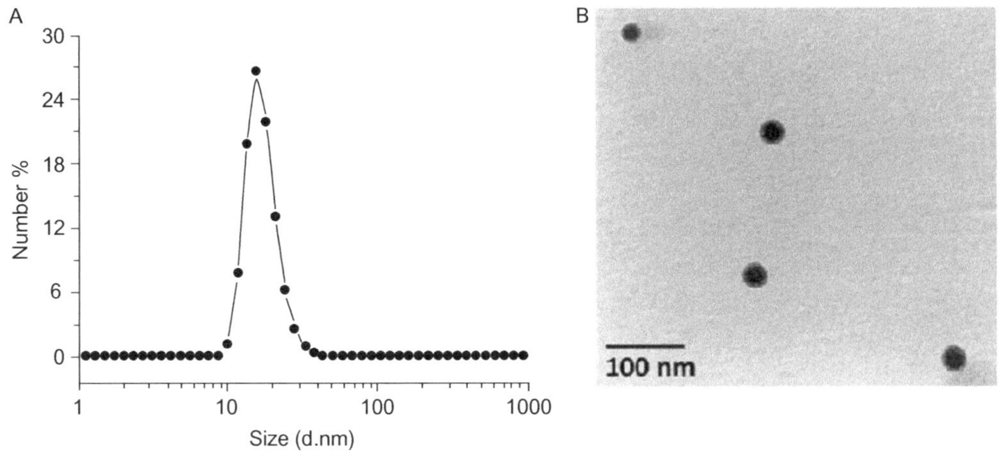

Fig. 13 Nanogel characterization by (A) DLS and (B) TEM (Molla et al., 2014).

4.2.2 Gravimetric Extraction Procedure

In this method, the organic solvent and the surfactant from the nanogel reaction mixture are removed by taking advantage of the density of nanogel, which is associated with aqueous solvent. Since the nanogel is associated with trace amounts of water, its density is more comparable to that of bulk organic phase, which is heptane in this case. This difference in density can be utilized to pellet the hydrogel from the bulk organic phase by using a high centrifugal force.

Equipment and chemicals

PBS (pH 7.4 buffer), 10-mL centrifuge tubes made of polypropylene material, centrifuge.

Procedure

(1) Take the nanogel solution formed after UV irradiation in a centrifuge tube.
(2) Centrifuge for 30 min at 15,000 × *g* at 4°C to pellet the nanogel.
(3) Decant the supernatant and then wash the pellet with 5 mL of heptane.
(4) Repeat steps 2 and 3 twice and resuspend the nanogel pellet in PBS solution or a suitable buffer of choice.
(5) To calculate the concentration of nanogel, take small amount of aqueous solution and lyophilize it to calculate the concentration of nanogel using the same volume of buffer as blank.

4.3 Activity Assay for the Released Enzyme

Protein quantification assay such as bicinconic acid assay (Pierce™ BCA Protein Assay protocol) can be utilized to estimate the amount of protein in the nanogel sample. In order to determine the activity of enzyme entrapped in the nanogel, GAA activity assay can be performed using *para*-nitrophenol-α-D-glucopyranoside as a substrate. The active enzyme will cleave the substrate to produce *para*-nitrophenolate product which has an absorbance maximum at 400 nm. The activity data for the enzyme released from the nanogel are shown in Fig. 14.

Equipment and chemicals

100 m*M* citrate buffer (pH 5), 200 m*M* sodium borate buffer (pH 10), *para*-nitrophenol-α-D-glucopyranoside (Sigma-Aldrich), nanogel solution with an estimated concentration of encapsulated protein, 96-well clear flat-bottom plates, plate reader for measuring absorbance.

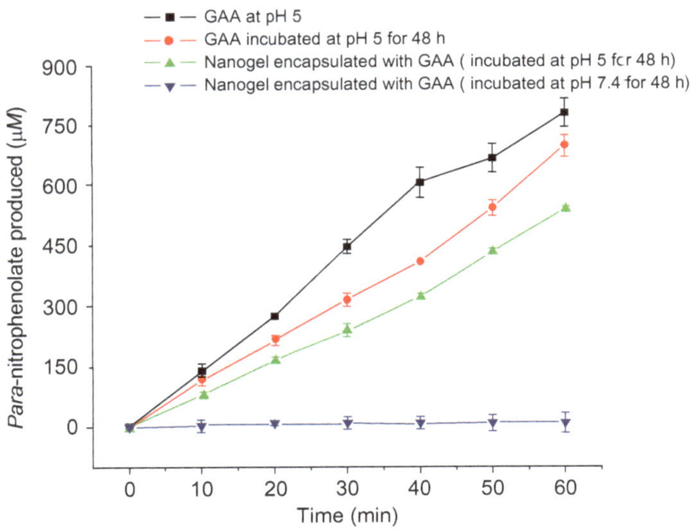

Fig. 14 Activity recovery of enzyme after release from the nanogel (Molla et al., 2014).

Procedure

(1) Make a nanogel sample stock solutions in (i) 100 m*M* citrate (pH 5) buffer and (ii) PBS (pH 7.4) having protein concentration 0.34 μ*M* and incubate it for 48 h at 4°C.

(2) Prepare a α-glucosidase stock solution in 100 m*M* Citrate (pH 5) buffer and incubate it for 48 h at 4°C.

(3) Take 50 μL of substrate solution in an ultracentrifuge tube and add 50 μL of nanogel/protein stock solution to it which was prepared in previous steps and mix gently.

(4) For activity measurement, take 10 μL of the reaction mixture (step 2) and add it to 290 μL of 200 m*M* sodium borate buffer (pH 10) and measure the absorbance at 400 nm.

(5) Take the activity measurement every 10 min for an hour to get the activity profile.

5. CONCLUSIONS

Nano-armoring of enzymes using polymers is a useful technique to formulate therapeutic proteins for systemic administration. While armoring enzymes using polymers serve the purpose of providing a protective shield from aggregation and degradation, it also offers an additional feature of surface functionalization thus, providing targeting capabilities to the enzymes. In this chapter, we have discussed three different methods to formulate enzyme–polymer complexes based on the type of interactions between them. Depending on the physical and chemical properties of the enzyme of interest and the therapeutic target, a suitable formulation method can be applied to provide a nano-armor for the enzyme. Covalent means of enzyme–polymer complexes will be a suitable technique for proteins that have surface accessible and reactive functional groups. Alternatively, electrostatic complexation methods can be employed for enzymes having a significant overall surface charge at physiological pH. Noncovalent enzyme entrapment methods, on the other hand, would serve better if the enzyme of interest lacks surface accessible reactive functional groups or if using covalent means is detrimental to maintaining the activity of the protein cargo.

ACKNOWLEDGMENTS

We thank the Army Research Office (W911NF-15-1-0568) and the National Institutes of Health (GM-065255 and CA-169140) for partial support of this research.

REFERENCES

Ashwinkumar, N., Maya, S., & Jayakumar, R. (2014). Redox-responsive cystamine conjugated chitin–hyaluronic acid composite nanogels. *RSC Advances*, *4*, 49547–49555.

Ayame, H., Morimoto, N., & Akiyoshi, K. (2008). Self-assembled cationic nanogels for intracellular protein delivery. *Bioconjugate Chemistry*, *19*, 882–890.

Azagarsamy, M. A., Alge, D. L., Radhakrishnan, S. J., Tibbitt, M. W., & Anseth, K. S. (2012). Photocontrolled nanoparticles for on-demand release of proteins. *Biomacromolecules*, *13*, 2219–2224.

Babu, K. R., Moradian, A., & Douglas, D. (2001). The methanol-induced conformational transitions of β-lactoglobulin, cytochrome c, and ubiquitin at low pH: A study by electrospray ionization mass spectrometry. *Journal of the American Society for Mass Spectrometry*, *12*, 317–328.

Balendiran, G. K., Dabur, R., & Fraser, D. (2004). The role of glutathione in cancer. *Cell Biochemistry and Function*, *22*, 343–352.

Brange, J., Andersen, L., Laursen, E. D., Meyn, G., & Rasmussen, E. (1997). Toward understanding insulin fibrillation. *Journal of Pharmaceutical Sciences*, *86*, 517–525.

Brocchini, S., Godwin, A., Balan, S., Choi, J.-w., Zloh, M., & Shaunak, S. (2008). Disulfide bridge based PEGylation of proteins. *Advanced Drug Delivery Reviews*, *60*, 3–12.

Chang, H. Y., & Yang, X. (2000). Proteases for cell suicide: Functions and regulation of caspases. *Microbiology and Molecular Biology Reviews*, *64*, 821–846.

Cohen, J. A., Beaudette, T. T., Tseng, W. W., Bachelder, E. M., Mende, I., Engleman, E. G., et al. (2008). T-cell activation by antigen-loaded pH-sensitive hydrogel particles in vivo: The effect of particle size. *Bioconjugate Chemistry*, *20*, 111–119.

Curatolo, L., Valsasina, B., Caccia, C., Raimondi, G. L., Orsini, G., & Bianchetti, A. (1997). Recombinant human IL-2 is cytotoxic to oligodendrocytes after in vitro self aggregation. *Cytokine*, *9*, 734–739.

Dai, C., Wang, B., Zhao, H., Li, B., & Wang, J. (2006). Preparation and characterization of liposomes-in-alginate (LIA) for protein delivery system. *Colloids and Surfaces. B, Biointerfaces*, *47*, 205–210.
Desnick, R. J., & Schuchman, E. H. (2012). Enzyme replacement therapy for lysosomal diseases: Lessons from 20 years of experience and remaining challenges. *Annual Review of Genomics and Human Genetics*, *13*, 307–335.
Dyal, A., Loos, K., Noto, M., Chang, S. W., Spagnoli, C., Shafi, K. V., et al. (2003). Activity of Candida rugosa lipase immobilized on γ-Fe2O3 magnetic nanoparticles. *Journal of the American Chemical Society*, *125*, 1684–1685.
Fishburn, C. S. (2008). The pharmacology of PEGylation: Balancing PD with PK to generate novel therapeutics. *Journal of Pharmaceutical Sciences*, *97*, 4167–4183.
Ghosh, S., Basu, S., & Thayumanavan, S. (2006). Simultaneous and reversible functionalization of copolymers for biological applications. *Macromolecules*, *39*, 5595–5597.
Ghosh, P., Yang, X., Arvizo, R., Zhu, Z.-J., Agasti, S. S., Mo, Z., et al. (2010). Intracellular delivery of a membrane-impermeable enzyme in active form using functionalized gold nanoparticles. *Journal of the American Chemical Society*, *132*, 2642–2645.
González-Toro, D. C., Ryu, J.-H., Chacko, R. T., Zhuang, J., & Thayumanavan, S. (2012). Concurrent binding and delivery of proteins and lipophilic small molecules using polymeric nanogels. *Journal of the American Chemical Society*, *134*, 6964–6967.
Heredia, K. L., & Maynard, H. D. (2007). Synthesis of protein-polymer conjugates. *Organic & Biomolecular Chemistry*, *5*, 45–53.
Hermeling, S., Crommelin, D. J. A., Schellekens, H., & Jiskoot W. (2004). Structure-immunogenicity relationships of therapeutic proteins. *Pharmaceutical Research*, *21*, 897–903.
Joralemon, M. J., McRae, S., & Emrick, T. (2010). PEGylated polymers for medicine: From conjugation to self-assembled systems. *Chemical Communications*, *46*, 1377–1393.
Knubovets, T., Osterhout, J. J., & Klibanov, A. M. (1999). Structure of lysozyme dissolved in neat organic solvents as assessed by NMR and CD spectroscopies. *Biotechnology and Bioengineering*, *63*, 242–248.
Leader, B., Baca, Q. J., & Golan, D. E. (2008). Protein therapeutics: A summary and pharmacological classification. *Nature Reviews. Drug Discovery*, *7*, 21–39.
Lee, A. L., Wang, Y., Ye, W.-H., Yoon, H. S., Chan, S. Y., & Yang, Y.-Y. (2008). Efficient intracellular delivery of functional proteins using cationic polymer core/shell nanoparticles. *Biomaterials*, *29*, 1224–1232.
Luckarift, H. R., Spain, J. C., Naik, R. R., & Stone, M. O. (2004). Enzyme immobilization in a biomimetic silica support. *Nature Biotechnology*, *22*, 211–213.
Matsumoto, N. M., González-Toro, D. C., Chacko, R. T., Maynard, H. D., & Thayumanavan, S. (2013). Synthesis of nanogel–protein conjugates. *Polymer Chemistry*, *4*, 2464–2469.
Mero, A., Ishino, T., Chaiken, I., Veronese, F. M., & Pasut, G. (2011). Multivalent and flexible PEG-nitrilotriacetic acid derivatives for non-covalent protein pegylation. *Pharmaceutical Research*, *28*, 2412.
Molla, M. R., Marcinko, T., Prasad, P., Deming, D., Garman, S. C., & Thayumanavan, S. (2014). Unlocking a caged lysosomal protein from a polymeric nanogel with a pH trigger. *Biomacromolecules*, *15*, 4046–4053.
Mueller, C., Capelle, M. A. H., Seyrek, E., Martel, S., Carrupt, P.-A., Arvinte, T., et al. (2012). Noncovalent PEGylation: Different effects of dansyl-, l-tryptophan-, phenylbutylamino-, benzyl- and cholesteryl-PEGs on the aggregation of salmon calcitonin and lysozyme. *Journal of Pharmaceutical Sciences*, *101*, 1995–2008.
Murthy, N., Xu, M., Schuck, S., Kunisawa, J., Shastri, N., & Fréchet, J M. (2003). A macromolecular delivery vehicle for protein-based vaccines: Acid-degradable protein-loaded microgels. *Proceedings of the National Academy of Sciences*, *100*, 4995–5000.

Nischan, N., & Hackenberger, C. P. R. (2014). Site-specific PEGylation of proteins: Recent developments. *The Journal of Organic Chemistry*, *79*, 10727–10733.
Pasut, G., & Veronese, F. M. (2012). State of the art in PEGylation: The great versatility achieved after forty years of research. *Journal of Controlled Release*, *161*, 461–472.
Poznansky, M. J., & Juliano, R. L. (1984). Biological approaches to the controlled delivery of drugs: A critical review. *Pharmacological Reviews*, *36*, 277–336.
Ryu, J.-H., Chacko, R. T., Jiwpanich, S., Bickerton, S., Babu, R. P., & Thayumanavan, S. (2010). Self-cross-linked polymer nanogels: A versatile nanoscopic drug delivery platform. *Journal of the American Chemical Society*, *132*, 17227–17235.
Standley, S. M., Kwon, Y. J., Murthy, N., Kunisawa, J., Shastri, N., Guillaudeu, S. J., et al. (2004). Acid-degradable particles for protein-based vaccines: Enhanced survival rate for tumor-challenged mice using ovalbumin model. *Bioconjugate Chemistry*, *15*, 1281–1288.
Vellodi, A. (2005). Lysosomal storage disorders. *British Journal of Haematology*, *128*, 413–431.
Ventura, J., Eron, S. J., González-Toro, D. C., Raghupathi, K., Wang, F., Hardy, J. A., et al. (2015). Reactive self-assembly of polymers and proteins to reversibly silence a killer protein. *Biomacromolecules*, *16*, 3161–3171.
Vermonden, T., Censi, R., & Hennink, W. E. (2012). Hydrogels for protein delivery. *Chemical Reviews*, *112*, 2853–2888.

CHAPTER SEVENTEEN

Nanoarmored Enzymes for Organic Enzymology: Synthesis and Characterization of Poly(2-Alkyloxazoline)–Enzyme Conjugates

Melanie Leurs, Joerg C. Tiller[1]
TU Dortmund, Dortmund, Germany
[1]Corresponding author: e-mail address: joerg.tiller@tu-dortmund.de

Contents

Abstract

The properties of enzymes can be altered significantly by modification with polymers. Numerous different methods are known to obtain such polymer–enzyme conjugates (PECs). However, there is no universal method to render enzymes into PECs that are fully soluble in organic solvents. Here, we present a method, which achieves such high

Methods in Enzymology, Volume 590
ISSN 0076-6879
http://dx.doi.org/10.1016/bs.mie.2017.01.008
© 2017 Elsevier Inc.
All rights reserved.

degree of modification of proteins that the majority of modified enzymes will be soluble in organic solvents. This is achieved by preparing poly(2-alkyloxazoline)s (POx) with an NH_2 end group and coupling this functional polymer via pyromellitic acid dianhydride onto the amino groups of the respective protein. The resulting PECs are capable of serving as surfactants for unmodified proteins, rendering the whole mixture organosoluble. Depending on the nature of the POx and the molecular weight and the nature of the enzyme, the PECs are soluble in chloroform or even toluene. Another advantage of this method is that the poly(2-alkyloxazoline) can be activated with the coupling agent and used for the enzyme conjugation without further purification. The POx–enzyme conjugates generated by this modification strategy show modulated catalytic activity in both, aqueous and organic, systems.

1. INTRODUCTION

Since modern biotechnology allows the design and large-scale production of enzymes, they are increasingly being used as therapeutics in medicine (Liu, Li, & Lu, 2015), as catalysts in water purification and for food processing (Kent, 2013), and even considered as regio- and enantioselective catalysts in chemical production processes (Bos & Roelfes, 2014; Faber, 2004). However, these biocatalysts are often not sufficiently active in organic solvents, which is essential for the synthesis of most fine chemicals.

One way to stabilize enzymes and render their solubility in organic solvents is the approach of chemical modification of their functional groups, preferably at the surface. In order to change as few functional groups as possible and achieve significant effect on enzyme stability and solubility, the end-on functionalization with synthetic polymers is one approach. Such polymer–enzyme conjugates (PECs) are well established in armoring enzymes against degradation by proteases and denaturation (Harris, Martin, & Modi, 2001; Moreno-Pérez et al., 2016).

The most investigated polymer for protein and enzyme modification is poly(ethyleneglycol) (PEG), and the modification is termed PEGylation. The first PEGylation was performed in 1977 by Abuchowski and coworkers (Abuchowski, McCoy, Palczuk, van Es, & Davis, 1977). Since then, practically, all functional groups of enzymes were PEGylated using a myriad of coupling methods (Herman, Hooftman, & Schacht, 1995; Hooftman, Herman, & Schacht, 1996; Roberts, Bentley, & Harris, 2012). The focus in more recent studies has been the PEGylation of thiol groups (Katsumi et al., 2016; Sawhney et al., 2016; Yoon, Shin, Kim, Kim, & Chung,

2016). Thereby, PEGylation research greatly focuses on the synthesis of therapeutically utilizable conjugates (Grigoletto et al., 2016), leading to commercial drugs (Fee & Van Alstine, 2006; Harris et al., 2001). PEGylation in those cases, for example, leads to a longer circulation time in the body and higher stability against proteases (Harris et al., 2001).

PEGylation is also known to increase solubility in organic solvents (Gaertner & Puigserver, 1988; Takahashi et al., 1984) and different cases show some catalytic activity in the organic reaction media (Castillo, Solá, Ferrer, Barletta, & Griebenow, 2008).

Also several other polymers were used for enzyme modification, as described by many authors of this volume. For example, polystyrene residues were attached to the enzymes, which leads to the assembly of big amphiphilic structures, the so-called giant amphiphiles (Boerakker et al., 2002; Hannink et al., 2001; Velonia, Rowan, & Nolte, 2002). Furthermore, smart and stimuli-responsive PECs were investigated (Cobo, Li, Sumerlin, & Perrier, 2015). Such conjugates with poly (*N*-isopropylacrylamide), for example, indicated temperature-switchable activities (Cao & Wang, 2016; Ivanov et al., 2003; Shiroya, Yasui, Fujimoto, & Kawaguchi, 1995; Zhou, He, & Wang, 2016). Modification with poly[(oligo ethylene glycol)methyl ether methacrylate] leads to temperature-dependent self-assembly of the conjugates (Moatsou et al., 2015).

An alternative, especially to PEG, is the polymer class of poly(2-alkyloxazoline)s (POx) which can be made by cationic ring opening polymerization of 2-alkyl-2-oxazolines. Due to the living character and the tolerance of different side chains to this method, a variety of polymer architectures can be achieved. Amphiphilic copolymers (Krumm et al., 2012) and amphiphilic polymer conetworks (Dech, Wruk, Fik, & Tiller, 2012; Schoenfeld et al., 2013; Sittko et al., 2015) are some examples as carriers for enzymes. Hydrophobicity of the polymer is controlled by choosing appropriate monomers (Hoogenboom, 2009). The polymer is also known to be benign to a variety of enzymes (Plothe et al., 2017). Additionally, the poly(2-oxazoline)s can easily be equipped with different functional groups by applying diverse polymerization initiation or termination reagents. Another advantage of this method is that it allows the generation of well-defined polymers with a low polydispersity, and the degree of polymerization can be preset by adjusting the initiator to monomer ratio. By the usage of a microwave reactor, various polymers with different molecular weights can be synthesized rapidly (Wiesbrock, Hoogenboom,

Abeln, & Schubert, 2004; Wiesbrock, Hoogenboom, Leenen, Meier, & Schubert, 2005).

In 1990, POx–enzyme conjugate was made for the first time by the modification of catalase with poly(2-methyloxazoline) (PMOx) and poly(2-ethyloxazoline) (PEtOx) (Miyamoto, Naka, Shiozaki, Chujo, & Saegusa, 1990). The highest organosolubility of such a POx–catalase conjugate has been stated as 0.077 mg/mL in benzene. Since then POxylation has been applied for therapeutics (Luxenhofer et al., 2012; Tong et al., 2013; Viegas et al., 2011) and for organic enzymology (Naka, Chujo, Miyamoto, & Saegusa, 1997; Naka, Miyamoto, Chujo, & Saegusa, 1999; Naka, Ohki, & Maeda, 1991). In the case of functionalization of a lipase from *Candida rugosa* with poly(2-isopropyloxazoline), activations of about 160 were achieved even if this conjugate is not fully soluble in the used solvent, toluene (Naka et al., 1997).

In this chapter, we describe synthesis of organosoluble poly(2-alkyloxazoline)–enzyme conjugates and their catalytic activities in both aqueous and organic solvents (Konieczny, Fik, Averesch, & Tiller, 2012; Konieczny, Krumm, Doert, Neufeld, & Tiller, 2014). The conjugates showed a higher activity when compared to the corresponding suspended native enzyme in organic solvents, but not when compared to the enzyme in aqueous media. The highest enzyme activation (5000-fold) was found for the PMOx30–laccase conjugate. α-Chymotrypsin modified with PMOx30 is 1146 more active than the corresponding native enzyme in dry chloroform. This was achieved by synthesizing different poly(2-alkyloxazoline)s with one NH_2 end group and conjugating the free amino groups to amino groups on the enzymes via the reactive bifunctional agent, pyromellitic acid dianhydride (PADA).

2. SYNTHESIS OF POLY(2-ALKYLOXAZOLINE)S WITH AN NH_2 END GROUP

Detailed protocols for the synthesis of poly(2-methyloxazoline) (Fig. 1) with 30 repeating units and the terminal benzyl and ethylene diamine groups are described below. All other polymers are prepared similarly by adjusting the reaction time and temperature according to the targeted molecular weight and the nature of the monomer (Konieczny et al., 2012, 2014; Krumm, Konieczny, Dropalla, Milbradt, & Tiller, 2013).

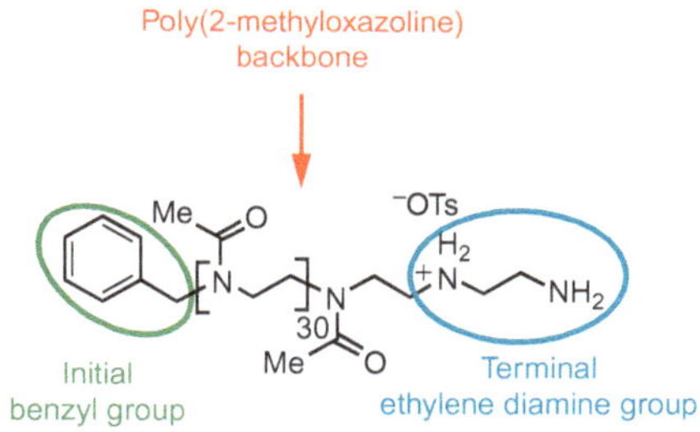

Fig. 1 Structure of poly(2-methyloxazoline) with 30 repeating units and terminal groups of benzyl and ethylene diamine.

2.1 Purification of Chemicals

Precautions:

For all syntheses, the usual precautions and safety equipment of classical chemical laboratories are necessary.

Equipment

- Distillation apparatus
- Schlenk flask
- Round-bottom flask
- Karl-Fischer titrator, e.g., TitroLine KF, Schott

Chemicals

- Chloroform (VWR)
- Activated alumina 90 (Merck)
- CombiCoulomat fritless (Karl-Fischer reagent, Merck)
- Molecular sieve 4 Å (Merck)
- 2-Methyl-2-oxazoline (Alfa Aesar)
- Calcium hydride (Merck)
- Argon

Protocol

1. Dry the solvent chloroform with activated alumina for 30 min at 600 mbar and distill it subsequently (40°C, 350 mbar) under argon atmosphere. This way, the chloroform contains less than 1 ppm of water (determined by Karl-Fischer titration). Store it in a Schlenk flask equipped with a septum at 4°C under argon over molecular sieves (4 Å).
2. Distill the commercially available monomer, 2-methyl-2-oxazoline (MOx, 40°C, 40 mbar) twice from CaH_2 under reduced pressure and argon atmosphere. Store it under argon and over molecular sieves (4 Å) in a round-bottom flask, closed with a septum, at –20°C.

2.2 Polymerization Procedure

Equipment

- Synthesis microwave reactor, e.g., CEM Discover equipped with compressed air cooling, magnetic stirrer, and infrared sensor for continuous temperature monitoring and automatic adjustment
- SCHOTT flask (DURAN®) with ethylene tetrafluoroethylene cap and PTFE-coated silicone sealing (50 mL)
- Beaker (3 L)
- Syringe (5 mL and 20 mL)

Chemicals

- Argon
- Distilled 2-methyl-2-oxazoline prepared as described in Section 2.1
- Dry chloroform prepared as described in Section 2.1
- Benzyl tosylate

Protocol

The polymers are synthesized in a microwave reactor (Hoogenboom, 2009).

1. Carry out preparation and reaction under argon atmosphere.
2. Use a 50 mL SCHOTT flask (DURAN®) with ethylene tetrafluoroethylene cap and PTFE-coated silicone sealing equipped with a stirring bar. Put it in a 3-L beaker floated with argon to perform preparation under argon atmosphere.
3. Flush a 20-mL syringe three times with argon and then fill it with argon. Inject the argon through the septum into the chloroform-containing vessel, take 20 mL of dry chloroform (4 mL per mL of the monomer) up, and inject it into the 50-mL SCHOTT flask (DURAN®).
4. Take 5 mL 2-methyl-2-oxazoline (58.75 mmol, 30 equiv.) out of the round-bottom flask with a 5-mL syringe following the protocol above and inject it into the chloroform.
5. Add 513.2 mg of the solid benzyl tosylate (1.96 mmol, 1 equiv.) and close the vessel with the ethylene tetrafluoroethylene cap with PTFE-coated silicone sealing. After closing, lift the vessel out of the beaker floated with argon.
6. Heat the reaction mixture to 100°C for 4 h under medium stirring in the closed flask in a CEM Discover synthesis microwave reactor equipped with compressed air cooling, magnetic stirrer, and infrared sensor for continuous temperature monitoring and automatic adjustment. Set the following parameters in the regulation software (Synergy 1.36):

temperature 100°C, ramp time 3 min, hold 240 min, Power Max off, stirring medium. The maximum demand is 300 W and gets automatically adjusted by the equipment.

2.3 Termination Procedure

Equipment

- Beaker (3 L)
- Syringe (5 mL)

Chemicals

- Ethylene diamine (ABCR)
- Poly(2-methyloxazoline) synthesized in Section 2.2

Protocol

1. Cool the reaction mixture down to 45°C after polymerization.
2. Take the flask out of the microwave reactor and put it in a 3-L glass beaker floated with argon. Generally, the reaction mixture turns slight yellow during polymerization.
3. Add 5 mL ethylene diamine (EDA) (74.9 mmol, 38 equiv.) under argon atmosphere with a syringe following the protocol described above for preparation of polymerization. Thereby, typically the yellow color of the reaction mixture gets brighter.
4. Close the vessel and put it into an oil bath at 45°C.
5. Continue stirring the reaction mixture at 45°C for 48 h.

2.4 Purification

Equipment

- Falcon tubes (50 mL)
- Centrifuge (Z300, Hermle)
- Rotary evaporator
- ZelluTrans cellulose membrane (molecular weight cut off 1000 g/mol)
- Brackets to close the dialysis tubes
- Dropping funnel (50 mL)
- Beaker 600 mL

Chemicals

- Diethylether (VWR)
- Chloroform (VWR)
- Methanol (AppliChem)

Protocol

1. Purify the raw polymer by repeated reprecipitation in about 200 mL diethylether/30 mL chloroform. Therefore, fill six 50-mL Falcon tubes with 35 mL diethylether. Then add 5 mL of the reaction mixture dropwise to each tube via a syringe. Thereby, the polymer precipitates. Close the tubes and mix thoroughly by shaking. Then adjust them to the same mass and centrifuge at 6000 rpm for 5 min. Decant the liquid, dry the solid by floating with compressed air, and dissolve it again in overall 30 mL chloroform. Repeat this protocol three times. Typically at the first reprecipitation step, a mucous light yellow solid separates, whereas during the last step it precipitates flaky and lighter.
2. Afterward dissolve the resulting polymer in 30 mL chloroform and dry it under reduced pressure with a rotary evaporator. Then dissolve it in about 15 mL methanol.
3. Prepare a 15-cm-long piece of a ZelluTrans cellulose membrane with a molecular weight cut off (MWCO) of 1000 g/mol as followed for conditioning to methanol: Put the membrane in 50 mL doubly distilled water in a beaker and add 50 mL methanol dropwise via a dropping funnel under slow stirring with a magnetic stirring bar. Discard 50 mL of the resulting solution and add again 50 mL methanol as described above. Repeat these steps five times.
4. Then perform a dialysis of the polymer against distilled methanol for 2 days. Therefore, seal the dialysis membrane by folding it twice tight at the bottom and fixing it with a bracket. Fill the polymer solution into the tube. Fold the upper side twice and seal it with another bracket equipped with a wire loop. Then hook it into a 600-mL beaker filled with methanol equipped with a magnetic stirring bar and stir slowly. Perform solvent exchange every 12 h.
5. After dialysis, pour the polymer solution out of the membrane and evaporate the solvent under reduced pressure to obtain a white to slightly yellow powder.

2.5 Analysis of the Poly(2-Methyloxazoline)

For subsequent enzyme conjugation, it is important that the used polymers are well defined and highly functionalized, especially with regard to the primary amino group at the end of the polymer chain. Because the latter is used for direct coupling of the polymer to the primary amino groups of the respective protein via a bifunctional linker, it is necessary that no dimerization occurs during termination of the POx and that the degree of termination is near complete.

To address these points, the resulting polymers were characterized with ^{1}H NMR spectroscopy, size exclusion chromatography (SEC), and ESI-MS.

Equipment

- NMR spectrometer, e.g., Bruker DRX-500 with a 5-mm sample head operating at 500.13 MHz
- GPC system, e.g., Viscotek GPCMax system equipped with a refractive index (RI) detector
- ESI-MS spectrometer, e.g., LTQ-FT-ICR-Ultra/Advion Triversa Nanomate
- PTFE syringe filter 0.2 μm pore size

Chemicals

- Terminated and purified PMOx from Section 2.5
- DMSO-d_6 (Amar Chemicals)
- Dimethyl formamide (DMF, VWR)
- Lithium bromide (Sigma-Aldrich)
- Doubly distilled water
- Acetonitrile (VWR)
- Formic acid (Merck)

2.5.1 ^{1}H NMR Measurements

^{1}H NMR measurements are performed to further characterize the polyoxazolines with regard to their degree of polymerization and degree of termination. The ^{1}H NMR spectra are recorded in DMSO-d_6 (70 mg polymer dissolved in 600 μL DMSO-d_6) using a Bruker DRX-500 spectrometer with a 5-mm sample head operating at 500.13 MHz. Fig. 2 shows the ^{1}H NMR spectrum of an EDA-terminated PMOx with a desired DP of 30.

The five aromatic protons of the benzyl group (1–3) are used as reference signal; this means the surface integral is set to 5. The degree of polymerization (DP_{NMR}) can be calculated referring to the surface integral signals generated by the polymer backbone (6–8, blue squares; 4, protons per repeating unit) as well as the signal from the polymer side chain (5, green squares; 3, protons per repeating unit) by the following formula:

$$\mathrm{DP(NMR)} = \frac{1}{2}\cdot\left(\frac{\Sigma\,\text{backbone protons}}{4} + \frac{\Sigma\,\text{sidechain protons}}{3}\right)$$

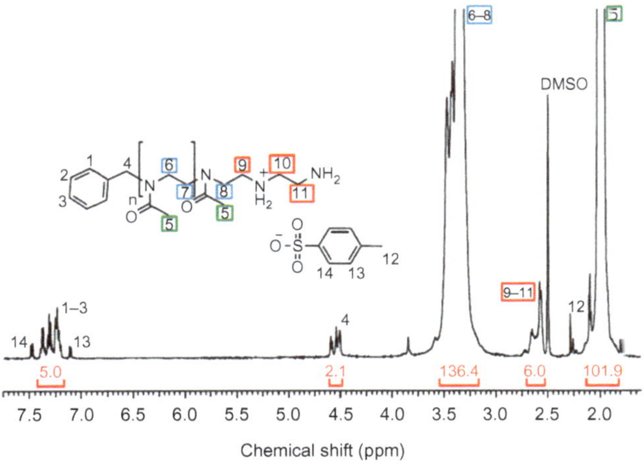

Fig. 2 ^{1}H NMR spectrum of an EDA-terminated PMOx in DMSO-d_6 at 500 MHz; the numbering refers to the protons; equivalent protons are numbered same.

Based on the obtained spectrum, the DP_{NMR} is calculated to be 34, which is in good agreement with the expected DP of 30.

The degree of termination (D_T) is calculated referring to the surface integral of the two CH_2 groups of the terminal EDA group and the last CH_2 group of the polymer backbone (9–11, red squares). Thereby, an integral of 6 refers to a functionalization of 100%, as shown in this example.

The basic ^{1}H NMR signals are listed below:

PMOx: ^{1}H NMR (DMSO-d_6): δ (ppm) = 3.65–3.10 (b, n·4H, $N(CH_2)_2$), 2.7–2.55 (b, 6H, CH_2–NH–CH_2–CH_2–NH_2), 2.10–1.85 (b, n·3H, NCO–CH_3).

2.5.2 SEC

Additionally, SEC is used to check if dimerization occurred during the termination of the PMOx and to calculate the polydispersity index (PDI). SEC was performed on a Viscotek GPCMax system equipped with an RI detector in saline DMF (20 m*M* LiBr) at 60°C with a flow rate of 0.7 mL/min. Two TSKgel GMHHR-M 7.8 × 300 mm columns and one precolumn were used. Calibration was performed using polystyrene standards.

Prepare the samples as described below:

1. Dissolve the PMOx30 (6 mg/mL) in 1.5 mL of DMF (20 m*M* LiBr).
2. Store at room temperature (RT) for 12 h and filter through a PTFE syringe filter with 0.2 μm pore size, before measurement.
3. Inject 100 μL of the sample onto the column.

The SEC measurements provide signals at molecular weights in the range of the desired DP of 30 ($M_n = 3400$, $M_w = 4200$, DP = 36) and no signals for the double or higher molecular weight can be found, which means that no dimer formation. Additionally, the PDI of 1.2 is a good value for a narrow size distribution which is typical for living polymerizations.

2.5.3 ESI-MS

A more accurate, although also more elaborate, method for analyzing the functional polymer is electrospray ionization mass spectrometry (ESI-MS). Fig. 3 depicts the ESI-MS spectrum of PMOx30. The measurements were performed on an LTQ-FT-ICR-Ultra coupled with an Advion TriversaNanomate. The nitrogen pressure was 28–63 mbar and the ionization tension is between 1.4 and 2.0 kV. The spectra were recorded in positive mode in a mass range between 225 and 2000 with a decay of 1,000,000.

Prepare the samples as follows:

1. Dissolve the PMOx30 (1 μg/mL, use a dilution series) in a mixture of 70 vol% doubly distilled water, 30 vol% acetonitrile, and 0.1 vol% formic acid.
2. The injection volume is 3 μL and the flow rate 200 nL/min.

The numbers on the peaks are the individual molecular weights. Their distance is exactly one repeating unit of the PMOx (85 g/mol). The overall mass of each major peak is exactly an x-fold of the repeating units plus the molecular weight of the benzyl group plus that of the EDA. Such a group of identical polymers which only differ in the number of repeating units is called a generation. The minor generation without given masses represents a

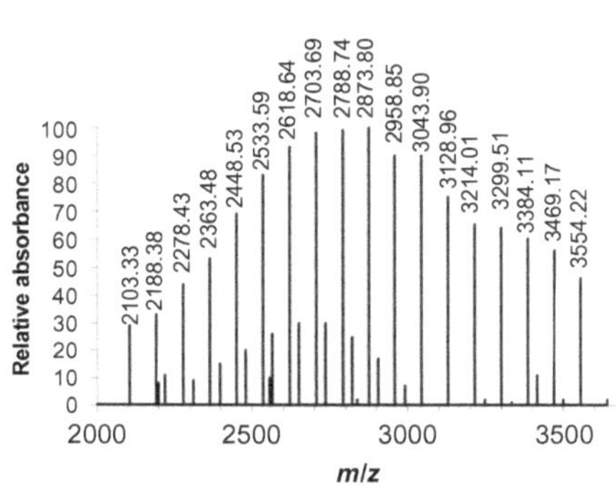

Fig. 3 ESI-MS spectrum of ethylene diamine-terminated PMOx30.

respective chain transfer side product with a proton as end group and an EDA group. No mass signals for products where two or more PMOx chains react with one EDA could be found, indicating that only one PMOx chain reacts with one EDA group, i.e., practically all polymer chains have an NH_2 end group.

3. CONJUGATION OF ENZYMES WITH POLY(2-ALKYLOXAZOLINE)S

We found that the most convenient way for enzyme functionalization with poly(2-alkyloxazoline)s is to bind the polymers with one terminal amino group via the highly reactive PADA as bifunctional linker to the primary amino groups of the respective protein (Konieczny et al., 2012, 2014). The schematic procedure of the polyoxazoline enzyme conjugation is shown in Fig. 4.

A great advantage of this method is that the polymer does not require purification after reaction with PADA. In typical literature procedures with other coupling agents such purification is necessary (Mero et al., 2008; Tong, Luxenhofer, Yi, Jordan, & Kabanov, 2010). The reason why this step

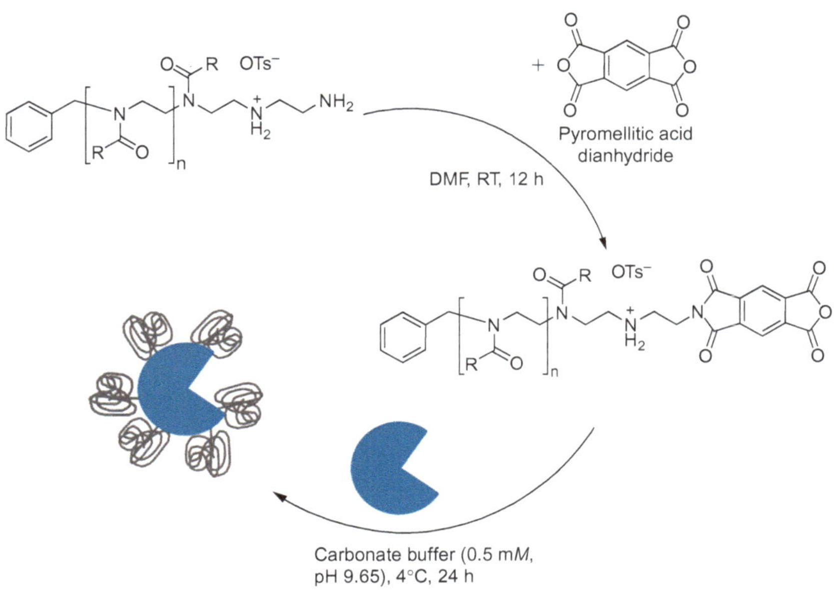

Fig. 4 Strategy for enzyme modification with POx and PADA.

is not required here is that PADA is highly reactive. In the first step, PADA reacts very fast with the amino group at the end of the POx. Due to diffusion limitations, this polymer derivative is less reactive than the free, unbound linker. This is why single functionalization reactions are favored to dimerization. Therefore, PADA can be added equimolar or even sub-stoichiometries to the amino group containing polyoxazoline. The SEC measurements of the polymers modified with PADA indicate no signals for the double or higher molecular weight signals when compared to that of the EDA-terminated polymer. Thus, the polymer does not form dimers and so the second anhydride group can react with the primary amino groups of the enzyme.

The polymer amount for the enzyme modification was generally adjusted to a tenfold molar excess referring to the primary amino groups of the enzyme. The PADA was used in a substoichiometric amount (0.8 equiv.) referring to the primary amino group at the end of the polymer chain to ensure that no free PADA remains in the reaction mixture. Such free PADA could also react with the enzymes' primary amino groups and, thus, block them for polymer functionalization or even lead to cross-linked enzymes. The detailed protocol for the conjugation of enzymes with PMOx30 is given below. All other modifications with different polymers were performed in the same manner, only adjusting the solvent mixtures and dialysis medium according to the respective polymer used (Konieczny et al., 2012, 2014).

Equipment

- Distillation apparatus
- Round-bottom flask (5 mL, 50 mL)
- Septum
- Syringes (5 mL)
- Beaker (600 mL, 3 L)
- Flask 100 mL
- Lever lid glass (10 mL)
- ZelluTrans cellulose membranes (MWCO 10,000 g/mol)
- Freeze-drying apparatus, e.g., Alpha 1–4 LD plus

Chemicals

- DMF (VWR)
- Phosphorus pentoxide (Sigma-Aldrich)
- Argon
- Terminated and purified PMOx from Section 2.5
- PADA (Sigma-Aldrich)

- Sodium bicarbonate (Merck)
- Sodium carbonate (Merck)
- Bidistilled water
- Lysozyme from hen egg white (AppliChem)
- Liquid nitrogen

3.1 Protocol for Lysozyme Conjugation With PMOx30

The conjugation of lysozyme (hen egg white) with PMOx30 is performed as follows:

1. Distill DMF under reduced pressure (53°C, 15 mbar) and stir with a magnetic stirring bar over phosphorus pentoxide under argon atmosphere to dry it and free it from amines.
2. Place a 5-mL round-bottom flask equipped with a magnetic stirring bar in a 3-L beaker floated with argon to perform preparation under argon atmosphere.
3. Add 5 mL dry and amine-free DMF with a syringe. Therefore, flush a 5-mL syringes three times with argon and then fill it with argon. Inject the argon through the septum into the DMF-containing vessel, take the DMF, and inject it into the round-bottom flask.
4. Dissolve 5.9 mg PADA (0.8 equiv. referring to the molar PMOx amount, 27.4 μmol) in the DMF. Wait until all crystals are dissolved. This can take up to 2 min.
5. Add 99.4 mg PMOx (10 equiv. referring to the enzymes' primary amino groups, 34.3 μmol) under argon atmosphere. Close the flask with a septum and allow the mixture to react for 12 h at RT under vigorous stirring.
6. No argon atmosphere is necessary for the following steps. Prepare a sodium carbonate buffer (10 m*M*) by adding 80.5 mg $NaHCO_3$ and 20.0 mg Na_2CO_3 in a flask and filling up to 100 mL with doubly distilled water. This buffer is stable and can be stored at RT.
7. Dilute the buffer with doubly distilled water to obtain a concentration of 0.5 *M*. Adjust the pH to 9.65 with aqueous HCl or aqueous NaOH.
8. After 12-h reaction time, dissolve 7 mg of the lysozyme (seven primary amino groups (Habeeb & Atassi, 1970), 0.49 μmol) in 1 mL of the sodium carbonate buffer (0.5 m*M*, pH 9.65) in a 10-mL lever lid glass. Then add all of the polymer/PADA/DMF reaction mixture. The solution should stay clear. Stir the mixture for 24 h at 4°C.

9. Dialyze the reaction mixture for 2 days in a 600-mL beaker filled with doubly distilled water (solvent exchange every 12 h) for 2 days using ZelluTrans cellulose membranes with an MWCO of 10,000 g/mol (tube with 10 cm in length). The membrane is used without further conditioning. The dialysis is performed as described above.
10. Transfer the complete solution from the dialysis tube to a 50-mL round-bottom flask, freeze it in liquid nitrogen for 10 min, and lyophilize it at 0.001 mbar and −56°C condenser temperature using a freeze-drying apparatus (e.g., Alpha 1–4 LD plus) to obtain a white powder.

4. CONJUGATE CHARACTERIZATION

4.1 Solubility

Due to the modification with POx, the enzyme polymer conjugates are rendered soluble in organic media like DMF, chloroform, toluene, or THF (Fig. 5; Table 1).

Solubility in this context is defined as transmission of more than 90% of the solvent containing 1 mg PEC per mL. The light transmission of the PEC solutions was measured spectrophotometrically at 760 nm using 0.2-cm cuvettes against the pure solvent. So far, the only enzyme that could not be rendered organosoluble by POXylation is glucose oxidase (EC 1.1.3.4, 150 kDa (Viegas et al., 2011)). This indicates that there might be a molecular weight limitation or the number of accessible amino groups on the enzyme surface.

Fig. 5 Pictures of 7 mg lysozyme in 5 mL DMF: *left*, native; *right*, modified with PMOx30 and PADA (Konieczny et al., 2012).

Table 1 List of Organosoluble Conjugates Consisting of Different POx Derivatives and Enzymes

Enzyme	Polymer	Solvent
Alcohol dehydrogenase (*Saccharomyces cerevisiae*, EC 1.1.1.1)	PMOx30	$CHCl_3$
Alcohol dehydrogenase (horse liver, EC 1.1.1.1)	PMOx30	$CHCl_3$
α-Chymotrypsin (bovine pancreas, EC 3.4.21.1)	PMOx30	$CHCl_3$, DMF
	PBuOx-b-PMOx	$CHCl_3$, THF, toluene
	PHeptOx30	Toluene
Deoxyribose-phosphate aldolase (*Escherichia coli*, EC 4.1.2.4)	PMOx30	$CHCl_3$
Collagenase (*Clostridium histolyticum*, EC 3.4.24.3)	PMOx30	$CHCl_3$
Laccase (*Trametes versicolor*, EC 1.10.3.2)	PMOx30	$CHCl_3$, DMF, DMSO
	PHeptOx30	THF
Lipase (*Candida rugosa*, EC 3.1.1.3)	PMOx30	$CHCl_3$, DMF
Lipase (*Candida antarctica*, EC 3.1.1.3)[a]	PMOx30	$CHCl_3$, DMF
	PMOx60	$CHCl_3$, DMF
	PEtOx40	$CHCl_3$, DMF
	PBuOx-b-PMOx	THF, toluene
	PHeptOx30	THF, toluene
Lipase (*Rhizomucor miehei*, EC 3.1.1.3)	PMOx30	$CHCl_3$
	PHeptOx30	Toluene
Lysozyme (hen egg white, EC 3.2.1.17)	PMOx30	$CHCl_3$, DMF
	PBuOx-b-PMOx	THF, MeOH, EtOH, $CHCl_3$
	PHeptOx30	$CHCl_3$, THF, toluene
Monooxygenase (*Bacillus megaterium*, EC 1.14.14.1)	PMOx30	$CHCl_3$

Table 1 List of Organosoluble Conjugates Consisting of Different POx Derivatives and Enzymes—cont'd

Enzyme	Polymer	Solvent
Papain (*Carica papaya*, EC 3.4.22.25)	PMOx30	DMF, $CHCl_3$
	PHeptOx30	THF, toluene
Peroxidase (horseradish, 1.11.1.7)	PMOx30	$CHCl_3$
Proteinase K (*Tritirachium album*, *EC* 3.4.21.64)	PMOx30	DMF, $CHCl_3$
Trypsin (bovine pancreas, EC 3.4.21.4)	PMOx30	$CHCl_3$, DMF
	PBuOx-b-PMOx	$CHCl_3$, THF
NADH dehydrogenase (*Bacillus subtilis*, EC 1.6.99.1)	PMOx30	$CHCl_3$

[a]Transmission more than 70%.
Solubility in this context is defined as transmission of more than 90% of a dispersion of PEC in 1 mL solvent containing 1 mg of enzyme if not indicated otherwise (calculated from the initial enzyme weight) (Konieczny et al., 2012, 2014).

4.2 Sodium Dodecyl Sulfate Polyacrylamide Gel Electrophoresis

The formation of conjugates is further confirmed by sodium dodecyl sulfate polyacrylamide gel electrophoresis (SDS-PAGE) and isoelectric focusing, as described below.

SDS-PAGE analytics are chosen as a tool to further monitor the success of protein modification. With this method, the molecular weight of natural and modified enzymes is determined. There should be increase in the molecular weight for every modified amino group. Thus, the degree of modification of the enzymes can be verified by this method. This is performed according to a literature procedure (Laemmli, 1970).

Equipment

- SDS-PAGE apparatus, e.g., Minigel-Twin System (Analytik Jena, Jena) connected to a EV231 power supply (Consort, Turnhout, Belgium)

Chemicals

- SDS-PAGE buffers (lower Tris buffer (4×), upper Tris buffer (4×), loading buffer, ammonium peroxodisulfate (APS) solution, running buffer (10×) prepared according to Laemmli (1970))
- POx–enzyme conjugates and the corresponding native enzyme
- Dithiothreitol (DTT, AppliChem)
- Tris–glycine gel (14%) with 1 mm gel thickness

- Roti®-Mark 10–150 (Roth) as protein ladder
- Ethanol (VWR)
- Doubly distilled water
- Glacial acid (Merck)
- Coomassie brilliant blue R250 (Roth)
- Sodium acetate trihydrate (Roth)
- Sodium thiosulfate (Merck)
- Silver nitrate (Roth)
- Formalin 37 wt% in water (Merck)
- Sodium carbonate (Merck)
- EDTA disodium dihydrate (Roth)

Protocol

1. Dissolve the native enzyme (5 μg in about 5 μL) and conjugate (30 μg in 25 μL) in water and mix them with the same amount of 2 × loading buffer, heat the mixture to 96°C for 10 min, and add 2 μL aqueous 1 *M* DTT per 10 μL of sample.
2. Load the mixtures on a 14% Tris–glycine gel (for composition see Table 2) with 1 mm gel thickness. Roti®-Mark 10–150 (6.5 μL) can be used as protein ladder.
3. The gel run can be performed in a Minigel-Twin System (Analytik Jena, Jena) connected to a EV231 power supply (Consort, Turnhout, Belgium). Use a constant voltage of 60 V for the run in the stacking gel. After the Bromophenol Blue band arrived in the resolving gel, change the current to 160 V and perform gel run till the Bromophenol Blue band has just left the gel.
4. In the case of a Coomassie Blue staining, shake the gel overnight in 100 mL of a solution of 40 vol% ethanol, 10 vol% glacial acetic acid,

Table 2 Composition of Stacking As Well As Resolving Gel for the SDS-PAGE

	Stacking Gel (4%) [μL]	Resolving Gel (14%) [μL]
Lower Tris, 4 ×	—	1500.0
Upper Tris, 4 ×	375.0	—
Doubly distilled water	900.0	2700.0
Acrylamide/bisacrylamide 40%	225.0	1800.0
APS	7.5	15.0
TEMED	3.0	7.5

and 290 mg/L Coomassie Brilliant Blue R250 in doubly distilled water for 12 h. Carry out destaining in 100 mL of a solution of 25 vol% ethanol and 8 vol% glacial acetic acid in doubly distilled water until the background of the gel gets transparent (De St. Groth, Webster, & Datyner, 1963).

5. Silver staining can be performed according to a modified protocol (Heukeshoven & Dernick, 1985) by the protocol below:
   (a) Incubate the gel for 30 min in the following solution

Ethanol	50 mL
Glacial acetic acid	10 mL
Doubly distilled water	ad 100 mL

   (b) Then incubate the gel for 20 min in a solution consisting

Ethanol	30 mL
Sodium acetate trihydrate	6.8 g
Sodium thiosulfate	0.127 g
Doubly distilled water	ad 100 mL

   (c) Afterward wash the gel twice for 20 min with 100 mL doubly distilled water, respectively.
   (d) Stain the gel for 30 min with a solution of

Silver nitrate	0.2 g
Formalin 37 wt% in water	28.8 μL
Doubly distilled water	ad 100 mL

   (e) Put 100 mL of doubly distilled water on the gel and remove it immediately.
   (f) Next slew it for 1 min in a solution with

Sodium carbonate	2.5 g
Doubly distilled water	ad 100 mL

(g) Carry out the developing step with

Sodium carbonate	2.5 g
Formalin 37 wt% in water	28.8 μL

until the desired coloration is achieved. This typically takes 5–10 min.

(h) Stop the developing by incubating 20 min with a solution of

EDTA disodium dihydrate	1.85 g
Doubly distilled water	ad 100 mL

Lysozyme, for example, has six lysine residues plus the N-terminus, which can be modified by PADA as well. Literature data show that all lysine residues are accessible for chemical modification (Habeeb & Atassi, 1970). The maximum possible molecular weight of lysozyme modified with PMOx30 is about 40 kDa (native enzyme 14.4 kDa). The resulting PMOx30–lysozyme conjugate (band LyCon30) shows a broad band from 24 to 48 kDa with a maximum at about 30 kDa on the SDS-PAGE (Fig. 6A). This strongly indicates that the lysozyme has been successfully modified but with varying

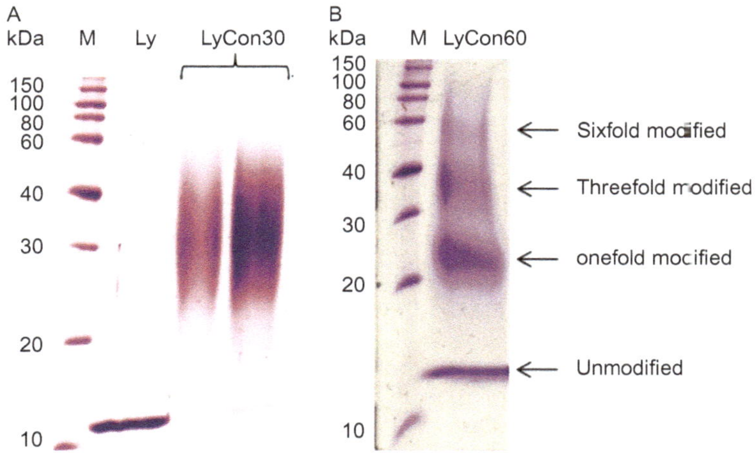

Fig. 6 SDS-PAGE gels of lysozyme and POxylated lysozyme: (A) *Ly*, native lysozyme; *LyCon30*, PMOx30–lysozyme conjugate; *M*, marker; (B) *LyCon60*, PMOx60–lysozyme conjugate; *M*; marker (Konieczny et al., 2012).

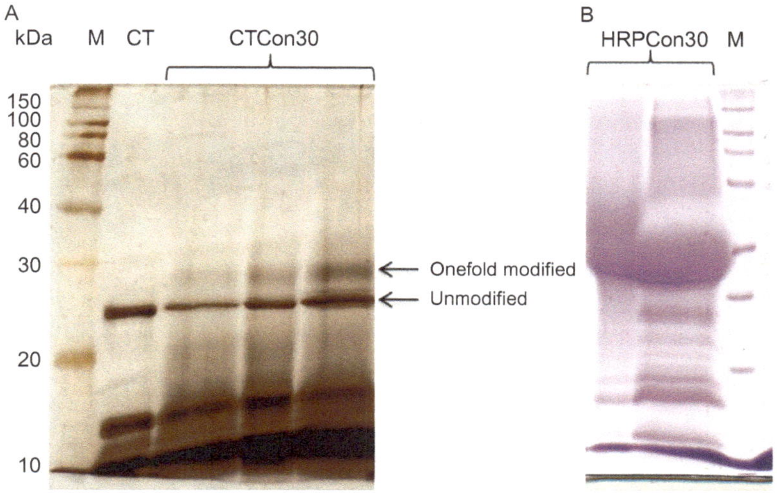

Fig. 7 SDS-PAGE gels of α-chymotrypsin and POxylated α-chymotrypsin and horseradish peroxidase: (A) *CT*, native α-chymotrypsin; *CTCon30*, PMOx30–α-chymotrypsin conjugate; *M*, marker; (B) *HRPCon30*, horseradish peroxidase conjugate; *M*, marker (Konieczny et al., 2012).

degrees of modification. Furthermore, it can be seen that there is almost no pristine lysozyme left.

The conjugates of lysozyme with PMOx60 show discrete bands in the SDS-PAGE, which represent different degrees of modification (Fig. 6B). Thereby, conjugates with up to six polymer chains were obtained and fully modified. However, the SDS-PAGE also shows that not all lysozyme molecules are modified to this extent.

The SDS gels of POxylated chymotrypsin and horseradish peroxidase clearly show functionalized enzyme, but also a significant amount of native enzyme (Fig. 7). Nevertheless, the samples are fully soluble in organic media such as chloroform. The gel of PMOx30–α-chymotrypsin conjugate also shows some discrete bands. Besides the native enzyme (lane CT, Fig. 7A), a new band at around 28–29 kDa appears which can be calculated as α-chymotrypsin with one to two polymer chains bound. A weak second band at about 31 kDa possibly originating from a conjugate of CT and three PMOx30 chains is also visible (Fig. 7A).

4.3 Isoelectric Focusing

Equipment

- Electrophoresis system, e.g., Multiphor II (Pharmacia Uppsala, Sweden)

Chemicals

- POx–enzyme conjugates and the corresponding native enzyme
- IEF gels Servalyt™ Precotes™ (Serva)
- Protein test mixture for p*I* determination, pH 3–10 (Serva)
- Ethanol (VWR)
- Doubly distilled water
- Glacial acid (Merck)
- Coomassie Brilliant Blue R250 (Roth)

Protocol

The success of the conjugation of enzymes with poly(2-oxazoline)s can also be revealed by performing isoelectric focusing. The IEF is able to distinguish between natural and modified enzymes. There is a shift in the isoelectric point (p*I*) for every modified amino group. We assumed that the more POx molecules were attached to the enzyme, the more decreases the p*I*.

Perform the IEF as follows:

1. Apply 24 μL containing 80 μg of the sample or rather 15 μg of the native enzyme on the gel. The used IEF gels can be purchased from Serva (Servalyt™ Precotes™) with a pH gradient from 3 to 10 and a thickness of 300 μm. Protein test mixture for p*I* determination, pH 3–10 (Serva) (12 μL), can be used as protein ladder.
2. The IEF analysis can be performed on a Multiphor II electrophoresis system (Pharmacia Uppsala, Sweden) with an EPS 3501 XL power supply (GE Healthcare, Munich). The electrophoresis program is 500 V, 3 mA for 150 Vh and 2000 V, 2 mA for 3500 Vh at 4°C.
3. Stain the gel with Coomassie Blue as described above.

The sample of a PMOx30–lysozyme conjugate shows the formation of several new bands on the IEF gel and only a small amount of unmodified lysozyme is visible. The p*I* decreased from 11.4 (Thomas, Vekilov, & Rosenberger, 1998) of the unmodified lysozyme to minimal 4.2 (Fig. 8). This leads to the conclusion that almost all lysozyme molecules are modified, but most likely to different degrees. Thus, the IEF measurement of the lysozyme conjugate is in good compliance with the SDS-PAGE results.

4.4 Size Exclusion Chromatography

Equipment

- HPLC system equipped with a diode array detector, e.g., LaChrom Elite (Hitachi)

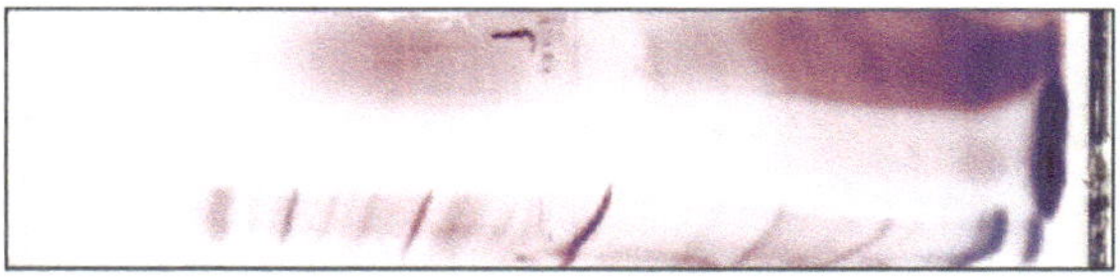

Fig. 8 Isoelectric focusing of lysozyme and the corresponding POx–enzyme conjugate: *A*, PMOx30–lysozyme conjugate modified in DMF; *Ly*, lysozyme; *M*, marker (Konieczny, Fik, Averesch, & Tiller, 2012).

Chemicals

- POx–enzyme conjugates and the corresponding native enzyme
- 0.1 *M* KH_2PO_4 buffer

Protocol

SEC for the enzymes and enzyme conjugates can be performed on a LaChrom Elite HPLC system equipped with a diode array detector in buffer (0.1 *M* KH_2PO_4 buffer, pH 7) at 30°C with a flow rate of 1 mL/min. A Biobasic SEC-300 7.8 × 300 mm column and one precolumn were used. The injection is performed with doubly distilled water. The detection was performed at 280 nm.

Prepare the samples as described below:

1. Dilute the 6 mg/mL of the sample in 1.5 mL KH_2PO_4 buffer (0.1 *M*, pH 7).
2. Filter samples through a PP syringe filter with 0.2 μm pore size before injection.
3. Inject 3 μL onto the column.

The measurements of lysozyme functionalized with different MOx showed conjugate formation (Fig. 9). There is a shift in elution time toward higher molecular weights. Also, in compliance with the SDS-PAGE analytics for the PMOx60-modified enzyme, different maxima, representing the different degrees of functionalization, appeared. For the PMOx30–lysozyme conjugate, a broad peak including different degrees of modification is seen. Moreover, as also seen in the SDS-PAGE, there is a small amount of native lysozyme. Thus, the SEC measurements fully support the SDS-PAGE analytics and illustrate that conjugate formation occurs.

4.5 Dynamic Light Scattering

Equipment

- Dynamic light scattering (DLS) spectrometer, e.g., Malvern Zetasizer nano S

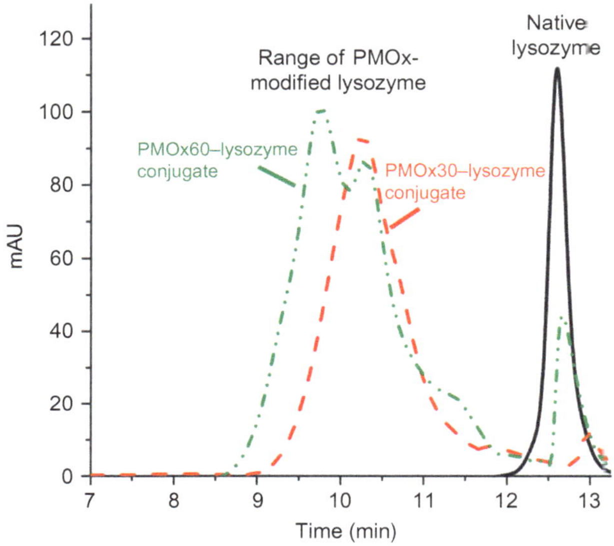

Fig. 9 Normalized SEC chromatograms of lysozyme and the corresponding POX–enzyme conjugates: lysozyme (*black*), PMOx30–lysozyme conjugate (*red*), PMOx60–lysozyme conjugate (*green*).

- Quartz glass cuvette
- PTFE and regenerated cellulose syringe filters with 0.2 μm pore size

Chemicals

- POx–enzyme conjugates and the corresponding native enzyme and POx
- Chloroform (VWR)
- Doubly distilled water

The SDS-PAGE analytics show that, in several cases, native enzyme is left after the modification step. Nevertheless, nearly all of these samples are soluble in organic media such as chloroform. We assumed that there must be some kind of noncovalent interactions which cause the solubility of these free enzymes on the organic media. To obtain a deeper insight into this aspect, we performed DLS measurements of different enzyme conjugates in chloroform and compared them to those of the unbound polymer and the free enzyme.

The DLS measurements are performed on a Malvern Zetasizer nano S at 20°C.

Prepare the samples as follows:

1. Dissolve the samples of PMOx (10 mg/mL) and the PMOx–enzyme conjugates (20 mg/mL) in chloroform and filter them through a PTFE syringe filter with 0.2 μm pore size before measurement.

2. Add 1.5 mL of the sample into a quartz glass cuvette and put it into the console.
3. Let it equilibrate for 2 min at 20°C and then start the measurement.
4. The sizes of free enzymes were measured in water following the same protocol as described above after filtering using a syringe filter with 0.2 μm pore size consisting of regenerated cellulose.

As seen in Fig. 10, in all cases the diameter of the PMOx30 conjugate increases in comparison to that of the corresponding native enzyme. The PMOx30–lysozyme conjugate was almost fully functionalized, *albeit* to different degrees. The diameter of these conjugates in chloroform was 7.8 nm. This can be calculated as the sum of the measured diameters of the native enzyme (3.6 nm) and that of the free PMOx (2 nm). Thus, the conjugates do not tend to aggregate to a great extent (Fig. 10).

In contrast, the PMOx30 conjugates of CT and HRP have bigger diameters than the sum of PMOx and native enzyme diameters. Thus, some kind of aggregate formation occurred. The SDS-PAGEs of the conjugates show

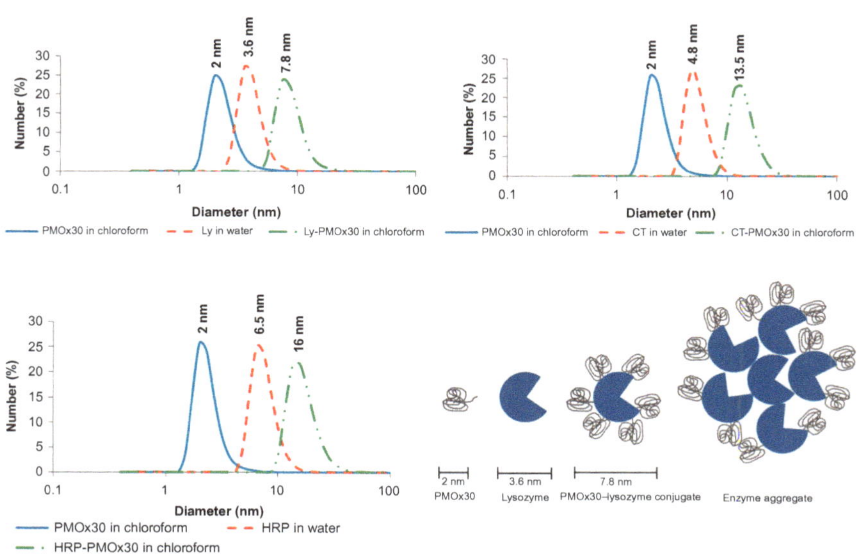

Fig. 10 DLS measurements of PMOx30–enzyme conjugates: lysozyme (Ly-PMOx30), α-chymotrypsin (CT-PMOx30), and horseradish peroxidase (HRP-PMOx30) (*green curves*) and the free PMOx30 (*blue curves*) in chloroform compared to the corresponding native enzymes (*red curves* in water). Schematic illustration of the diameters of PMOx30, lysozyme, and the PMOx30–lysozyme conjugate corresponding to the DLS measurements and the desired enzyme aggregates of POx–enzyme conjugates and native enzymes.

that there is unmodified and functionalized enzyme in the mixture, which is still soluble in chloroform. With regard to the DLS measurements, it seems that the functionalized enzymes form some clusters which incorporate or solubilize the unmodified enzyme molecules, i.e., they first act as surfactants for the latter (Fig. 10).

5. ACTIVITY EVALUATION OF POLY(2-ALKYLOXAZOLINE)–ENZYME CONJUGATES

Equipment

- Gas chromatography system, e.g., a Clarus 500 GC from PerkinElmer
- UV/vis spectrometer, e.g., Specord 210 (Analytik Jena)

Chemicals

- POx–enzyme conjugates and the corresponding native enzyme and POx
- *N*-Acetyl-L-phenylalanine ethylester (APEE, Sigma-Aldrich)
- 1-Propanol (Merck)
- *n*-Tetradecane (Merck)
- 2,6-Dimethoxyphenol (DMP, Fisher Scientific)
- Lauric acid (Sigma-Aldrich)
- 1-Octanol (Sigma-Aldrich)

Protocols

The activity of POx–α-chymotrypsin conjugates as well as the unmodified enzyme in water was determined by the transesterification reaction of APEE to *N*-acetyl-L-phenylalanine propylester using 1-propanol (Savin, Bruns, Thomann, & Tiller, 2005). One unit (U) is defined as a formation of 1 μmol *N*-acetyl-L-phenylalanine propylester per minute.

1. Prepare a solution consisting of 60 m*M* APPE, 1 *M* 1-propanol, and 0.002 m*M* tetradecane as internal standard in dry chloroform.
2. Dissolve 6 mg of the conjugates or native enzyme in dry chloroform.
3. Add 1 mL of substrate solution.
4. Incubate the mixture at 37°C.
5. Take a sample every 24 h.
6. The product formation is measured quantitatively by gas chromatography. The equipment used is a Clarus 500 GC from PerkinElmer equipped with a CP-Sil 8 CB (Varian) column. Nitrogen is used as a carrier gas. The following temperature program is suitable for measurement: 50°C for 2 min, heating at a rate of 20 K/min up to 150°C

then heating at a rate of 10 K/min to 200°C and after that 20 K/min to 280°C.

Laccase activity in chloroform was measured with a modified reported procedure (Wan et al., 2010). One unit is defined as the increase in absorbance of 0.001 per minute at 37°C.

1. Prepare a reaction mixture composed of 1 mL DMP (26 mmol/L) in 1 mL water-saturated chloroform containing 5.3 mg/mL of the laccase conjugate and 5.8 mg/mL unmodified laccase.
2. Measure the absorbance spectrophotometrically at 468 nm under constant stirring at 37°C.

The lipase activity of the native enzyme formulation and the conjugates in organic solvents was determined by esterification of 1-octanol with lauric acid at 28°C. One U is defined as a formation of 1 μmol of the ester per minute.

1. Prepare a solution of 50 m*M* of lauric acid in 10 mL of dry chloroform and a solution containing 100 m*M* of 1-octanol in 10 mL of dry chloroform.
2. Add 0.004 m*M* tetradecane as internal standard to the 1-octanol solution.
3. Dissolve 5 mg of the POx–lipase conjugate or native enzyme in 1 mL dry chloroform.
4. Add 500 μL of the lauric acid solution to the enzyme solution and afterward add 500 μL of the 1-octanol solution.
5. Incubate the mixture at 28°C.
6. Take 45 μL samples every 30 min.
7. The product formation is measured quantitatively by gas chromatography. The equipment used is a Clarus 500 GC from PerkinElmer equipped with a CP-Sil 8 CB (Varian) column. Nitrogen is used as a carrier gas. The following temperature program is suitable for measurement: 100°C for 1 min, heating at a rate of 20 K/min up to 250°C, and holding 250°C for 2 min.

The measurements reveal that the activity of the conjugated enzymes in chloroform is higher than that of the corresponding native enzyme (Table 3). For example, the activity of the PMOx30–α-chymotrypsin conjugate is about 1146-fold and the one of PMOx30–laccase is nearly 5000-fold higher than that of the corresponding native enzyme in chloroform. The residual activity in aqueous solution (e.g., for PMOx30–laccase conjugate 474 U/mg) is always lower than the original enzymatic activity (e.g., for laccase 14,800 U/mg). Considering the activity drop in water upon

Table 3 Activities of Some POx–Enzyme Conjugates and the Corresponding Unmodified Enzyme in Dry Chloroform (Konieczny et al., 2014)

Enzyme	Polymer	Activity of POx–Enzyme Conjugate	Activity of Native Enzyme
α-Chymotrypsin (bovine pancreas, EC 3.4.21.1)	PMOx30	0.55 U/g	0.00048 U/g
Laccase (*Trametes versicolor*, EC 1.10.3.2)	PMOx30	83 U/mg[a]	0.017 U/mg[a]
Lipase (*Candida antarctica*, EC 3.1.1.3)	PMOx30	17.18 U/mg	0.582 U/mg
	PMOx60	3.04 U/mg	
	PEtOx40	0.365 U/mg	
Peroxidase (horseradish, 1.11.1.7)	PMOx30	0.98 U/mg	0.52 U/mg

[a]Water-saturated chloroform was used as solvent.

POxylation, the highest activation for the still active PMOx30–laccase conjugate is even 153,000 (Konieczny et al., 2014).

6. CONCLUSIONS

The technique for the conjugation of enzymes and poly-(2-alkyloxazolines) (POx), called POXylation, using amino group-terminated polymers and PADA as bifunctional coupling agent leads to highly functionalized and organosoluble PECs. The polymer-conjugated enzymes act as surfactants for the solubilization of unmodified proteins. This way, even incomplete conversion leads to fully soluble enzyme conjugates in organic solvents. The solubility can be controlled by the nature of the versatile 2-alkyl-2-oxazoline monomers, which are readily available and quickly prepared in the lab without the safety equipment required for working for instance with PEG monomers. Another advantage of the described method is the activation of the amino-functionalized polymer with PADA, which does not require additional purification typical for such kind of modification.

Activity measurements reveal that although the residual activities of the conjugates in aqueous solution are lower than the original enzymatic activities, the former are in many cases superior to the native enzyme powder in organic solvents.

Additionally, the enzyme conjugates can also be used to generate highly selective and active artificial metalloenzymes (Konieczny, Leurs, & Tiller, 2015; Leurs, Spiekermann, & Tiller, 2016). The concept behind this is that their structure offers the possibility to apply them as polymer amphiphilic nanocontainers with the protein inside surrounded by a more hydrophobic POx shell (Ho et al., 2012; Kapil, Singh, & Das, 2015) for dissolving inorganic salts in organic solvents, and thus forcing the catalytically active salt to attach to the protein, which represents the hydrophilic core of the nanocontainer. An artificial enzyme composed of a laccase–PMOx conjugate with an osmate salt at its active site acts as chiral catalyst for the asymmetric dihydroxylation of alkenes in chloroform. The use of former thereby leads to highly enantioselective product formation (up to 99.4% *ee* for 1-phenyl-1,2-ethanediol for styrene dihydroxylation) (Konieczny et al., 2015; Leurs et al., 2016) that even exceed the classical Sharpless catalysts (97% *ee*) (Kolb, VanNieuwenhze, & Sharpless, 1994; Sharpless et al., 1992)).

ACKNOWLEDGMENTS

We thank Thorsten Moll for the SEC measurements. All polymers were synthesized using CEM Discover microwaves. The research leading to these results has received funding from the Ministry of Innovation, Science and Research of North Rhine-Westphalia within the CLIB-Graduate Cluster Industrial Biotechnology, contract no: 314-108 001 08.

REFERENCES

Abuchowski, A., McCoy, J. R., Palczuk, N. C., van Es, T., & Davis, F. F. (1977). Effect of covalent attachment of polyethylene glycol on immunogenicity and circulating life of bovine liver catalase. *Journal of Biological Chemistry*, *252*(11), 3582–3586.

Boerakker, M. J., Hannink, J. M., Bomans, P. H. H., Frederik, P. M., Nolte, R. J. M., Meijer, E. M., et al. (2002). Giant amphiphiles by cofactor reconstitution. *Angewandte Chemie International Edition*, *41*(22), 4239–4241.

Bos, J., & Roelfes, G. (2014). Artificial metalloenzymes for enantioselective catalysis. *Current Opinion in Chemical Biology*, *19*, 135–143.

Cao, Y., & Wang, Y. (2016). Temperature-mediated regulation of enzymatic activity. *ChemCatChem*, *8*, 2740–2747. http://dx.doi.org/10.1002/cctc.201600406.

Castillo, B., Solá, R. J., Ferrer, A., Barletta, G., & Griebenow, K. (2008). Effect of PEG modification on subtilisin Carlsberg activity, enantioselectivity, and structural dynamics in 1,4-dioxane. *Biotechnology and Bioengineering*, *99*(1), 9–17.

Cobo, I., Li, M., Sumerlin, B. S., & Perrier, S. (2015). Smart hybrid materials by conjugation of responsive polymers to biomacromolecules. *Nature Materials*, *14*(2), 143–159.

Dech, S., Wruk, V., Fik, C. P., & Tiller, J. C. (2012). Amphiphilic polymer conetworks derived from aqueous solutions for biocatalysis in organic solvents. *Polymer*, *53*(3), 701–707.

De St. Groth, S. F., Webster, R. G., & Datyner, A. (1963). Two new staining procedures for quantitative estimation of proteins on electrophoretic strips. *Biochimica et Biophysica Acta*, *71*, 377–391.

Faber, K. (2004). *Biotransformations in organic chemistry* (6th ed.). Berlin, Heidelberg: Springer-Verlag.
Fee, C. J., & Van Alstine, J. M. (2006). PEG-proteins: Reaction engineering and separation issues. *Chemical Engineering Science, 61*(3), 924–939.
Gaertner, H. F., & Puigserver, A. J. (1988). Peptide synthesis catalyzed by polyethylene glycol-modified chymotrypsin in organic solvents. *Proteins: Structure, Function, and Bioinformatics, 3*(2), 130–137.
Grigoletto, A., Maso, K., Mero, A., Rosato, A., Schiavon, O., & Pasut, G. (2016). Drug and protein delivery by polymer conjugation. *Journal of Drug Delivery Science and Technology, 32*(Pt B), 132–141.
Habeeb, A. F. S. A., & Atassi, M. Z. (1970). Enzymic and immunochemical properties of lysozyme. Evaluation of several amino group reversible blocking reagents. *Biochemistry, 9*(25), 4939–4944.
Hannink, J. M., Cornelissen, J. J. L. M., Farrera, J. A., Foubert, P., De Schryver, F. C., Sommerdijk, N. A. J. M., et al. (2001). Protein–polymer hybrid amphiphiles. *Angewandte Chemie International Edition, 40*(24), 4732–4734.
Harris, J. M., Martin, N. E., & Modi, M. (2001). Pegylation. *Clinical Pharmacokinetics, 40*(7), 539–551.
Herman, S., Hooftman, G., & Schacht, E. (1995). Poly(ethylene glycol) with reactive endgroups: I. Modification of proteins. *Journal of Bioactive and Compatible Polymers, 10*(2), 145–187.
Heukeshoven, J., & Dernick, R. (1985). Simplified method for silver staining of proteins in polyacrylamide gels and the mechanism of silver staining. *Electrophoresis, 6*(3), 103–112.
Ho, C. H., Thiel, M., Celik, S., Odermatt, E. K., Berndt, I., Thomann, R., et al. (2012). Conventional and microwave-assisted synthesis of hyperbranched and highly branched polylysine towards amphiphilic core-shell nanocontainers for metal nanoparticles. *Polymer, 53*, 4623–4630.
Hooftman, G., Herman, S., & Schacht, E. (1996). Review: Poly(ethylene glycol)s with reactive endgroups. II. Practical consideration for the preparation of protein-PEG conjugates. *Journal of Bioactive and Compatible Polymers, 11*(2), 135–159.
Hoogenboom, R. (2009). Poly(2-oxazoline)s: A polymer class with numerous potential applications. *Angewandte Chemie International Edition, 48*(43), 7978–7994.
Ivanov, A. E., Edink, E., Kumar, A., Galaev, I. Y., Arendsen, A. F., Bruggink, A., et al. (2003). Conjugation of penicillin acylase with the reactive copolymer of N-isopropylacrylamide: A step toward a thermosensitive industrial biocatalyst. *Biotechnology Progress, 19*(4), 1167–1175.
Kapil, N., Singh, A., & Das, D. (2015). Cross-β amyloid nanohybrids loaded with cytochrome C exhibit superactivity in organic solvents. *Angewandte Chemie, 127*(22), 6592–6595.
Katsumi, H., Nishikawa, M., Hirosaki, R., Okuda, T., Kawakami, S., Yamashita, F., et al. (2016). Development of PEGylated cysteine-modified lysine dendrimers with multiple reduced thiols to prevent hepatic ischemia/reperfusion injury. *Molecular Pharmaceutics, 13*, 2867–2873. http://dx.doi.org/10.1021/acs.molpharmaceut.6b00557.
Kent, J. A. (2013). *Handbook of industrial chemistry and biotechnology*. New York: Springer US, Springer Science+Business Media.
Kolb, H. C., VanNieuwenhze, M. S., & Sharpless, K. B. (1994). Catalytic asymmetric dihydroxylation. *Chemical Reviews, 94*(8), 2483–2547.
Konieczny, S., Fik, C. P., Averesch, N. J. H., & Tiller, J. C. (2012). Organosoluble enzyme conjugates with poly(2-oxazoline)s via pyromellitic acid dianhydride. *Journal of Biotechnology, 159*, 195–203.

Konieczny, S., Krumm, C., Doert, D., Neufeld, K., & Tiller, J. C. (2014). Investigations on the activity of poly(2-oxazoline) enzyme conjugates dissolved in organic solvents. *Journal of Biotechnology*, *181*, 55–63.

Konieczny, S., Leurs, M., & Tiller, J. C. (2015). Polymer enzyme conjugates as chiral ligands for sharpless dihydroxylation of alkenes in organic solvents. *ChemBioChem*, *16*(1), 83–90.

Krumm, C., Fik, C. P., Meuris, M., Dropalla, G. J., Geltenpoth, H., Sickmann, A., et al. (2012). Well-defined amphiphilic poly(2-oxazoline) ABA-triblock copolymers and their aggregation behavior in aqueous solution. *Macromolecular Rapid Communications*, *33*(19), 1677–1682.

Krumm, C., Konieczny, S., Dropalla, G. J., Milbradt, M., & Tiller, J. C. (2013). Amphiphilic polymer conetworks based on end group cross-linked poly(2-oxazoline) homo- and triblock copolymers. *Macromolecules*, *46*(9), 3234–3245.

Laemmli, U. K. (1970). Cleavage of structural proteins during the assembly of the head of bacteriophage T4. *Nature*, *227*(5259), 680–685.

Leurs, M., Spiekermann, P. S., & Tiller, J. C. (2016). Optimization of and mechanistic considerations for the enantioselective dihydroxylation of styrene catalyzed by osmate-laccase-poly(2-methyloxazoline) in organic solvents. *ChemCatChem*, *8*(3), 593–599.

Liu, Y., Li, J., & Lu, Y. (2015). Enzyme therapeutics for systemic detoxification. *Advanced Drug Delivery Reviews*, *90*, 24–39.

Luxenhofer, R., Han, Y., Schulz, A., Tong, J., He, Z., Kabanov, A. V., et al. (2012). Poly(2-oxazoline)s as polymer therapeutics. *Macromolecular Rapid Communications*, *33*(19), 1613–1631.

Mero, A., Pasut, G., Via, L. D., Fijten, M. W. M., Schubert, U. S., Hoogenboom, R., et al. (2008). Synthesis and characterization of poly(2-ethyl 2-oxazoline)-conjugates with proteins and drugs: Suitable alternatives to PEG-conjugates? *Journal of Controlled Release*, *125*(2), 87–95.

Miyamoto, M., Naka, K., Shiozaki, M., Chujo, Y., & Saegusa, T. (1990). Preparation and enzymic activity of poly[(N-acylimino)ethylene]-modified catalase. *Macromolecules*, *23*(13), 3201–3205.

Moatsou, D., Li, J., Ranji, A., Pitto-Barry, A., Ntai, I., Jewett, M. C., et al. (2015). Self-assembly of temperature-responsive protein–polymer bioconjugates. *Bioconjugate Chemistry*, *26*(9), 1890–1899.

Moreno-Pérez, S., Orrego, A. H., Romero-Fernández, M., Trobo-Maseda, L., Martins-DeOliveira, S., Munilla, R., et al. (2016). Chapter three—Intense PEGylation of enzyme surfaces: Relevant stabilizing effects. In K. Challa Vijaya (Ed.), *Vol. 571*. *Methods in enzymology* (pp. 55–72). New York: Elsevier, Academic Press.

Naka, K., Chujo, Y., Miyamoto, M., & Saegusa, T. (1997). Effect of modifier on enzymatic function of poly[(N-acylimino)ethylene]-modified lipases in organic solvents. *Journal of Macromolecular Science, Part A*, *34*(1), 35–48.

Naka, K., Miyamoto, M., Chujo, Y., & Saegusa, T. (1999). Effect of solvent polarity on enzymatic function of poly [(<i>N</i>-acylimino)ethylene] modified lipase. *Proceedings of the Japan Academy, Series B*, *75*(3), 49–53.

Naka, K., Ohki, A., & Maeda, S. (1991). Amphiphilic block copolymer-horseradish peroxidase aggregate as a new polymer-enzyme hybrid in organic solvents. *Chemistry Letters*, *20*(8), 1303–1306.

Plothe, R., Sittko, I., Lanfer, F., Fortmann, M., Roth, M., Kolbach, V., et al. (2017). Poly(2-ethyloxazoline) as matrix for highly active electrospun enzymes in organic solvents. *Biotechnology and Bioengineering*, *114*, 39–45. http://dx.doi.org/10.1002/bit.26043.

Roberts, M. J., Bentley, M. D., & Harris, J. M. (2012). Chemistry for peptide and protein PEGylation. *Advanced Drug Delivery Reviews*, *64*(Suppl.), 116–127.

Savin, G., Bruns, N., Thomann, Y., & Tiller, J. C. (2005). Nanophase separated amphiphilic microbeads. *Macromolecules*, *38*(18), 7536–7539.

Sawhney, P., Kumar, S., Maheshwari, N., Guleria, S. S., Dhar, N., Kashyup, R., et al. (2016). Site-specific thiol-mediated PEGylation of streptokinase leads to improved properties with clinical potential. *Current Pharmaceutical Design*, *22*(38), 5868–5878.

Schoenfeld, I., Dech, S., Ryabenky, B., Daniel, B., Glowacki, B., Ladisch, R., et al. (2013). Investigations on diffusion limitations of biocatalyzed reactions in amphiphilic polymer conetworks in organic solvents. *Biotechnology and Bioengineering*, *110*(9), 2333–2342.

Sharpless, K. B., Amberg, W., Bennani, Y. L., Crispino, G. A., Hartung, J., Jeong, K. S., et al. (1992). The osmium-catalyzed asymmetric dihydroxylation: A new ligand class and a process improvement. *Journal of Organic Chemistry*, *57*, 2768–2771.

Shiroya, T., Yasui, M., Fujimoto, K., & Kawaguchi, H. (1995). Control of enzymatic activity using thermosensitive polymers. *Colloids and Surfaces B: Biointerfaces*, *4*(5), 275–285.

Sittko, I., Kremser, K., Roth, M., Kuehne, S., Stuhr, S., & Tiller, J. C. (2015). Amphiphilic polymer conetworks with defined nanostructure and tailored swelling behavior for exploring the activation of an entrapped lipase in organic solvents. *Polymer*, *64*, 122–129.

Takahashi, K., Nishimura, H., Yoshimoto, T., Okada, M., Ajima, A., Matsushima, A., et al. (1984). Polyethylene glycol-modified enzymes trap water on their surface and exert enzymic activity in organic solvents. *Biotechnology Letters*, *6*(12), 765–770.

Thomas, B. R., Vekilov, P. G., & Rosenberger, F. (1998). Effects of microheterogeneity in hen egg-white lysozyme crystallization. *Acta Crystallographica Section D*, *54*(2), 226–236.

Tong, J., Luxenhofer, R., Yi, X., Jordan, R., & Kabanov, A. V. (2010). Protein modification with amphiphilic block copoly(2-oxazoline)s as a new platform for enhanced cellular delivery. *Molecular Pharmaceutics*, 7(4), 984–992.

Tong, J., Yi, X., Luxenhofer, R., Banks, W. A., Jordan, R., Zimmerman, M. C., et al. (2013). Conjugates of superoxide dismutase 1 with amphiphilic poly(2-oxazoline) block copolymers for enhanced brain delivery: Synthesis, characterization and evaluation in vitro and in vivo. *Molecular Pharmaceutics*, *10*(1), 360–377.

Velonia, K., Rowan, A. E., & Nolte, R. J. M. (2002). Lipase polystyrene giant amphiphiles. *Journal of the American Chemical Society*, *124*(16), 4224–4225.

Viegas, T. X., Bentley, M. D., Harris, J. M., Fang, Z., Yoon, K., Dizman, B., et al. (2011). Polyoxazoline: Chemistry, properties, and applications in drug delivery. *Bioconjugate Chemistry*, *22*(5), 976–986.

Wan, Y.-Y., Lu, R., Xiao, L., Du, Y.-M., Miyakoshi, T., Chen, C.-L., et al. (2010). Effects of organic solvents on the activity of free and immobilised laccase from Rhus vernicifera. *International Journal of Biological Macromolecules*, *47*(4), 488–495.

Wiesbrock, F., Hoogenboom, R., Abeln, C. H., & Schubert, U. S. (2004). Single-mode microwave ovens as New reaction devices: Accelerating the living polymerization of 2-ethyl-2-oxazoline. *Macromolecular Rapid Communications*, *25*(22), 1895–1899.

Wiesbrock, F., Hoogenboom, R., Leenen, M. A. M., Meier, M. A. R., & Schubert, U. S. (2005). Investigation of the living cationic ring-opening polymerization of 2-methyl-, 2-ethyl-, 2-nonyl-, and 2-phenyl-2-oxazoline in a single-mode microwave reactor. *Macromolecules*, *38*(12), 5025–5034.

Yoon, A., Shin, J. W., Kim, S., Kim, H., & Chung, J. (2016). Chicken scFvs with an artificial cysteine for site-directed conjugation. *PLoS One*, *11*(1), e0146907.

Zhou, J. Q., He, T., & Wang, J. W. (2016). The microbial transglutaminase immobilization on carboxylated poly(N-isopropylacrylamide) for thermo-responsivity. *Enzyme and Microbial Technology*, *87–88*, 44–51.

CHAPTER EIGHTEEN

Preparation and Applications of Dendronized Polymer–Enzyme Conjugates

Andreas Küchler, Daniel Messmer, A. Dieter Schlüter, Peter Walde[1]
ETH Zürich, Zürich, Switzerland
[1]Corresponding author: e-mail address: peter.walde@mat.ethz.ch

Contents

Abstract

Dendronized polymer–enzyme conjugates are large, water-soluble macromolecular structures built from a linear, fully synthetic, dendronized polymer (denpol), and several copies of enzyme molecules covalently bound to the peripheral functional groups of the denpol. Since denpol chains comprise repeating units with regularly branched side chains (dendrons), denpols have a cylindrical shape and are much thicker than conventional linear polymers. Depending on the dendron generation and chemical structure, denpols may have a large number of functional groups on their surface,

Methods in Enzymology, Volume 590
ISSN 0076-6879
http://dx.doi.org/10.1016/bs.mie.2017.01.014
© 2017 Elsevier Inc.
All rights reserved.

exposed to the aqueous medium in which they are dissolved. Enzymes (and also other molecules) can be attached to these functional groups, for example, via a stable bis-aryl hydrazone (BAH) bond. The dendronized polymer scaffold might also serve as a nanoarmor and stabilize the delicate enzymes. One of the denpols which can be used for the preparation of denpol–enzyme conjugates is *de*-PG2. It has a poly(methacrylate) backbone and consists of second-generation dendrons with four peripheral amino groups in each repeating unit. The synthesis of *de*-PG2 and the preparation of a *de*-PG2 conjugate carrying BAH-linked proteinase K (proK), as an example, are described here for applications in the field of enzyme immobilization on solid surfaces. The nanoarmored enzyme–polymer conjugate indicated high stability and retention of enzymatic activity.

ABBREVIATIONS

AIBN 2,2′-azobisisobutyronitrile
BHT 3,5-di-*tert*-butyl-4-hydroxytoluene
DCM dichloromethane
DMAP 4-dimethylaminopyridine
DMF *N*,*N*-dimethylformamide
DMSO dimethyl sulfoxide
EDC 1-ethyl-3-(3-dimethylaminopropyl)carbodiimide
EtOAc ethyl acetate
FRP free-radical polymerization
GPC gel permeation chromatography (=size-exclusion chromatography)
MAC methacryloyl chloride
MeCN acetronitrile
MeOH methanol
MES 2(*N*-morpholino)ethanesulfonic acid
M_n number-average molar mass
MOPS 3(*N*-morpholino)propanesulfonic acid
MWCO molecular weight cut off
NMP *N*-methyl-2-pyrrolidone
NMWL nominal molecular weight limit
PDI polydispersity index
PMMA poly(methyl methacrylate)
P_n number-average degree of polymerization
r.u. repeating unit
RAFT reversible addition–fragmentation chain transfer
R_f retention factor (TLC)
RT room temperature (≈25°C)
S-4FB *N*-succinimidyl 4-formylbenzoate
S-HyNic *N*-succinimidyl 6-hydrazinonicotinate acetone hydrazone
TEA triethylamine
THF tetrahydrofuran
TLC thin-layer chromatography

1. INTRODUCTION

Dendronized polymers (denpols) are linear, cylindrical polymers, which carry a dendritic side chain in each repeating unit (r.u.) (Chen & Xiong, 2010; Frauenrath, 2005; Paez, Martinelli, Brunetti, & Strumia, 2012; Rosen et al., 2009; Schlüter, 2005; Schlüter & Rabe, 2000). This so-called dendron can be of first, second, third, fourth (e.g., Guo et al., 2009), or even higher generation (Yu, Schlüter, & Zhang, 2014). With increasing generation (i.e., number of branching points), the denpol molecules get thicker and stiffer, and they increasingly differ in general structure and properties from conventional linear polymers, bottle-brush polymers, or dendrimers. In Fig. 1, the chemical structure of a second-generation denpol is shown, abbreviated as *de*-PG2 (**1**), "*de*" specifying that the protecting group used during synthesis has been removed, i.e., the denpol is deprotected. This type of denpol has a poly(methacrylate) backbone,

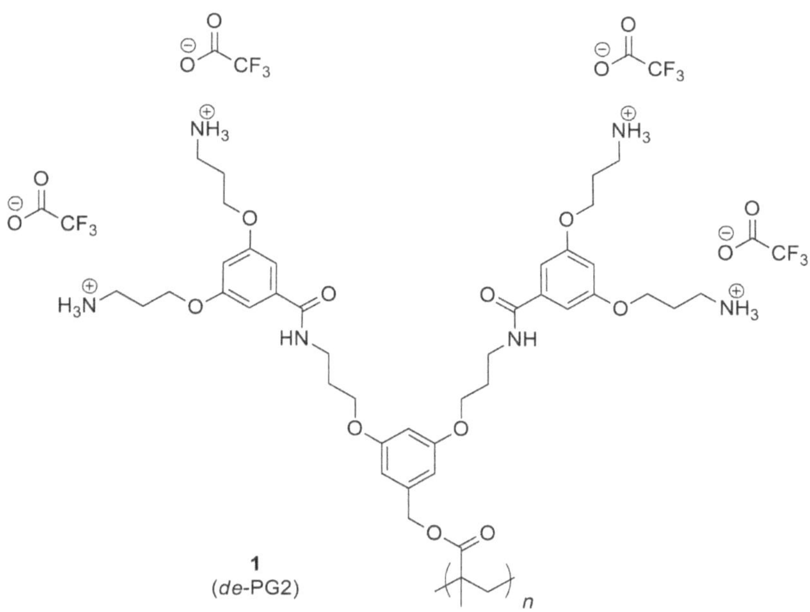

1
(*de*-PG2)

Fig. 1 Chemical structure of the dendronized polymer (denpol) *de*-PG2 (**1**) as trifluoroacetate salt. For the description in the text, $n = 2000$.

1,3,5-trisubstituted benzene as dendron branching unit, and peripheral amino groups. These groups increase in number with increasing generation.

In *de*-PG2, each r.u. carries four amino groups. At acidic or neutral pH values, the amino groups are protonated, and therefore, the denpols of type *de*-PGX are water soluble due to their polycationic nature. The pK_a value of the ammonium ion is expected to be about 9–10, as in the case of poly (L-lysine) (de Kruijff, Rietveld, Telders, & Vaandrager, 1985). Under slightly alkaline aqueous conditions, some of the amino groups can be modified, for example, with a fluorescent dye or enzyme molecules (Grotzky, Nauser, Erdogan, Schlüter, & Walde, 2012). In the latter case, denpol–enzyme conjugates can be obtained in such a way that several copies of one type of enzyme (or several copies of two—or possibly even more—different types of enzymes) can be bound along one and the same denpol chain (see Fig. 2; Grotzky et al., 2012; Küchler, Adamcik, Mezzenga, Schlüter, & Walde, 2015; Küchler, Bleich, Sebastian, Dittrich, & Walde, 2015). This type of conjugate differs from those enzyme–polymer conjugates in which one or several polymer (or denpol) chains are bound to one and the same enzyme molecule (Fuhrmann et al., 2013; Moreno-Pérez et al., 2016; Veronese, Mero, & Pasut, 2009).

Denpol–enzyme conjugates prepared from *de*-PG2 can be applied for the immobilization of enzymes on silicate surfaces, for example, on microscopy glass coverslips, inside glass micropipettes, or on microfluidic chips (Gustafsson, Küchler, Holmberg, & Walde, 2015; Küchler, Adamcik, et al., 2015; Küchler, Bleich, et al., 2015). Immobilization is achieved by simple noncovalent adsorption of the conjugates from aqueous solution. The dendronized polymer scaffold might also serve as nanoarmor and protect the enzyme against denaturation or deactivation, which could be a significant advantage of the current approach.

In the following, a protocol for the synthesis of *de*-PG2 (**1**, Fig. 1) is given, together with a detailed description of the stepwise preparation of a conjugate between *de*-PG2 and the enzyme proteinase K (proK, EC 3.4.21.64), as an elaborated example. The enzyme is covalently bound to *de*-PG2 via a stable bis-aryl hydrazone (BAH) unit, resulting in a denpol–enzyme conjugate which is abbreviated as *de*-PG2-BAH-proK (Fig. 2). The synthesis of *de*-PG2-BAH-proK involves (i) the partial modification of *de*-PG2 with *N*-succinimidyl 6-hydrazinonicotinate acetone hydrazone (S-HyNic) to yield *de*-PG2-HyNic, (ii) the partial modification of proK with *N*-succinimidyl 4-formylbenzoate (S-4FB) to yield proK-4FB, and

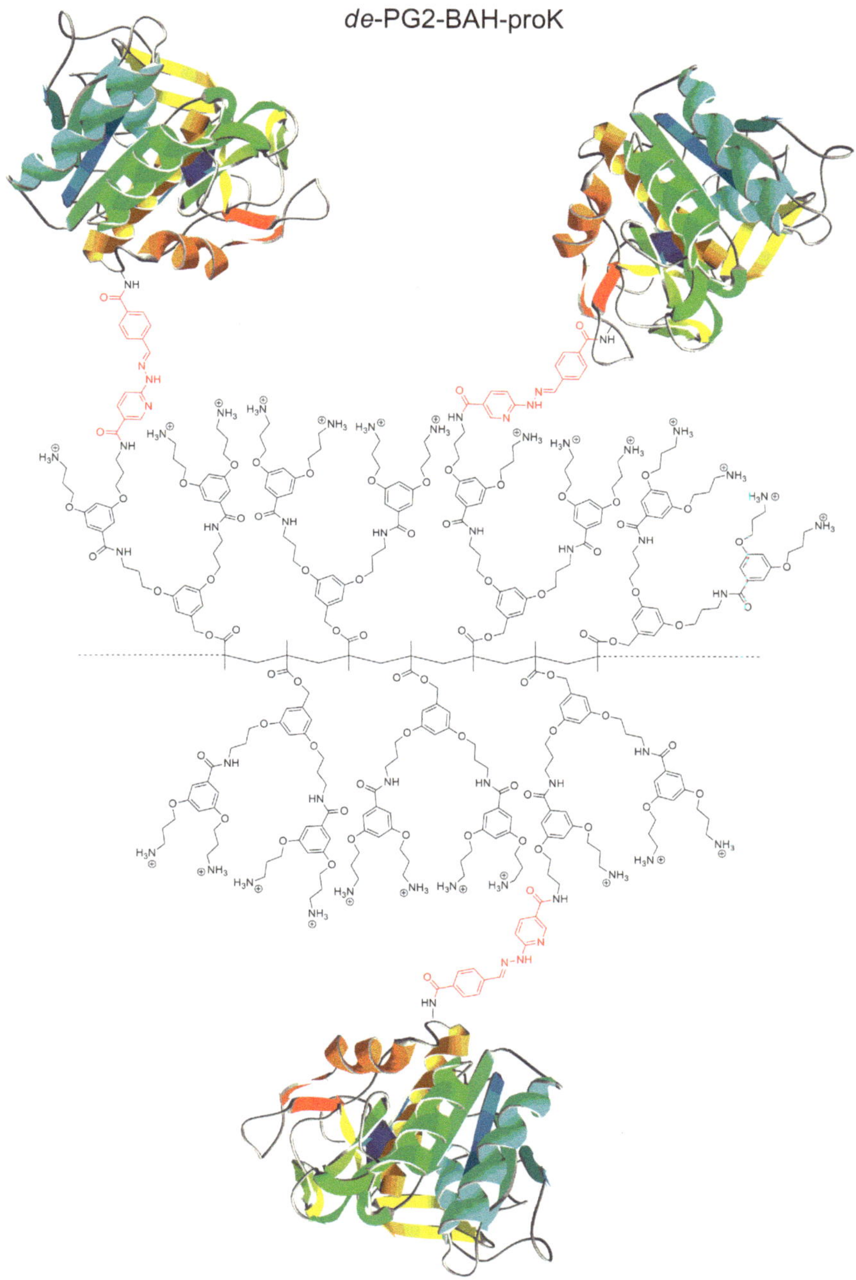

Fig. 2 Schematic representation of a part of the denpol–enzyme conjugate *de*-PG2-BAH-proK, illustrating that on the periphery of the water-soluble second-generation denpol *de*-PG2 (**1**) several copies of the enzyme proteinase K (proK) are covalently bound through a stable bis-aryl hydrazone (BAH) linker (see Küchler, Bleich, et al., 2015; proK PDB entry: 1IC6). The denpol and proK are not drawn to scale. The approximate diameter of *de*-PG2 is 2 nm (Küchler, Adamcik, et al., 2015), and the size of proK is about 4 nm (Betzel, Pal, & Saenger, 1988).

finally (iii) the reaction between *de*-PG2-HyNic and proK-4FB. After purification by ultrafiltration, the amount of proK bound to the denpol can be determined through a spectrophotometric quantification of the amount of BAH at $\lambda = 354$ nm (molar absorption coefficient $\varepsilon_{354nm} = 2.9 \times 10^4\ M^{-1}\ cm^{-1}$; Grotzky, Manaka, Kojima, & Walde, 2011; Hermanson, 2013; Solulink, 2016).

2. THE DENPOL *de*-PG2

The synthesis of denpol *de*-PG2 (**1**) involves several steps as outlined in Figs. 3 and 4. The procedures have been elaborated previously (Guo et al., 2009) and are given in detail in the following, with indication of how the reagents and solvents were prepared, how the products were purified, and how the actual synthesis has to be carried out to obtain *de*-PG2 as pure compound.

2.1 Synthesis of Macromonomer MG1 and Dendronization Agent DG1

2.1.1 Materials and General Methods

Dendritic precursor **5** and dendronization agent DG1 (**9**) (Fig. 3) are commercially available from Synwit Technology Co. Ltd. (Beijing), however, only in batches of 25 g or larger; their syntheses are therefore described later. 3-Bromopropylamine *tert*-butylcarbamate (**3**) is available from various suppliers but can also be prepared from cheap precursors as described later. Methacryloyl chloride (MAC) (Sigma Aldrich, 3,5-di-*tert*-butyl-4-hydroxytoluene (BHT) stabilized) was distilled from hydroquinone shortly before use and stored at −18°C under N_2 in a Schlenk flask; 2,2′-azobisisobutyronitrile (AIBN) (Fluka) was recrystallized from methanol (MeOH) before use; solvents for column chromatography (dichloromethane (DCM), hexane, ethyl acetate (EtOAc)) were obtained in technical grade (EGT Chemie AG) and distilled by rotary evaporation before use. 1,4-Dioxane (Fisher, BHT stabilized) was distilled directly before use. Dry tetrahydrofuran (THF) and dry DCM were prepared by storing reagent-grade solvents (Sigma Aldrich) over freshly activated 4 Å molecular sieves (Sigma Aldrich) for at least 16 h prior to use. Column chromatography was performed using silica gel (SiliaFlash P60, Silicycle). All other reagents and solvents were reagent grade or better and used as received. Thin-layer chromatography (TLC) was carried out on precoated

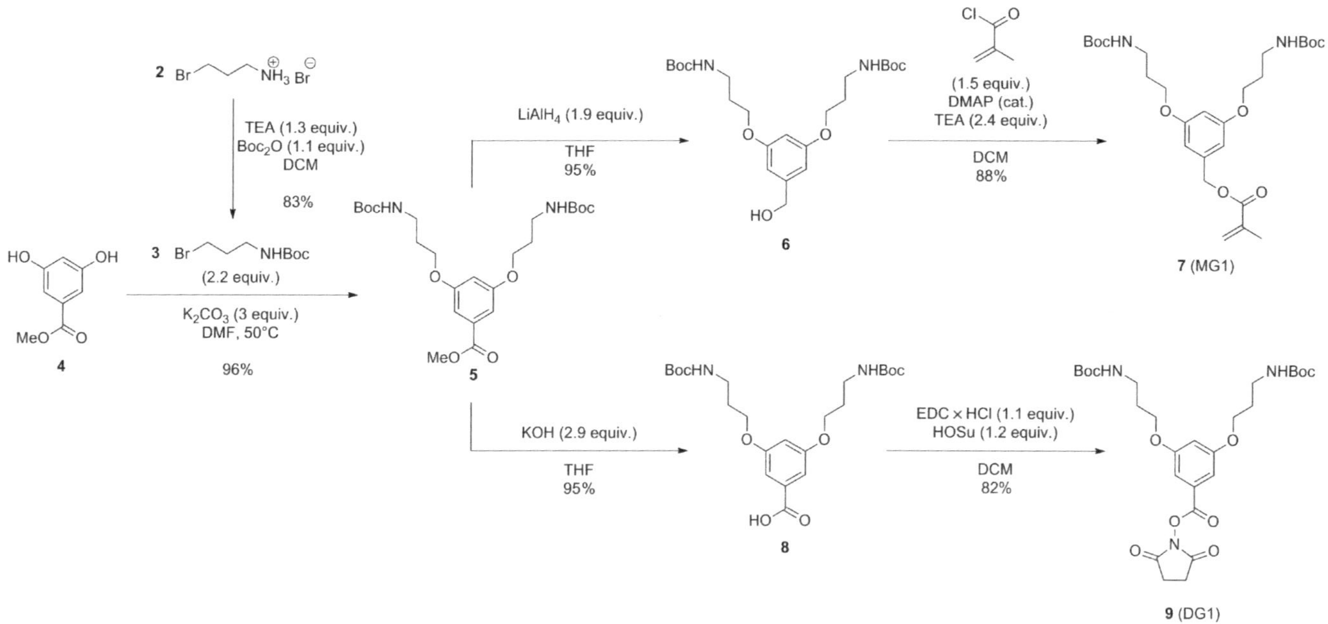

Fig. 3 Scheme for the chemical synthesis of the macromonomer MG1 (**7**) and of the dendronization agent DG1 (**9**).

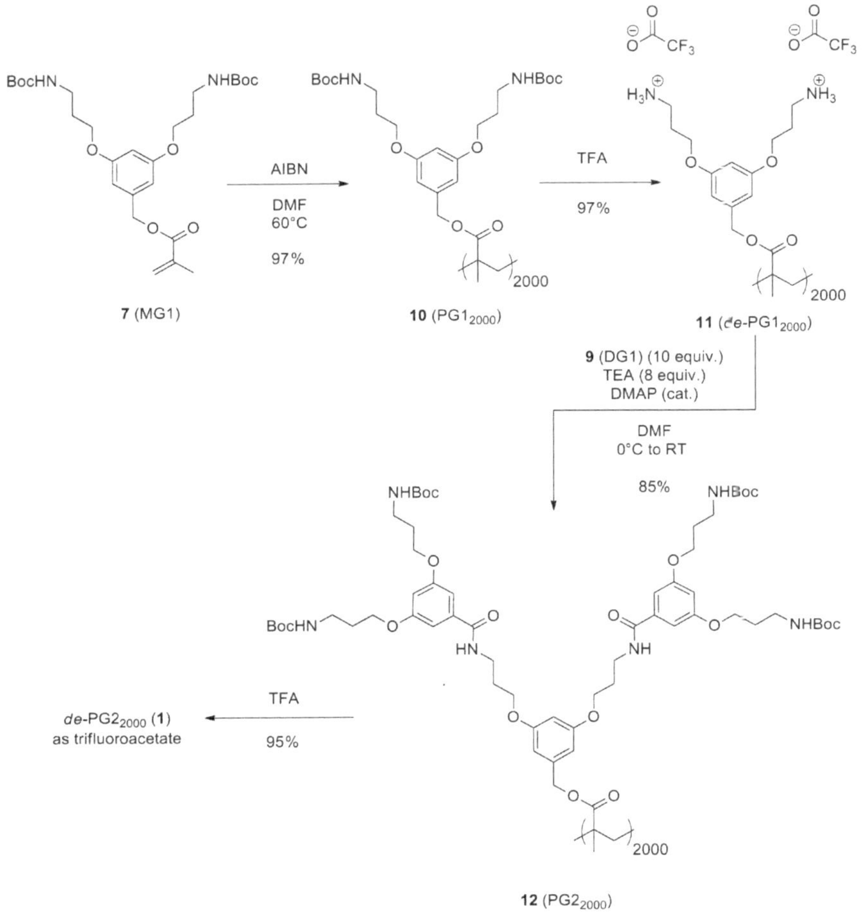

Fig. 4 Scheme for the synthesis of *de*-PG2 (**1**) from the macromonomer MG1 (**7**) and the dendronization agent DG1 (**9**). The water-soluble denpol *de*-PG2 (**1**) is obtained as trifluoroacetate salt. In accordance with the description in the text, details are given for the synthesis of *de*-PG2 (**1**) with $P_n = 2000$.

aluminum-backed plates (Macherey-Nagel, TLC Silica Gel 60 F_{254}). Gel permeation chromatography (=size-exclusion chromatography) (GPC) was performed on a PL-GPC 220 (Polymer Laboratories) equipped with two PL-Gel Mix-B LS columns (Polymer Laboratories) operated at 45°C with 0.1 g L^{-1} LiBr in *N,N*-dimethylformamide (DMF) as eluent. Molecular weights were determined employing universal calibration with monodisperse PMMA standards (Polymer Laboratories). ^1H NMR spectra were measured on an AVANCE™ 300 MHz instrument (Bruker), equipped with a SampleXpress® autosampler.

Where dry solvent use is indicated, reactions were conducted under an atmosphere of dry N_2, and reaction vessels were dried by three cycles of heating with a heat gun, then flushing of the vessel with N_2.

Solvents were removed by rotary evaporation in vacuum at 40°C unless otherwise noted. Drying in high vacuum and lyophilization was performed using a Alpha 2-4 LD plus lyophilizer (Martin Christ Gefriertrocknungsanlagen GmbH) coupled to an RC5 double-stage rotary vane pump (Vacuubrand), supplying 10^{-2} to 10^{-1} mbar.

The reactions are summarized in Fig. 3 with indication of an example of the isolated yields achieved.

Safety comments

General considerations

- Before conducting any experiments, material safety datasheets (MSDS) of all chemicals should be consulted.
- Syntheses should be conducted in a dedicated synthetic chemical laboratory, with proper media (cooling water, dry N_2), ventilation (fume cabinets), and safety infrastructure (emergency shower, eye shower, fire safety).
- Standard personal protective equipment (gloves, safety goggles, lab coat) must be worn at all times when handling chemicals.
- A respirator is recommended for handling of silica during preparation of chromatography columns.
- In case of exposure or spillage, the guidelines outlined in the compound's MSDS must be followed.
- Chemicals and solvents must be disposed of properly.

Particular hazards

- Many solvents used in the syntheses outlined later (DCM, 1,4-dioxane, DMF, hexane, MeOH, THF) are carcinogenic and/or pose other long-term health risks. Chronic exposure must be avoided.
- AIBN is classified as an explosive; heating and friction should be avoided, and plastic spatulae should be used in handling the pure solid.
- MAC is sensitive to moisture and heat as well as volatile and highly toxic upon inhalation. Strict adherence to proper techniques for handling of compounds under inert atmosphere is therefore required. Exposure must be avoided at all cost, and any equipment in contact with MAC must be treated with a 1:1 mixture of MeOH and 1 *M* aq.

NaOH inside a well-vented fume hood before regular cleaning. Leftover MAC must be quenched by dropwise addition to a great excess of a 1:1 mixture of MeOH and 1 *M* aq. NaOH which is cooled in an ice bath. The resulting mixture is stirred at room temperature (RT) overnight before disposal.

- $LiAlH_4$ is highly sensitive to moisture and oxygen, and its solutions may ignite on contact with air. Strict adherence to proper techniques for handling of compounds under inert atmosphere is therefore required. Equipment in contact with $LiAlH_4$ solutions must be rinsed with Et_2O or THF prior to standard cleaning. Due to the high risk of ignition and the compound's reactivity with CO_2, a suitable extinguisher (dry powder) must be at hand.
- Pure TFA is highly corrosive and quite volatile; contact with skin and mucous membranes must be avoided.

2.1.2 Synthesis of 3-Bromopropylamine tert-Butylcarbamate (3)

(1) 3-Bromopropylamine hydrobromide (**2**, 0.99 g, 4.5 mmol, 1.0 equiv.) and di-*tert*-butyl dicarbonate (1.16 g, 5.3 mmol, 1.17 equiv.) are suspended in DCM (25 mL) in a 50-mL round-bottom flask containing a magnetic stir bar.

(2) Triethylamine (TEA) (0.76 mL, 5.5 mmol, 1.22 equiv.) is added dropwise via a syringe; the initial suspension clears up in the process.

(3) The mixture is stirred at RT for 18 h.

(4) The clear solution is diluted with EtOAc (100 mL) and transferred to a 250-mL separation funnel, together with deionized water (20 mL).

(5) The organic phase is washed with 1 *M* aqueous HCl (100 mL), twice with sat. aq. $NaHCO_3$ (100 mL each), and finally with brine (100 mL).

(6) The organic phase is dried over $MgSO_4$, filtered, and solvents are removed.

(7) Column chromatography with EtOAc/hexane 1:9 (v/v) affords the desired product **3** as colorless oil (e.g., 0.89 g, 83%).

2.1.3 Synthesis of 5

(1) A mixture of methyl 3,5-dihydroxybenzoate (**4**, 280 mg, 1.7 mmol, 1.00 equiv.) and anhydrous potassium carbonate (782 mg, 5.7 mmol, 3.4 equiv.) in DMF (20 mL) in a 50-mL round-bottom flask is stirred at RT for 15 min.

(2) 3-Bromopropylamine *tert*-butylcarbamate (**3**, 880 mg, 2.2 equiv.) in DMF (5 mL) is added. The mixture is heated to 40°C and stirred for 13 h under N_2.

(3) The reaction mixture is cooled to RT, transferred into a 250-mL separation funnel, and diluted with EtOAc (100 mL). The organic phase is washed twice with water (100 mL each) and once with brine (100 mL).

(4) The organic phase is dried over $MgSO_4$, filtered, and then the solvents are removed.

(5) The resulting white solid is purified by flash chromatography (EtOAc/hexane 1:2).

(6) Evaporation of solvents followed by drying in high vacuum affords the desired product **5** as a white powder (e.g., 771 mg, 96%).

2.1.4 Synthesis of 6

(1) A 1-L three-neck round-bottom flask is equipped with a large magnetic stir bar, a valve connected to a Schlenk line supplying vacuum and dry N_2, a 250-mL pressure-equalizing dropping funnel and a septum.

(2) Compound **5** (40.06 g, 83 mmol, 1.0 equiv.) is added to the round-bottom flask in N_2 counter flow and dissolved in dry THF (450 mL), which is transferred into the setup using a double-tipped cannula.

(3) The flask is cooled to 0°C by immersion in an ice/water bath.

(4) A $LiAlH_4$ solution (2.4 *M* in THF, 65 mL, 156 mmol, 1.9 equiv.) is added to the dropping funnel via syringe and diluted with dry THF (100 mL) added via double-tipped cannula.

(5) The diluted $LiAlH_4$ solution is added continuously under vigorous stirring over a period of 30 min.

(6) After completion of the addition, the reaction mixture is stirred for 30 min, then the ice bath is removed, and the mixture is stirred at RT for 4 h.

(7) The flask is again cooled to 0°C in an ice–water bath.

(8) Excess $LiAlH_4$ is quenched by dropwise addition of EtOAc (250 mL), then dropwise addition of deionized water (200 mL).

(9) The resulting slurry is further diluted with EtOAc (250 mL) and water (200 mL); the mixture is transferred into a 1-L separation funnel, then washed with water (2 × 800 mL) and brine (400 mL).

(10) The organic phase is dried over $MgSO_4$, filtered, and concentrated by rotary evaporation.

(11) The resulting oil is precipitated by layering with hexane and briefly sonicating the flask.

(12) Vacuum filtration, followed by washing with hexane and drying in high vacuum, affords the desired product **6** as a white solid (e.g., 36.2 g, 95%).

2.1.5 Synthesis of MG1 (7)

(1) A 1-L three-neck round-bottom flask is equipped with a large magnetic stir bar, a valve, a 250-mL pressure-equalizing dropping funnel and a septum. The valve is connected to a Schlenk line supplying vacuum and dry N_2.

(2) Compound **6** (36 g, 79 mmol, 1 equiv.) and 4-dimethylaminopyridine (DMAP) (362 mg, 3 mmol, 0.04 equiv.) are added to the round-bottom flask in N_2 counter flow.

(3) Dry DCM (400 mL) is added to the round-bottom flask via double-tipped cannula, and TEA (33 mL, 191 mmol, 2.4 equiv.) is added via syringe.

(4) The round-bottom flask with the resulting solution is cooled to 0°C in an ice/water bath.

(5) Freshly distilled MAC (11.5 mL, 119 mmol, 1.5 equiv.) is added to the dropping funnel and diluted with dry DCM (100 mL).

(6) The MAC solution is added dropwise under vigorous stirring.

(7) After addition is complete, the ice bath is removed, and the mixture is stirred for 5 h.

(8) The resulting mixture is diluted with DCM (300 mL) and transferred to a 1-L separation funnel.

(9) The organic phase is washed twice with sat. aq. $NaHCO_3$ (300 mL each), then once with brine (200 mL).

(10) The separated organic phase is dried over $MgSO_4$, filtered, and then concentrated to a yellow oil by rotary evaporation at 30°C.

(11) The oil is purified by flash chromatography (DCM/acetone 15:1).

(12) Product fractions are concentrated by rotary evaporation at 30°C to a colorless oil and precipitated in a 500-mL round-bottom flask by layering with hexane (100 mL) and briefly immersing the flask into an ice-cooled ultrasonication bath.

(13) The solid is isolated by vacuum filtration and washing with copious amounts of hexane; drying in high vacuum affords the desired product **7** (MG1) as a white powder (e.g., 36.3 g, 88%).

(14) Compound **7** (MG1) which is not immediately used for polymerization reactions is stored under N_2 and at $-18°C$ in a Schlenk flask wrapped in aluminum foil.

2.1.6 Synthesis of 8

(1) In a 1-L round-bottom flask equipped with a magnetic stir bar, KOH (5.74 g, 102 mmol, 2.9 equiv.) is dissolved in a mixture of MeOH (360 mL) and water (120 mL).

(2) Compound **5** (30.43 g, 35.9 mmol, 1.0 equiv.) is added, and the flask is equipped with a reflux condenser.

(3) The suspension is stirred at reflux for 15 h.

(4) The resulting clear solution is cooled to RT, then concentrated to a volume of ca. 150 mL by rotary evaporation.

(5) The mixture is transferred to a 1-L beaker, diluted with deionized water (100 mL), and cooled to 0°C in an ice/water bath.

(6) A 1:1 mixture of acetic acid and deionized water is added dropwise under vigorous stirring until the pH reaches a value of approximately 5 (monitored using universal pH indicator paper, pH 1–11, Macherey-Nagel). A white solid precipitates in the process.

(7) The resulting white powder is collected by vacuum filtration, washed with water, and dried in high vacuum, affording the desired product **8** (e.g., 27.69 g, 93.7%).

2.1.7 Synthesis of DG1 (9)

(1) A 50-mL round-bottom Schlenk flask is equipped with a septum, a magnetic stir bar and connected to a Schlenk line providing vacuum and dry N_2.

(2) Compound **8** (500 mg, 1.07 mmol, 1.0 equiv.) and *N*-hydroxysuccinimide (139 mg, 1.2 mmol, 1.13 equiv.) are added in N_2 counter flow and dissolved in dry DCM (20 mL) which is added via syringe.

(3) The solution is cooled to 0°C by immersion into an ice/water bath.

(4) EDC·HCl (262 mg, 1.37 mmol, 1.28 equiv.) is added. The resulting mixture is stirred at 0°C for 1 h.

(5) The ice bath is removed, and the slurry is stirred at RT for 14 h.

(6) The resulting solution is diluted with EtOAc (100 mL) and washed successively with sat. aq. NH_4Cl (100 mL), deionized water (100 mL), brine (75 mL), then dried over $MgSO_4$ and filtered.

(7) The solvents are removed, and the resulting oil is purified by column chromatography (EtOAc/hexane 1:1).

(8) Removal of solvents affords a colorless oil which is precipitated by layering with hexane (10 mL) and brief ultrasonication. The resulting solid is collected by vacuum filtration, washed with hexane, and dried in high vacuum, affording the desired product **9** (DG1) as a white powder (e.g., 496 mg, 82%).

2.2 Polymerization of MG1 and Dendronization With DG1

The denpols of the type PGX currently (2016) are not commercially available. The synthesis of PG1 by free-radical polymerization (FRP) is described later. PG1 can also be synthesized by RAFT polymerization as described elsewhere (Yu, Schlüter, & Zhang, 2012). The latter method allows for more precise control of chain length and chain length distribution below a number-average degree of polymerization (P_n), of ca. 1000. In the enzyme conjugation experiments described further later, FRP-derived denpols of $P_n = 2000$ are employed as an example. The two synthetic protocols only differ in that RAFT employs an additional reagent (e.g., cumyl dithiobenzoate) to control chain length and distribution. While FRP is operationally slightly simpler and somewhat less sensitive to dissolved oxygen than RAFT, it suffers from occasional cross-linking side reactions, which lead to insoluble denpol networks.

The reactions are summarized in Fig. 4 with indication of an example of the isolated yields achieved.

2.2.1 Synthesis of $PG1_{2000}$ (10) by FRP

(1) A 20-mL Schlenk tube is equipped with a septum, a magnetic stir bar and connected to a Schlenk line providing high vacuum (10^{-2} mbar or better) and dry N_2.

(2) Compound **7** (MG1) (2 g, 3.8 mmol, 1 equiv.) and freshly recrystallized AIBN (10 mg, 0.06 mmol, 0.016 equiv.) are added to the Schlenk tube.

(3) The solids are dissolved in dry DMF (3 mL).

(4) The resulting clear solution is degassed by applying freeze/pump/thaw cycles until no more gas evolves on thawing.

(5) The Schlenk tube is backfilled with dry N_2, the valve is closed, and the tube is immersed into an oil bath preheated to 60°C. The solution is gently stirred overnight.

(6) The Schlenk tube is opened to air, DCM (15 mL) is added, and the tube is shaken several hours to dissolve the viscous product.
(7) The dissolved denpol is purified by column chromatography (DCM as eluent).
(8) Denpol fractions are combined, and the solvent is removed.
(9) The resulting denpol film is dissolved in 1,4-dioxane (10 mL) and lyophilized, affording the denpol **10** as a white powder (e.g., 1.95 g, 97%), which is stored at −18°C in the dark. GPC, e.g., $M_n = 1.1 \times 10^6\ \mathrm{g\ mol^{-1}}$, corresponding to $P_n = 2000$; polydispersity index (PDI) = 4.3; i.e., denpol **10** is abbreviated as $PG1_{2000}$.

Comments: The denpol **10** ($PG1_{2000}$) is not retained by silica in the way typical for small molecules; it elutes from the column with the solvent front; there is, however, significant tailing. The elution is best monitored by TLC (EtOAc/hexane 1:1; $R_f \approx 0$) to avoid contamination with unreacted **7** (MG1). The plates should be stained with ninhydrin and heated thoroughly to visualize the denpol spots.

2.2.2 Deprotection of $PG1_{2000}$ (10) to Yield de-$PG1_{2000}$ (11)

(1) Compound **10** (400 mg) is added to a 50-mL round-bottom flask equipped with a magnetic stir bar and cooled with an ice/water bath.
(2) Cold TFA (5 mL) is added dropwise under vigorous stirring.
(3) After completed addition, the flask is closed with a stopper and the resulting solution is stirred at 0°C for 10 min, then at RT overnight.
(4) The flask is cooled in an ice/water bath, and MeOH (20 mL) is added slowly.
(5) The resulting solution is concentrated by rotary evaporation at 35°C; a second aliquot of MeOH (20 mL) is added, and the solvents are again removed by rotary evaporation at 35°C.
(6) The resulting film is further dried by rotary evaporation for 30 min at 35°C and <10 mbar.
(7) The denpol is dissolved in deionized water and lyophilized, affording the deprotected first-generation denpol **11** (*de*-$PG1_{2000}$) as an off-white powder (e.g., 410 mg, 97%), which is stored at −18°C in the dark.

Comments: In most instances, the reaction mixture turns milky white during or shortly after step (2); in this case, cold MeOH is added dropwise to the reaction mixture at 0°C until a clear solution results. Initial failure of lyophilization during step (7) is likely related to trace amounts of MeOH; repetition of steps (5)–(7) with elongated vacuum drying is usually successful.

2.2.3 Synthesis of $PG2_{2000}$ (12) and de-$PG2_{2000}$ (1)

(1) Deprotected denpol **11** (*de*-$PG1_{2000}$) (0.374 g, 0.7 mmol (r.u.), 1.0 equiv.) is added to a 50-mL flask and dissolved in dry DMF (4 mL).

(2) After complete dissolution, the flask is cooled by placing it in an ice/water bath.

(3) TEA (0.75 mL, 5.4 mmol, 8 equiv.) is added slowly, followed by DMAP (51 mg).

(4) DG1 (**9**, 2.304 g, 4.1 mmol, 6 equiv.) is added. The ice bath is removed, the flask is closed with a stopper, wrapped in aluminum foil, and stirred at RT.

(5) Another portion of DG1 (**9**, 768 mg, 1.4 mmol, 2 equiv.) is added after 2 days of stirring by addition of the solid to the flask which is cooled by immersion in an ice/water bath. The reaction mixture is then stirred again at RT.

(6) A last aliquot of DG1 (**9**, 768 mg, 1.4 mmol, 2 equiv.) is added as described in the previous step after an additional 1 day of stirring.

(7) After further 2 days of stirring, the resulting solution is loaded onto a short silica column and eluted with DCM.

(8) Denpol fractions are combined, and the solvent is removed by rotary evaporation.

(9) The resulting polymer is dissolved in 1,4-dioxane (10 mL) and lyophilized, affording the second-generation denpol **12** ($PG2_{2000}$) as a white powder (706 mg, 85%).

The deprotection procedure for **12** ($PG2_{2000}$) follows the steps outlined for **10** ($PG1_{2000}$); no modification is necessary.

Comments: Like mentioned for **10** ($PG1_{2000}$) earlier, the denpol **12** ($PG2_{2000}$) is not retained by silica and elutes with the solvent front, with significant tailing. The elution is best monitored by TLC (EtOAc/hexane 1:1; $R_f \approx 0$) to avoid contamination with unreacted **9** (DG1). The plates should be stained with ninhydrin and heated thoroughly to visualize the denpol spots.

For all denpols, storage under an inert atmosphere, exclusion of light and refrigeration is advisable. Additionally, the deprotected denpols *de*-$PG1_{2000}$ **(11)** and *de*-$PG2_{2000}$ **(1)** are hygroscopic and the lyophilizates show some tendency to become insoluble over long periods of storage; it is therefore advisable to (a) store deprotected polymers such as to exclude moisture or (b) store the deprotected polymers as aqueous solutions, which have proven to be stable over the long term. Deprotected denpols are rendered usually irreversibly insoluble when exposed to incompatible solvents. Compatible solvents include: H_2O, MeOH, EtOH, DMF, DMSO, NMP; solvents of lesser polarity such as MeCN act as precipitants.

2.3 Characterization of *de*-PG2

With the procedure shown in Fig. 4 and described earlier, so far various batches of *de*-PG2 have been obtained as trifluoroacetate salt, differing mainly in average degree of polymerization. The Boc-protected PG2 was analyzed by ^{1}H NMR spectroscopy (Grotzky et al., 2013; Guo et al., 2009; Yu et al., 2014), and the completeness of the dendronization step (usually >99%) was determined by labeling of unreacted amines with Sanger's reagent according to the published method (Shu, Gössl, Rabe, & Schlüter, 2002). The average molar mass and the PDI were determined by analytical size-exclusion chromatography (Grotzky et al., 2013; Guo et al., 2009; Yu et al., 2014). Due to the large thickness of *de*-PG2, a statistical analysis of the length of PG2 deposited from solution onto mica by atomic force microscopy is also possible (Guo et al., 2009). For an aqueous solution of *de*-$PG2_{1000}$ (pH 5.0), the UV absorption spectrum shows an absorption maximum at $\lambda = 285$ nm ($\varepsilon_{285\text{nm}} = 5.0 \times 10^3\ M^{-1}$ r.u. cm^{-1}) (Fornera, Balmer, Zhang, Schlüter, & Walde, 2011).

3. THE DENPOL–ENZYME CONJUGATE *de*-PG2-BAH-proK

The denpol–enzyme conjugate *de*-PG2-BAH-proK (Fig. 2) is synthesized from *de*-PG2 (**1**, Fig. 1) and proK by first partially modifying *de*-PG2 with S-HyNic to yield *de*-PG2-HyNic, and partially modifying proK with S-4FB to yield proK-4FB, followed by reaction of *de*-PG2-HyNic with proK-4FB (Fig. 5) and purification of the conjugate by ultrafiltration. The experimental details given lead to the formation of a stable *de*-PG2-BAH-proK conjugate (Küchler, Bleich, et al., 2015). Deviations from the experimental conditions described may lead to altered properties of the conjugate (different stability or activity). The activity of proK is measured spectrophotometrically with succinyl-L-alanyl-L-alanyl-L-prolyl-L-phenylalanyl *para*-nitroanilide (Suc-AAPF-*p*NA) as chromogenic substrate. The different reaction steps for obtaining *de*-PG2-BAH-proK are outlined for *de*-PG2 (**1**, Fig. 1) with $P_n = 2000$ (Küchler, Bleich, et al., 2015).

It is useful to prepare the following buffer and stock solutions. MES and MOPS are available in high purity, for example, from AppliChem.

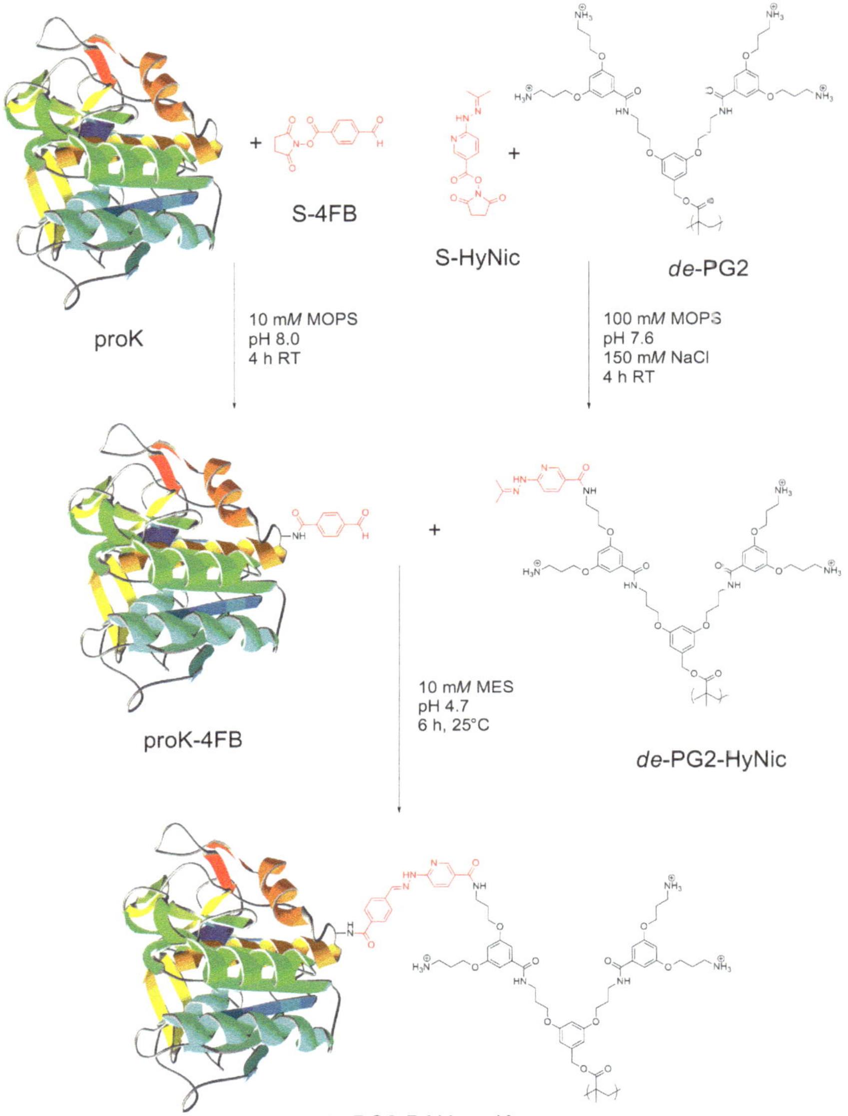

Fig. 5 Scheme of the stepwise synthesis of the denpol–enzyme conjugate *de*-PG2-BAH-proK from *de*-PG2 (**1**) and proteinase K (proK): Modification of proK with S-4FB to yield proK-4FB, modification of *de*-PG2 with S-HyNic to yield *de*-PG2-HyNic, and reaction of *de*-PG2-HyNic with proK-4FB to yield *de*-PG2-BAH-proK. Shown is the conjugation at one of the repeating units of the denpol.

Buffer solutions: 100 m*M* MES, 150 m*M* NaCl, pH 4.7; 10 m*M* MES, pH 4.7; 100 m*M* MES, pH 5.0; 10 m*M* MES, pH 5.0; 10 m*M* MOPS, 150 m*M* NaCl, pH 7.6; 10 m*M* MOPS, pH 7.0; 10 m*M* MOPS, pH 8.0.

(a) *de-PG2$_{2000}$ stock solution:* Denpol *de*-PG2$_{2000}$ (**1**, 1 mg) is dissolved in MOPS buffer (500 μL, 100 m*M* MOPS, 150 m*M* NaCl, pH 7.6). The precise r.u. concentration can be determined spectrophotometrically, for example, by using a NanoDrop ND-1000 ($\varepsilon_{285\text{nm}}$(*de*-PG2$_{2000}$) = $5.0 \times 10^3\ M^{-1}$ r.u. cm^{-1}) (Fornera et al., 2011). A typical *de*-PG2 concentration is 1.5 m*M* r.u.

(b) *proK stock solution*: Enzyme proK (10 mg, from Roche Applied Science) is dissolved in 1 mL MOPS buffer (10 m*M*, pH 8.0). The precise protein concentration is determined spectrophotometrically (for example, by using a NanoDrop ND-1000), taking into account $\varepsilon_{280\text{nm}}$(proK) = $4.1 \times 10^4\ M^{-1}\ \text{cm}^{-1}$ (Ebeling et al., 1974).

(c) *S-HyNic stock solution*: S-HyNic (2.90 mg, from Solulink, or synthesized as described by Grotzky et al., 2012) is dissolved in dry DMF (500 μL; S-HyNic concentration 20 m*M*).

(d) *S-4FB stock solution*: S-4FB (4.94 mg, from Solulink, or synthesized as described by Grotzky et al., 2012) is dissolved in dry DMF (200 μL; S-4FB concentration 100 m*M*).

(e) *4-Nitrobenzaldehyde stock solution*: 4-Nitrobenzaldehyde (7.6 mg, from Fluka) is dissolved in dry DMF (1.0 mL; 4-nitrobenzaldehyde concentration 50 m*M*).

(f) *2-Hydrazinopyridine stock solution*: 2-Hydrazinopyridine dihydrochloride (9.1 mg, from Sigma Aldrich) is dissolved in MES buffer (1.0 mL, 10 m*M* MES, pH 4.7; 2-hydrazinopyridine concentration 50 m*M*).

(g) *Suc-AAPF-pNA stock solution*: Suc-AAPF-*p*NA (15.62 mg, from Bachem) is dissolved in dry DMF (500 μL; Suc-AAPF-*p*NA concentration 50 m*M*).

Safety comment: Standard personal protective equipment (gloves, safety goggles, lab coat) must be worn at all times when handling chemicals.

3.1 Synthesis and Characterization of *de*-PG2-HyNic

(1) 500 μL *de-PG2$_{2000}$ stock solution* is added to a 1.5-mL polypropylene reaction tube (750 nmol, 1 equiv. r.u.).

(2) 22.5 μL *S-HyNic stock solution* is added to the reaction tube (300 nmol, 0.4 equiv. per r.u.). The solution is mixed by inversion of the reaction tube and kept at RT.

(3) After 4 h, the reaction mixture is transferred to a 100-kDa NMWL ultrafiltration device (Amicon Ultra-0.5) and concentrated at 14,000 × *g* (=13,100 rpm, Eppendorf 5415 D centrifuge) to a final volume of about 100 μL.

(4) The concentrate is diluted by addition of 400 μL MES buffer (10 m*M* MES, pH 4.7), mixed inside the ultrafiltration device, and concentrated again at 14,000 × *g* (=13,100 rpm, Eppendorf 5415 D centrifuge).

(5) The filtrate obtained in step 4 is analyzed with a NanoDrop ND-1000 UV/vis spectrophotometer, and steps 4 and 5 are repeated until no absorbance is measured in the filtrate at 220 nm.

(6) The obtained *de*-$PG2_{2000}$-HyNic concentrate is collected in a 1.5-mL reaction tube and stored at 4°C until further use.

(7) The HyNic quantification reagent is prepared by dilution of 10 μL of the *4-nitrobenzaldehyde stock solution* with 990 μL of MES buffer (100 m*M* MES, pH 5.0).

(8) 5 μL of *de*-$PG2_{2000}$-HyNic concentrate is diluted with 95 μL HyNic quantification reagent, and a UV/vis absorption spectrum is recorded immediately after mixing with a spectrophotometer (e.g., a JASCO V-670, with a ultramicrocuvette 105.201-QS, volume 0.1 mL, path 1 cm, from Hellma). The absorbance at 390 nm is measured, $A_{390nm}(0\ h)$.

(9) The quantification solution is recovered from the cuvette, transferred into a 0.5-mL polypropylene reaction tube, and kept at 40°C for 1 h.

(10) After cooling the quantification solution to RT, the UV/vis absorption spectrum of the solution is remeasured, $A_{390nm}(1\ h)$.

(11) The molar concentration of HyNic units in the *de*-$PG2_{2000}$-HyNic concentrate, *c*(HyNic), is calculated by using Eq. (1):

$$c(\mathrm{HyNic}) = \frac{(A_{390\,\mathrm{nm}}(1\,\mathrm{h}) - A_{390\,\mathrm{nm}}(0\,\mathrm{h})) \times 20}{2.4 \times 10^{4}\,M^{-1}\,\mathrm{cm}^{-1} \times 1\,\mathrm{cm}} \quad (1)$$

(12) The *de*-PG2 r.u. quantification reagent is prepared by dissolving 0.4 mg trypan blue (from Acros Organics) in MilliQ water.

(13) Calibration solutions of *de*-$PG2_{2000}$ are prepared at concentrations of 0–150 μ*M* r.u.; an aliquot of the obtained *de*-$PG2_{2000}$-HyNic solution is then diluted to a r.u. concentration of approximately

100 μ*M*, as estimated from the UV/vis absorbance at 285 nm, neglecting the absorbance of the HyNic moiety ($\varepsilon_{285\text{nm}}(de\text{-PG2}_{2000}) = 5.0 \times 10^3\ M^{-1}\ \text{r.u.}\,\text{cm}^{-1}$) (Fornera et al., 2011).

(14) 10 μL of the diluted *de*-PG2$_{2000}$-HyNic solution, 110 μL MES buffer (100 m*M* MES, 150 m*M* NaCl, pH 4.7), and 5 μL trypan blue solution are mixed in a 1.5-mL polypropylene reaction tube. The same procedure is followed for the calibration solutions.

(15) The reaction solutions are kept at 37°C for 1 h.

(16) After cooling to RT, the reaction solutions are centrifuged at 16,100 × *g* for 20 min. (=14,050 rpm, Eppendorf 5415 D centrifuge).

(17) 100 μL of the supernatant are carefully removed, and the UV/vis absorption spectrum is recorded (e.g., with a Analytik Jena Specord S600, ultramicrocuvette 105.201-QS, volume 0.1 mL, path 1 cm, from Hellma).

(18) The absorbance of the calibration solutions at 580 nm is plotted against the known r.u. concentrations, and the r.u. concentration of the *de*-PG2$_{2000}$-HyNic solution is read from the calibration curve using the measured absorbance at 580 nm.

3.2 Synthesis and Characterization of proK-4FB

(1) 2 mL of a 50-μ*M* proK solution (100 nmol, 1 equiv.) are prepared from the *proK stock solution* in a 5-mL polypropylene reaction tube using MOPS buffer (10 m*M*, pH 8.0).

(2) 120 μL of the *S-4FB stock solution* (12 μmol, 120 equiv., see Küchler, Bleich, et al., 2015) are added, and the solution is mixed by gentle inversion of the reaction tube.

(3) After 4 h at RT, the reaction mixture is loaded into an ultrafiltration device (Amicon Ultra-4, 10 kDa NMWL), diluted to 4 mL with MES buffer (10 m*M*, pH 4.7), and concentrated to approximately 1 mL at 2500 × *g* (=4000 rpm, Hermle Z 320 K centrifuge).

(4) The filtrate obtained in step 3 is analyzed with a NanoDrop ND-1000 UV/vis spectrophotometer, and steps 3 and 4 are repeated until no absorbance is measured in the filtrate at 220 nm.

(5) The obtained proK-4FB concentrate is collected in a 1.5-mL reaction tube and stored at 4°C until further use.

(6) The 4FB quantification reagent is prepared by dilution of 10 μL of the *2-hydrazinopyridine stock solution* with 980 μL of MES buffer (10 m*M* MES, pH 4.7).

(7) 10 μL of the proK-4FB concentrate is added to the 4FB quantification solution, and a UV/vis spectrum is measured immediately after mixing. The absorbance at 350 nm is measured, $A_{350\text{nm}}(0\text{ h})$.

(8) After 3 h at RT, the UV/vis spectrum of the reaction solution is re-measured, $A_{350\text{nm}}(3\text{ h})$, and the molar concentration of 4FB moieties in the proK-4FB concentrate, c(4FB), is calculated using Eq. (2):

$$c(4\text{FB}) = \frac{(A_{350\,\text{nm}}(3\text{h}) - A_{350\,\text{nm}}(0\text{h})) \times 100}{2.45 \times 10^4\,M^{-1}\,\text{cm}^{-1} \times 1\,\text{cm}} \quad (2)$$

(9) The proK concentration in the proK-4FB concentrate is determined with the Bradford protein assay (Sigma Aldrich) as described by the supplier (Zor & Selinger, 1996), using solutions with known proK concentrations for the calibration.

3.3 Synthesis and Characterization of *de*-PG2-BAH-proK

(1) In a 1-mm path length quartz cuvette, *de*-$PG2_{2000}$-HyNic and proK-4FB are added to MES buffer (10 mM, pH 4.7) to obtain a final concentration of 100 μM HyNic and 50 μM 4FB.

(2) The solution in the cuvette is mixed by gentle inversion of the cuvette, and the cuvette is mounted in a Peltier-controlled holder (temperature set to 25°C), for example, in a JASCO V-670 UV/vis spectrophotometer.

(3) UV/vis spectra are recorded immediately after mixing the reaction solution and in intervals of 5 min over a time of 6 h.

(4) The reaction mixture is recovered from the cuvette, loaded into an ultrafiltration device (Sartorius Centrisart I, 300 kDa MWCO), diluted to 2.5 mL with MOPS buffer (10 mM MOPS, pH 7.0), and concentrated to approximately 500 μL at 2500 × g (=4000 rpm, Hermle centrifuge Z 320 K).

(5) The filtrate obtained in step 4 is analyzed with a NanoDrop ND-1000 UV/vis spectrophotometer, and steps 4 and 5 are repeated until no absorbance is measured in the filtrate at 220 nm.

(6) The obtained *de*-$PG2_{2000}$-BAH-proK concentrate is collected in a 1.5-mL reaction tube and stored at 4°C until further use.

(7) The amount of bound proK on the *de*-$PG2_{2000}$ is calculated from the recorded UV/vis spectra using the known absorption coefficient of the formed BAH bond ($\varepsilon_{354\text{nm}}$(BAH) = $2.9 \times 10^4\,M^{-1}\,\text{cm}^{-1}$; Solulink,

2016) and the known concentration of *de*-PG2$_{2000}$ as determined earlier (Küchler, Bleich, et al., 2015).

3.4 Enzymatic Activity of proK and *de*-PG2-BAH-proK

The catalytic efficiency of proK in the denpol–enzyme conjugate *de*-PG2-BAH-proK (Fig. 2) can be analyzed according to the Michaelis–Menten model. To evaluate the effects of the conjugation of proK on the catalytic activity, the results can be compared to the kinetic constants obtained for native proK (see later).

(1) The concentration of proK in the *de*-PG2$_{2000}$-BAH-proK solution is determined using UV/vis spectrophotometry. The proK concentration is calculated from the absorption of the BAH bond ($\varepsilon_{354\text{nm}}$(BAH) = $2.9 \times 10^4\ M^{-1}\ \text{cm}^{-1}$; Solulink, 2016), assuming the same concentration for the BAH bond and proK.

(2) A solution of *de*-PG2$_{2000}$-BAH-proK is prepared at a concentration of 1.65 μ*M* proK in MOPS buffer (10 m*M* MOPS, pH 7.0). Additionally, a 0.5-μ*M* proK solution is prepared from a fresh *proK stock solution*.

(3) Dilutions of Suc-AAPF-*p*NA at concentrations of 3, 5, 7.5, 10, 12.5, 18, and 25 m*M* are prepared in dry DMF.

(4) 980 μL of MOPS buffer (10 m*M* MOPS, pH 7.0) is placed in a polystyrene semimicrocuvette (path 1 cm), and a blank spectrum is acquired on a UV/vis diode-array spectrophotometer (e.g., a Analytik Jena Specord S600).

(5) 10 μL of the *de*-PG2$_{2000}$-BAH-proK solution and 10 μL of one of the Suc-AAPF-*p*NA solutions are added to the cuvette. Immediately after mixing of the solutions, UV/vis spectra are recorded with a diode-array spectrometer at intervals of 5 s for a total of 1 min.

(6) The catalytic activity is evaluated using the absorption of the *p*-nitroaniline product at 410 nm ($\varepsilon_{410\text{nm}} = 8.8 \times 10^3\ M^{-1}\ \text{cm}^{-1}$; Erlanger, Kokowsky, & Cohen, 1961). The kinetic constants of native proK and proK in *de*-PG2$_{2000}$-BAH-proK, k_{cat} and K_M, are obtained by a nonlinear fit of the initial reaction velocities, v_i, against the substrate concentrations according to the Michaelis–Menten equation, Eq. (3), S being the substrate Suc-AAPF-*p*NA and E being the enzyme proK:

$$v_i = \frac{k_{cat} \times [\text{E}]_{tot} \times [\text{S}]}{K_M + [\text{S}]} \tag{3}$$

The data obtained for K_M and k_{cat} for proK, proK-4FB, and *de*-PG2-BAH-proK (10 m*M* MOPS, pH 7.0, 25°C) are given in the legend of Fig. 6.

4. SYNTHESIS AND CHARACTERIZATION OF DENPOLS CARRYING OTHER TYPES OF ENZYMES

The general procedure for the denpol–enzyme conjugate synthesis as described for *de*-PG2$_{2000}$-BAH-proK can be adapted to synthesize conjugates containing other enzymes (Grotzky et al., 2013, 2012; Küchler, Adamcik, et al., 2015). For each enzyme, the experimental conditions have to be optimized accordingly which also depends on the desired enzyme loading density.

In addition to conjugates carrying several copies of an enzyme on the same polymer chain, a special class of denpol–enzyme conjugates has been synthesized consisting of polymer chains decorated with *several copies of two different types of enzymes*. In a sequential conjugation approach, a denpol modified with the HyNic linker was reacted with the two 4FB-modified enzymes, first attaching the larger of the two enzymes to the denpol. The second, smaller enzyme was then attached to the denpol–enzyme conjugate in a second step, exploiting the reactivity of the remaining free HyNic groups on the denpol–enzyme conjugate (horseradish peroxidase (HRP, EC 1.11.1.7) and superoxide dismutase (EC 1.15.1.1), see Grotzky et al., 2012; glucose oxidase (GOD, EC 1.1.3.4) and HRP, see Küchler, Adamcik, et al., 2015).

5. IMMOBILIZATION OF *DE*-PG2-BAH-proK ON SILICATE SURFACES

The denpol–enzyme conjugates described earlier can be used for a simple and efficient immobilization of the enzymes on different glass surfaces. The ability of the polycationic denpol to adsorb on glass surfaces when applied in a dilute aqueous solution allows a single-step immobilization of the enzymes bound to the denpol. The adsorption process of such denpol–enzyme conjugates on silicate surfaces has been analyzed in situ using quartz crystal microbalance with dissipation monitoring and the transmission interferometric adsorption sensor (Gustafsson et al., 2015; Küchler, Adamcik, et al., 2015; Küchler, Bleich, et al., 2015).

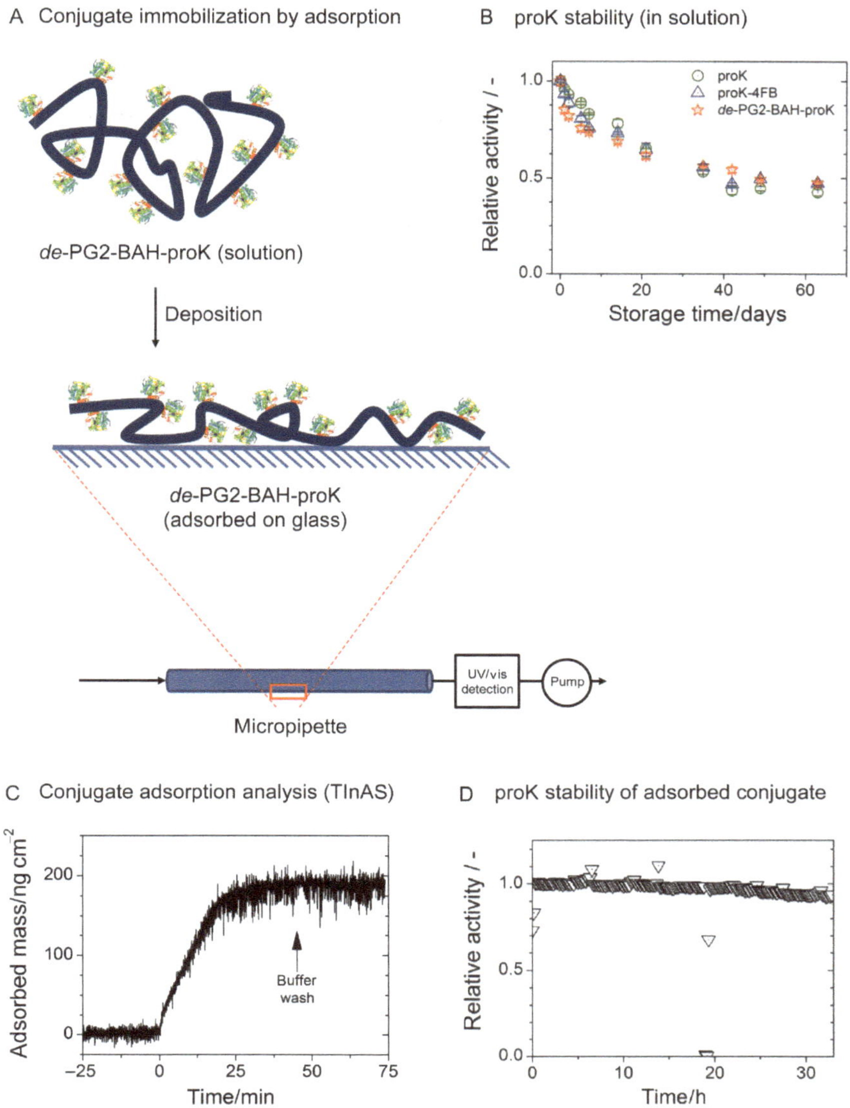

Fig. 6 Application of the denpol–enzyme conjugate *de*-PG2-BAH-proK for the simple immobilization of proK on silicate surfaces. (A) Schematic representation of the immobilization process through simple adsorption from aqueous solution onto the inner glass wall of a micropipette, used as flow reactor. (B) Stability of proK, proK-4FB, and *de*-PG2-BAH-proK during storage at 4°C at pH 7.0. (C) Monitoring of the adsorption of *de*-PG2-BAH-proK on a silicate surface with the transmission interferometric adsorption sensor (TInAS). (D) Stability of immobilized *de*-PG2-BAH-proK inside glass micropipettes under continuous operation. For C and D, the proK activity was measured with

(Continued)

5.1 Immobilization on Glass Slides

For the immobilization of proK on the surface of planar glass slides, round microscopy glass coverslips (8 mm diameter, total surface 1 cm^2) were used as solid support (Küchler, Bleich, et al., 2015).

(1) The coverslips are cleaned with ethanol in a sonication bath three times for 10 min and dried in a stream of nitrogen.

(2) The clean coverslips are kept in 2-mL polypropylene reaction tubes, and the surface of the coverslips is wetted by addition of 1 mL MOPS buffer (10 m*M* MOPS, pH 7.0).

(3) The MOPS buffer is removed from the polypropylene tubes and immediately replaced by 1 mL of *de*-PG2-BAH-proK solution (1 μ*M* proK concentration).

(4) After 1 h of incubation at RT, the *de*-PG2-BAH-proK solution is removed from the reaction tubes and the coverslips are washed three times with MOPS buffer.

(5) The coverslips coated with *de*-PG2-BAH-proK are stored in MOPS buffer (10 m*M* MOPS, pH 7.0) at 4°C until further use.

5.2 Immobilization Inside Glass Micropipettes

In a similar procedure to the immobilization of proK on planar glass surfaces, the adsorption of *de*-PG2-BAH-proK inside glass micropipettes allowed the production of enzymatic flow reactors (Küchler, Bleich, et al., 2015).

(1) Glass micropipettes (intraMARK 200 μL nominal volume, length 14 cm, inner diameter 1.6 mm) are cleaned with ethanol in a sonication bath (three times 10 min) and dried in a stream of nitrogen.

(2) The clean micropipettes are connected to a 1-mL syringe using poly(tetrafluoroethylene) tubing, and the inner surface of the micropipette is wetted by aspiration of MOPS buffer (10 m*M* MOPS, pH 7.0).

(3) After dispensing the MOPS buffer from the micropipette, a solution of *de*-PG2-BAH-proK is aspirated (1 μ*M* proK concentration).

Fig. 6—Cont'd Suc-AAPF-*p*NA as proK substrate at pH 7.0. The Michaelis–Menten kinetic constants (K_M and k_{cat}) were 185 ± 14 μ*M* and 172 ± 8 s^{-1}, 138 ± 15 μ*M* and 138 ± 8 s^{-1}, and 66 ± 7 μ*M* and 23 ± 1 s^{-1} for dissolved proK, proK-4FB, and *de*-PG2-BAH-proK, respectively. These mainly indicating a decrease of k_{cat} for denpol-bound proK. *Panels (B)–(D): Reprinted with permission from Küchler, A., Bleich, J. N., Sebastian, B., Dittrich, P. S., & Walde, P. (2015). Stable and simple immobilization of proteinase K inside glass tubes and microfluidic channels.* ACS Applied Materials & Interfaces, 7, *25970–25980. Copyright (2015) American Chemical Society.*

(4) After incubation for 1 h at RT, the *de*-PG2-BAH-proK solution is dispensed and the micropipette is washed three times with MOPS buffer.
(5) The micropipette with the inner surface coated with *de*-PG2-BAH-proK is filled with MOPS buffer (10 m*M* MOPS, pH 7.0) and stored at 4°C until further use.

6. ENTRAPMENT OF DENPOL–ENZYME CONJUGATES INSIDE PHOSPHOLIPID VESICLES

Since the denpol–enzyme conjugates of the type described earlier are water-soluble hybrid structures between a denpol and several copies of one type (or different types) of enzymes, it is possible to entrap these conjugates in micrometer-sized giant vesicles for developing vesicle-confined enzymatic reactions. This possibility has been explored with *de*-PG1-BAH-HRP, i.e., a conjugate between the first-generation denpol *de*-PG1 (**11**) and HRP, and phospholipid vesicles formed from POPC (1-palmitoyl-2-oleoyl-*sn*-glycero-3-phosphocholine) (Grotzky et al., 2013). The entrapment was achieved by using one of the known methods (Walde, Cosentino, Engel, & Stano, 2010), the "water droplet transfer method," which involves the initial formation of a water-in-oil emulsion of micrometer-sized aqueous droplets in a water-immiscible organic solvent, followed by the transfer of these droplets through a phospholipid monolayer interface into an aqueous solution to form the giant vesicles. For a proof of the successful entrapment of the denpol–enzyme conjugate, the conjugate was labeled with fluorescein and the giant vesicles containing the fluorescently labeled conjugate were analyzed by confocal laser scanning microscopy (Grotzky et al., 2013). Retention of HRP activity within the giant vesicles was demonstrated by adding to the giant vesicles hydrogen peroxide (for oxidizing HRP) and the fluorogenic HRP substrate amplex red, which upon oxidation yields inside the vesicles resorufin which is fluorescent (Grotzky et al., 2013).

7. CONCLUSIONS AND OUTLOOK

Immobilizing enzymes on silicate surfaces for bioanalytical or biosynthetic applications through simple adsorption of a previously prepared conjugate between a dendronized polymer and several copies of the enzyme(s) of interest is an alternative approach to the more established enzyme immobilization methods (e.g., Hanefeld, Gardossi, & Magner, 2009; Küchler,

Yoshimoto, Luginbühl, Mavelli, & Walde, 2016; Mateo, Palomo, Fernandez-Lorente, Guisan, & Fernandez-Lafuente, 2007; Sheldon & van Pelt, 2013). Conjugates like *de*-PG2-BAH-proK (Fig. 2) or *de*-PG2-BAH-HRP adhere to silicate surfaces due to many weak interactions between surface-exposed functional groups of some of the many enzyme molecules of the conjugates and/or between the numerous free amino groups of the denpol and the surface. With respect to the latter mode of interaction, strong adherence of enzyme-free *de*-PG2 (**1**) to silicate surfaces was demonstrated and applied in a related approach for the stepwise surface immobilization of enzymes via the biotin–avidin system: first adsorption of partially biotinylated *de*-PG2, then adsorption of avidin, and finally adsorption of partially biotinylated enzymes (HRP, GOD, or β-galactosidase (EC 3.2.1.23), Fornera, Bauer, Schlüter, & Walde, 2012; Fornera, Kuhn, et al., 2012). In both cases—enzyme immobilization with *de*-PG2 and the biotin–avidin system, or enzyme immobilization through the preparation of *de*-PG2-BAH enzyme—the denpol *de*-PG2 (**1**) plays a key role. Its synthesis certainly is time consuming (see earlier), which means that *de*-PG2 (**1**) is not thought for large-scale applications, but rather for small-scale devices, e.g., micropipettes or microfluidic chips (Fornera, Bauer, et al., 2012; Fornera, Kuhn, et al., 2012; Küchler, Adamcik, et al., 2015; Küchler, Bleich, et al., 2015).

Concerning the denpol–enzyme conjugates described in detail earlier, the optimal conditions for their synthesis have to be determined individually for each enzyme of interest. The details provided for proK hopefully are a useful guideline for further exploring the preparation and application of conjugates with other enzymes.

Finally, the possibility of varying the degree of polymerization and of varying the chemical structure of the denpol (including the branching unit and the generation) should be considered, as well as the possibility of using other linking units than BAH, for preparing an improved denpol–enzyme conjugate, possibly with even higher enzyme stability than obtained so far in the case of proK (Fig. 6) or HRP (Gustafsson et al., 2015; Küchler, Adamcik, et al., 2015). One challenging application of denpol–enzyme conjugates is in the field of localized enzymatic cascade reactions (Küchler et al., 2016; Schoffelen & van Hest, 2013). The type of denpol–enzyme conjugates described here offers the unique possibility of colocalizing different types of enzymes on one and the same denpol chain (Grotzky et al., 2012; Küchler, Adamcik, et al., 2015), which may be of advantage not only for surface-confined reactions (Gustafsson et al., 2015; Küchler, Adamcik, et al., 2015) but also for enzymatic cascade reactions inside vesicular compartments (Grotzky et al., 2013).

ACKNOWLEDGMENTS

The elaboration of the synthesis of the type of mentioned denpols by many collaborators (in particular Afang Zhang, Yifei Guo, and Baozhong Zhang) and the contributions to the synthesis of the denpol–enzyme conjugates mentioned (Andrea Grotzky, Julian Bleich, and Nicolas Gheczy) are highly acknowledged.

REFERENCES

Betzel, C., Pal, G. P., & Saenger, W. (1988). Three-dimensional structure of proteinase K at 0.15 nm resolution. *European Journal of Biochemistry*, *178*, 155–171.

Chen, Y., & Xiong, X. (2010). Tailoring dendronized polymers. *Chemical Communications*, *46*, 5049–5060.

de Kruijff, B., Rietveld, A., Telders, N., & Vaandrager, B. (1985). Molecular aspects of the bilayer stabilization induced by poly(L-lysines) of varying size in cardiolipin liposomes. *Biochimica et Biophysica Acta*, *820*, 295–304.

Ebeling, W., Hennrich, N., Klockow, M., Metz, H., Orth, H. D., & Lang, H. (1974). Proteinase K from *Tritirachium album* limber. *European Journal of Biochemistry*, *47*, 91–97.

Erlanger, B. F., Kokowsky, N., & Cohen, W. (1961). The preparation and properties of two new chromogenic substrates of trypsin. *Archives of Biochemistry and Biophysics*, *95*, 271–278.

Fornera, S., Balmer, T. E., Zhang, B., Schlüter, A. D., & Walde, P. (2011). Immobilization of peroxidase on SiO_2 surfaces with the help of a dendronized polymer and the avidin-biotin system. *Macromolecular Bioscience*, *11*, 1052–1067.

Fornera, S., Bauer, T., Schlüter, A. D., & Walde, P. (2012). Simple enzyme immobilization inside glass tubes for enzymatic cascade reactions. *Journal of Materials Chemistry*, *22*, 502–511.

Fornera, S., Kuhn, P., Lombardi, D., Schlüter, A. D., Dittrich, P. S., & Walde, P. (2012). Sequential immobilization of enzymes in microfluidic channels for cascade reactions. *ChemPlusChem*, *77*, 98–101.

Frauenrath, H. (2005). Dendronized polymers—Building a new bridge from molecules to nanoscopic objects. *Progress in Polymer Science*, *30*, 325–384.

Fuhrmann, G., Grotzky, A., Lukić, R., Matoori, S., Yu, H., Zhang, B., et al. (2013). Sustained gastrointestinal activity of dendronized polymer-enzyme conjugates. *Nature Chemistry*, *5*, 582–589.

Grotzky, A., Altamura, E., Adamcik, J., Carrara, P., Stano, P., Mavelli, F., et al. (2013). Structure and enzymatic properties of molecular dendronized polymer-enzyme conjugates and their entrapment inside giant vesicles. *Langmuir*, *29*, 10831–10840.

Grotzky, A., Manaka, Y., Kojima, T., & Walde, P. (2011). Preparation of catalytically active, covalent α-polylysine-enzyme conjugates via UV/vis-quantifiable bis-aryl hydrazone bond formation. *Biomacromolecules*, *12*, 134–144.

Grotzky, A., Nauser, T., Erdogan, H., Schlüter, A. D., & Walde, P. (2012). A fluorescently labeled dendronized polymer-enyzme conjugate carrying multiple copies of two different types of active enzymes. *Journal of the American Chemical Society*, *134*, 11392–11395.

Guo, Y., van Beek, J. D., Zhang, B., Colussi, M., Walde, P., Zhang, A., et al. (2009). Tuning polymer thickness: Synthesis and scaling theory of homologous series of dendronized polymers. *Journal of the American Chemical Society*, *131*, 11841–11854.

Gustafsson, H., Küchler, A., Holmberg, K., & Walde, P. (2015). Co-immobilization of enzymes with the help of a dendronized polymer and mesoporous silica nanoparticles. *Journal of Materials Chemistry B*, *3*, 6174–6184.
Hanefeld, U., Gardossi, L., & Magner, E. (2009). Understanding enzyme immobilisation. *Chemical Society Reviews*, *38*, 453–468.
Hermanson, G. T. (2013). *Bioconjugate techniques* (3rd ed.) (p. 762). Amsterdam: Elsevier.
Küchler, A., Adamcik, J., Mezzenga, R., Schlüter, A. D., & Walde, P. (2015). Enzyme immobilization on silicate glass through simple adsorption of dendronized polymer-enyzme conjugates for localized enzymatic cascade reactions. *RSC Advances*, *5*, 44530–44544.
Küchler, A., Bleich, J. N., Sebastian, B., Dittrich, P. S., & Walde, P. (2015). Stable and simple immobilization of proteinase K inside glass tubes and microfluidic channels. *ACS Applied Materials & Interfaces*, *7*, 25970–25980.
Küchler, A., Yoshimoto, M., Luginbühl, S., Mavelli, F., & Walde, P. (2016). Enzymatic reactions in confined environments. *Nature Nanotechnology*, *11*, 409–420.
Mateo, C., Palomo, J. M., Fernandez-Lorente, G., Guisan, J. M., & Fernandez-Lafuente, R. (2007). Improvement of enzyme activity, stability and selectivity via immobilization techniques. *Enzyme and Microbial Technology*, *40*, 1451–1463.
Moreno-Pérez, S., Orrego, A. H., Romero-Fernández, M., Trobo-Maseda, L., Martins-DeOliveira, S., Munilla, R., et al. (2016). Intense PEGylation of enzyme surfaces: Relevant stabilizing effects. *Methods in Enzymology*, *571*, 55–72.
Paez, J. I., Martinelli, M., Brunetti, V., & Strumia, M. C. (2012). Dendronization: A useful synthetic strategy to prepare multifunctional materials. *Polymers*, *4*, 355–395.
Rosen, B. M., Wilson, C. J., Wilson, D. A., Peterca, M., Imam, M. R., & Percec, V. (2009). Dendron-mediated self-assembly, diassembly, and self-organization of complex systems. *Chemical Reviews*, *109*, 6275–6540.
Schlüter, A. D. (2005). A covalent chemistry approach to giant macromolecules with cylindrical shape and an engineerable interior and surface. *Topics in Current Chemistry*, *245*, 151–191.
Schlüter, A. D., & Rabe, J. R. (2000). Dendronized polymers: Synthesis, characterization, assembly at interfaces, and manipulation. *Angewandte Chemie, International Edition*, *39*, 864–883.
Schoffelen, S., & van Hest, J. C. M. (2013). Chemical approaches for the construction of multi-enzyme reaction systems. *Current Opinion in Structural Biology*, *23*, 613–621.
Sheldon, R. A., & van Pelt, S. (2013). Enzyme immobilisation in biocatalysis: Why, what and how. *Chemical Society Reviews*, *42*, 6223–6235.
Shu, L., Gössl, I., Rabe, J. P., & Schlüter, A. D. (2002). Quantitative aspects of the dendronization of dendronized linear polystyrenes. *Macromolecular Chemistry and Physics*, *18*, 2540–2550.
Solulink Inc. (2016). At http://www.solulink.com/.
Veronese, F. M., Mero, A., & Pasut, G. (2009). Protein PEGylation, basic science and biological applications. In F. M. Veronese (Ed.), *Pegylated protein drugs: Basic science and clinical applications* (pp. 11–31). Basel: Birkäuser.
Walde, P., Cosentino, K., Engel, H., & Stano, P. (2010). Giant vesicles: Preparations and applications. *ChemBioChem*, *11*, 848–865.
Yu, H., Schlüter, A. D., & Zhang, B. (2012). Synthesis of dendronized polymers by a "$n+2$" approach. *Macromolecules*, *45*, 8555–8560.
Yu, H., Schlüter, A. D., & Zhang, B. (2014). Synthesis of high generation dendronized polymers and quantification of their structure perfection. *Macromolecules*, *47*, 4127–4135.
Zor, T., & Selinger, Z. (1996). Linearization of the Bradford assay increases its sensitivity: Theoretical and experimental studies. *Analytical Biochemistry*, *236*, 302–308.

CHAPTER NINETEEN

Nanoarmoring of Enzymes by Interlocking in Cellulose Fibers With Poly(Acrylic Acid)

Caterina M. Riccardi, Rajeswari M. Kasi, Challa V. Kumar[1]
University of Connecticut, Storrs, CT, United States
[1]Corresponding author: e-mail address: challa.kumar@uconn.edu

Contents

Methods in Enzymology, Volume 590
ISSN 0076-6879
http://dx.doi.org/10.1016/bs.mie.2017.01.009
© 2017 Elsevier Inc.
All rights reserved.

Abstract

A simple method for interlocking glucose oxidase (GOx) and horseradish peroxidase (HRP) in cellulose fibers using poly(acrylic acid) (PAA) as an armor around the enzyme, without any need for activation of the cellulose support, is reported here. The resulting enzyme paper is an inexpensive, stable, simple, wearable, and washable biosensor. PAA functions as a multifunctional tether to interlock the enzyme molecules around the paper fibers so that the enzymes are protected against thermal/chemical denaturation and not released from the paper when washed with a detergent. The decreased conformational entropy of the interlocked enzyme protected by the nanoarmor is likely responsible for increased enzyme stability to heat and chemical denaturants (retained ≥70 percent enzyme activity after washing with urea or SDS for 30 min) and the polymer protects the enzyme against inactivation by proteases, bacteria, inhibitors, etc. The kinetics of the interlocked enzyme were similar to that of the enzyme in solution. The V_{max} was 6(±0.5) m*M* per minute before washing, then increased slightly to 9(±1.4) m*M* per minute after washing with water. The K_m was 22(±6.4 m*M*), which was slightly higher compared to GOx in solution (25–27 m*M*). Because the surface area of the paper does not limit the enzyme loading, about 20% of enzyme was successfully loaded onto the paper (0.2 g enzyme per gram of paper), and ≥95% of the enzyme was retained after washing. Interlocking works with other enzymes such as laccase, where ≥60% of the enzyme activity is retained. This novel methodology provides a low cost, simple, modular approach of achieving high enzyme loadings in ordinary filter paper, not limited by cellulose surface area, and there has been no need for complex methods of enzyme engineering or toxic methods of activation of the solid support to prepare highly active biocatalysts.

1. INTRODUCTION

Paper-based enzyme devices hold high promise for a wide range of applications such as diagnostic assays (Parolo & Merkoçi, 2012; Pelton, 2009; Yetisen, Akram, & Lowe, 2013), biosensors (Liana, Raguse, Gooding, & Chow, 2012; Martinez, Phillips, Butte, & Whitesides, 2007), and biofuel cells (Wang et al., 2014). The specific properties of paper, such as its hydrophilicity, flexibility, insolubility in water, no toxicity, and abundance make it an attractive support for enzymes (Credou & Berthelot, 2014). A serious limitation of using paper for enzyme loading, which is mostly composed of cellulose with a chemical formula of $(C_6H_{10}O_5)_n$, has been the expensive chemical modification that is needed for the covalent linking of the enzyme to the support. This serious roadblock has been overcome by the current approach of interlocking of enzymes in the fibrous network of paper, which is novel, efficient, simple, inexpensive, and amenable to laboratories with limited resources.

Cellulose, the main component of paper, is hydrophilic, odorless, chiral, biodegradable, composed of D-glucose units, and insoluble in most organic solvents and dilute mineral acids. There are few approaches for convenient loading of enzymes onto a cellulose matrix such as physical adsorption (Sulaiman, Mokhtar, Naim, Baharuddin, & Sulaiman, 2014) and covalent crosslinking of the enzyme to the cellulose surface after chemical activation of the solid surface (Edwards, Prevost, Condon, & French, 2011). Physical adsorption is known to involve weak interfacial forces and the associated leaching of the enzyme from the surface into solution during exposure and storage. Chemical crosslinking limits enzyme loading to the accessible surface area of the matrix as well as the extent of chemical modification that has been achieved on the cellulose surface. As of now, there is no universal method for enzyme immobilization onto cellulose at high loading capacity that does not require any chemical activation of the support. We devised a simple, novel approach for the interlocking of enzymes onto cellulose fibers without any surface modification/activation of the support. In our approach, the enzyme is locked into cellulose fibers by covalent linking with a synthetic polymer such that the enzyme–polymer nanogels are directly embedded within the fibrous network (interlocking). This approach virtually eliminates long-term leaching of the enzyme and significantly enhances the stability of the interlocked enzyme. The method described here can be used for any enzyme and any polymer molecule that has the appropriate crosslinking functionalities.

1.1 The Enzymes: Glucose Oxidase and Peroxidase

Glucose oxidase (GOx, from *Asperigillus niger*) and horseradish peroxidase (HRP, from horseradish) were chosen as model systems. GOx, which catalyzes the oxidation of D-glucose to hydrogen peroxide and gluconic acid, is a widely studied enzyme for its possible usage in glucose biosensors (Oliveira, e Silva, de Souza, Martins, & Coltro, 2014; Soni & Jha, 2015; Zhang, Lyu, Ge, & Liu, 2014; Zhang et al., 2012) and biofuel cells (Wang et al., 2014). The second enzyme, HRP, then uses the peroxide produced from the GOx reaction and converts iodide to elemental iodine, which is brown, which is produced in proportion to the concentration of glucose present in the sample (Martinez et al., 2007). Thus, the extent of brown color produced upon loading the test sample, as measured by an ordinary scanner, is then proportional to glucose present and after calibration of the sensor, one can obtain the concentration of glucose in the sample.

1.2 The Polymer: Poly(Acrylic Acid)

Poly(acrylic acid) (PAA, $Mw = 8000$) is a negatively charged (Lützenkirchen, van Male, Leermakers, & Sjöberg, 2011), flexible, hydrophilic polymer (Kadajji & Betageri, 2011), which is commercially available in large quantities (Buchholz, 2000). Conjugation of enzymes to PAA drastically improves enzyme stability against thermal and proteolytic degradation (Mudhivarthi et al., 2012; Riccardi et al., 2014), and these conjugates are useful for paper-based enzyme devices.

1.3 Need for Nanoarmoring of Enzymes

The stabilization of enzymes against inactivation is a major challenge that often limits their application in the laboratory or industry. There are many methods for enzyme–polymer conjugation, and one such method is the wrapping of the enzyme with a long chain synthetic polymer which can enhance enzyme stability by lowering its conformational entropy.

Stabilizing the enzyme while also maintaining enzyme activity is possible by recognizing the thermodynamics of enzyme denaturation. The native (N, folded) and denatured (D, unfolded) states are thermodynamically related via the free energy changes. In order to retain enzyme function under harsh conditions, the native state must be maintained, and the transition from native to denatured state must be inhibited. In this method, the free energy of the denatured state is increased (destabilized) and the free energy of the native state is lowered (stabilized) (Scheme 1, left). In order to increase the free energy of the denatured state, its conformational entropy is reduced by wrapping the enzyme with a synthetic polymer (right). In our case, PAA is used to wrap the enzyme and curtail the number of conformational states that the denatured state can achieve. The restricted conformational space of the denatured state is therefore, lower in free energy than a complete random coil. This is called the "entropy control" method, which has been successfully used in our laboratory to improve enzyme stability (Deshapriya & Kumar, 2013; Mudhivarthi, Bhambhani, & Kumar, 2007).

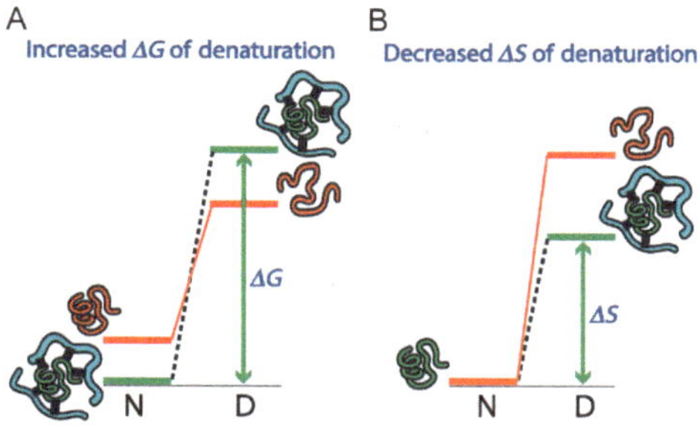

Scheme 1 (A) Enzymes can be stabilized by raising the free energy of denatured state (D) with respect that of the native state (N). (B) The increase in free energy of D may be achieved by decreasing its conformational entropy upon wrapping with a synthetic polymer.

The many available amine groups on enzymes provide a multipoint attachment by covalent linking to the polymer's carboxylic acid groups, and the covalent conjugation is readily achieved with the standard 1-ethyl-3-(3-(dimethylamino)propyl) carbodiimide (EDC) chemistry, under mild conditions at room temperature. Wrapping of the enzyme with the polymer "armors" the enzyme in its native conformation, and increases the energy needed to unfold the enzyme. At the same time, the highly charged nature of the polymer prevents enzyme aggregation over time, so the enzyme remains active. A similar strategy can also be designed with a cationic polymer such as poly(L-lysine) and its amine side chains can be attached to the COOH groups on most enzymes using the carbodiimide chemistry. Because of the polymer's net charge, it is also known to protect the enzyme from proteolytic degradation and inhibit inhibition by inhibitors that have the same charge as the polymer. These nanoarmored enzymes can then function even under unfavorable conditions when compared to the unmodified enzyme, at a minimal cost, low toxicity, while also maintaining excellent activity but enhanced stability.

1.4 Rationale for Using Cellulose

Paper-based devices have been widely used in a variety of applications such as in diagnostic assays, biosensors, and biofuel cells. Paper is mostly made of cellulose fibers (Scheme 2), and cellulose is attractive due to its biodegradability, hydrophilicity, insolubility in water, flexibility, filtration capabilities, and its high abundance make it an attractive support for enzymes (Nery and Lauro, 2016). The cellulosic fibers are benign to most enzymes and most

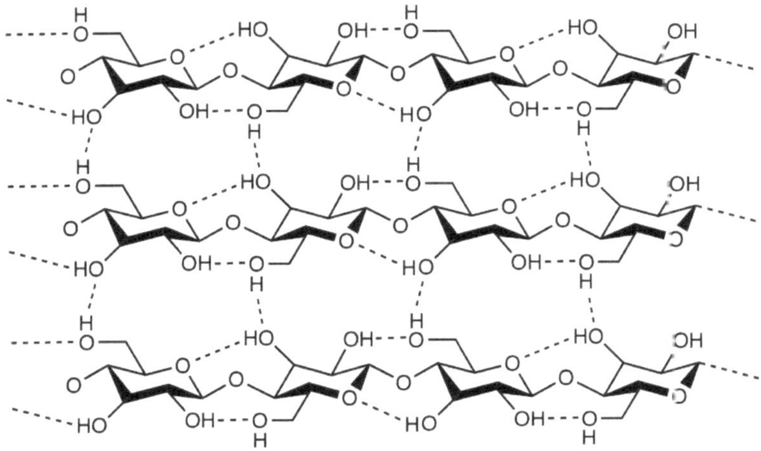

Scheme 2 The fibers of cellulose are shown earlier where the cellulose polymer molecules are aligned and hydrogen bonds between and within the cellulose molecules hold them together.

enzymes cannot degrade, hydrolyze, or oxidize it. Cellulosic paper is also extremely inexpensive for commercial applications of bound enzymes.

1.5 Concept of Interlocking as a Novel Method for Sensor Fabrication

There are many ways of trapping enzymes in a cellulose matrix, such as covalent crosslinking or adsorption but these methods often result in severe enzyme deactivation, or loss of enzyme when subjected to washing. The method of "interlocking" enzymes in cellulose is a new approach, and it is analogous to two chain links stuck together, but not joined to one another. The only way to remove one link would be to break the link open. Similarly, the enzyme–PAA conjugate is one link, while the cellulose is the other link. The enzyme–PAA conjugate is wrapped around the nanofibers in cellulose. Together, the enzyme is "interlocked" and cannot be washed away. At the same time, the enzyme is nanoarmored by the polymer so that the enzyme's stability is enhanced as well without compromising its biological activity. Thus, two positive outcomes are achieved in one step.

1.6 General Approach to Enzyme Interlocking and Key Results

Because of the nonspecific nature of the EDC chemistry that occurs between the amine groups of the enzyme and the carboxylic acid groups of the PAA polymer, there is minimal work needed to create the nanoarmored interlocked enzyme–PAA. Most enzymes have multiple lysine groups on their surface and PAA has a large number of carboxyl groups and it is also available commercially providing several molecular weights as desired. Large enzyme molecules would need longer PAA molecules to wrap them with, while smaller enzyme molecules might work well even with shorter PAA chains, and hence, one needs to choose an appropriate molecular weight polymer for interlocking. Typically, the polymer length can be about 10–15 times longer than the average diameter of the enzyme.

The interlocking approach via EDC chemistry can be broken down into two convenient steps: (1) activation of the polymer with EDC to prevent any enzyme–enzyme crosslinking and (2) addition of the enzyme to the activated polymer followed by deposition onto the cellulose surface to initiate interlocking. The reaction proceeds at room temperature until dry, and the enzyme paper is ready in a few minutes. This approach is also amenable to the inkjet technology as well as the 3D printing where the polymer, EDC, and enzyme solutions can be deposited on the paper in a predefined sequence and pattern. Ideally, no interactions between the cellulose fibers and the enzyme/polymer are required to interlock enzymes in paper. This is advantageous, because any such interactions might distort the enzyme's structure, and the absence of such interactions or when the interactions are very weak there is no threat to the enzyme structure. Enzymes can therefore be bound to paper without significant distortion of enzyme structure, and the bound enzyme cannot be washed away due to the "interlock" technology. Thus, the "interlock" technology and paper might provide a general protocol to bind enzymes to solid substrates by a universal approach, without loss of activity but improved stability.

Intricate structures in 2- or 3-dimensions on porous substrates can be made, where the aqueous solutions can wet the fibrous matrix. All manipulations are carried out at room temperature, using buffered solutions, desired pH and ionic strength conditions, thus preserving the native structure of the enzyme without adverse effects. The enzymes interlocked into the fibrous network of the paper are stable against washing, leaching is prevented, they remain active, and are stable against denaturants. Because the pores of the paper matrix are not blocked by this approach, the medium can be used to separate particulate material before analysis. The sample can be spotted on one side of the paper and activity assayed from the other side, so that any debris would not affect the analytical result. On the other hand, the paper matrix may also be exploited for performing any further chemistries on the sample by simply adding multiple paper layers to the sensor.

Thus, paper based, interlocked enzymes have a high potential for analytical applications. Benign, one step, simple interlocking of one or more enzymes in the cellulose matrix is reported here.

2. EQUIPMENT AND REAGENTS

2.1 General Equipment

1. *Safety*: Goggles, laboratory coat, latex, or nitrile gloves, ventilation hood. Caution when weighing out SDS (dust mask is recommended), as it is a fine powder and extremely irritating when inhaled.
2. *Waste management*: General aqueous waste receptacle.
3. Filter paper (Whatman 1 cellulose, typical 180 μm thickness, 87 g m^{-2} weight, and with 11 μm pore size, 185 mm diameter).
4. Deionized water.
5. A UV–visible spectrophotometer.
6. A benchtop mini centrifuge.
7. Microcentrifuge tubes (1.5 mL).
8. Volumetric micropipettes (20, 100, 200, and 1000 μL).
9. Beakers (100–1000 mL).
10. Stir/hot plate.
11. Analytical balance.
12. pH meter.

2.2 Enzyme Interlocking on Paper

1. Glucose oxidase (*Asperigillus niger*, GOx, EC 1.1.3.4, Sigma-Aldrich).
2. Peroxidase from horseradish (HRP, EC 1.11.1.7, Calzyme).
3. 1-Ethyl-3-(3-(dimethylamino)propyl) carbodiimide (EDC, TCI America).
4. Poly(acrylic acid) (PAA, $Mw = 8000$, Sigma-Aldrich). *Tip*: Other molecular weights of PAA have been used in our laboratory, and all different samples behaved similarly. Synthesis concentrations, however, must be adjusted according to the molecular weight, as higher molecular weight polymer solutions become more viscous and these will not coat the paper evenly. The conditions of interlocking need to be adjusted to the type of enzyme used and the molecular weight of PAA, on a case-by-case basis.
5. Phosphate buffer (200 m*M*, pH 7.0).
6. Household kitchen wax paper.

7. Precision cutter.
8. Heat press.

2.3 Characterization by Gel Electrophoresis

1. Agarose (high gel temperature, USBiological Life Sciences).
2. Bromophenol blue (Sigma-Aldrich).
3. Acetic acid (Sigma-Aldrich).
4. Tris-base (Sigma-Aldrich).
5. Microwave.
6. Glycerol (Sigma-Aldrich).
7. Gel electrophoresis apparatus (Gibco Model 200, Life Technologies Inc.).
8. Brilliant blue R250 (Sigma-Aldrich).

2.4 Characterization by SDS-PAGE

1. Sodium dodecyl sulfate (Sigma-Aldrich).
2. Tris-base (Sigma-Aldrich).
3. Bromophenol blue (Sigma-Aldrich).
4. β-Mercaptoethanol (BME, Sigma-Aldrich).
5. Acrylamide (Sigma-Aldrich).
6. Glycine (Sigma-Aldrich).
7. *N′N′*-bis-methylene-acrylamide (Sigma-Aldrich).
8. Hydrochloric acid (Sigma-Aldrich).
9. Ammonium persulfate (APS, Sigma-Aldrich).
10. Isopropanol (Sigma-Aldrich).
11. Acetic acid (Sigma-Aldrich).
12. Tetramethylethylenediamine (Sigma-Aldrich).
13. Molecular weight marker (broad range protein ladder, Thermo Fischer Scientific).

2.5 Characterization by Circular Dichroism

1. A circular dichroism (CD) spectropolarimeter (JASCO model J715).
2. 0.05-cm path length quartz cuvette.

2.6 Characterization of Cellulose-Interlocked Enzymes by Laser Confocal Fluorescence Microscopy

1. 1-Ethyl-3-(3-(dimethylamino)propyl) carbodiimide (EDC, TCI America).

2. Fluoresceinamine Isomer 1 (FA, Sigma-Aldrich).
3. 5(6)Carboxy-X-rhodamine *N*-succinimdyl ester (ROX, Sigma-Aldrich).
4. Acetone (Sigma-Aldrich).
5. Sodium bicarbonate (Sigma-Aldrich).
6. Dimethylsuldocide (DMSO, Sigma-Aldrich).
7. Dialysis membrane (Spectra/Por Biotech Membranes, MWCO 15K).
8. Laser confocal microscope (Nikon A1R Confocal Microscope).

2.7 Colorimetric Enzymatic Activity Assay

1. Potassium iodide (Sigma-Aldrich).
2. Glucose (Sigma-Aldrich).
3. Color scanner (Canon, CanoScan LIDE 200).
4. ImageJ image processing software (free online) with histogram package.

2.8 Enzyme Loading on Paper by Bradford Assay

1. Brilliant blue G-250 (Sigma).
2. Ethanol (Sigma-Aldrich).
3. Phosphoric acid (Sigma-Aldrich).
4. Whatman 1 filter paper, standard grade.

3. METHODS

3.1 Enzyme Conjugation to PAA and Characterization by Agarose Gel Electrophoresis

Agarose gel electrophoresis was done to verify complete conjugation of the enzymes to PAA in solution (Fig. 1A). The nanoarmored GOx–HRP–PAA (Lane 3) showed increased electrophoretic mobility toward the positive electrode when compared to GOx/HRP (Lane 1) and GOx/HRP/PAA (Lane 2) controls. The increased negative charge is due to the covalent attachment of the enzyme (NH_2 groups) to the negatively charged PAA (COOH groups), which also indicates that no unconjugated enzyme remained after the EDC reaction was complete (Scheme 3). Normally EDC reaction is 40%–50% complete, but we achieved complete conjugation because of the possible optimal ratio of the enzymes to the polymer and high EDC concentration and conditions followed. Another reason is interlocking itself is expected to be efficient because not 100% of the available carboxyls or amine groups need to react. A substantial fraction should

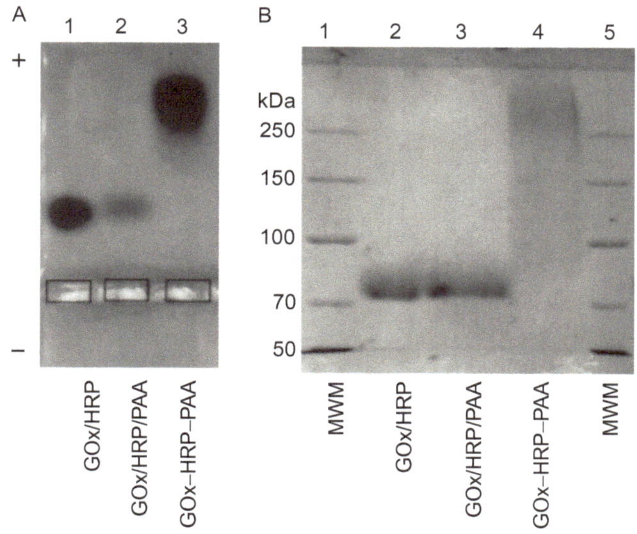

Fig. 1 (A) Agarose gel of the GOx–HRP–PAA conjugate in solution (Lane 3, concentration ratio of 16:3:10,000 μ*M* GOx:HRP:PAA), corresponding controls of GOx/HRP (Lane 1, simple mixture in solution at a concentration ratio of 16:3 μ*M* GOx:HRP) and GOx/HRP/PAA physical mixture in solution prior to EDC treatment (Lane 2). (B) SDS-PAGE of the GOx–HRP–PAA conjugate (Lane 4) shows the increased molecular weight compared to the corresponding controls of GOx/HRP (Lane 2) and GOx/HRP/PAA physical mixtures (Lane 3). Standard molecular weight markers are shown (Lanes 1 and 5). *Reproduced from Riccardi, C. M., Mistri, D., Hart, O., Anuganti, M., Lin, Y., Kasi, R. M., & Kumar, C. V. (2016). Covalent interlocking of glucose oxidase and peroxidase in the voids of paper: Enzyme–polymer "spider webs".* Chemical Communications, *52, 2593–2596.*

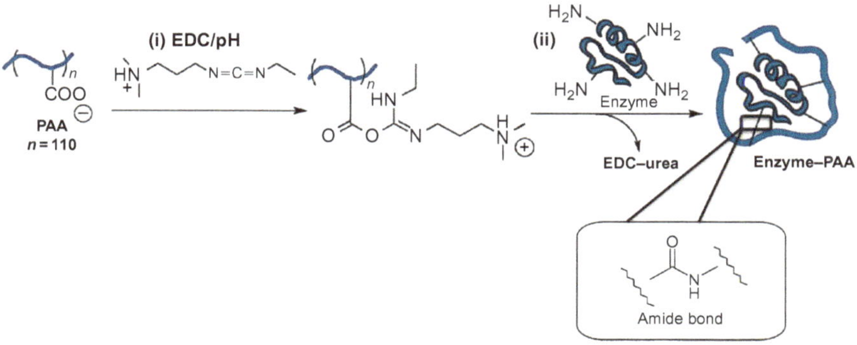

Scheme 3 Chemical reaction between amine and carboxyl groups catalyzed by EDC.

react to securely interlock the components in the fibrous matrix. Thus, there is a significant margin of error that is allowed in the extent of crosslinking needed for complete interlocking of the components.

1. Prepare the agarose gel (0.125 g agarose) by microwaving a solution of agarose (0.5%, w/v) in Tris-acetate buffer (40 m*M*, pH 7.0, 25 mL) for 1 min on high setting. After microwaving, the agarose should all have dissolved.
2. Leave the agarose to cure in the gel mold for 30 min.
3. Meanwhile, prepare samples by mixing 5–6 μ*M* with loading buffer (10 μL, 50% (v/v) glycerol, and 0.1% (w/v) bromophenol blue).
4. Load 15 μL of sample in each well. Once loaded, run the agarose gel at 100 V for 30 min.
5. Stain the gel with brilliant blue R250 (0.1%, w/v) for 4 h, and then destain the gel with acetic acid (10%, v/v) overnight.

3.2 Characterization of Enzyme–PAA Conjugates by SDS-PAGE

Covalent attachment and formation of the GOx–HRP–PAA conjugate and verification of the increase in molecular weight were also confirmed by SDS-PAGE. Because of the highly crosslinked nature of the nanoarmored enzymes, the conjugate is not expected to move as freely through the gel (Fig. 1B). A 7% polyacrylamide separating gel was used with a 7% stacking gel. By comparing to the standard molecular weight markers (Lanes 1 and 5), the GOx/HRP exhibits a sharp band at ~70 kDa (Lane 2). Without the addition of EDC, the physical mixture GOx/HRP/PAA also exhibits the same molecular weight as GOx/HRP (Lane 3). It is evident that the nano-armored GOx–HRP–PAA had a broad and high molecular weight band (Lane 4, 90–250 kDa) with no unconjugated enzyme present.

3.3 Characterization of Enzyme–PAA Conjugates by CD

CD is an excellent way of tracking overall changes in enzyme secondary and tertiary structure. One might question how the enzyme structure is affected by the covalent attachment of PAA to its surface. The ellipticity of the nano-armored GOx–HRP–PAA in solution was found by measuring the CD spectra with a JASCO model J715 spectropolarimeter. All spectra are an average of 10 accumulations, measured with a 1 nm data pitch, and scan speed of 50 nm/min. Cuvette path lengths and sample concentrations were 0.05 cm and ~3 μ*M* (UV-CD, 190–260 nm). The GOx has a flavin adenine dinucleotide (FAD) cofactor which plays a role in the oxidation–reduction

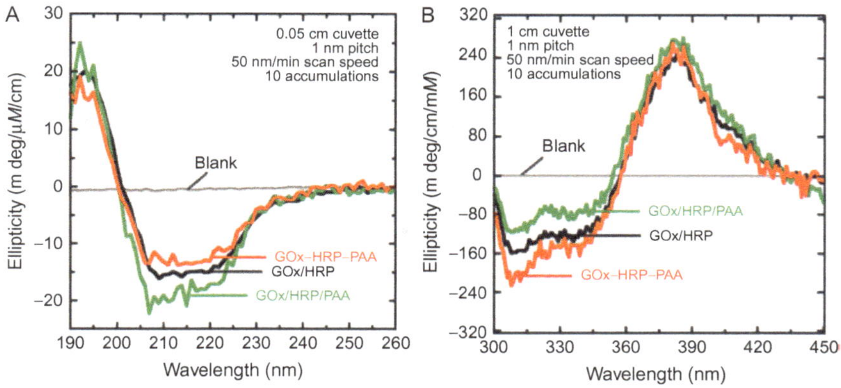

Fig. 2 (A) UV and (B) Soret circular dichroism (CD) of GOx–HRP–PAA conjugates and corresponding GOx/HRP and GOx/HRP/PAA physical mixture controls. *Reproduced from Riccardi, C. M., Mistri, D., Hart, O., Anuganti, M., Lin, Y., Kasi, R. M., & Kumar, C. V. (2016). Covalent interlocking of glucose oxidase and peroxidase in the voids of paper: Enzyme–polymer "spider webs".* Chemical Communications, 52, *2593–2596.*

reaction, and its molar ellipticity was measured using 1 cm cuvette path length and ~15 μ*M* enzyme (Soret-CD, 300–450 nm). All ellipticities were normalized by dividing with path length and sample concentration. The CD spectra of all samples remained the same after conjugation to PAA (Fig. 2) with some changes in intensities. GOx–HRP–PAA in solution retained 80% ellipticity at 209 nm when compared to unmodified GOx/HRP physical mixture. The GOx/HRP/PAA control also retained 100% ellipticity. In conclusion, the spectra strongly suggest that the enzyme structure is mostly retained after nanoarmoring with PAA.

3.4 Method for Enzyme Interlocking in Paper

In this approach, circular assay "wells" were prepared on filter paper by coating with wax. The enzyme–polymer conjugate (GOx–HRP–PAA) was subsequently formed by crosslinking the carboxylate groups of the PAA to the lysine residues of GOx and HRP by the water-soluble carbodiimide, EDC (Scheme 4A). The resulting interlocked enzyme–PAA nanoarmored conjugate can be colorimetrically assayed and activity can be quantified by a color scanner and ImageJ software (Scheme 4B and C).

The significance of using a colorimetric assay was a model for a portable quick and disposable test strip, where minimal sample preparation is required in order to obtain a reading. Additionally, a colorimetric assay is useful for enzyme characterization, monitoring how the enzyme is active in cellulose,

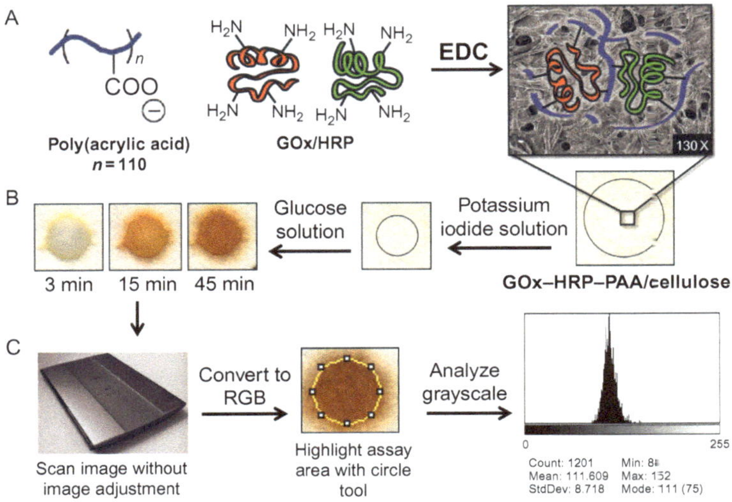

Scheme 4 Synthesis and analysis of GOx–HRP–PAA biosensors. (A) Conjugation of the GOx and HRP to PAA is shown via EDC chemistry. The final enzyme–PAA conjugate interlocked within the filter paper (cartoon not to scale). (B) Addition of potassium iodide to the filter paper develops the color once glucose is added. (C) The resulting color intensity on the test strip is scanned and processed with ImageJ. *Reproduced from Riccardi, C. M., Mistri, D., Hart, O., Anuganti, M., Lin, Y., Kasi, R. M., & Kumar, C. V. (2016). Covalent interlocking of glucose oxidase and peroxidase in the voids of paper: Enzyme–polymer "spider webs".* Chemical Communications, 52, *2593–2596.*

and how its stability is enhanced when compared to that of the enzyme in solution.

1. Cut circles (8 mm diameter) into wax paper using a precision cutter.
2. Place the cut wax paper on top of a single sheet of Whatman filter paper and apply heat press (280 °C, 20 s). The wax-coated filter paper wells have an inner diameter of ~3.5 mm. *Tip*: The well's boundary uniformity can be verified by wetting with distilled water and then observing if the water diffused into the waxed areas. The wax edges should be sharp and well defined.
3. Prepare a diluted stock solution of PAA (37 m*M*). *Tip*: Because the viscosity of PAA is very high, using a micropipette is not ideal. Instead, use a volumetric syringe to dilute the PAA. Remove any air bubbles from the syringe by pressing the plunger down quickly.
4. Prepare stock solutions of GOx ($\varepsilon_{280\ \mathrm{nm}} = 267{,}000$) and HRP ($\varepsilon_{403\ \mathrm{nm}} = 100{,}000$) each at about 16 mg/mL. Centrifuge the enzyme

solutions to remove any large impurities and obtain the enzyme concentrations using a UV–vis.

5. *Reaction mixture concentrations*: Mix PAA (10 m*M*) in phosphate buffer (200 m*M*, pH 7.0), deionized water, and EDC (100 m*M*) in a microcentrifuge tube and stir for 10 min to activate the PAA. *Tip*: EDC is hygroscopic, so it should be weighed immediately before use.
6. Premix the GOx and HRP to be added to ensure uniform distribution of enzyme in the conjugate, add the enzyme mixture to the PAA solution so that final reaction concentrations are 16 μ*M* for GOx and 3.2 μ*M* for HRP.
7. Immediately transfer 2.5 μL of the activated GOx/HRP/PAA/EDC to the wells in the wax-coated filter paper, where the crosslinking occurs.
8. *Control solutions*: Prepare samples as above (1) without EDC to create physical mixture (GOx/HRP/PAA), (2) without EDC or PAA for the adsorbed enzyme (GOx/HRP), and (3) GOx–HRP crosslinked in the absence of PAA.
9. Air dry for about 1 h at room temperature.
10. Wash the resulting GOx–HRP–PAA/cellulose and controls, three times in 25 mL of distilled water with gentle stirring to remove any unreacted PAA and EDC urea.
11. Dry for 1 h at room temperature.

3.5 Characterization of Cellulose-Interlocked Enzymes by Laser Confocal Fluorescence Microscopy

The nanoarmored enzymes interlocked in the filter paper were visualized with multichannel laser confocal microscopy. This showed the overall distribution of the enzyme and polymer in the cellulose matrix. GOx was labeled with a red dye 5(6)carboxy-X-rhodamine N succinimidyl ester (ROX–GOx) and the PAA with a green dye fluoresceinamine isomer 1 (FA–PAA). By exciting and monitoring at two different wavelengths, the area with both red (ROX–GOx) and green (FA–PAA) indicates the location of the nanoarmored enzymes in the cellulose.

The confocal microscope images show that both the enzyme (red) and the polymer (green) are localized on the cellulose fibers, indicating that the conjugate might be forming within the small intrafiber voids in the cellulose (Fig. 3A). Ambient lighting is shown in the bottom left quadrant while the overlay of all three quadrants is shown in the bottom right. The physical mixture control ROX–GOx/FA–PAA/cellulose shows a decrease

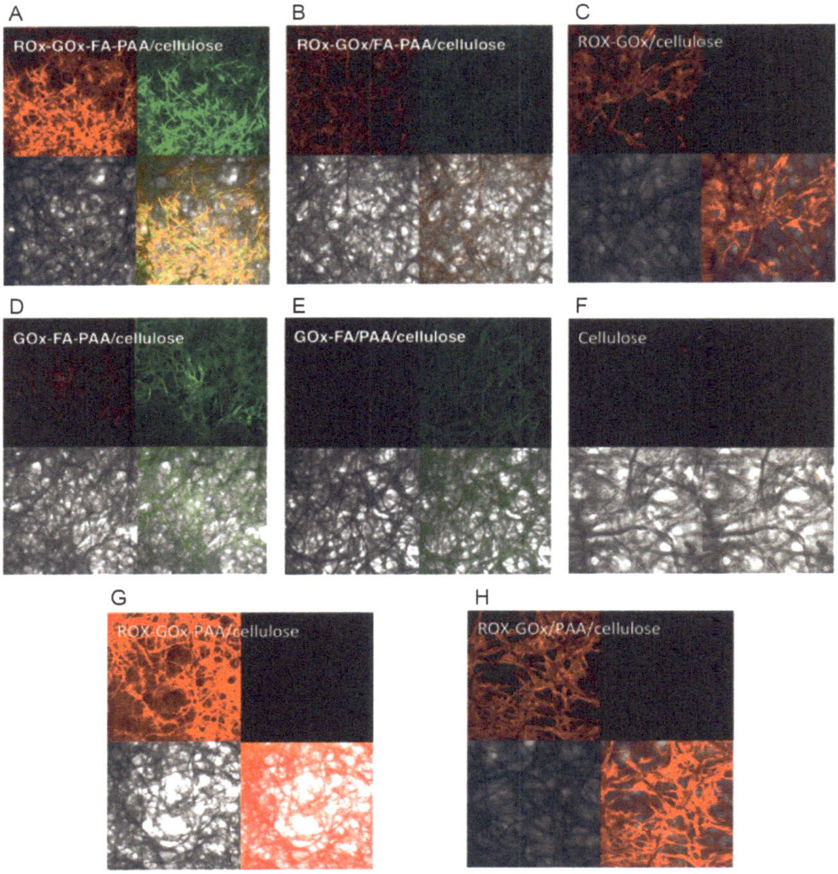

Fig. 3 Confocal fluorescence microscopy images of (A) ROX–GOx–FA–PAA/cellulose conjugate, (B) physical mixture ROX–GOx–FA–PAA/cellulose, and (C) ROX–GOx/cellulose. Additional controls include (D) the conjugate GOx–FA–PAA/cellulose, (E) physical mixture GOx/FA–PAA/cellulose, (G) conjugate ROX–GOx–PAA/cellulose, and (H) the physical mixture ROX–GOx/PAA/cellulose. (F) Bare cellulose. *Reproduced from Riccardi, C. M., Mistri, D., Hart, O., Anuganti, M., Lin, Y., Kasi, R. M., & Kumar, C. V. (2016). Covalent interlocking of glucose oxidase and peroxidase in the voids of paper: Enzyme–polymer "spider webs".* Chemical Communications, 52, *2593–2596*

in fluorescence intensity (Fig. 3B), indicating that the enzyme and polymer are both mostly washed away. The physical adsorption of the enzyme ROX–GOx on cellulose (Fig. 3C) shows decreased fluorescence intensity compared to Fig. 3A and B, indicating that the enzyme is mostly washed off. The conjugate made with unlabeled GOx and FA–PAA (Fig. 3D) shows a high intensity of green fluorescence from the interlocked conjugate, while the physical mixture of this same sample shows decreased green fluorescence (Fig. 3E). The bare cellulose control is shown (Fig. 3F) indicating no red or green fluorescence. The conjugate made with labeled ROX–GOx and unlabeled PAA is shown (Fig. 3F) indicating high red fluorescence from the labeled enzyme, while its corresponding physical mixture control (Fig. 3G) shows decreased fluorescence. In all cases, the physical mixtures show decreased fluorescence, indicating that the enzyme and polymer are both washing off from the cellulose paper, whereas the conjugates are retained.

3.5.1 Fluorescent Labeling of PAA Using Fluoresceinamine Isomer 1

1. Prepare a stock solution of PAA (9 m*M*) in deionized water.
2. Add EDC (1 *M*) and stirred for 10 min.
3. The FA stock can be prepared in DMSO. Add FA (10 m*M*) slowly to the PAA solution with stirring, and react for 4 h in the dark at room temperature.
4. Centrifuge the PAA solution for 30 min (10,000 rpm) to remove any unreacted solid FA.
5. Remove the pellet, and then add equal volume of acetone to the FA–PAA solution. This will cause the FA–PAA to precipitate and form a globular gel.
6. Centrifuge for 10 min (10,000 rpm).
7. Remove the supernatant containing acetone/water/unreacted dye.
8. Resuspend the FA–PAA in deionized water. Repeat acetone precipitation from Step 5 three times. This will be the purified FA–PAA stock.

3.5.2 Fluorescent Labeling of GOx With ROX

1. Prepare ROX in DMSO (10 mg/mL).
2. Add the ROX (100 μL) slowly to a solution containing GOx (20 mg/mL) in bicarbonate buffer (0.2 *M*, pH 8.3).
3. Stir for 4 h in the dark at room temperature, and then dialyze against phosphate buffer (pH 7.0, 20 m*M*) for 4 h (switch fresh buffer out every hour) to remove the unreacted fluorescent dye.

3.5.3 Laser Confocal Fluorescence Microscopy

1. Nikon A1R Confocal Microscope parameters
   a. 10 times dry objective lens.
   b. Filter paper samples are placed directly on the microscope stage.

c. The FA dye is excited by a 488 nm argon laser (power level 2.4) and the fluorescence emission is monitored at 525 nm, with a PMT high voltage of 125.
d. ROX dye is excited by a 561 nm argon laser (power level 2.0) and monitored at 595 nm, with a PMT high voltage of 120.
e. The instrument pinhole was 2.17 μm.

3.6 Characterization by Colorimetric Activity Assay

An enzymatic assay using potassium iodide and glucose was used to monitor enzymatic activity and quantify enzyme retention within cellulose (Scheme 4B and C). GOx catalyzes the oxidation of β-D-glucose by molecular oxygen to form hydrogen peroxide and gluconic acid. HRP then catalyzes the oxidation of the iodide to molecular iodine, which is brownish orange colored (Scheme 5). It was quantitatively found that the brown color appears gradually and the reaction is complete after about 20–30 min (Fig. 4). A scanner is used to collect the images so that the color intensity appearance over time can be monitored using ImageJ imaging software.

Nearly 100% of the color intensity was retained for the GOx–HRP–PAA/cellulose after washing, indicating that the enzyme was retained within the paper. Respective controls (without EDC, indicated by a slash) of GOx/HRP/cellulose and GOx/HRP/PAA/cellulose physical mixtures, and GOx–HRP–PAA conjugate drop cast on cellulose 1 day after synthesis, were washed under the same conditions. After 1 day, no further conjugation could occur because of EDC hydrolysis, and no entanglement within the pores was expected. Less than 30% of the enzyme was retained in all three controls.

1. To each circular "well" add 2 μL of potassium iodide solution, prepared in deionized water (0.3 *M*). *Tip*: The potassium iodide solution can be stored at 4 °C, in the dark for at least 1 week.
2. Dry for 15 min.
3. With the scanner hooked up and ready to scan, add the glucose solution (2–40 m*M* in deionized water).

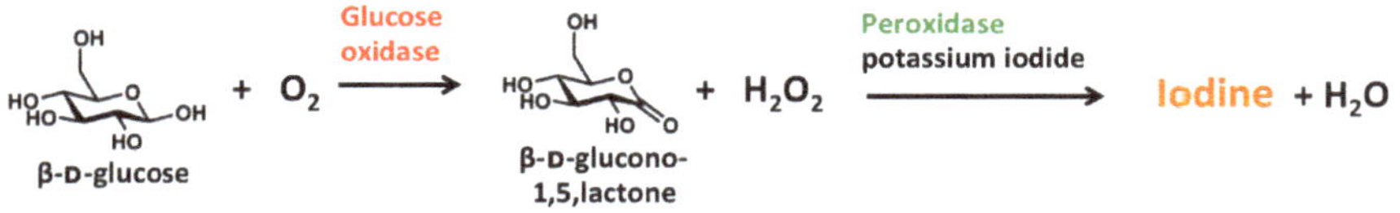

Scheme 5 Colorimetric enzyme assay by the addition of glucose oxidase and potassium iodide to the interlocked and nanoarmored enzymes–PAA conjugate.

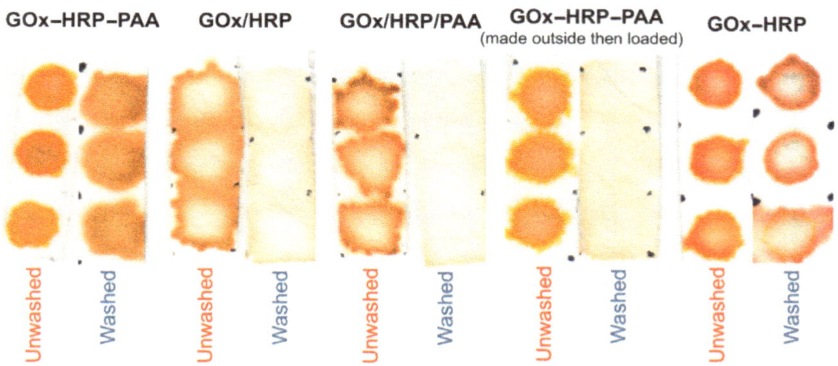

Fig. 4 Synthesis validation for GOx–HRP–PAA/cellulose conjugate interlocked with paper fibers. All samples were washed to remove any unbound material. Controls are GOx/HRP and GOx/HRP/PAA physical mixtures, GOx–HRP–PAA conjugate deposited on the filter paper 1 day after synthesis, and GOx–HRP crosslinked in the absence of PAA. *Reproduced from Riccardi, C. M., Mistri, D., Hart, O., Anuganti, M., Lin, Y., Kasi, R. M., & Kumar, C. V. (2016). Covalent interlocking of glucose oxidase and peroxidase in the voids of paper: Enzyme–polymer "spider webs".* Chemical Communications, 52, *2593–2596.*

4. Scan the yellow to brown color (conversion of potassium iodide to elemental iodine) with the scanner every 2 min (600 dpi, color setting). *Tip*: The scanner may have a color adjustment setting—this should be turned off.
5. Open the resulting image with ImageJ, and convert to gray scale.
6. Highlight one "well" with the circle tool and analyze the gray scale color intensity using the histogram tool under the "analyze" dropdown.
7. Record the "mean" average color intensity and subtract this value from 255. This value increases as more iodine is produced.

3.7 Enzyme Loading on Cellulose by Bradford Assay

Because the conjugate is forming around the cellulose fibers, and the conjugate size can increase as both the enzyme and polymer concentrations increase, the cellulose surface area therefore does not limit the maximum loading of enzyme in the cellulose. Finding the enzyme loading on cellulose can be useful, as this interlocking and armoring method can easily be extended to other enzymes and other suitable supports. The loading will also indicate the extent of conjugation and how much of the enzyme washes off, which is significant to know, especially when using more expensive enzymes. Enzyme concentration can easily be quantified by using the

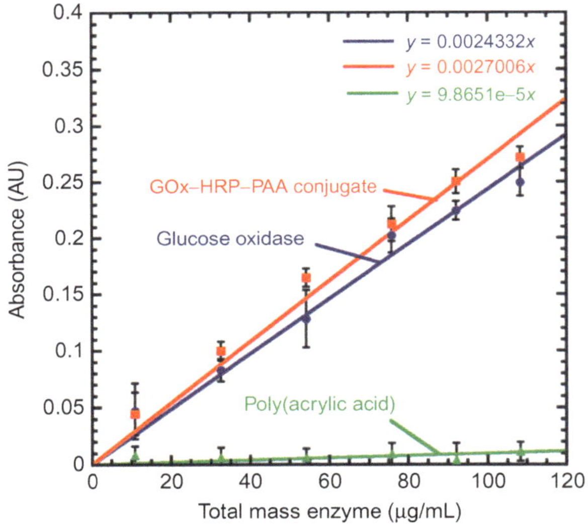

Fig. 5 Calibration graph of enzyme standards (1, 3, 5, 7, 8.5, and 10 µg enzyme in 0.1 mL) by the net absorbance at 595 nm of after adding Bradford reagent. Enzyme standards of GOx (*blue*), GOx–HRP–PAA conjugate (*red*), and control of PAA without enzyme (*green*) are shown. Net absorbance was calculated by subtracting the absorbance of buffer and reagent from the mixture of enzyme and reagent. *Reproduced from Riccardi, C. M., Mistri, D., Hart, O., Anuganti, M., Lin, Y., Kasi, R. M., & Kumar, C. V. (2016). Covalent interlocking of glucose oxidase and peroxidase in the voids of paper: Enzyme–polymer "spider webs".* Chemical Communications, 52, *2593–2596.*

Bradford reagent and measuring the absorbance at 595 nm with respect to increasing enzyme concentration. A calibration graph was made for the Bradford reagent, and the unknown enzyme concentration in the washes was calculated from the calibration plot (Fig. 5).

As total enzyme mass was increased during the interlocking process, it was found that enzyme percent loading did not saturate in the range tested, confirming the system was not limited by cellulose surface area (Fig. 6A). Loading increased as more enzyme was added. The enzyme percent retention after washing also increased with increasing enzyme mass, with the highest retention for the 26 mg GOx sample that exhibited less than 1% loss of enzyme during subsequent washes (Fig. 6B). Because crosslinking probability increases as more GOx is added, high enzyme retention is achieved with more crosslinking until no more enzyme–polymer can be accommodated in the voids.

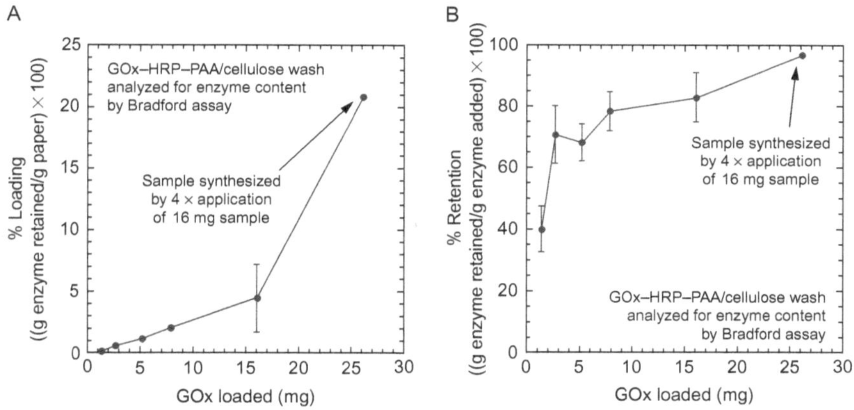

Fig. 6 (A) Percent loading of enzyme per gram of paper and (B) the percent enzyme retained within the paper after washing are shown. Error bars are provided but some are smaller than the points. *Reproduced from Riccardi, C. M., Mistri, D., Hart, O., Anuganti, M., Lin, Y., Kasi, R. M., & Kumar, C. V. (2016). Covalent interlocking of glucose oxidase and peroxidase in the voids of paper: Enzyme–polymer "spider webs".* Chemical Communications, 52, *2593–2596.*

3.7.1 Bradford Reagent Preparation and Calibration of Enzyme Concentration

1. In a 200 mL beaker, dissolve brilliant blue G-250 (100 mg) in ethanol (95%, w/v, 50 mL). *Tip*: Make sure at this point that most of the brilliant blue is dissolved.
2. To this solution, add phosphoric acid (85%, w/v, 100 mL) and then dilute to 1 L with deionized water. Final concentrations of the reagent are 0.01% (w/v) brilliant blue G-250, 4.7% (w/v) ethanol, and 8.5% (w/v) phosphoric acid.
3. Filter the solution through Whatman 1 filter paper before each use until the color is light brown (usually 2–3 times). This solution can be stored up to 6 months at 4 °C.
4. Prepare enzyme standards containing either GOx–HRP–PAA or GOx at 1, 3, 5, 7, 8.5, and 10 μg enzyme in 0.1 mL. Respective control of only PAA is made at identical PAA mass present in GOx–HRP–PAA standards.
5. To these standards, add 1 mL of Bradford reagent and mix.
6. Measure the absorbance of the solution at 595 nm 10 min after adding the reagent with respect to total enzyme concentration (Fig. 5). *Blank solution*: Phosphate buffer (20 m*M*, pH 7.0) and reagent.

3.7.2 Enzyme Loading by Bradford Assay

1. During synthesis of the GOx–HRP–PAA/cellulose, the GOx concentration was increased (16, 30, 60, 90, 180 μ*M*).
2. The EDC concentration is also increased to 1.1 *M* so that all carboxylate groups on PAA were activated.
3. Wash the GOx–HRP–PAA/cellulose samples three times in 25 mL of water for 1 h to remove any free enzyme. *Tip*: Measure the exact volume using a graduated cylinder after each wash is complete to account for slight evaporation.
4. Analyze the wash for enzyme concentration by its absorbance at 450 nm after adding the Bradford reagent. The absorbance of the wash should fall within the calibration curve, dilute as necessary.
5. Calculate the total enzyme concentration in the cellulose, taking into account volume dilutions.
6. Enzyme percent loading ((g enzyme retained/g cellulose) × 100) is calculated by dividing the mass of enzyme retained in the paper by the total mass of the cellulose, and then multiplying by 100 (Fig. 6A).
7. The percent enzyme retained (Fig. 6B) within the paper is calculated by dividing the mass of enzyme retained by the original mass of enzyme added, and then multiplying by 100. In all experiments, total filter paper weight was about 270 mg.

3.8 Porosity of the Cellulose-Interlocked Enzymes by Mercury Intrusion Porosimetry

Maintaining porosity of the cellulose paper matrix is important for minimizing low diffusion rate of the substrate to the enzymes in the matrix. In order to quantify how much the porosity changes due to interlocking, mercury intrusion porosimetry was done on the GOx–HRP–PAA/cellulose sample with the highest enzyme loading (26 mg GOx). This ensures that, even with such high loadings, the porosity is still maintained.

Four samples of GOx–HRP–PAA/cellulose and four control samples of bare cellulose were sent according to the specifications needed by Micromeritics, Georgia, USA. The measurements were conducted automatically on an AutoPore IV 9500V1.09. The samples were sealed in a penetrometer, weighed, and subjected to mercury intrusion. The GOx–HRP–PAA/cellulose had a pore volume of 0.65 mL/g, a total pore area of 14 m^2/g, a median pore diameter of 3.2 μm, and a porosity of 21%. The corresponding control of bare cellulose had a pore volume of 1.4 mL/g, a total pore area of

7.3 m^2/g, a median pore diameter of 12 μm, and a porosity of 48%. These data along with the laser confocal microscopy strongly suggest that the porosity of the cellulose is diminished but largely maintained.

3.9 Stability of the Cellulose-Interlocked Enzymes Against Denaturation

The nanoarmoring and interlocking of enzymes in cellulose is expected to stabilize the native state, and increase the enzyme's resistance against denaturation, while also giving them the advantage of being washable. Washable paper-based systems are useful in applications such as wearable sensors, or flow-through systems such as wastewater treatment. In such cases, the enzyme must remain active in harsh conditions.

Thus, the GOx–HRP–PAA/cellulose was washed in denaturants such as urea (5 *M*) or SDS (20%, w/v) for 30 min, and then rinsed three times with deionized water for an additional 30 min. It was found that the GOx–HRP–PAA/cellulose retained 103 ± (8)% of its activity after washing in urea, and 70 ± (8)% activity after washing with SDS (Fig. 7A). All other controls of GOx/HRP/cellulose (Fig. 7B) and GOx/HRP/PAA/cellulose (Fig. 7C) retained 30% or less activity after washing in these conditions. Because a 30-min cycle is a typical washing cycle, the nanoarmored and interlocked enzyme can be considered a washable system.

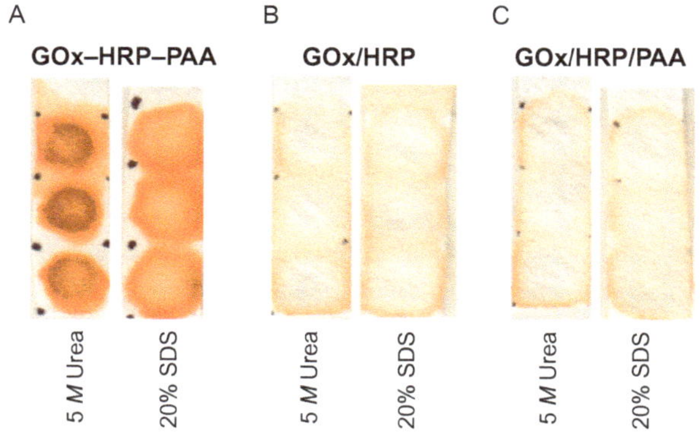

Fig. 7 Resistance to chemical denaturation of interlocked (A) GOx–HRP–PAA/cellulose and controls of (B) GOx/HRP/cellulose and (C) GOx/HRP/PAA/cellulose. The samples were all washed in urea or SDS for 30 min and then rinsed three times with distilled water for an additional 30 min.

Fig. 8 Synthesis validation for (C) nanoarmored laccase interlocked with paper fibers. All samples were washed to remove any unbound material. Controls are (A) adsorbed laccase and (B) laccase physical mixtures (unpublished work).

3.10 Modular Approach for Enzyme Interlocking

Thus far, we have shown that this method of nanoarmoring by covalent conjugation to PAA improves the stability of enzymes in solution such as hemoglobin, catalase, cytochrome *c*, lysozyme, and others (Mudhivarthi et al., 2012; Riccardi, Kasi, & Kumar, 2015; Riccardi et al., 2014; Thilakarathne, Briand, Zhou, Kasi, & Kumar, 2011). The interlocking and nanoarmoring method is therefore expected to work with other enzymes and other polymers that have the required functional groups.

We recently interlocked laccase in cellulose, an enzyme that reduces a variety of phenols (Fig. 8). Even though our methods have not yet been optimized, the interlocked laccase remained in the filter paper even after extensive washing (Fig. 8C), while the respective controls of adsorbed and physical mixture of laccase was washed away (Fig. 8A and B) In an effort to enhance the stability and loading of laccase in cellulose, and because laccase only has five available lysines to conjugate to PAA, we have pre-interlocked BSA as a first coating on the filter paper, then added another layer of laccase with BSA and/or PAA. This new approach of using BSA as "packing peanuts" appears to help increase the loading and retention of laccase compared to when only PAA is used (only about 60% of the laccase activity is retained when interlocked only with PAA, but activity retention increased to greater than 60% when BSA was added. The interlocked laccase was also significantly more stable than the free enzyme (the half-life of laccase interlocked with BSA instead of PAA was $\geq$20 days, while laccase in solution at room temperature has a half-life of less than 3 days).

4. CONCLUSIONS

In conclusion, the nanoarmoring of enzymes with PAA and their interlocking within the cellulose matrix of ordinary paper is an excellent way of trapping enzymes, without the need for direct covalent attachment or surface modification. This is the first example where the physical interlocking of the conjugate in the cellulose allows for a washable system that remains active even under strong denaturing conditions with long-term stability and reusability after washing.

ACKNOWLEDGMENTS

This work was supported by NSF EAGER award DMR-1441879 and by the University of Connecticut VPR Research Excellence award. We would like to thank K. Cole, K. Benson, and K. Kadimisetty for technical help.

REFERENCES

Buchholz, F. L. (2000). Polyacrylamides and poly(acrylic acids). In *Ullmann's Encyclopedia of Industrial Chemistry*. Wiley-VCH Verlag GmbH & Co, KGaA Retrieved from http://onlinelibrary.wiley.com/doi/10.1002/14356007.a21_143/abstract.

Credou, J., & Berthelot, T. (2014). Cellulose: From biocompatible to bioactive material. *Journal of Materials Chemistry B*, *2*(30), 4767.

Deshapriya, I. K., & Kumar, C. V. (2013). Nanobio interfaces: Charge control of enzyme/inorganic interfaces for advanced Biocatalysis. *Langmuir*, *29*(46), 14001–14016.

Edwards, J. V., Prevost, N. T., Condon, B., & French, A. (2011). Covalent attachment of lysozyme to cotton/cellulose materials: Protein verses solid support activation. *Cellulose*, *18*(5), 1239–1249.

Kadajji, V. G., & Betageri, G. V. (2011). Water soluble polymers for pharmaceutical applications. *Polymers*, *3*(4), 1972–2009.

Liana, D. D., Raguse, B., Gooding, J. J., & Chow, E. (2012). Recent advances in paper-based. *Sensors*, *12*(9), 11505–11526.

Lützenkirchen, J., van Male, J., Leermakers, F., & Sjöberg, S. (2011). Comparison of various models to describe the charge–pH dependence of poly(acrylic acid). *Journal of Chemical & Engineering Data*, *56*(4), 1602–1612.

Martinez, A. W., Phillips, S. T., Butte, M. J., & Whitesides, G. M. (2007). Patterned paper as a platform for inexpensive, low-volume, portable bioassays. *Angewandte Chemie International Edition*, *46*(8), 1318–1320.

Mudhivarthi, V. K., Bhambhani, A., & Kumar, C. V. (2007). Novel enzyme/DNA/inorganic nanomaterials: A new generation of biocatalysts. *Dalton Transactions*, (47), 5483–5497.

Mudhivarthi, V. K., Cole, K. S., Novak, M. J., Kipphut, W., Deshapriya, I. K., Zhou, Y., et al. (2012). Ultra-stable hemoglobin–poly(acrylic acid) conjugates. *Journal of Materials Chemistry*, *22*(38), 20423–20433.

Nery, E. W., & Lauro, T. K. (2016). Evaluation of enzyme immobilization methods for paper-based devices—A glucose oxidase study. *Journal of Pharmaceutical and Biomedical Analysis*, *117*, 551–559.

Oliveira, K. A., e Silva, P. B. M., de Souza, F. R., Martins, F. T., & Coltro, W. K. T. (2014). Kinetic study of glucose oxidase on microfluidic toner-based analytical devices for clinical diagnostics with image-based detection. *Analytical Methods*, *6*(14), 4995–5000.

Parolo, C., & Merkoçi, A. (2012). Paper-based nanobiosensors for diagnostics. *Chemical Society Reviews*, *42*(2), 450–457.

Pelton, R. (2009). Bioactive paper provides a low-cost platform for diagnostics. *Trends in Analytical Chemistry*, *28*(8), 925–942.

Riccardi, C. M., Cole, K. S., Benson, K. R., Ward, J. R., Bassett, K. M., Zhang, Y., et al. (2014). Toward "stable-on-the-table" enzymes: Improving key properties of catalase by covalent conjugation with poly(acrylic acid). *Bioconjugate Chemistry*, *25*(8), 1501–1510.
Riccardi, C., Kasi, R., & Kumar, C. V. (2015). In *Stable on the table, edible, paper-based biosensors: Glucose oxidase and poly(acrylic acid) nanogels for metal sensing Abstracts of Papers, 250th ACS National Meeting & Exposition, Boston, MA, United States, August 16–20, 2015 (p. PHYS–526)*, American Chemical Society.
Soni, A., & Jha, S. K. (2015). A paper strip based non-invasive glucose biosensor for salivary analysis. *Biosensors and Bioelectronics*, *67*, 763–768.
Sulaiman, S., Mokhtar, M. N., Naim, M. N., Baharuddin, A. S., & Sulaiman, A. (2014). A review: Potential usage of cellulose nanofibers (CNF) for enzyme immobilization via covalent interactions. *Applied Biochemistry and Biotechnology*, *175*(4), 1817–1842.
Thilakarathne, V., Briand, V. A., Zhou, Y., Kasi, R. M., & Kumar, C. V. (2011). Protein polymer conjugates: Improving the stability of hemoglobin with poly(acrylic acid). *Langmuir*, *27*(12), 7663–7671.
Wang, Y., Ge, L., Wang, P., Yan, M., Yu, J., & Ge, S. (2014). A three-dimensional origami-based immuno-biofuel cell for self-powered, low-cost, and sensitive point-of-care testing. *Chemical Communications*, *50*(16), 1947–1949.
Yetisen, A. K., Akram, M. S., & Lowe, C. R. (2013). Paper-based microfluidic point-of-care diagnostic devices. *Lab on a Chip*, *13*(12), 2210–2251.
Zhang, Y., Lyu, F., Ge, J., & Liu, Z. (2014). Ink-jet printing an optimal multi-enzyme system. *Chemical Communications*, *50*(85), 12919–12922.
Zhang, L., Wu, S., Buthe, A., Zhao, X., Jia, H., Zhang, S., et al. (2012). Poly(ethylene glycol) conjugated enzyme with enhanced hydrophobic compatibility for self-cleaning coatings. *ACS Applied Materials & Interfaces*, *4*(11), 5981–5987.

AUTHOR INDEX

Note: Page numbers followed by "*f*" indicate figures, and "*t*" indicate tables.

C

G

L

M

N

O

P

Q

R

S

T

U

V

W

X

Y

Z

SUBJECT INDEX

Note: Page numbers followed by "*f*" indicate figures, "*t*" indicate tables, and "*s*" indicate schemes.

N

R

S

FSC
www.fsc.org
MIX
Papier | Fördert
gute Waldnutzung
FSC® C083411

Kolibri 360 GmbH
Ferdinand-Jühlke-Straße 7
99095 Erfurt, Deutschland
produktsicherheit@kolibri360.de